THE THEORY OF PROBABILITY

THE THEORY OF PROBABILITY

An Inquiry into the Logical and Mathematical Foundations of the Calculus of Probability

By HANS REICHENBACH

PROFESSOR OF PHILOSOPHY IN THE UNIVERSITY OF CALIFORNIA AT LOS ANGELES

ENGLISH TRANSLATION BY

ERNEST H. HUTTEN

AND

MARIA REICHENBACH

Second Edition

UNIVERSITY OF CALIFORNIA PRESS

BERKELEY AND LOS ANGELES · 1949

UNIVERSITY OF CALIFORNIA PRESS
BERKELEY AND LOS ANGELES
CALIFORNIA

CAMBRIDGE UNIVERSITY PRESS
LONDON, ENGLAND

PRINTED IN THE UNITED STATES OF AMERICA
BY MACKENZIE & HARRIS, INC., SAN FRANCISCO, CALIF.

PREFACE TO THE GERMAN EDITION

Les mathématiciens ont autant besoin
d'être philosophes que les philosophes
d'être mathématiciens.

G. W. Leibniz
(Lettre á Malebranche, mars, 1699)

The philosophical problems of the concept of probability have time and again occupied the minds of philosophers and mathematicians. Recently they have been brought to the fore with even greater emphasis: first, because of the prominence of the concept of probability in modern physics, where it has gradually replaced the concept of causality; second, because of the development of the modern philosophy of nature, which has analyzed the concept of probability for its own sake. With the incorporation of the results of symbolic logic, the new philosophy of nature has developed, in the meantime, from a critical investigation of the thinking of mathematical physics to a scientific theory of knowledge. It has now reached the stage at which it begins to replace the era of metaphysical constructions in philosophy by the establishment of a philosophical science. Abundant results have already been reached in the investigation of the space-time problem, in the logical criticism of mathematics, in the analysis of the problem of causality, and in the general criticism of scientific concepts. The problem of probability, however, has resisted with peculiar persistence all attempts at solution. Yet it has become obvious, as never before, that this problem, because of its intimate relation to the problem of induction, contains the nucleus of every theory of knowledge.

The philosophical theory of the probability problem, more than any other, is based upon mathematical analysis. Previous attempts to solve the problem were bound to break down because the mathematical basis of the calculus of probability had not yet been developed in the rigorous form on which philosophical criticism depends. Philosophical analysis was the starting point for a new mathematical construction of the calculus of probability. Further inquiry showed that the new mathematical construction made possible the long-sought transition from two-valued logic to probability logic, that is, to a logic with a continuous scale of truth values. Thus a final theory of probability that satisfies both mathematical and logical requirements can now be presented.

In a period in which the value of speculative philosophy is seriously questioned, it will not be regarded as surprising that a philosophical theory expects new insights from the construction and elucidation of mathematical theories.

I hope to demonstrate that, conversely, the mathematical theory can be furthered by philosophical points of view. The unification of the calculus of probability with the calculus of logic that determines the form of this exposition and, furthermore, the inclusion of all problems of the theory of probability in a general calculus of probability, as well as the theory of the order of probability sequences developed in this connection, seem to be results that may be of value to the mathematician also. He may be interested likewise in the continuous generalization of the concept of probability sequence, leading from the familiar geometrical probabilities to continuous probability sequences. The complete elaboration of the mathematical structure, which is necessary for the philosophical aim, makes this book at the same time a mathematical textbook on the calculus of probability. I hope that clarification of fundamental concepts will make the presentation particularly adequate for this purpose. Nothing makes the penetration into a new mathematical field more difficult than the improper treatment of fundamental concepts. But the so-called mathematical difficulties vanish when the meanings of the concepts whose relations make up the mathematical calculus are clarified.

I was fortunate to have had the opportunity, since 1927, of repeatedly presenting the ideas of this book to my students at the University of Berlin. Many details and examples were developed in class discussion, and the congenial atmosphere greatly helped me in working out solutions.

Only a few of the many names that I should like to mention may be included. I owe valuable help to C. Hempel, A. Becker, M. Strauss, O. Helmer, and my old friends K. Grelling and W. Dubislav. V. Bargmann assisted in the mathematical part and elaborated numerous details and proofs. I thank my friend R. Carnap of Prague for reading the introduction to symbolic logic. E. Kokott assisted in drawing the figures. C. Hempel and V. Bargmann also aided in the correction of proofs.

HANS REICHENBACH

Istanbul,
August, 1934

PREFACE TO THE ENGLISH EDITION

IT IS NOW more than fifteen years since this book was written in the German version. During this time the theory of probability presented in it has been much discussed, and the attacks made upon it have given me the opportunity to think over its content again and again. I am grateful to my opponents because their criticism has compelled me to check every item of my theory. Many details have thus been improved in this presentation. In particular, the exposition of the problems of application and induction has been rearranged and supplemented by many additions. The English version is therefore designated as the second edition of the book.

But, on the whole, I found that I had nothing to change in my views. The theory has stood up to the test of a critical discussion in all its fundamentals. The generalization of the customary theory, which identifies probability sequences with random sequences, into a comprehensive system that embraces sequences of all types of order, including continuous probability sequences, has turned out to be an adequate integration of the pursuit of the mathematician. The theory of the application of probability concepts to physical reality and the analysis of the problem of induction have shown that an empiricist solution of these problems can be given. It seems to me that the various attacks launched against this part of my theory have helped very much to make its significance clear.

The attacks came mainly from two sides, which I should like to distinguish as the rationalists and the unshakable Humeans. Both agree so far as they regard my justification of induction as unsatisfactory, but they do so for different reasons. The rationalists find this justification too weak because they wish to find a stronger one, which in some way or other bases induction and probability on a rational belief. The unshakable Humeans also maintain that I did not justify induction; but they do so because they believe, like Hume, that any such justification is impossible and that Hume's position represents the final stage in the analysis of the problem. The more I read of these arguments, the more convinced I became that my conception is correct; and I should like to explain in what points I see the merits of this conception.

The first achievement, I think, consists in the demonstration that the frequency interpretation of probability can be carried through for all uses of the term "probable". The difficulties that others had found in interpreting the probabilities of single events as limits of frequencies disappear when statements about single cases are regarded, not as assertions, but as posits. When, in the face of these results, there remain philosophers who maintain that there is a second concept of probability, which is not reducible to frequency notions, I must answer that there is no need for such a concept. Whoever wishes to reserve the right of using private meanings of a nonverifiable pattern, or of a structure useless for predictions, may do so. What has been shown, however, is that the frequency interpretation of probability leads to a meaning of the term that makes the usage of language conform to human behavior. I do not think there are other grounds that a philosophical theory could adduce for its claim to provide an adequate interpretation of terms.

The second achievement I find in the fact that, when the frequency interpretation is used, all nondeductive methods of the calculus of probability reduce to one kind of inference: the inference of induction by enumeration. Since all inductive methods of science, including the theory of indirect evidence and the formation of scientific theories, are interpretable in terms of inferences supplied by the calculus of probability, this result establishes the thesis that all forms of inductive inference are reducible to one form, to the inference of induction by enumeration. This thesis was implicitly contained in some older forms of a theory of induction, in particular, in Hume's theory. But so far it had not been given a proof.

As the third achievement, I should like to mention the fact that the theory of probability has been freed from all forms of a rational belief in synthetic statements, whether they appear in the form of a synthetic *a priori*, or an animal faith, or a belief in a uniformity of the world, or a principle of insufficient reason. All such notions are remnants of a philosophy of rationalism, which holds that human reason has access to knowledge of the physical world by other means than sense observation. What has been shown is that a theory of probability can be built up without the use of such notions. There is no such thing as inductive self-evidence. It is unfortunate that this discovery of David Hume has so often been forgotten, and that so many attempts have been made to reintroduce inductive self-evidence in new forms whenever some older forms had been proved to be untenable.

The fourth achievement of the theory I see in its successful justification of the inference of induction by enumeration, and therefore of all forms of inductive inference, in spite of its renunciation of arguments based on synthetic self-evidence. The inductive inference is regarded as permissible, not because it appears self-evident, but because it represents an instrument of prediction so devised that it must lead to success if success is attainable.

This conception of a justification has been attacked by various means. Although it is now generally admitted that it would be asking too much to require a proof that the inductive inference must lead to true conclusions, other requirements for a justification were advanced with the intention of ruling out my form of a justification. For instance, a justification of a method has been defined as a proof that there exists some inductive evidence, or probability, that the method will reach its aim. On the basis of such a definition it is easy to prove that what I have given is not a justification of induction.

It should be clear, however, that the argument I have set forth for the use of induction cannot be deprived of its rigor by narrowing down the meaning of words. That it is advisable to use a certain means if it can be shown to be the best means for a desired end, and if nothing is known about the attainability of this end, appears to be an argument exempt from doubt; and it seems to me that the word "justification" has always been applied to such situations. When Magellan sailed along the American shore with the intention of finding a western passage, he had no inductive evidence that there was one; but his enterprise was justified because it was a means to reach his aim if the aim was attainable. It is in this sense of the word that a justification of induction has been given.

What makes this justification of induction appear weak is the adherence to rationalistic standards, the creed that we can do more for the finding of predictions than prepare everything for the case where success is attainable. I do not think that any such creed is compatible with the principles of empiricism, nor that it is an indispensable backbone for action in this recalcitrant world. The search for certainty is a desire deeply rooted in human nature; yet we must not infer that we should submit to it. I prefer a philosophy that teaches us to walk without the crutches of any kind of faith. It is of the essence of empiricism that it refuses to recognize any form of rationalized belief. That such refusal need not lead to skepticism, that actions anticipating the future can be justified without any reference to belief, has been shown through my analysis of the problem of induction. I do not think that a justification of the inductive method of prediction may be called weak if it proves that employing this method is the best we can do for the attainment of the aim. This proof is the result to which I came through the construction of my theory of probability and induction some fifteen years ago. It is with the intention of submitting this result to the judgment of the English-speaking world that I present this English edition of my book.

HANS REICHENBACH

University of California,
Los Angeles,
May, 1948

CHANGES AND ADDITIONS IN THE ENGLISH EDITION

THROUGHOUT the book, numerous minor changes were made; but only the major changes will be mentioned. In chapter 1 the last paragraphs of § 2 were cut off and added to § 3, which was shortened in other parts. The introduction to symbolic logic, given in chapter 2, was abbreviated and adjusted to the results of a detailed exposition of symbolic logic given in another publication (see footnote, p. 12). In chapter 3 and the following chapters the symbolic notation was adapted to the English language; in particular, the German symbol W was replaced by the symbol P, standing for "probability". Some changes and additions were made in §§ 12 and 22. A new section was inserted as § 24; furthermore, the appendix to chapter 3, which contains exercises, was added. From §§ 1 to 23 the numbers of the English and the German sections coincide; from §§ 25 to 46 the numbers of the English sections exceed by one those of the German sections. In chapter 4, § 30 was supplemented by references to recent publications on the problem of randomness. Chapters 5 and 6 remained unchanged, except for some additions in § 45 and the addition of § 47. From §§ 48 to 62 the numbers of the English sections exceed by two those of the German sections. In chapters 7 and 8 changes were made in §§ 49, 57, and 62.

Chapters 9–11 take the place of the German chapters 9–10. These chapters were greatly changed in the order of the material presented, so that no simple rule for the comparison of sections can be given; and many changes and additions were made in the interior of the sections. The following sections of these chapters are new: §§ 70–74, 81–88. The appendix of the German edition, which contains some mathematical extensions and proofs, was omitted.

The changes made in chapters 9–11 may be summarized as follows. A distinction is made between primitive knowledge and advanced knowledge, and the problems of induction and of the meaning of limit statements are given different treatments for the two cases. In advanced knowledge, which presupposes the use of previous inductions, both the inductive inference and the meaning of limit statements can be accounted for in terms of probability concepts. In primitive knowledge, which does not make use of previous inductions, the meaning of limit statements can be defined by the help of a "finitization", which eliminates infinite sequences; the use of infinite sequences then appears as a simplification introduced for technical reasons. The inductive inference, as far as it is applied in primitive knowledge, is made legitimate by my justification of induction.

A further distinction is made between the object-language interpretation and the metalanguage interpretation of probability. The latter is used for the construction of probability logic and is thus also called the logical interpretation. At the time the German book was written, little was known about the significance of this distinction. The omission of the distinction in the German book had no bearing upon my theory of induction or on the mathematical parts of the book; the truth tables of probability logic given in the German edition were not affected by the omission either. But since the distinction has turned out to be relevant from many logical aspects, it has been carried through in the new presentation. The method of derivation in probability logic was revised and greatly elaborated.

The exposition of inductive methods in their various forms was widened, and a section on the inductive introduction of all-statements and the probability of hypotheses was added. Recent publications on the theory of statistical estimation were included in the frame of the discussion.

ACKNOWLEDGMENTS

THE SIGNIFICANCE of the distinction between object-language and metalanguage interpretation of probability was called to my attention through criticisms by R. Carnap and C. Hempel. The necessity of a minor correction of my axiom system of the calculus of probability was pointed out to me by J. C. C. McKinsey and S. C. Kleene (p. 56). Some results communicated to me by N. Dalkey and H. Schott were the starting point for the analysis of complete probability systems (§ 24). In the course of oral discussions with O. Helmer, I recognized the fact that the method of anticipative posits can be extended to advanced knowledge (§ 88). E. Nagel's criticism induced me to be more explicit on the subject of the probability of hypotheses (§ 85). The University of California Press and its staff were of great help in giving the book its printed form. For mathematical details and the reading of the proofs I had the assistance of J. Freund, A. Birnbaum, and H. Edmundson.

I wish to express my sincere gratitude to the translators of the book, Ernest H. Hutten and my wife Maria Reichenbach. Their work has shown that the problem of translating German into English is not insoluble if it is attacked with the courage to deviate from sentence structure in favor of preserving the meaning. As a result of this method, the English formulation is often more precise than the German original. Whenever the deviations appeared to them to demand too great a venture, I did not hesitate to make use of an author's right to rephrase his text. Any criticism that the translation is too free should therefore not be directed against my translators, for whose active contribution to my book I am deeply obliged.

HANS REICHENBACH

University of California,
Los Angeles,
September, 1948

CONTENTS

Chapter 1. Introductory Considerations

Chapter 11. Induction

Chapter 1

INTRODUCTORY CONSIDERATIONS

INTRODUCTORY CONSIDERATIONS

§ 1. The Probability Concept of the Language of Everyday Life

The word "probable" is frequently used in everyday language; more often, however, the concept is employed without being explicitly expressed. We must restrict to mere probability not only statements of comparatively great uncertainty, like predictions about the weather, where we would cautiously add the word "probable", but also statements of so high a degree of probability that we do not consider it necessary to mention the unavoidable uncertainty, or statements of the probability character of which we are not even conscious. Sometimes we express the uncertainty by a gesture or the accentuation of words. If we expect the plumber to come for some repairs we may communicate the news to the family by a shrug of the shoulders. But in many instances even this symbolic gesture is missing. Thus, when we go to the station to catch a certain train, it does not occur to us that because of an accident the train might for once be late. Even a desultory consideration of the statements of daily life shows clearly that a great number of them owe their character of "certainty" to a confusion of certainty with a high degree of probability. On close inspection, finally, it becomes evident that there are no statements of absolute certainty, if the statements are not to designate empty logical relations but to assert the existence of specific facts.

Incidentally, it would be a mistake to believe that the concept of probability concerns only statements about the future. In many statements about the past, we evidently use the concept of probability. The historian considers it very probable that Nero ordered the burning of Rome; he believes it less probable that Henrietta of England, who lived at the court of Louis XIV as the Duchess of Orleans, was murdered; he regards it as improbable that Bacon was the author of Shakespeare's plays. Even events of the past can only be asserted as probable.

Although we *apply* the concept of probability in daily life as a matter of course, we find it difficult to say what we *mean* by the concept "probable". We know that a probability statement neither asserts nor denies the facts that it designates, but we do not restrict it to an assertion of the mere possibility of an event, since we make distinctions in the degree of probability. For instance, we regard Nero's responsibility for the burning of Rome as more probable than Bacon's authorship of Shakespeare's plays. But we have only a vague notion of the *applicability* of the concept without being able to explain its *meaning*. Since, furthermore, probability statements appear to be grounded in the insufficiency of human knowledge, they seem to be more or less subjective. It is impossible to know with certainty who the Man in the Iron Mask was; his identity, however, is an objective fact, and if those who knew his origin had left a trustworthy account, they would have spared us the uncertainty. To know how the weather will be tomorrow is not possible for us; but we hope that future meteorology will predict the weather of the next day with certainty, or at least with the same certainty that the arrival of trains is predicted today.

What can be the significance, for philosophical investigation, of a concept whose interpretation is vague and whose origin seems to be rooted in the inadequacy of human knowledge? The analysis of the probability concept of everyday language has, indeed, been rather fruitless for philosophical investigation. Its inefficiency has manifested itself in the philosophical critique of the probability concept carried through in traditional philosophy. Philosophers have been satisfied to construe probability as an uncertainty originating in the imperfection of human knowledge, and to connect the concept of probability with that of possibility; this was virtually all that philosophy could discover so long as it restricted its studies to the probability concept of everyday language. Thus the first line of development of a theory of probability—which, incidentally, goes back to Aristotle—did not supply any significant results.

Philosophers even tried to eliminate the concept of probability from science and to restrict it to prescientific language—which was an evasion rather than a solution of the problem. It cannot be admitted that everyday language uses concepts essentially different from those of scientific language. The analysis of science has shown that there is no sharp borderline between scientific and prescientific statements, that the concepts of daily life are absorbed by the language of science, in which they take on a more concise form and a clearer content without being abandoned. Thus it has become evident that criticism of fundamental concepts in their scientific formulation is more fruitful than reflection about their naïve usage; that scientific expression, through its precise wording, leads to a clearer interpretation of concepts and a deeper understanding of their meanings.

It may be recalled that the analysis of the concepts of time and space remained futile so long as the philosophical discussion did not extend beyond the use of these concepts in daily life. Only with the elaboration of the scientific theory of space and time, carried through in non-Euclidean geometry and the theory of relativity, did it become possible to uncover the ultimate nature of space-time concepts and to achieve a more profound understanding of their application to daily life. Thus we now have a better knowledge of what an architect means when he specifies lengths and widths in the plan of a building, and what a watchmaker does when he synchronizes a number of watches. Another example is logic, the fundamentals of which, though continuously applied in everyday language, remained unclarified until mathematicians undertook the analysis and formulation of logical relations. The probability concept, therefore, can be studied successfully only within the realm of its scientific application.

From this point of view a second line of development of the probability concept—its evolution within the exact sciences—may be traced.

§ 2. The Historical Development of the Scientific Concept of Probability

The scientific evolution of the probability concept, which began with the construction of the mathematical theory of games of chance about the middle of the seventeenth century, is a striking example of the materialistic origin of intellectual developments. Wealthy noblemen, who spent their ample hours of leisure in the excitement of gambling, supplied the stimulus that induced ingenious mathematicians to construct the mathematical theory of probability. The thrill of complication had made the rules of games of chance so involved that some farsighted knights of the green table turned to eminent mathematicians like Pascal and Fermat for an exact calculation of all chances. The mathematicians undertook the task with increasing interest as they discovered the consistent logical structure of a mathematical theory. It is surprising that even Jacob Bernoulli developed his profound mathematical theorem in the pursuit of a detailed mathematical calculation of a variety of games of chance.

If we compare the status of knowledge of the concept of probability as obtained from the theory of games of chance with the status resulting from the discussion of the probability concept of everyday language, we notice remarkable progress. Only in one respect did the analysis of probability concepts remain on the naïve level: The idea that human ignorance is the source of the concept of probability was taken over by mathematicians, who recognized that probability statements about the die or the roulette are used only because it is impossible to portray the individual cast of the die or the turning of the wheel by mathematical laws. But one new insight was opened by games of chance: the discovery that probability statements can be transformed into statements of a high degree of certainty if they are applied to a great number of homogeneous cases, that is, if they are transformed into statistical statements. The relation between probability and frequency was made clear for the first time in the mathematical study of games of chance. The implications of this epistemological discovery will be discussed in a later chapter.

About a decade after the fundamental inquiries of Pascal and Fermat into the theory of games of chance, the calculus of probability was first applied to social statistics. The problems of life insurance were investigated, for at that time many of the middle class wished to secure a lifelong income with a single outlay of capital. The calculus of probability was applied, also, to statistics on the efficacy of medical treatments, the reliability of testimony in courts, and so on. The rationalism of the eighteenth century embraced such ideas with fervor. By the application of such rational methods to all the questions of daily life, Condorcet hoped to achieve the ideal that "our reason cease to be the slave of our impressions".

In continuation of these developments, the calculus of probability was applied to the mathematical sciences, particularly to astronomy and geodesy, where the theory of errors presented statistical questions. The construction of suitable methods, which is connected primarily with the names of Laplace and Gauss, bears witness to mathematical genius; it linked the calculus of probability to methods of mathematical analysis. What is more important, epistemologically speaking, is the fact that for the first time the probability concept found an application in the exact sciences. The study of games of chance had helped to clarify the meaning of probability. They became models by which the theorems of probability were explained. Even today, games of chance are significant in the mathematical calculus of probability because they represent easily understandable applications of probability laws. In social statistics and in the theory of errors, however, the field of application itself became the subject of scientific interest.

The indispensability of the probability concept for the natural sciences became even more apparent when a new field of application was opened—the kinetic theory of gases and liquids. Whereas in the theory of errors a higher precision of observational results, an improvement in the numerical aspect of physical knowledge, had been achieved, the appearance of the probability concept in the kinetic theory meant nothing less than its penetration into the concept of natural law, for which new perspectives were suddenly revealed. The statistical gas theory asserted that certain laws that had formerly been considered to be strict physical laws were statistical laws, that is, laws of a probability character. This holds, in particular, for Boltzmann's interpretation of the second principle of thermodynamics, the epistemological implications of which were not completely understood until our day. Boltzmann's theory implied that certain laws that had previously been regarded as strict laws of nature are not different from the statistical laws of games of chance, and that the law of great numbers, which was uncovered in the theory of games of chance, represents a general type of physical law. Thus the concept of probability was related to that of causality, and the concept of the statistical law of nature took its place beside that of the causal law of nature.

The fundamental significance of the probability concept was not generally acknowledged even then. Its application to the gas theory was regarded by many scholars as an expedient necessitated by human ignorance, by the impossibility of following the movements of individual gas molecules with the methods of physics. The adherents of this conception upheld the postulate that the events of the microcosm are controlled by strict causal laws, and they looked upon the laws of probability as a crude reconstruction of the macrocosm to which we resort because of the insufficiency of human observations and calculations. A similar conception was customary in social statistics, in which it seemed obvious that statistical laws are the product of an omission of causal considerations that in principle can be carried through for each individual case separately. Even in Boltzmann's gas theory, therefore, the ascendancy was gained by an interpretation that belittled the significance of the probability concept and from the very outset hindered insight into the epistemological position of the concept.

At the same time, however, there developed a second conception that granted much greater significance to the concept of probability. Since certain laws, formerly regarded as strict causal relations, had been revealed as probability laws, it seemed possible that the same fate awaited all the so-called strict laws of physics. According to this conception, the causal interpretation of nature represents a rather crude form of description that is necessitated by the inaccuracy of human observations and is made possible by the concerted action of many elementary processes in macrocosmic phenomena. Whereas the first conception asserts the primacy of the causality concept over the probability concept, the second conversely maintains the primacy of the probability concept over the causality concept; the microcosm seems to be governed only by probability laws, whereas for the macrocosm there result statistical regularities that we take for causal laws and from which the idea of a strict causal determination has been incorrectly extrapolated.

By these considerations the theory of the probability concept was connected with a third line of development—the problem of causality—which was to join the other two in giving the concept of probability a leading position. Their merging, however, did not occur until our day.

The idea of the strict causal connection of all natural phenomena is justly called a product of Western physical science, for it was the science of the modern age that, through the consistent application of the principle of causality, especially in mathematical physics, brought the principle to the fore and made it the basic principle of a knowledge of nature. The philosophical criticism of the idea of causality could be attempted only after the principle had been sufficiently elaborated.

The decisive turn was made in the criticism of David Hume, to whom we owe the greatest discoveries in regard to the logical structure of the causality concept. Hume recognized that the causality relation establishes a mere coördination of events and that all metaphysical ideas of an intrinsic connection of events constitute anthropomorphisms having no objective meaning. It is only a relation of the form *if-then*, in the meaning of a concurrence without exception, that is asserted in the laws of nature. Bacon had emphasized that this relation is established by means of inductive inference. Hume, however, saw that the peculiar structure of inductive inference calls for serious criticism. He considered the inference in its simplest form—induction by enumeration: observing repeatedly that an event A is accompanied by an event B, we infer that the concurrence will always take place. Although he saw that there are differences between scientific manipulation and everyday life, and that the scientist insists upon precise analysis of all factors involved in a phenomenon before he applies the inference, Hume realized clearly that such qualifications cannot eliminate the inductive inference from science and that all scientific inferences ultimately presuppose the legitimacy of induction by enumeration. Nor does the method of scientific experiment change the situation. Through experiment the scientist produces conditions that supply particularly instructive trains of events. That, however, under the same conditions the same thing will always happen is an indispensable assumption, in which inductive inference is applied.

I do not intend to inquire now into the strange and enigmatic nature of inductive inference, which will be considered explicitly in later sections of the book. I only wish to point out that Hume's insight into the problematic nature of inductive inference opened the path for the connection of causal with probability methods. The relation of inductive inference to probability is obvious, for we never claim that the inductive conclusion is certain; to what extent it can at least be called probable is a question I shall attempt to answer in the later analysis.

Strangely enough, Hume's early contribution toward a connection of causality and probability was forgotten in later philosophical developments. It was primarily Kant's unfortunate doctrine of the apriority of causality that misdirected subsequent analysis of the problem. Although Kant, according to his own words, "was awakened from his dogmatical slumber" by Hume's criticism, he was unable to do justice to Hume's questioning of the legitimacy of inductive inference; and, in fact, he never attempted to apply his general theory of the *a priori* to a specific solution of Hume's problem. Otherwise he would have seen that even the assumption of the *a priori* validity of the principle of causality cannot make inductive inferences dispensable for the discovery of individual causal relations. If, for instance, after repeated observation of the deviation of the magnetic needle by an electric current, we proceed to the assertion that always and everywhere the electric current will have the same effect on magnetic needles, it does not help our inductive inference to know that there are causal relations between the determining factors of physical occurrences. The statement that the observed relation is generally valid, i.e., that we already have sufficient knowledge of the determining factors of the problem, still depends on inductive inference.

Later, when the transition from thermodynamics to a statistics of gases was completed and the statistical law had found its place beside the causal law as a second type of natural law, critical investigations of the causality concept were connected with the analysis of the probability concept. These considerations originated in an inquiry into the peculiar relation existing between natural law and reality. The law never portrays the actual occurrence completely, but represents an idealization in which only certain prominent factors are considered, whereas an infinite number of other factors are neglected. Without such a schematization, natural events would be too complex for interpretation.

This method, however, gives rise to a peculiar problem: we calculate the expected effect on the assumption of certain ideal conditions, although we know that the real conditions do not correspond completely to the ideal ones. The problem finds its solution in the discovery that an application of natural laws to reality is never expressed in certainty statements, but only in probability statements. It is through the coördination of the ideal structure to reality that the probability concept enters into the physical sciences. The existence of ideal conditions, or even of approximately ideal ones, cannot be asserted with certainty, but only with probability, though the probability may be rather high if a sufficient tolerance is admitted for the conditions. The expected effect can, therefore, be predicted only with probability. It is possible to improve the probability of a prediction by a more precise analysis of the conditions, but we can never rid ourselves of probability statements.

We might attempt the assumption that the probability of a prediction will approach certainty indefinitely with a more exact analysis of the determining factors; this assumption, in fact, represents the strict form in which the hypothesis of causality is to be expressed. The very formulation reveals that the assumption cannot be regarded as *a priori* necessary, but that it is a matter of experience whether an increase in the probability of predictions is possible. It may be that the contrary is true, that the increase in the probability of predictions is restricted to remain below a certain limit that is lower than certainty. Such a restriction would represent the transition to a more general form of natural law. This generalization, which I developed in earlier papers[1] in the sense of a possibility, was later recognized as an actuality by quantum mechanics, and was formulated in Heisenberg's principle of indeterminacy.

[1] H. Reichenbach, "Die Kausalstruktur der Welt und der Unterschied von Vergangenheit und Zukunft," in *Ber. d. bayer. Akad., math.-phys. Kl.*, 1925, p. 138.

Thus the historical development of physics led to the result that the probability concept is fundamental in all statements about reality. Strictly speaking, we cannot make a single statement about reality the validity of which can be asserted with more than probability. Only a theory of the probability concept, therefore, can supply an exhaustive analysis of the structure of statements about reality. This is why the theory of probability stands today in the focus of investigations that, within the frame of a scientific philosophy, are concerned with clarification of the nature of knowledge.

§ 3. Remarks about the Plan of the Book

From the preceding historical account it will be evident to the scientific philosopher that a satisfactory analysis of the probability concept can be carried through only in connection with a study of the probability concept as it developed in the mathematical calculus of probability. Only in the mathematical theory did the concept acquire the precise formulation that reveals its logical structure. The concepts that are applied in the process of knowledge are not always understood by the investigator from the beginning; only through continuous application do they assume clearer meanings until they reach a stage of determinateness that makes philosophical analysis fruitful.

This book is not meant to analyze the connection of the probability concept with the problem of causality, but will be restricted to a treatment of the mathematical calculus of probability. Nonetheless, all the results may be transferred to every application of the probability concept in science or daily life. In spite of the derivation of ingenious theorems, the calculus of probability as it has been formulated by mathematicians does not, however, possess the degree of logical strictness and clarity that is necessary for philosophical analysis. It is my intention in this work to present a construction of the calculus of probability that is mathematically as well as logically satisfactory, and then, returning to the logical and epistemological problems, to show that all the questions on the nature and application of the probability concept can be answered satisfactorily.

The solution here presented for the problem of the application of the probability concept to physical reality will look very different from the traditional conceptions. It will be shown that the analysis of probability statements referring to physical reality leads into an extension of logic, into a *probability logic*, that can be constructed by a transcription of the calculus of probability; and that statements that are merely probable cannot be regarded as assertions in the sense of classical logic, but occupy an essentially different logical position. Within this wider frame, transcending that of traditional logic, the problem of probability finds its ultimate solution.

The results of this inquiry demand a readjustment of traditional conceptions of the foundations of knowledge. It is not surprising, however, that the analysis of probability should lead to far-reaching results. The concept of probability is not the instrument of some narrow scientific discipline; it is the fundamental concept on which all knowledge of nature is based, the interpretation of which determines the formulation of any theory of knowledge. Hence there is good reason for attributing so much significance to the analysis of probability. What is sought is not only an interpretation of a concept of mathematics and mathematical physics but the interpretation of the knowledge of nature, the answer to the question of the ultimate meaning of statements about the physical world.

The construction of the calculus of probability presented in this book has the form of an axiomatic system. Some assumptions are assembled in the beginning and employed as axioms, and all the other theorems of the calculus are derived from them. This procedure has a mathematical advantage in that it presents the content of the mathematical discipline of probability in a logically ordered form. Moreover, it is of great help to a philosophical analysis because it exhibits clearly the logical problems of the theory and compels one to formulate exhaustively all presuppositions.

Expositions given in textbooks of probability are often written with the intention of making the theorems of the calculus plausible to the reader. Certain presuppositions, therefore, are not mentioned but are regarded as "understood"; and it is assumed that this kind of presentation is the safest means of instilling belief in the theorems of probability. The method cannot be said to clarify the reasons why the theorems should be accepted, since the philosophical problems are usually contained in the presuppositions that are "understood". If an assumption is treated as a matter of course, one should always suspect that it serves to conceal certain philosophical problems. By uncovering hidden assumptions, the axiomatic procedure accomplishes something philosophical: it raises unconscious laws of reasoning to the level of conscious assumptions. Formulation, therefore, is an important means of philosophical analysis; mathematical presentation becomes a tool of the philosopher when it is directed toward exhaustive formulation.

Mathematicians have developed a method that entails the most precise formulation, namely, formalization. Formalization is the introduction of a set of symbols that permit us to abandon conversational language entirely and to express thought relations by relations of algorismic symbols. The value of the method consists in its avoidance of a language whose structure, originating from everyday needs, cannot express the logic of philosophical problems. Even if the manipulation of the algorism offers difficulties to the untrained student, his mental effort will be repaid by a clearer understanding of the subject and an abbreviation of the process of subsequent learning.

An algorism of this kind will be developed for the exposition of the mathematical calculus of probability. The task may be simplified greatly by the adoption of the algorism of symbolic logic, with the addition of a sign for probability. The algorism presents the fundamentals of the calculus of probability in a clear and instructive way; it is expedient, too, for the formulation of complicated theorems and for practical applications.

The usefulness of a symbolism that goes beyond that of mathematics is not widely recognized; in particular, mathematicians have not always been on friendly terms with logical symbolism. The student of probability, therefore, may not be familiar with the symbolic technique. For this reason the following chapter offers a short introduction to symbolic logic. The presentation is restricted to the parts of the technique that are used in the construction of the probability calculus given in this book. If the brevity of the exposition leaves unanswered questions in the mind of the reader, he may consult an exhaustive presentation on symbolic logic by the author.[1]

The logic presented in chapter 2 is called *deductive logic.* Its theorems are necessarily true; and if its methods of derivation are applied to true statements, the resulting conclusions will be true. This part of logic is to be distinguished from *inductive logic,* which cannot guarantee the truth of its conclusions, and which will be presented in chapter 11.

[1] H. Reichenbach, *Elements of Symbolic Logic* (New York, 1947). Hereafter referred to as *ESL.*

Chapter 2

INTRODUCTION TO SYMBOLIC LOGIC

INTRODUCTION TO SYMBOLIC LOGIC

§ 4. The Calculus of Propositions

In its first part, the calculus of propositions, symbolic logic treats operations performed with propositions. A proposition is a sign combination that is either true or false. The terms "sentence" and "statement" will be regarded as synonymous with "proposition". Examples are given by the propositions: "Berlin is situated on the Spree"; "Two times two is four"; "If potassium is thrown into water, it begins to burn"; "Gold is lighter than water". The last sentence is, certainly, a false proposition, but that does not deprive it of its propositional character. Symbols arranged in a meaningless combination, however, do not form a proposition. If someone says, "Light is a prime number", or "Two times two equals and", he does not make a statement, not even a false one; he only combines symbols without constructing a meaningful combination. The meaningless combination must therefore be distinguished, as a third category, from true or false sign combinations; only the last two sign combinations are propositions.

Propositions are denoted by variables a, b, ... The most important operations with propositions, namely, propositional operations, are

$\bar{a}$	non-a (negation)
$a \vee b$	a or b (disjunction, logical sum)
$a.b$	a and b (conjunction, logical product)
$a \supset b$	a implies b (implication)
$a \equiv b$	a is equivalent to b (equivalence, logical equality)

By means of a propositional operation, a *compound* statement is constructed from the *elementary* statements a, b, ... Thus by the operation of conjunction the compound statement $a.b$ is constructed from the elementary statements a and b.

The terms "and" and "or" will be used in the description; such use is not circular, however, since it is not intended to define the terms, i.e., to replace them by others. They will be considered as primitive concepts the meaning of which is known. Explaining their meaning more precisely is not a definition, but a characterization of the propositional operations.

The statement $\bar{a}$ is true if a is false.

The statement $a \vee b$ is true if a is true and b is false, or if a is false and b is true, or if a and b are true. This operation represents the *inclusive* "or", not the *exclusive* "or", which conversational language expresses by "either-or", and which excludes the case that a and b are true. For the exclusive "or" the symbol $a \wedge b$ will be used, but not often, since it can be replaced by other symbols, for instance, by the combination

$$(a \vee b) \,.\, \overline{a.b} \tag{1}$$

The inclusive "or" does not state that the case a *and* b must hold; it states only that this case may also hold. The name "logical sum" is used for the inclusive "or" to indicate a certain correspondence of the or-sign to the plus-sign of arithmetic. If, for example, things of a different kind are added, the or-connection of the respective concepts corresponds to the addition of the numbers. Thus 8 male persons plus 6 female persons equals 14 persons—this equation is true because the term "person" is the or-combination of the terms "male person" and "female person".

The statement $a.b$ is true if both a and b are true. The name "logical product" used for this operation indicates a certain correspondence of the and-sign to the multiplication sign of arithmetic. If the numerical values belonging to concepts of different kinds are multiplied, the concepts enter into an and-combination. Thus 3 meters times 4 kilograms equals 12 meter kilograms. The concept "meter kilogram" may be conceived as the logical pair of the concepts "meter" and "kilogram", that is, it denotes an object that is characterized by the concepts "meter" and "kilogram".

The statement $a \supset b$ is true if a is true and b is true, or if a is false and b is true, or if a is false and b is false. This characterization of the implication, or if-then statement, which will later be discussed in detail, may seem unusual, but it corresponds to linguistic usage so far as it excludes only the second of the four possible combinations

$$a.b \qquad a.\bar{b} \qquad \bar{a}.b \qquad \bar{a}.\bar{b} \tag{2}$$

An implication is therefore regarded as true if one of the three other combinations holds. In the expression $a \supset b$, the term a is called the *implicans* and b the *implicate.*

The statement $a \equiv b$ is true if a is true and b is true, or if a is false and b is false. The equivalence establishes the same truth value for a and b. The equivalence relation plays a role in logic that resembles the relation of equality in arithmetic and is therefore sometimes called a logical equality. It is used in logical formulas to state that certain symbol combinations are logically equivalent, that is, have the same truth value. Assertions of this kind represent a great part of the logical formulas that are derived in the calculus of propositions, though not every logical formula is an equivalence. It is easily seen that an equivalence means the same as implications in both directions, that is, $a \equiv b$ holds if both $a \supset b$ and $b \supset a$ hold (formula 7a). Whereas the first implication excludes the second of the four combinations (2), the second implication excludes the third combination, so that only the first and the fourth remain, corresponding to the foregoing characterization of the equivalence.

The given characterizations of the propositional operations are expressed in the truth table 1, in which the *truth value* $V(a)$ of a statement a, which is either truth or falsehood, is denoted, respectively, by the letters T and F.

TABLE 1

A. NEGATION

$V(a)$	$V(\bar{a})$
T	F
F	T

B. SUM, PRODUCT, IMPLICATION, EQUIVALENCE

$V(a)$	$V(b)$	$V(a \vee b)$	$V(a.b)$	$V(a \supset b)$	$V(a \equiv b)$
T	T	T	T	T	T
T	F	T	F	F	F
F	T	T	F	T	F
F	F	F	F	T	T

Table 1A refers to the *monadic* operation of negation; the argument column (left) contains the truth values T and F of the elementary proposition a; the functional column (right), the truth values of the compound proposition $\bar{a}$. Table 1B refers to the *binary* operations, which connect two propositions; the table has two argument columns (left) and determines, in each of the functional columns, the truth values T or F of the respective compound proposition.

The truth tables can be read in two directions. When read from right to left, they state, for a compound statement given as true, the possible combinations of elementary propositions. Thus if $a \vee b$ is true, table 1B furnishes the result that a is true and b is true, or a is true and b is false, or a is false and b is true. When read from left to right, the tables state, for given elementary propositions, the truth value of the compound proposition. For instance, if a is true and b is true, table 1B states that $a \vee b$ is true. This distinction of directions leads to two possible interpretations of the truth tables. If only the direction from right to left is used, we speak of the *connective* interpretation; if both directions are used, we speak of the *adjunctive* interpretation.

In the adjunctive interpretation, the truth values of the elementary propositions determine the truth value of the compound proposition. Since in this relation no reference is made to the meaning of the propositions, the propositional operations of the adjunctive interpretation are called *truth-functional.*

In the connective interpretation no such determination is possible. Thus if both a and b are known to be true, this interpretation admits of no statement whether $a \vee b$ is true; the answer to the question depends on further information. However, if $a \vee b$ is known to be true, this information is regarded as sufficient for the statement of the disjunction of the T-cases of the compound statement, in the example, for the statement that a is true and b is true, or a is true and b is false, or a is false and b is true.

An example of an adjunctive "or" is the statement, "The speech of the prime minister will be transmitted by station KFI or station KNX". If we observe that station KFI transmits the speech, the statement is regarded as verified. An illustration of a connective "or" is the statement, "A man suffering from severe diabetes takes insulin injections or will soon die". We do not regard this statement as verified if we see a man taking insulin injections; the totality of knowledge about the nature of diabetes is involved in the statement. However, the disjunction of the T-cases is employed in this instance; of the combinations (2) the statement excludes only the last combination. Since the meaning of the propositional operations depends on the interpretation, a distinction is made between *adjunctive operations* and *connective operations.*

Conversational language employs both kinds of propositional operations. The two examples cited illustrate the two meanings of the inclusive "or"; the exclusive "or", too, is used in both meanings. The "and" is used almost exclusively in the adjunctive sense. The implication, however, is usually meant in the connective sense. Thus we do not regard it as a verification of the statement, "If you take this medicine your cold will disappear", when the medicine is taken and the cold disappears. Such a coincidence, observed in only one instance, might be due to chance. We would regard it as even less a verification if the medicine were not taken and the cold disappeared—a combination that represents one of the T-cases of the implication.

The adjunctive implication rarely occurs in conversational language. Moreover, it leads to peculiar consequences that have been called the paradoxes of implication. Every false sentence implies adjunctively any sentence, and every true sentence is adjunctively implied by any sentence. For instance, the sentence, "The earth is a flat disk implies the earth does not revolve around the sun", is true in the adjunctive sense of the word "implies" because the implicans is false; and the sentence, "The earth is a flat disk implies the earth revolves around the sun", is also true, for the same reason. The implication, "Sugar is sweet implies water is wet", is true because the implicate is true. These examples lose their paradoxical character if we realize that the word "implies" is used in a meaning different from the one accepted for conversational language. The adjunctive implication does not connect meanings, but simply adjoins two propositions according to rules referring to their truth values. Similarly, the equivalence of conversational language is usually connective.

Symbolic logic employs adjunctive operations throughout. In doing so it makes use of the scientist's privilege of defining his own simplified concepts. Such an attitude does not mean that the definition of connective operations is regarded as unnecessary; it is only postponed to a later stage. It is possible to construct connective operations from adjunctive ones, though such a definition is rather involved.[1]

The four binary operations of table 1*B* (p. 17) are not the only possible operations. This can be seen from the fact that the functional columns contain only certain arrangements of the values *T* and *F*. Any other possible arrangement of the values *T* and *F* would also define an operation. There are $2^4 = 16$ possible arrangements all together, and thus 16 binary operations can be defined. One of them is the exclusive "or", which is distinguished from the inclusive "or" in that the first line of its column contains an *F* instead of a *T*. The operations of table 1*B*, however, are sufficient for all purposes, since the meaning of other operations can be constructed through a combination of these symbols.

We do not even need all the four operations of table 1*B*, for some are reducible to others. Thus the statement $\bar{a} \vee b$ has the same arrangement of *T*'s and *F*'s in its column as the statement $a \supset b$; therefore, the latter statement may be defined by the first, and written

$$a \supset b =_{Df} \bar{a} \vee b \tag{3}$$

The sign $=_{Df}$ expresses *equality by definition*. On its left is the new symbol, or *definiendum*; on its right, a certain combination of old symbols, called *definiens*. An expression like (3) is called a *definition*. A definition is often used to replace a long symbol combination by a short new symbol, which is called an *abbreviation*.

Among the 16 possible arrangements of *T*'s and *F*'s in a column are two that must be discussed separately. The first contains a *T* in every line, and thus represents a combination of elementary statements that is true no

[1] See *ESL*, §§ 7, 9, and chap. viii. In the meaning of the term "adjunctive", the term "extensional" is sometimes used; but, since the meaning of "extensional" is not clearly defined, the term will not be used here. See *ibid*., p. 31.

matter what truth values the elementary statements have. The truth value of the compound statement is thus independent of the truth values of the elementary statements; it is always T. Such a combination is called a *tautology*, or an *analytic statement*. An example of a tautology is given by the statement

$$a \supset b \equiv \bar{a} \vee b \tag{4}$$

Its tautological character is proved by *case analysis*, that is, by assuming for a and b successively the values T and F, and then applying the truth tables for each of the operations occurring in the formula. A T is then found for each of the four possible cases.

The construction of tautologies is the very objective of logic. Since a tautology is true for all truth values of the elementary propositions of which it is composed, it is necessarily true. Tautological character, therefore, expresses *logical necessity*. All logical formulas are tautologies.

The second arrangement to be considered is that in which the column of the truth table contains only the sign F. Such a formula is false for every combination of the truth values of the elementary statements and is therefore called a *contradiction*. It follows from the given characterization that a contradiction is the negation of a tautology. By the negation of (4), for example, we obtain the contradiction

$$\overline{a \supset b \equiv \bar{a} \vee b} \tag{5}$$

This definition explains why a contradiction is called necessarily false. Thus, to say that the formula

$$a \equiv \bar{a} \tag{6}$$

is contradictory means that it is false no matter what truth value the proposition a has.

Statements that are neither analytic nor contradictory are called *synthetic statements*. Though they do not reveal the truth values of their elementary statements, they express a restricting condition for these truth values: the synthetic statement $a \vee b$ excludes the case that both a and b are false. Thus they inform us about physical objects and situations. A tautology, on the contrary, does not exclude any case and thus does not inform us about anything; it is therefore an *empty* statement. The term "empty" must be carefully distinguished from the term "meaningless" used previously. A meaningless symbol combination is neither true nor false, but the empty statement of a tautology is true. Its negation, therefore, is not meaningless either; it is false. Thus (6) is not meaningless but false.

In the list on pages 21–22 a number of useful logical formulas are presented. The proof of their tautological character can be given by case analysis with the help of the truth tables.

A formula written on a separate line should be taken to mean that the formula is asserted to be true. Thus a symbol is saved for assertion. In order to avoid parentheses, a rule of binding force is introduced, expressed by the following listing of operations:

strongest binding force $\cdot \vee \supset \equiv$ weakest binding force

The "and" has the strongest binding force, the equivalence the weakest. The bar of negation, which indicates the *scope* of the negation, connects like parentheses. Thus $a \vee b.c$ means the same as $a \vee (b.c)$; that is why parentheses are omitted in this formula, whereas they cannot be omitted in $(a \vee b).c$.

Tautologies in the Calculus of Propositions

CONCERNING ONE PROPOSITION

1a.	$a \equiv a$	rule of identity (1a–1c)
1b.	$a \vee a \equiv a$	
1c.	$a.a \equiv a$	
1d.	$\bar{\bar{a}} \equiv a$	rule of double negation
1e.	$a \vee \bar{a}$	*tertium non datur*
1f.	$\overline{a.\bar{a}}$	rule of contradiction
1g.	$a \supset \bar{a} \equiv \bar{a}$	*reductio ad absurdum*

SUM

2a.	$a \vee b \equiv b \vee a$	commutativity of "or"
2b.	$a \vee (b \vee c) \equiv (a \vee b) \vee c \equiv a \vee b \vee c$	associativity of "or"

PRODUCT

3a.	$a.b \equiv b.a$	commutativity of "and"
3b.	$a.(b.c) \equiv (a.b).c \equiv a.b.c$	associativity of "and"

SUM AND PRODUCT

4a.	$a.(b \vee c) \equiv a.b \vee a.c$	1st distributive rule
4b.	$a \vee b.c \equiv (a \vee b).(a \vee c)$	2d distributive rule
4c.	$(a \vee b).(c \vee d) \equiv a.c \vee b.c \vee a.d \vee b.d$	twofold distribution (4c–4d)
4d.	$a.b \vee c.d \equiv (a \vee c).(b \vee c).(a \vee d).(b \vee d)$	
4e.	$a.(a \vee b) \equiv a \vee a.b \equiv a$	redundance of a term

NEGATION, PRODUCT, SUM

5a.	$\overline{a.b} \equiv \bar{a} \vee \bar{b}$	breaking of negation line (5a–5b)
5b.	$\overline{a \vee b} \equiv \bar{a}.\bar{b}$	
5c.	$a.(b \vee \bar{b}) \equiv a$	dropping of an always-true factor
5d.	$a \vee b.\bar{b} \equiv a$	dropping of an always-false term
5e.	$a \vee \bar{a}.b \equiv a \vee b$	redundance of a negation

IMPLICATION, NEGATION, PRODUCT, SUM

6a.	$a \supset b \equiv \bar{a} \vee b$	dissolution of implication
6b.	$a \supset b \equiv \overline{a . \bar{b}}$	
6c.	$a \supset b \equiv \bar{b} \supset \bar{a}$	contraposition
6d.	$a \supset (b \supset c) \equiv b \supset (a \supset c) \equiv a . b \supset c$	symmetry of premises
6e.	$(a \supset b) . (a \supset c) \equiv a \supset b . c$	merging of implications
6f.	$(a \supset c) . (b \supset c) \equiv a \vee b \supset c$	
6g.	$(a \supset b) \vee (a \supset c) \equiv a \supset b \vee c$	
6h.	$(a \supset c) \vee (b \supset c) \equiv a . b \supset c$	

EQUIVALENCE, IMPLICATION, NEGATION, PRODUCT, SUM

7a.	$(a \equiv b) \equiv (a \supset b) . (b \supset a)$	dissolution of equivalence
7b.	$(a \equiv b) \equiv a . b \vee \bar{a} . \bar{b}$	
7c.	$\overline{a \equiv b} \equiv (a \equiv \bar{b})$	negation of equivalence
7d.	$(a \equiv b) \equiv (\bar{a} \equiv \bar{b})$	negation of equivalent terms

ONE-SIDED IMPLICATIONS

8a.	$a \supset a \vee b$	addition of an arbitrary term
8b.	$a . b \supset a$	implication from "both" to "any"
8c.	$a \supset (b \supset a)$	arbitrary addition of an implication
8d.	$\bar{a} \supset (a \supset b)$	
8e.	$a . (a \supset b) \supset b$	inferential implication
8f.	$(a \supset b) \supset (a \supset b \vee c)$	addition of a term in the implicate
8g.	$(a \supset b) \supset (a . c \supset b)$	addition of a factor in the implicans
8h.	$(a \vee c \supset b) \supset (a \supset b)$	dropping of a term in the implicans
8i.	$(a \supset b . c) \supset (a \supset b)$	dropping of a factor in the implicate
8j.	$(a \supset b) . (c \supset d) \supset (a . c \supset b . d)$	derivation of a merged implication
8k.	$(a \supset b) . (c \supset d) \supset (a \vee c \supset b \vee d)$	
8l.	$(a \supset b) . (b \supset c) \supset (a \supset c)$	transitivity of implication
8m.	$(a \equiv b) . (b \equiv c) \supset (a \equiv c)$	transitivity of equivalence

The statements and formulas so far considered belong in the *object language*, the language in which we speak about physical objects. When we speak about symbols, however, we use another language, which is called the *metalanguage*. The assertion that a given statement is true belongs in the metalanguage and thus is one level higher than the original, or object, language. Similarly, the terms "tautology", "synthetic", and so on belong in the metalanguage. The distinction between the different levels of language is one of the important discoveries of modern symbolic logic.[2]

The distinction finds a significant application in the construction of connective operations. When we write the tautology

$$\overline{a \vee b} \supset \bar{a} \tag{7}$$

the implication of the formula, although adjunctive, supplies a "reasonable" implication that is free from the paradoxes mentioned above; it is an implication of logical necessity and may therefore be regarded as a connective implication. Connective operations can be defined by the help of the metalanguage; thus a tautological implication is a connective implication. Similarly, a tautological equivalence like (4) supplies a connective equivalence; it expresses the relation of *having the same meaning*. In fact, the definition (3) is admissible only because (4) is a tautological equivalence. However, only analytic connective operations can be constructed thus. Synthetic connective operations, especially a synthetic connective implication, require further means for their definition.

§ 5. The Method of Derivation

The superiority of symbolic logic over the older forms of logic consists in the fact that logical symbolism can be used, like mathematical symbolism, for the derivation of other formulas from given ones. All derivations can be reduced to the application of two rules:

1. Rule of substitution. For a propositional variable in a logical formula it is permissible to substitute any other propositional variable, or a propositional constant, or any compound expression composed of such variables or constants.

Thus in formula (4, § 4), $c \vee d$ may be substituted for a, so that the formula

$$c \vee d \supset b \equiv \overline{c \vee d} \vee b \tag{1}$$

results. This formula, like (4, § 4), is a tautology.

2. Rule of inference. If it is known or has been proved that both a and $a \supset b$ are true, then b may be asserted.

[2] The transition from a sign to a sign for that sign is often expressed by the use of quotation marks. In fact, many of our previous statements should be quoted. Thus I should write, "The statement '$a \vee b$' is true". I omit the quotes for the sake of simplicity and follow the widely accepted rule that their function is, in general, taken over by italics; thus symbols in italics are the names of the corresponding roman symbols. There are a few exceptions to this rule, which, however, are easily understood.

Thus the two statements a and $a \supset b$ may be omitted and b asserted separately. The rule of inference may be symbolized by the schema

$$\frac{\begin{array}{c} a \\ a \supset b \end{array}}{b} \tag{2}$$

In traditional logic the rule of inference is called *modus ponens.*

It is important to realize that the two rules cannot be expressed by the symbols of the calculus, because they are not formulated *in* the calculus but speak *about* the calculus. They refer to a procedure—the procedure of derivation—that is applied to statements in the object language, and therefore they belong in the metalanguage. If a symbolic notation of these rules is desired, it could be constructed only in the metalanguage; in fact, schema (2) can be regarded as a symbolism belonging in the metalanguage. A complete symbolization, however, would include symbols for such terms as "propositional variable".

A *rule* is not a statement; it is a *directive,* i.e., it does not state a matter of fact, but has a character similar to that of a command. It differs from a command only so far as it does not order what is to be done, but grants a permission.[1] The phrases "may be substituted" and "may be asserted", used in the two rules, indicate this directive character. A rule, therefore, cannot be proved to be true or false; these concepts do not apply to directives. Instead, a directive requires a *justification.* It must be shown that the directive serves the purpose for which it is established, that it is a means to a specific end. The procedure of derivation, to which the rules refer, is intended to supply true statements; a justification of the rules is therefore achieved if it can be shown that the rules will always provide true statements. The proof, which is easily given, can be formulated in a *metatheorem.* Thus we can prove the metatheorem that if the rules of substitution and inference are applied to true statements the resulting statements are true. Unlike a rule, a metatheorem is a statement (in the metalanguage) and thus is true or false.

That the metatheorem supplies a justification of the rule is a consequence of the aim of derivation. If we had the aim of deriving false formulas, the two rules would not be a suitable means and therefore would not be justifiable. Or let us consider the aim of deriving tautologies. For this aim the rule of inference, in the form given, is not justifiable; the rule must be restricted to the use of tautological premises, since only then will it supply a tautological conclusion. A justification can be given only with respect to a certain aim.

That the rules of deductive logic require a justification has often been overlooked. The oversight seems to be connected with the fact that the justifica-

[1] A permission has the imperative character of a command, though in a weakened form. See *ESL*, p. 343.

tion is easily given. It is quite the contrary with inductive logic, in which the problem of justification is extremely difficult (see § 91). The justification of induction has been discussed since the time of Hume. If some philosophers believed they could deny the existence of the problem, they were misled by the erroneous conception that deductive logic could be established without a justification of its rules. The recognition of all rules as directives makes it evident that a justification of the rules is indispensable and that justifying a rule means demonstrating a means-end relation.

For the purpose of derivation it is advisable to introduce some further, or *secondary*, rules, which can be derived from the two fundamental rules. One is the *rule of replacement*, which permits the replacement of a propositional expression by another that is tautologically equivalent. Thus we can replace the expression $a \supset b$ by the expression $\bar{a} \vee b$ and proceed from the formula

$$(a \equiv b) \equiv (a \supset b) . (b \supset a) \tag{3}$$

to the formula

$$(a \equiv b) \equiv (\bar{a} \vee b) . (b \supset a) \tag{4}$$

This rule is frequently applied in derivations. A replacement differs from a substitution in that it need not be done in all the places where the replaced expression occurs; furthermore, it may be applied to compound expressions, as in the example given, and is not restricted to a replacement of elementary variables.

During the procedure of derivation, the meaning of the formulas need not be understood; it is sufficient to consider the formulas as combinations of signs with which certain operations are performed. This treatment of formulas is called the *formal conception* of the system. Only when we refer the formulas to physical objects must we understand them, that is, know the meaning of the symbols. We then apply *material thinking* and say that the system is given an *interpretation*, or employed in an *interpreted conception*.

It should be noted that the formal conception, during a derivation, is restricted to the object language. The metalanguage must be used in an interpreted conception. Thus, in order to make an inference, one must know whether the given formulas satisfy the conditions expressed in the rule of inference. Formal manipulations with formulas require material thinking in the metalanguage.

The method of derivation enables us to derive many of the formulas in the list (pp. 21–22) from others. It has been shown that all formulas of the calculus of propositions are derivable from a few axioms. The axioms are:

$$\begin{aligned} & a \vee a \supset a \\ & a \supset a \vee b \\ & a \vee b \supset b \vee a \\ & (a \supset b) \supset (c \vee a \supset c \vee b) \end{aligned} \tag{5}$$

The tautological character of the axioms follows by case analysis. Every formula derivable from them is also a tautology. The implication in the axioms is regarded as defined by (3, § 4).

§ 6. The Calculus of Functions

In the calculus of propositions, the propositions are regarded as wholes and therefore represent undivided units for all operations. The part of logic that analyzes the inner structure of propositions and employs it for operations is called the calculus of functions.

Consider the proposition "Aristotle was a Greek". It is symbolized in the calculus of propositions by the single letter a. But in the calculus of functions we refer to the fact that the proposition has an inner structure; and we separate the subject Aristotle, about whom something is said, from what is said about him, namely, his property of being a Greek. As the symbol of this relation we employ that of the mathematical function and symbolize the proposition in the form

$$f(x) \tag{1}$$

The argument sign x corresponds to the subject about which we speak; the function sign f corresponds to the property holding for it. The symbol (1) indicates the inner structure, or the form, of the proposition.

Since f and x are variables, the expression (1) can be given various meanings by suitable specialization of these variables. If only f is specialized, say, as meaning the property of being a Greek, the expression (1) is not yet a proposition; it can be true or false, depending on what we substitute for the variable x. Thus (1) is true if we put "Aristotle" for x, and false if we put "Goethe" for x. The sign f is called a *propositional function;* the combination $f(x)$ is called a *functional.* If we wish to indicate the variable in a propositional function, we write $f(\hat{x})$; the circumflex distinguishes this expression from the functional $f(x)$. The expression $f(\hat{x})$, therefore, means the same as f. In the same meaning as the term *propositional function*, the word *predicate* is often used. The object correlate of the propositional function, the *property* denoted by it, is called a *situational function.* For instance, the property, or situational function, of redness is denoted by the predicate, or propositional function, "red".

There are two ways of constructing a proposition from a propositional function, or from a functional. The first has been mentioned. We substitute a constant x_1 for the variable x, that is, we go from $f(x)$ to $f(x_1)$. This is the method of *specialization.* It is not permissible, however, to substitute for x a value that makes $f(x)$ meaningless. Thus if f means being a Greek, and we put for x the number 4, the resulting expression, "The number 4 is a Greek", is meaningless. Besides the method of specialization there is a second procedure for constructing a proposition from a propositional function: the method of *binding the variables*,[1] which has two subdivisions.

The first form of binding the variables is the method of generalization, which employs the all-operator, written (x). We thus have

$$(x)\, f(x) \tag{2}$$

We read this expression in the form, "For all x, $f(x)$". It is a statement, not a propositional function, since it is either true or false. If, for instance, f means the property of being a Greek, (2) is false; if f is to mean the property of being a part of nature, (2) is true. Further examples of true all-statements obtain if we put a more complicated expression for $f(x)$, for instance, an implication. Thus we arrive at an expression of the form

$$(x)[f(x) \supset g(x)] \tag{3}$$

If we understand by f the property of being a man, by g the property of being mortal, (3) states that all men are mortal, and thus is true. In fact, we may substitute for x whatever we like, for instance, Mount Everest, since it is correct to say, "If Mount Everest is a man it is mortal".

The second form of binding the variables is constructed by the help of the existential operator, written $(\exists x)$. From $f(x)$ we thus construct the statement

$$(\exists x) f(x) \tag{4}$$

which is read, "There is an x such that $f(x)$". Like (2), the expression (4) is a statement, because it is either true or false. If we understand by f the property of being a Greek, (4) is true; if we interpret f as meaning the property of being an inhabitant of the moon, (4) is false.

The binding of variables thus supplies a statement of a peculiar inner structure. The statement contains a variable x, but as a whole it does not depend on the variable. Such a variable is therefore called a *bound variable* in contradistinction to a *free variable*. To illustrate the nature of a bound variable, we may refer to the variable of a definite integral, which is confined to the inside of the expression, whereas the whole expression does not depend on it.

The expression (3) is called a *general implication*. Although the implication sign occurring in it is of an adjunctive nature, the general implication is free, to a certain extent, from the paradoxes of the *individual implication*. The condition that the adjunctive implication hold for all x excludes certain paradoxical applications. Thus the foregoing example concerning the earth and its revolution around the sun is ruled out when we require for a reasonable implication that it be valid for all things. The statement, "For all x, if x is

[1] This method is frequently called "quantification"; the operators then are called "quantifiers".

a flat disk it revolves around the sun", is not true, even if we interpret the implication as adjunctive. A reasonable individual implication includes a reference to generality; the meaning of implication in conversational language is constructed by a *transfer of meaning from the general to the particular case.* Thus the implication, "If you heat this piece of iron it will expand", is a reasonable individual implication because it is a special case of a valid general implication.

The general implication is therefore an instrument for the definition of a *synthetic connective* implication, also called *physical* implication, which differs from the tautological, or *analytic connective,* implication (7, § 4) in that it expresses a *physical,* not a *logical,* relation. The complete definition of this kind of implication, however, requires some further means; in particular, the case of an always-false implicans and of an always-true implicate must be excluded, since otherwise the paradoxes of the individual adjunctive implication reappear. For this definition, which is given in the frame of a general theory of connective operations, the reader may consult another publication by the author.[2]

The all-operator can be regarded as a generalization of the "and", the existential operator as a generalization of the "or". For a finite range of the variables we thus have the tautological equivalences

$$(x)f(x) \equiv f(x_1) . f(x_2) \ . \ . \ . \ f(x_n) \tag{5}$$

$$(\exists x)f(x) \equiv f(x_1) \vee f(x_2) \vee \ . \ . \ . \ \vee f(x_n) \tag{6}$$

These relations, however, cannot be used in defining the operators, since the operations "and" and "or" are not defined for an infinite number of terms. The operators have, therefore, an independent meaning and can be regarded as generalizations of the operations "and" and "or" for an infinite number of terms.

So far, propositional functions of one variable, or one-place functions, have been considered. There are also propositional functions of several variables, or many-place functions. Their object correlates, i.e., the situational functions denoted by them, are often called *relations,* in contradistinction to *properties,* i.e., situational functions of one variable. Many-place functions are written in the form $f(\hat{x},\hat{y})$, $f(\hat{x},\hat{y},\hat{z})$, and so on. The circumflex notation is convenient for such functions because the symbol f would not indicate the number of variables.

Many-place functions frequently play a part in the language of everyday life. Thus the proposition "Peter is the brother of Paul" contains the propositional function "$\hat{x}$ is the brother of $\hat{y}$", which is symbolized by $f(\hat{x},\hat{y})$; the sentence itself is symbolized by the functional $f(x,y)$. "To be brother" there-

[2] *ESL,* chap. viii.

fore denotes a binary relation. A propositional function of three variables $f(\hat{x},\hat{y},\hat{z})$ occurs in the example, "The mother gives an apple to the child", where the variables x,y,z are represented by the expressions "the mother", "an apple", and "the child", and the term "gives" is symbolized by the function sign f. Another example of a ternary relation is the relation denoted by the word "between". "$\hat{y}$ is situated between $\hat{x}$ and $\hat{z}$" contains a propositional function $f(\hat{x},\hat{y},\hat{z})$.

The two procedures for constructing propositions can be applied also to propositional functions of several variables; the individual variables can be specialized or bound separately. Thus $f(x,\hat{y})$ is a propositional function containing one variable y. The two variables can also be bound in a different way. From $f(x,y)$, for instance, we can construct the proposition

$$(x)(\exists y)f(x,y) \tag{7}$$

Another binding of the variables is given by

$$(\exists x)(y)f(x,y) \tag{8}$$

A third is

$$(\exists y)(x)f(x,y) \tag{9}$$

Statements (7) and (9) are not identical because the order of the operators is relevant. This fact may be demonstrated by two examples in which we replace $f(x,y)$ by the more complicated expression

$$g(x) \supset g(y) . h(x,y) \tag{10}$$

If we understand by g "to be a natural number", by h "to be smaller", then (7) holds, since it then means, "For every natural number x there is a natural number y such that x is smaller than y". Statement (9) means, in this interpretation, "There is one number y such that all x are smaller than y". The statement, however, is false, since there is no greatest number. We obtain an example satisfying both formulas if we understand by h the relation "to be not smaller". Statement (9) then says, "There is a natural number y such that all natural numbers are not smaller than y". Such a number y exists: it is the number zero. Statement (7) is then *a fortiori* satisfied. It is obvious that (9) is the stronger statement; (9) implies (7), but not vice versa (16*a*). Whereas (7) represents the meaning of "every", (9) formulates the meaning of "all". The distinction between "all" and "every" is therefore expressed in terms of the order of operators. Two operators of the same kind, however, are commutative (15*a*, 15*b*).

An important connection between all-statements and existential statements is expressed by the formulas

$$\overline{(x)f(x)} \equiv (\exists x)\overline{f(x)} \tag{11}$$

$$\overline{(\exists x)f(x)} \equiv (x)\overline{f(x)} \tag{12}$$

They exhibit an important feature of the notation of the calculus of functions. The binding of variables is symbolized by a prefixed operator in order to express the two forms of negation resulting from the possibilities of negating the whole expression and the function alone. The expression

$$\overline{(x)f(x)} \tag{13}$$

therefore must be distinguished from the expression

$$(x)\overline{f(x)} \tag{14}$$

The two statements are distinguished by the scope of the negation, which in (13) includes the operator, whereas it does not in (14). The expression (14) is the stronger statement, that is, (14) implies (13), but not vice versa (13*c*).

The expression $f(x)$ in (2) and (4) is called the *operand* of the statement; the whole expression, including the operator and the operand, is called the *scope* of the operator. Not always does the scope of an operator extend over the whole formula. Thus, in the expression

$$(x)f(x) \supset (\exists y)g(y) \tag{15}$$

the scope of the all-operator is the implicans, and that of the existential operator is the implicate. Whereas (2) and (4) are *one-scope formulas*, (15) is not.

In (7) the operand of the existential operator is the expression $f(x,y)$, whereas the operand of the all-operator is given by the expression $(\exists y)f(x,y)$. This distinction explains the difference in the meanings of (7) and (9). In spite of this distinction, formulas like (7) and (9), in which the operators are assembled at the beginning and extend over the rest of the formula, are regarded as one-scope formulas.

For reasons of expediency, a combination of an operator with a proposition will be regarded as meaningful, and as meaning the same as the proposition. Thus we have

$$(x)a \equiv a \tag{16}$$

$$(\exists x)a \equiv a \tag{17}$$

Then a is called a constant. These relations are useful, in particular, when a is a functional expression that does not contain the variable x.

It is often expedient to express generality, not by means of an operator, but through a free variable. For this purpose the convention is introduced that if an expression containing a free variable is asserted, it is meant to hold for all values of the free variable. Thus the expression

$$f(x) \vee \overline{f(x)} \tag{18}$$

is true for all values of x and can be asserted; the variable x is then a free variable. The free variable, therefore, has the same meaning as an all-operator the scope of which is the whole formula. So it is permissible, in formulas containing free variables, to put an all-operator referring to these variables before the whole formula as its scope; this is called the *rule for free variables.*

Free variables have been used in the calculus of propositions, since the letters a, b, and so on, occurring in tautologies, express free variables. Similarly, the letter f in (18) expresses a free variable. The use of free variables is possible when the scope of the generalization is not restricted to parts of the formula. Mathematical notation makes frequent use of free variables. In a mathematical identity like

$$(x + y)^2 = x^2 + 2xy + y^2 \tag{19}$$

the letters x and y represent free variables the range of which is the domain of numbers. In conversational language, a free variable is expressed by the word "any". The means of expression, however, are limited for free variables; the difference between the expressions (13) and (14) is not expressible in terms of free variables.

The concept of tautology can be defined for functions, too. A tautology is a formula that is true for all values of the argument and of the functional variables. The derivation of tautologies, however, is more complicated, since a method of case analysis for functions cannot be carried through generally, although it can be applied, in a generalized form, to certain kinds of formulas. A table of tautologies in functions is presented below.

The axiomatization of the calculus of functions has shown that, besides the axioms (5, § 5) of the calculus of propositions, it is sufficient to employ formulas (14*a*) and (14*b*) as axioms. All tautologies in functions are then derivable.

The functional calculus so far developed is called the *simple calculus of functions.* It is distinguished from the *higher calculus of functions*, in which functional variables are bound by operators or occur as arguments of functions of a higher type. The higher calculus will not be presented in this book because it is not required for the exposition of the theory of probability.

Tautologies in the Calculus of Functions

FORMULAS CONCERNING FUSION OR DIVISION OF OPERANDS

9*a*. $(x)[f(x) . g(x)] \equiv (x)f(x) . (x)g(x)$

9*b*. $(x)f(x) \vee (x)g(x) \supset (x)[f(x) \vee g(x)]$

9*c*. $(x)[f(x) \vee g(x)] \supset (x)f(x) \vee (\exists x)g(x)$

9*d*. $(x)[f(x) \supset g(x)] \supset [(x)f(x) \supset (x)g(x)]$

9*e*. $(x)[f(x) \supset g(x)] \supset [(\exists x)f(x) \supset (\exists x)g(x)]$

9*f*. $(x)[f(x) \equiv g(x)] \supset [(x)f(x) \equiv (x)g(x)]$

9*g*. $(x)[f(x) \equiv g(x)] \supset [(\exists x)g(x) \equiv (\exists x)g(x)]$
9*h*. $(x)f(x) \cdot (x)[f(x) \supset g(x)] \supset (x)g(x)$

10*a*. $(\exists x)f(x) \cdot g(x) \supset (\exists x)f(x) \cdot (\exists x)g(x)$
10*b*. $(\exists x)[f(x) \lor g(x)] \equiv (\exists x)f(x) \lor (\exists x)g(x)$
10*c*. $(\exists x)[f(x) \supset g(x)] \equiv (x)f(x) \supset (\exists x)g(x)$
10*d*. $[(\exists x)f(x) \supset (\exists x)g(x)] \supset (\exists x)[f(x) \supset g(x)]$
10*e*. $[(\exists x)f(x) \supset (x)g(x)] \supset (x)[f(x) \supset g(x)]$
10*f*. $(\exists x)f(x) \cdot (x)g(x) \supset (\exists x)[f(x) \cdot g(x)]$

11*a*. $(x)[a \cdot f(x)] \equiv a \cdot (x)f(x)$
11*b*. $(x)[a \lor f(x)] \equiv a \lor (x)f(x)$
11*c*. $(x)[a \supset f(x)] \equiv a \supset (x)f(x)$
11*d*. $(x)[f(x) \supset a] \equiv (\exists x)f(x) \supset a$
11*e*. $(x)[f(x) \equiv a] \supset [(x)f(x) \equiv a]$
11*f*. $[(x)a] \equiv a$

12*a*. $(\exists x)[a \cdot f(x)] \equiv a \cdot (\exists x)f(x)$
12*b*. $(\exists x)[a \lor f(x)] \equiv a \lor (\exists x)f(x)$
12*c*. $(\exists x)[a \supset f(x)] \equiv a \supset (\exists x)f(x)$
12*d*. $(\exists x)[f(x) \supset a] \equiv (x)f(x) \supset a$
12*e*. $[(\exists x)f(x) \equiv a] \supset (\exists x)[f(x) \equiv a]$
12*f*. $[(\exists x)a] \equiv a$

FORMULAS CONCERNING NEGATION OF OPERATORS

13*a*. $\overline{(x)f(x)} \equiv (\exists x)\overline{f(x)}$
13*b*. $\overline{(\exists x)f(x)} \equiv (x)\overline{f(x)}$
13*c*. $(x)\overline{f(x)} \supset \overline{(x)f(x)}$
13*d*. $\overline{(\exists x)f(x)} \supset (\exists x)\overline{f(x)}$

FORMULAS OF SUBALTERNATION

14*a*. $(y)f(y) \supset f(x)$
14*b*. $f(x) \supset (\exists y)f(y)$
14*c*. $(x)f(x) \supset (\exists x)f(x)$

FORMULAS CONCERNING THE ORDER OF OPERATORS

15*a*. $(x)(y)f(x,y) \equiv (y)(x)f(x,y)$
15*b*. $(\exists x)(\exists y)f(x,y) \equiv (\exists y)(\exists x)f(x,y)$

16*a*. $(\exists x)(y)f(x,y) \supset (y)(\exists x)f(x,y)$
16*b*. $(\exists x)(y)f(x) \cdot g(y) \equiv (y)(\exists x)f(x) \cdot g(y)$
16*c*. $(\exists x)(y)[f(x) \lor g(y)] \equiv (y)(\exists x)[f(x) \lor g(y)]$
16*d*. $(\exists x)(y)[f(x) \supset g(y)] \equiv (y)(\exists x)[f(x) \supset g(y)]$
16*e*. $(x)(y)[f(x,y) \lor g(x,y)] \supset (\exists x)(y)f(x,y) \lor (x)(\exists y)g(x,y)$

17a. $(x)(y)f(x,y) \supset (x)f(x,x)$
17b. $(\exists x)(y)f(x,y) \supset (\exists x)f(x,x)$

§ 7. The Calculus of Classes

The calculus of functions leads to the introduction of a new and important concept: the concept of *class*. If a propositional function $f(\hat{x})$ is given, all arguments x that satisfy $f(\hat{x})$ can be incorporated in one class, the class F. Every propositional function thus defines a class. Vice versa, every class can be regarded as defined by a propositional function. The arguments x for which $f(x)$ is true are called *members*, or *elements*, of the class F. For the expression "x is a member of F", we write

$$x \in F \tag{1}$$

The symbol $\in$ corresponds to the copula of conversational language, for instance, to "is a" in "Saddle Peak is a mountain".

Because of the equivalence of propositional functions and classes it is not necessary to consider the concept of class as a primitive concept. We can rather conceive (1) as defined by $f(x)$:

$$x \in F =_{Df} f(x) \tag{2}$$

It should be noted that we do not thus define the concept of class, but the total expression, "x is a member of the class". This is not a deficiency of the theory, since the concept of class is never used independently. All statements referring to classes are translatable into statements in which the total combination (1) occurs. It suffices, therefore, that the combination has a meaning. The definition (2) is called a *definition in use* of the concept of class. The term "set" is used in the same sense as the term "class". The class F is also called the *extension* of the propositional function $f(\hat{x})$.

Similarly, we can go from many-place functions to the corresponding class. The extension of a function $f(\hat{x},\hat{y})$ is given by the class of the couples $x.y$ that satisfy the propositional function. In analogy with (2) we write

$$x.y \in F =_{Df} f(x,y) \tag{3}$$

From functions of three variables we can go in a similar way to the class of triplets. The procedure can be extended to functions of more variables.

The calculus of classes is concerned with operations that can be performed with classes. We obtain these operations by performing operations with the corresponding propositional functions and then transferring the results to classes.

To begin with propositional functions of one argument: since the range of the arguments that make $f(x)$ meaningful is a definite one, the class of the arguments x that make $f(x)$ false is as well determined as the class of the arguments that make $f(x)$ true. The class of the arguments that make $f(x)$ false is called *complementary class,* or *complement* of F, denoted by $\bar{F}$. In symbolic notation the definition reads

$$x \,\epsilon\, \bar{F} =_{Df} \overline{x \,\epsilon\, F} \tag{4}$$

Thus the class of the nonprime numbers is a well-determined class.

The introduction of the complementary class can be conceived as an operation by which a certain class is constructed from a given class. This operation is analogous to the operation of negation applied to propositions. Correspondingly, we can define operations between classes that are analogous to the other operations of the calculus of propositions. We then construct a new class H from the two given classes F and G by the corresponding operation. This procedure can be carried through in two different ways: the new class can be constructed as a class of members x, so that it is of the same kind as the original classes; or the couples x,y, which are constructed from members of the two classes F and G, can be used as members of the new class.

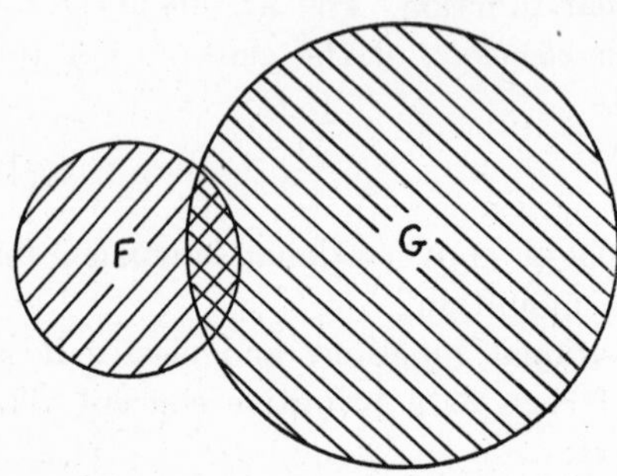

Fig. 1. Joint class and common class: joint class, according to (5), of classes F and G is whole shaded area; common class, according to (6), is double-shaded area.

Beginning with the first procedure, we define

$$x \,\epsilon\, F \vee G =_{Df} (x \,\epsilon\, F) \vee (x \,\epsilon\, G) \qquad \text{joint class, or disjunct} \tag{5}$$

$$x \,\epsilon\, F \,.\, G =_{Df} (x \,\epsilon\, F) \,.\, (x \,\epsilon\, G) \qquad \text{common class, or conjunct} \tag{6}$$

The classes so constructed are illustrated in figure 1. The joint class is produced by joining the two classes F and G; their common members, however, are counted only once. For instance, if two societies F and G sponsor a congress, persons who are members of either of the societies are entitled to take part, no matter whether the same person belongs to both societies. The class of the persons entitled to take part in the congress is therefore the joint class of the two societies. The common class is composed of persons who belong to both societies.

The second procedure mentioned provides classes the elements of which are the couples x,y. Two definitions, analogous to the preceding definitions, can be used:

$$x,y \,\epsilon\, F_{\hat{x}} \vee G_{\hat{y}} =_{Df} (x \,\epsilon\, F) \vee (y \,\epsilon\, G) \qquad \text{couple disjunct} \tag{7}$$

$$x,y\ F_{\hat{x}} . G_{\hat{y}} =_{Df} (x \,\epsilon\, F) \,.\, (y \,\epsilon\, G) \qquad \text{couple conjunct, or combination class} \tag{8}$$

The circumflex over the subscripts has a similar meaning as in the expression $f(\hat{x})$; it effects a sort of binding of the variable, and substitutions for the other letters x and y can be made without a change in the subscripts. The difference between these definitions and the two preceding ones becomes clear when the corresponding propositional functions are considered. The definitions (5) and (6) are equivalent to the transition from the two propositional functions $f(\hat{x})$ and $g(\hat{x})$ to the new propositional function of one variable:

$$f(\hat{x}) \vee g(\hat{x}) \tag{9}$$

$$f(\hat{x}) . g(\hat{x}) \tag{10}$$

The definitions (7) and (8), however, are equivalent to the transition from the same propositional functions to the new propositional functions of two variables:

$$f(\hat{x}) \vee g(\hat{y}) \tag{11}$$

$$f(\hat{x}) . g(\hat{y}) \tag{12}$$

To the function (11) corresponds the class defined in (7); to the function (12), the class defined in (8). An example of a couple conjunct, or combination class, is the class of possible telephone connections between subscribers in two cities. The corresponding couple disjunct is represented by the class of telephone connections that any of the parties in either city can have with any party in the entire country, that is, the class of telephone connections for which at least one party is a subscriber in one of the two cities.

A third form of operations will now be considered. The formation of classes defined in (7) and (8) can be specialized by restricting membership to certain couples x,y that are selected by a coupling relation $e(x,y)$ that establishes a one-one correspondence. The one-one correspondence may be symbolized by adding a subscript i to the variables, such that x_i,y_i denotes a couple of corresponding members. Thus the following definitions, analogous to (7) and (8), obtain:

$$x_i,y_i \in F_{\hat{x}_i} \vee G_{\hat{y}_i} =_{Df} (x_i \in F) \vee (y_i \in G) \quad \text{narrower couple disjunct} \tag{13}$$

$$x_i,y_i \in F_{\hat{x}_i} . G_{\hat{y}_i} =_{Df} (x_i \in F) . (y_i \in G) \quad \text{narrower couple conjunct} \tag{14}$$

The construction of the classes (13) and (14) is equivalent to the transition from the propositional functions $f(\hat{x})$ and $g(\hat{y})$ to the new propositional function of two variables, defined in terms of the coupling relation $e(x,y)$:

$$f(\hat{x}_i) \vee g(\hat{y}_i) \tag{15}$$

$$f(\hat{x}_i) . g(\hat{y}_i) \tag{16}$$

To eliminate the subscript i, the coupling relation $e(\hat{x},\hat{y})$ can be substituted; then the functions result:

$$e(\hat{x},\hat{y}) . [f(\hat{x}) \vee g(\hat{y})] \tag{17}$$

$$e(\hat{x},\hat{y}) . f(\hat{x}) . g(\hat{y}) \tag{18}$$

In application to probability relations, the notation by means of the subscript will be used.

An example of a narrower couple disjunct is the class of married couples one part of which belongs to family F or family G. The class of married

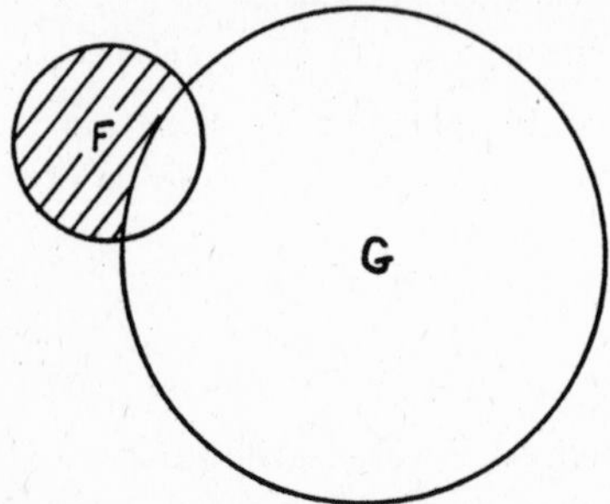

Fig. 2. Implication class of F and G: according to (19), all points of plane, except shaded area, belong to implication class.

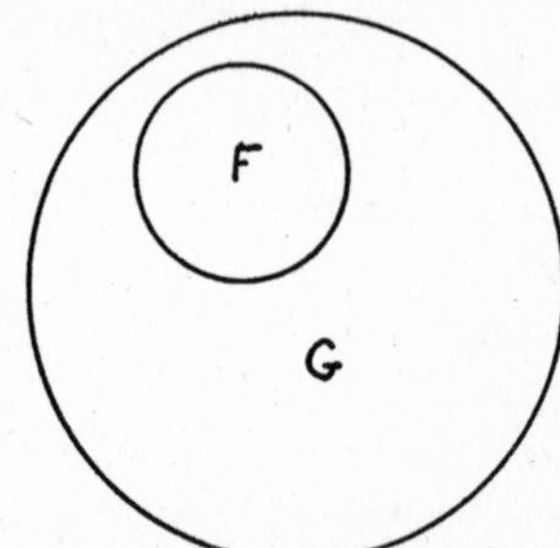

Fig. 3. Relation of class inclusion: F is included in G, according to (20*a*), (20*b*), and (21).

couples of which one part is white and the other colored, the class of marriages between whites and Negroes, is an example of a couple conjunct in the narrower sense.

The classes (5) and (6) can be conceived as special cases of the classes (13) and (14), resulting when the coupling relation $e(x,y)$ is the identity. Note that the couple disjunct and the couple conjunct can be defined even for a single class F.

For operations with classes, we use the same symbols as for operations with propositions. This simplification is possible because of the isomorphism between the calculus of classes and the calculus of propositions. The use of capital letters indicates that the operations standing between them are class operations and thus constitute classes, not propositions. The method can be extended to include further propositional operations; and, moreover, the transformations holding for propositions can be transferred to classes. Thus the implication class can be defined by the formula

$$x \, \epsilon \, F \supset G =_{Df} x \, \epsilon \, \bar{F} \vee G \tag{19}$$

This means that the implication class $F \supset G$ is the joint class of $\bar{F}$ and G. Assume that in figure 2 the domain of all meaningful arguments is represented by the points of the whole plane of the figure; then the implication class (19) is represented by all points of the plane except the shaded area.

A particularly important case results if the implication class is identical with the entire domain of all meaningful arguments. We then have

$$(x)(x \in F \supset G) \tag{20a}$$

or, in another notation,

$$(x)[(x \in F) \supset (x \in G)] \tag{20b}$$

Because of (2) this relation is identical with the general implication (3, § 6). An illustration of the resulting relation between the classes F and G is presented in figure 3. The class F is included in the class G; this relation is called *class inclusion*. For it the symbolization

$$F \subset G \tag{21}$$

is used. It should be noted that (21) is a proposition, not a class, since it means the same as (20*a*) or (20*b*). For this reason the usual implication sign, which would establish a class, cannot be used. Its place is taken by the reversed implication sign, which, when placed between capital letters, has the same meaning as the usual implication sign between small letters.

The class F included in G is also called a subclass of G. Since (21) is also true when we put F for G, every class is its own subclass. This terminology is inescapable because every proposition implies itself. Examples of class inclusion are found in biological classifications: the class of lions is a subclass of the class of mammals. This notation, now generally accepted in symbolic logic, distinguishes clearly between the relations of class inclusion and class membership, which are confused in traditional logic.

The class of all things is called *universal class* and is denoted by V. Its complement, the class that contains nothing and thus is empty, is the *null class*, denoted by Λ. The null class is a subclass of every class; this follows because a false proposition implies every proposition.

Two classes F and G are called identical when the relation

$$(x)[x \in F \equiv x \in G] \tag{22}$$

holds. We express the identity of classes by the symbol

$$F = G \tag{23}$$

which is defined by (22). Like (21), the expression (23) is a proposition, not a class. The symbols for class inclusion and class identity enable us to write propositions in class notation. Thus for (21) we can write

$$(F \supset G) = V \tag{24}$$

The following relation is a tautology:

$$(F \vee \bar{F}) = V \tag{25}$$

Classes were introduced as extensions of propositional functions. It is often convenient to regard propositions as degenerate cases of functions, and to speak of the extension of a proposition. In correspondence with the relations (16 and 17, § 6), the universal class will be regarded as the extension of a true proposition, the null class as the extension of a false proposition.

For class identity, or identity of extension, it is sufficient that (22) is true. In order that the meanings of the symbols F and G be identical (identity of intension), it is required, moreover, that (22) be either analytic or contain a synthetic connective equivalence.

§ 8. Axiomatic Systems

Whereas the formulas of logic are tautologies, science is constructed from synthetic statements. The aim of the scientist is to make, not empty statements, but statements that inform us about the physical world; and the assertion that certain synthetic statements are true is the very task of science. The proof of the truth is based on experience and observation, sources of knowledge that play no part in logic. A mere collection of true synthetic statements, however, would not be called science; they must be ordered logically so as to form a deductive system. Some synthetic statements, called *axioms*, are placed at the top, and all other statements are derived from them by logical methods. The ideal form of a science is therefore the *axiomatic system*, an aim that individual sciences have attained more or less successfully.

Derivations from synthetic premises are made by the same methods that are used for the derivation of tautologies. For the purpose of such derivations it is always permissible to add tautologies to synthetic premises; as tautologies are empty, their addition does not enlarge the empirical content of the axioms and thus cannot adulterate it. The process of derivation employs the rules of substitution, inference, replacement, free variables, and so on, that were presented above. By means of a logical device it is possible to reduce a derivation from synthetic axioms to a derivation of tautologies. For this purpose the conjunction of synthetic axioms is conceived as a statement a. If b is the theorem to be derived, the implication $a \supset b$ is then tautological.[1] The derivation of synthetic statements from synthetic axioms can then be replaced by the derivation of tautologies of the particular form $a \supset b$, where a represents the given set of axioms.

This conception of a derivation from synthetic premises shows clearly that the contribution of logic to science consists in the construction of tautologies, namely, of formulas of the form $a \supset b$; the truth of the axioms a, however, is not a matter of logic. The contribution of mathematics to science

[1] The only precaution to be taken is that a must not contain free variables; the expression a must therefore be *closed* by means of binding the variables through all-operators. See *ESL*, pp. 105, 144, 237.

is to be interpreted in the same way. Mathematics, too, represents a method of deriving consequences from given sets of axioms, whereas the truth of the axioms is irrelevant for mathematics. Thus mathematics is concerned only with tautological implications. I therefore agree with Russell in regarding mathematics as a branch of logic, a branch of a complicated structure that has grown from the treatment of axiom systems of practical significance.

Strangely enough, the assertion that mathematics deals only with tautologies has been regarded by some mathematicians as a disparagement of mathematical investigation. Such a judgment, of course, results from a misunderstanding, for the given definition of mathematics in no way diminishes the value of mathematical inquiry. The construction of complicated tautologies, as expressed in mathematical theorems, rather demands an extraordinary display of ingenuity and sagacity. Even if the tautology itself does not assert anything, the statement that a certain complicated symbol combination is a tautology can represent a discovery of the highest value.

The given definition of mathematics determines the construction of the calculus of probability, which will be given in axiomatic form. Disregarding the question of the truth of the axioms, I shall first present the conclusions that can be inferred from them. In fact, the mathematical calculus of probability is nothing but the deductive system that deals with the relations between axioms and theorems. The truth of the axioms will be discussed later.

The axiom systems of mathematics, apart from being synthetic, differ from logical formulas in containing certain symbols that are not used in the purely logical calculus and for which a meaning is not specified. Like the *truth* of the axioms, the *meaning* of the unspecified symbols is not determined by logic or mathematics: the logical relations of an axiomatic system can be developed without anything being known about the meaning of the symbols. Thus the axioms of geometry contain symbols for the concepts "point", "straight line", "plane", and so on; it is not necessary, however, to make use of the meaning of these symbols for the derivation of geometrical theorems. The derivation can be carried through in the *formal conception* of the axiomatic system. The formal conception differs from the one employed in § 5 in that the meanings of the logical symbols are regarded as known; only the non-logical, unspecified symbols are used formally, i.e., without reference to their meanings.

In order to proceed to an *interpreted conception,* we assign meanings to the unspecified symbols. Thus the symbol "point", used in the axiomatic system of geometry, is taken to mean a small piece of matter; "straight line", a light ray, and so on. Since the meanings are not determined by the axiomatic system, they are arbitrary to a certain degree. In fact, the unspecified symbols may be used in different meanings, that is, may be coördinated to various kinds of objects. We therefore speak of different interpretations of the unspecified symbols. For instance, "point" may mean a number triplet; "straight line", a pair of linear equations; "plane", a linear equation. The interpretation then differs from the usual interpretation of the axiomatic system and results in analytic geometry. The rules that coördinate an interpretation to the unspecified symbols are called *coördinative definitions.*

The interpretation, however, is not completely arbitrary. The objects coördinated to the unspecified symbols must satisfy the relations postulated in the axioms; only if this condition is fulfilled is an *admissible interpretation* constructed. If the coördinated objects are of an empirical nature, for instance, pieces of matter and light rays, the proof that the interpretation is admissible is given by reference to physical laws. If the coördinated objects are mathematical constructions, for instance, number triplets and equations, the admissibility of the interpretation is proved by mathematical laws. The plurality of interpretations explains why the same axiomatic system can be applied to various subjects. All these subject matters are, however, *isomorphous*, that is, they have the same logical structure, a consequence expressed in the fact that they constitute admissible interpretations of the same axiomatic system.

Only when the unspecified symbols are interpreted can the axioms be called true or false. So long as no interpretation is given, the axioms are neither true nor false, but represent definitions of the unspecified symbols. However, since the meaning of the latter is not completely determined by the axioms, the definitions merely set up certain relations that are to hold between the unspecified symbols; they define only certain structural properties for these symbols. Such definitions are called *implicit definitions.* They cannot be written in the form of *explicit definitions* (see p. 19); for implicit definitions a separation of the *definiendum* from a *definiens* is not possible.

The incomplete character of implicit definitions, so far as the meaning of the unspecified symbols is concerned, is demonstrated by the plurality of interpretations. When implicit definitions include more than one unspecified symbol, a further peculiarity results, since the question whether a certain physical object may be regarded as an interpretation of one of the symbols cannot be answered until the interpretations of the other unspecified symbols are added. If, for instance, "points" are taken to mean small objects, "straight lines" to mean light rays, and so on, a system of objects is obtained that fits the implicit definitions of these concepts in the geometrical axiom system. If, however, "straight lines" are understood to be a pair of linear equations, this interpretation cannot be combined with the coördinative definition "small object" for point. Implicit definitions are, therefore, of a logical nature very different from that of explicit definitions. They are, so to say, hollow forms that can be filled with different materials.

The coördination of an interpretation to an axiomatic system is also called the construction of a model. Thus the system of formulas of analytic geometry represents a model of the axiomatic system of Euclidean geometry. Geometricians have also developed models of non-Euclidean systems of axioms, as, for instance, the so-called Klein model of the Bolyai-Lobachevski geometry.

After the construction of an axiomatic system, the question of its consistency must be considered. If it is possible to coördinate to an axiomatic system another system in the sense of an interpretation, the consistency of the first system is reduced to that of the second. Thus the use of analytic geometry as a model of Euclidean geometry proves that the latter is consistent in the same sense as arithmetic, since analytic geometry represents a part of arithmetic. In other words, if the consistency of arithmetic is established, the consistency of Euclidean geometry is established likewise. Such reductions have great practical significance, but their ultimate value depends on the reasons that can be adduced for the consistency of certain of the simplest systems of axioms, such as the systems of logic and arithmetic. A discussion of these problems, however, is beyond the province of this book. Suffice it to say that the consistency of the calculus of propositions and of the simple calculus of functions can be proved, whereas the consistency of the higher calculus of functions, and with it that of mathematics, is still an unsolved problem.

Chapter 3

ELEMENTARY CALCULUS OF PROBABILITY

ELEMENTARY CALCULUS OF PROBABILITY

§ 9. The Probability Implication

The investigation of the concept of probability begins with an analysis of the logical structure of probability statements. The problem, which so far has not been given sufficient attention in the mathematical calculus of probability, is amenable to precise solution with symbolic methods. Symbolic logic has devised means of characterizing the logical form of a statement without regard to its content; these methods can be extended to include a characterization of probability statements. The formalization of the probability statement, in fact, is one of the first objectives in the philosophy of probability.

To consider a typical probability statement: when a die is thrown, the appearance of face 1 is to be expected with the probability $\frac{1}{6}$. This statement has the logical form of a relation. It is not asserted unconditionally that face 1 will appear with the probability $\frac{1}{6}$; the assertion, rather, is subject to the condition that the die be thrown. If it is thrown, the occurrence of face 1 is to be expected with the probability $\frac{1}{6}$; this is the form in which the probability statement is asserted. No one would say that the probability of finding a die on the table with face 1 up has the value $\frac{1}{6}$, if the die had not been thrown. Probability statements therefore have the character of an implication; they contain a first term and a second term, and the relation of probability is asserted to hold between these terms. This relation may be called *probability implication*. It is represented by the symbol

$$\underset{p}{\not\supset}$$

This is the only new symbol that the probability calculus adds to the symbols of the calculus of logic. Its connection with *logical implication* is indicated by the form of the symbol: a bar is drawn across the sign of logical implication. Whereas the logical implication corresponds to statements of the kind, "If a is true, then b is true", the probability implication expresses statements of the kind, "If a is true, then b is probable to the degree p".

The terms between which the probability implication holds are usually events. Let x be the event, "The die is thrown", and y the event, "The die has come to rest on the table"; then a probability implication between the two events is asserted. We recognize at once that this requires a more exact formulation. We speak of a definite probability only when the event is characterized in a certain manner, namely, as an event y in which face 1 is up. This means that the event y is regarded as belonging to a certain class B. We are dealing with a class, since the individual features of the event y are disregarded in the statement. It does not matter on what part of the table the die lies, or in which direction its edges point; only the attribute of having face 1 up is considered. Thus the event y is characterized only as to whether it can be said to belong to the class B. The same applies to the event x, since we do not consider with what force the die is thrown or what angular momentum is imparted to it; we demand only that x be a throw of the die, that it belong to a certain class A. Therefore we write the probability statement in the form

$$x \in A \underset{p}{\Rightarrow} y \in B \qquad (1)$$

This formulation, however, requires modification. We must express the fact that the elements of the classes are given in a certain order, for instance, in the order of time. In other words, the event x belongs to the discrete sequence of the events $x_1, x_2, \ldots x_i \ldots$, while at the same time the event y belongs to a corresponding sequence $y_1, y_2, \ldots y_i \ldots$ There is a one-one correspondence between the elements of the two sequences, expressed by equality of subscripts, and we assert only a probability implication between the corresponding elements x_i, y_i, so that we write, instead of (1),

$$x_i \in A \underset{p}{\Rightarrow} y_i \in B \qquad (2)$$

The coördination of the event sequences is necessary for the following reason. We do not wish to say that the probability implication holds, for instance, between the event x_i of throwing the die and the event y_{i+1} of obtaining a certain result. When we merely state that the event x_i of throwing the die occurs, we have not yet asserted that the event x_{i+1} of throwing the die will also occur and that, therefore, a probability for the occurrence of the event y_{i+1} exists.

However, even (2) does not completely represent the form of the probability statement; we must add the assertion that the same probability implication holds for each pair x_i, y_i. This generalization is expressible by two all-operators, meaning, "for all x_i and for all y_i". Using an abbreviation, we can reduce the two all-operators to one by placing only the subscript i in the parentheses of the operator. Thus the probability statement is written

$$(i)\ (x_i \in A \underset{p}{\Rightarrow} y_i \in B) \qquad (3)$$

This expression represents the final form of the probability statement: *The probability statement is a general implication between statements concerning a class membership of the elements of certain given sequences.*

To illustrate this formulation of the probability statement: a relation of the kind described is employed in dealing with the probability of a case of influenza leading to death. We do not speak unconditionally of the probability of the death of the patient, but only of the probability resulting from the fact that he has contracted influenza. Here again are two classes—the class of influenza cases and the class of fatal cases—and the probability implication is asserted to hold between them. If x_i is interpreted as a result of medical diagnosis, A as an influenza case, y_i as the state of the patient after one week of illness, and B as the death of the patient, then this example of a probability statement from the field of medicine has the form (3).

Another example is the probability of hitting a target during a rifle match. Here x_i represents the single shot, y_i the hit scored at the target, B the class of hits within a certain range, and A the class to which the rifleman belongs according to his ability. The probability of a hit will be different according to the contestant's degree of skill. Here again the probability is determined only when the classes A and B are chosen.

An example from physics is the bombardment of nitrogen by α-rays, or helium nuclei. There is a certain probability that a helium nucleus will eject a hydrogen nucleus from the nitrogen atom. Let A represent the class of α-rays, x_i the hit of an individual helium nucleus, and y_i the event produced by it. The event results in the occasional emission of a hydrogen nucleus, that is, it belongs to the class B. Although it is not possible to observe directly the causal connection between the helium nucleus and the released hydrogen nucleus, we assume, nevertheless, a one-one correspondence between x_i and y_i. Using a very weak radioactive preparation that rarely emits helium nuclei, we can employ the temporal coincidence observed for the α-rays and the hydrogen rays as a criterion of the correspondence.

In the previous examples, x_i and y_i stand in the relation of cause to effect, but other instances can easily be found in which y_i represents the cause and x_i the effect. In this case we carry out a reverse inference, from the effect to the probability of a certain cause, for example, in investigating the cause of a cold. And there are other examples for which the relation x_i to y_i is of a still different type. There exists a probability that a certain position of the barometer indicates rain, but there is no direct causal connection between the two events. In other words, one is not the cause of the other. Rather, the two events are effects produced by a common cause, which leads to their concatenation in terms of probabilities. It is easily seen that these examples also conform to the logical structure of (3).

The analysis presented shows that the probability implication can be regarded as a relation between classes. The class A will be called the *reference class;* the class B, the *attribute class*. It is the probability of the attribute B that is considered with reference to A. It must be added, however, that the probability relation between the two classes A and B is determined only after the elements of the classes are put into a one-one correspondence and ordered in sequences. For instance, the probability implication holding between the birth and the subsequent death of an infant—the rate of infant mortality—differs from one country to another, that is, it differs according to the sequence of events for which the statistics are tabulated. Even for an individual die there exists a particular pair $x_i y_i$ of sequences, and it is an assertion derived from experience that the probability remains the same for different dice. Therefore, strictly speaking, the probability implication must be regarded as a three-term relation between two classes and a sequence pair. The pair of sequences provides the *domain* with respect to which the probability implication assumes a determinate degree. Later the conception is extended to combinations of more than two sequences. The significance of the order of sequences is the subject of chapter 4.

Because of the equivalence that exists between classes and propositional functions, formula (3) may be expressed in a somewhat different way. According to (2, § 7), we may use instead of the statement $x \epsilon A$ the corresponding propositional functional $f(x)$ and, similarly, instead of $y \epsilon B$, the corresponding propositional functional $g(y)$. Then we must express the one-one correspondence between the sequences of x and y by a one-one functional $e(x,y)$ in order to determine for each x the corresponding value y. Thus (3) assumes the form

$$(x)(y)[f(x).e(x,y) \underset{p}{\Rightarrow} g(y)] \tag{4}$$

In this form it is not necessary to employ the subscript i, if the order of the elements is regarded as understood.

A special kind of probability implication is included in the general form (3) or (4). It may happen that the sequences coincide and that the elements x_i and y_i are identical. The function $e(x,y)$ then reduces to the identity relation. We thus obtain, instead of (3) and (4),

$$(i)\ (x_i \epsilon B \underset{p}{\Rightarrow} x_i \epsilon B_k) \tag{5}$$

$$(x)\ [f(x) \underset{p}{\Rightarrow} g(x)] \tag{6}$$

Since it refers to a probability implication within the same sequence, this form will be called an *internal probability implication*. It is employed in many important problems of probability, particularly in social statistics. Examples are the probability that an inhabitant of Bavaria suffers from goiter, or that a new-born baby is a boy. In such cases x_i is not represented by an event but by a person or an object that may possess the two properties B and B_k simultaneously. In more strictly statistical applications, the internal form of the probability statement prevails to so high a degree that it is usually made the basis of the probability calculus. Yet it would not be advisable to restrict the probability statement to this special form, since there are numerous other cases in which the more general types (3) or (4) are used. In particular, the application of the probability concept to the causal connection of events would be impossible if it were not based on the more general form of the probability statement as given above.

§ 10. The Abbreviated Notation

The form of the probability statement as given in (3, § 9) is rather complicated. An abbreviated notation, therefore, will be used for the development of the calculus of probability. Abbreviation is possible because certain properties of formula (3, § 9) occur in all probability statements in a similar manner, and can be suppressed in a simplified notation.

The probability statement has been written, so far,

$$(i)\ (x_i \epsilon A \underset{p}{\ni} y_i \epsilon B) \tag{1}$$

This formula will be abbreviated to the form

$$(A \underset{p}{\ni} B) \tag{2}$$

The transition from the abbreviated to the detailed notation is controlled by the following rule:

RULE OF TRANSLATION. *For every capital letter K substitute the expression* $x_i \epsilon K$, *using for different capital letters different variables* x_i, y_i . . . , *with the subscript i, but the same variable* x_i *for the capital letters* K_1, K_2 . . . . *In front of all parentheses containing capital letters place the symbol i within an all-operator.*

The method of abbreviation, as is seen from the rule, amounts to leaving out the specification of the sequence pair, an omission that is permissible because in probability statements the elements of the sequence pair never occur as free, but always as bound, variables. In the abbreviated notation, parentheses play the part of the all-operator; therefore, brackets must be used if generalization is not to be indicated. Furthermore, the difference between the two kinds of negation that exist for general statements is expressed as follows: in one case the negation bar is placed only above the expression written within parentheses; in the other it is extended above the parentheses. We thus define

$$(\overline{A \underset{p}{\ni} B}) =_{Df} (i)\ (\overline{x_i \epsilon A \underset{p}{\ni} y_i \epsilon B}) \tag{3}$$

$$\overline{(A \underset{p}{\ni} B)} =_{Df} \overline{(i)\ (x_i \epsilon A \underset{p}{\ni} y_i \epsilon B)} \tag{4}$$

The use of parentheses for the expression of the generalization applies also to formulas not containing the sign of the probability implication, and allows us to go from a class to a statement. Thus, $A \supset B$ is a class, and $(A \supset B)$ is a statement; according to the rule of translation, this statement has the form (20*b*, § 7) and is therefore identical with $A \subset B$. Adding parentheses to a class symbol means, in this notation, that the class is identical with the universal class and thus leads to the meaning expressed explicitly in (24, § 7).

If compound classes are used, like the class $A \supset B$, the rule of translation leads to the simple result: different capital letters mean narrower couple classes; equal capital letters with different subscripts mean simple classes. Couple classes containing implication or equivalence signs are interpreted by analogy with (13 and 14, § 7). The subscripts headed by circumflexes are dispensable for couple classes because their function is taken over by the difference of the capital letters. Class inclusion for different capital letters, i.e., for narrower couple classes, means a relation similar to the one illustrated in figure 3, § 7, for which the two circles are drawn in different planes, one on top of the other; corresponding points represent the couples of elements. Since for all practical purposes the narrower couple classes behave like simple classes, it is permissible to forget about the distinction for technical manipulations. The treatment of the general probability implication is technically not different from that of the internal probability implication.

A further abbreviation may be introduced. For many applications, particularly in mathematical calculations, we must solve the probability implication (2) for the degree p. We denote the degree p by $P(A,B)$, reading this symbol as "the probability from A to B". Some writers call this "the relative probability of B with respect to A". But in the present notation, the natural order, from the known to the unknown element of the relation, is used, thus introducing the same order of terms that is used in the implication $a \supset b$. The expression "probability from A to B" has the same grammatical form as the geometrical expression "distance from A to B", which also designates a relation. The order shows clearly that probabilities are treated as relations, in correspondence with the definition given in § 9. The calculus of probability in its usual form includes absolute as well as relative probabilities. The word "absolute" must be interpreted merely as an abbreviated notation, applying when the first term, the reference class, is dropped as being understood. Thus when it is said that there is the absolute probability $\frac{1}{6}$ for a face of the die, it is understood that the reference class is represented by the throwing of the die. This suppression of a first term has led to some confusion.

Instead of (2), then, the equation is written

$$P(A,B) = p$$

The p-symbol is a *numerical functor*, that is, a functional variable the special values of which are numbers.[1] It leads to statements only when it is used within mathematical equations. The P-symbol need not be considered as a primitive symbol; it can be reduced to the symbol of the probability implication by the definition

$$[P(A,B) = p] =_{Df} (A \underset{p}{\ni} B) \tag{5}$$

The symbol $P(A,B)$ itself is not defined—only the expression $P(A,B) = p$. This is permissible since the symbol $P(A,B)$ never occurs alone, but only in such equations. Thus a mere *definition in use* is given for $P(A,B)$. The equality sign used with this symbol represents arithmetical equality, i.e., equality between numbers. In the foregoing account of symbolic logic the sign was not explained because the rather complicated connection between logic and arithmetic could not be demonstrated. It may suffice to say that mathematical equality can be reduced to the basic logical operations.[2] The negation of a statement of mathematical equality is denoted by the inequality sign $\neq$. The notation by means of the P-symbol is called *mathematical notation;* that in terms of the $\underset{p}{\ni}$-symbol, *implicational notation.*

Another abbreviation is now introduced. Sometimes we omit the statement of the degree of probability and write

$$(A \ni B) \tag{6}$$

This relation is called *indeterminate probability implication.* Since it is not permissible simply to drop one constituent within a formula, a definition must be used to connect (6) with the symbols previously defined:

$$(A \ni B) =_{Df} (\exists p)\,(A \underset{p}{\ni} B) \tag{7}$$

The abbreviation (6) therefore means, "There is a p such that there exists between A and B a determinate probability implication of the degree p".

Passing from (6) to the detailed notation we obtain, according to the rule of translation,

$$(A \ni B) =_{Df} (\exists p)\,(i)\,(x_i \in A \underset{p}{\ni} y_i \in B) \tag{8}$$

The all-operator is placed after the existential operator, so that (8) represents the stronger form in the sense of (9, § 6).

The value p is often written within separate parentheses behind the probability implication:

$$(\exists q)\,(A \underset{q}{\ni} B)\,.\,(q = p) \tag{9}$$

---

[1] See *ESL*, p. 312.

[2] It is an identity of classes of a higher type. See *ibid.*, § 44.

This is merely a more convenient way of writing and has the same meaning as (2). We need this form because we shall later obtain for the probability degree p expressions that are too involved to be written as subscripts of the symbol of the probability implication. The resulting parentheses in the expression $(q = p)$ do not indicate an all-operator for the detailed notation because they do not contain capital letters.

The abbreviations given in this section will be useful in the following presentation of the theory of probability. In particular, it is an advantage that even in the abbreviated notation the symbols of the propositional operations can be manipulated according to the rules of the propositional calculus, although these symbols are placed between class symbols (that is, between capital letters) and thus represent class operations. This is possible because of the isomorphism of the two calculi (see § 7).

§ 11. The Rule of Existence

The *formal structure* of probability statements has been explained, but nothing has been said so far about their *meaning*. The laws of the probability implication can be completely developed, however, without interpretation. Discussion of the problem of interpretation will be deferred to a later section.

As a consequence, a method cannot yet be provided whereby we can determine whether, if two classes are given, a probability implication holds between them; in other words, we cannot yet ascertain the *existence* of a probability implication. However, this impossibility need not disturb us at this point. We assume the existence of some probability implications to be given; and we deal only with the question of how to derive new probability implications from the given ones. This operation exhausts the purpose of the probability calculus.

The existence of a probability implication I regard, in general, as a synthetic statement that cannot be proved by the calculus. The calculus can only transfer the existence character; with its help we can infer, from the known existence of certain probability implications, the existence of new ones. The property of transference by the calculus is, in part, directly expressed by the form of the axioms; some of the axioms, such as III and IV, directly assert the existence of new probability implications if certain others are given. However, these particular cases of transference do not suffice; for the transfer property will be required in a more general manner, as will be seen later. We must be able to assert that whenever the numerical value of a probability implication is determined by given probability implications, this probability implication does exist. It will become obvious (§ 17) that this existence is not self-evident, but must be asserted separately. The following postulate is therefore introduced.

RULE OF EXISTENCE. *If the numerical value p of a probability implication* $(A \underset{p}{\ni} B)$, *provided the probability implication exists, is determined by given probability implications according to the rules of the calculus, then this probability implication* $(A \underset{p}{\ni} B)$ *exists.*

The rule of existence is not an axiom of the calculus; it is a rule formulated in the metalanguage, analogous to the rule of inference or the rule of substitution (see § 5). It must be given an interpretation even in the formal treatment of the calculus. There must exist a formula that can be demonstrated in the calculus and that expresses the probability under consideration as a mathematical function of the given probabilities, with the qualification that the function be unique and free from singularities for the numerical values used. This is what is meant by the expression, "determined according to the rules of the calculus".[1]

§ 12. The Axioms of Univocality and of Normalization

From the discussion of the logical form we turn to the formulation of the laws of the probability implication. As explained above, an interpretation of probability is not required for this purpose. The laws will be formulated as a system of axioms for the probability implication—that is, as a system of logical formulas that, apart from logical symbols, contains only the symbol of the probability implication. Among the logical symbols, the logical implication occurs, and is thus used in formulating the laws of the probability implication.

The system to be constructed is called the system of axioms of the probability calculus. The name is justified by the fact that it is possible to derive from these axioms the formulas that are actually used in all applications of the probability calculus. When, at a later stage, an interpretation of probability is presented by means of statements about statistical frequencies, it will be possible to give another foundation to the axioms, showing that they are derivable from the given interpretation of probability. For the present, however, no use is made of the connection between probabilities and frequencies; instead, the axiom system is regarded as a system of formulas by which the properties of the probability concept are determined. By this procedure the axiomatic system of the probability calculus assumes a function comparable to that of the axiomatic system of geometry, which, in a similar way, determines *implicitly* the properties of the basic concepts of geometry, that is, of the concepts "point", "line", "plane", and so on (see § 8).

[1] The rule of existence can be replaced only incompletely by axioms. See footnote, p. 61.

We begin with the first two groups of axioms:

I. UNIVOCALITY $(p \neq q) \supset [(A \underset{p}{\twoheadrightarrow} B) . (A \underset{q}{\twoheadrightarrow} B) \equiv (\bar{A})]$

II. NORMALIZATION 1. $(A \supset B) \supset (\exists p) (A \underset{p}{\twoheadrightarrow} B) . (p = 1)$

2. $\overline{(\bar{A})} . (A \underset{p}{\twoheadrightarrow} B) \supset (p \geqq 0)$

Group II will be discussed first. The degree of probability is asserted by II,2 to be a positive number, including 0 as an extreme case. That p cannot be greater than 1 is not incorporated into the axioms because it will be derived as a theorem in § 13. The normalization to values in the interval from 0 to 1, end points included, is restricted to the case where the class A is not empty. The condition is expressed by the term $\overline{(\bar{A})}$, which means, according to the rule of translation (see p. 49), $\overline{(i)(\overline{x_i \epsilon A})}$, or, what is the same, $(\exists i)(x_i \epsilon A)$. The significance of this condition will be explained presently.

Axiom II,1 establishes a connection between the logical implication and the probability implication. Whenever a logical implication exists between A and B, there exists also a probability implication of the degree 1; the converse does not hold, however. It follows from a simple consideration that the reverse relation cannot be maintained. For the demonstration we use the formula corresponding to II,1:

$$(A \supset \bar{B}) \supset (\exists p) (A \underset{p}{\twoheadrightarrow} B) . (p = 0) \tag{1}$$

the necessity of which seems clear, though the exact derivation will be given later.

Formula (1) states that whenever an impossibility exists, a probability implication of the degree 0 exists also. For this case it is easy to illustrate why the reverse condition cannot be required. For instance, if we prick a sheet of paper with a needle, the probability (at least for a mathematical idealization of the problem) of hitting a given point is equal to 0; nevertheless a certain point is hit each time. Thus the probability 0 does not entail impossibility. Consequently, in order to remain free of contradictions, we must assert that certainty does not follow from the probability 1. Rather, certainty and the probability 1 stand in the relation of a narrower to a more comprehensive concept; certainty is a special case of the probability 1 (see § 18).

The relation of the two concepts is thus made clear in a very simple manner; the mysterious conception, which is occasionally voiced, that certainty and the probability 1 are incomparable concepts is untenable. On the contrary, the relation between the logical and the probability implication as expressed by II,1 represents an important relation holding between the two concepts, which connects the logic of the probability implication with classical logic. At this point the axiom system of probability differs from that of geometry. The concepts "point", "line", "plane", and so on, occurring in geometry, are of a type different from that of logical concepts; for that reason they can never assume the meaning of logical concepts, even for a special case.

The formulation of the univocality axiom I is clarified by the preceding remarks on the connection of the logical and the probability implication. It is obvious that the univocality of the degree of probability must be demanded. At first sight we might try to formulate univocality by

$$\overline{(A \underset{p}{\Rightarrow} B).(A \underset{q}{\Rightarrow} B).(p \neq q)} \tag{2}$$

However, this formula leads to contradictions. They result from the fact that in II,1 the logical implication was considered to be a special case of the probability implication. Certain properties of the logical implication prevent the assertion of (2) with complete generality. This is due to an above-mentioned property of the logical adjunctive implication, according to which a false proposition implies any proposition. In logic this fact is expressed by the *reductio ad absurdum*

$$(A \supset B).(A \supset \bar{B}) \equiv (\bar{A}) \tag{3}$$

Formula (3) is a generalization of (1*g*, § 4). It is proved by transforming the left side of (3) by means of (6*a*, § 4), applying (4*c*, § 4) and using (5*d* and 5*c*, § 4). Addition of the parentheses, meaning extension to an all-statement, is of course always permissible for tautologies. Logic thus admits an ambiguity of logical implication, but this case is restricted to the condition $(\bar{A})$. The ambiguity is transferred to the probability implication, since (3) with II,1 and (1) lead to the relation

$$(\bar{A}) \supset (\exists p)(\exists q)(A \underset{p}{\Rightarrow} B).(A \underset{q}{\Rightarrow} B).(p = 1).(q = 0) \tag{4}$$

In case of $(\bar{A})$ being true, the right side of the formula is valid, in contradiction to (2). Instead of (2) we therefore write axiom I, which brings the ambiguity of the probability implication into a form analogous to the ambiguity of logical implication. The condition $p \neq q$ must be written in front of I, since the expressions in brackets, contrary to (3), do not show whether we are dealing with different probability degrees.

In order to clarify I, it may be remarked that this axiom has the same meaning as the following implications:

$$(A \underset{p}{\Rightarrow} B).(A \underset{q}{\Rightarrow} B).(p \neq q) \supset (\bar{A}) \tag{5}$$

$$(\bar{A}) \supset (A \underset{p}{\Rightarrow} B).(A \underset{q}{\Rightarrow} B) \tag{6}$$

These two formulas result when formula (7a, § 4) is used to dissolve the equivalence in I into implications going in both directions. In this case the expression $(p \neq q)$ is dropped at the left side of (6); the condition is redundant because (6) holds likewise if the condition is not satisfied, that is, if $p = q$. From (6) is derived

$$(\bar{A}) \supset (A \underset{p}{\ni} B) \tag{7}$$

Since p can be chosen completely at random, the formula states that for the case $(\bar{A})$ any degree of probability may be asserted to hold between A and B. Formula (7) goes beyond (4) so far as it extends the ambiguity to any chosen degree of probability, including even values greater than 1 or smaller than 0.[1]

The ambiguity thus admitted is harmless because it applies only to the case in which the first sequence does not contain a single element x_i belonging to the class A. This follows because, according to the translation rule,

$$(\bar{A}) =_{Df} (i)(\overline{x_i \epsilon A}) \tag{8}$$

In the case $(\bar{A})$, therefore, the probability cannot be used to determine expectations of the events B because the event A is never realized, and so the plurality of values cannot lead to practical inconveniences. It seems reasonable, in such a case, to consider the probability implication between A and B with respect to the sequence pair x_iy_i as not defined at all and, therefore, to allow the assertion of any value for the degree of probability. This generalization of the probability concept extends it beyond practical needs; the extension is required because we wish to incorporate in the probability concept—as a special case—the logical implication as it is formulated in symbolic logic. The univocality, however, is always guaranteed if at least a single element x_i of the sequence belongs to the class A; it does not matter whether the corresponding y_i belongs to the class B. For, using the tautological equivalence provided by the propositional calculus,

$$a.b \supset c \equiv \overline{a.b} \vee c \equiv \bar{a} \vee \bar{b} \vee c \equiv \bar{a} \vee c \vee \bar{b} \equiv \overline{a.\bar{c}} \vee \bar{b} \equiv a.\bar{c} \supset \bar{b} \tag{9}$$

and substituting

$$\begin{aligned} &\text{for } a\text{: } (A \underset{p}{\ni} B).(A \underset{q}{\ni} B) \\ &\text{for } b\text{: } (p \neq q) \\ &\text{for } c\text{: } (\bar{A}) \end{aligned} \tag{10}$$

we derive from (5) the formula

$$(A \underset{p}{\ni} B).(A \underset{q}{\ni} B).\overline{(\bar{A})} \supset (p = q) \tag{11}$$

[1] The latter extension is necessary because otherwise the system of axioms would lead to contradictions, as J. C. C. McKinsey and S. C. Kleene have pointed out. See my note on probability implication in *Bull. Amer. Math. Soc.*, Vol. 47, No. 4 (1941), p. 265. It is for this reason that in this article I introduced for axiom II,2 the condition $\overline{(\bar{A})}$, which the German edition of this book does not contain.

When the double negation is removed and the translation rule (p. 49) and formula (13, § 6) are applied, we obtain

$$(A \underset{p}{\ni} B) . (A \underset{q}{\ni} B) . (\exists i) (x_i \epsilon A) \supset (p = q) \tag{12}$$

This means that the univocality of the degree of probability is guaranteed if there is at least one element x_i that belongs to the class A.

It is a result of the axioms I and II that the probability implication assumes the function of an extension of logical implication, the general implication introduced in (3, § 6). The latter is to be regarded as a special case of a probability implication, as we may recognize particularly from the form (6, § 9). This conception permits a more precise formulation of the concept of physical law, which was interpreted above as a general implication (§ 6). Closer inspection reveals that general implications that are absolutely certain can occur only if they are tautologies. The uncertainty of synthetic implications originates from the fact that any conceptual formulation of a physical event represents an idealization; the application of the idealized concept can possess only the character of probability (p. 8). The expression, "It follows according to a physical law", must therefore be represented, strictly speaking, not by a general implication but by a probability implication of a high degree (see § 85). Upon this fact rests the great importance of the probability implication: all laws of nature are probability implications.

There is an important difference between logical implication and probability implication. To the general implication $(A \supset B)$ corresponds an individual implication $a \supset b$, as defined by the truth tables $1B$ (§ 4). For probability implication such an individual relation is not used; the expression $A \underset{p}{\ni} B$, therefore, need not be considered as a meaningful expression. Only in a fictitious sense can the degree of probability, holding for the entire sequence, be transferred to the individual case. Like the meaning of an individual connective implication of the synthetic kind (see § 6), that of an individual probability implication is constructed by a *transfer of meaning from the general to the particular case.* This transfer makes understandable why a frequency interpretation of the degree of probability can be applied to single events, though only in a fictitious sense. The problem will be considered later (see § 72).

§ 13. The Theorem of Addition

A well-known theorem of the probability calculus is that the probability of a logical sum is determined by the arithmetical sum of the individual probabilities, provided the events are mutually exclusive. For instance, the probability of obtaining face 1 or 2 by throwing a die is calculated to be $\frac{1}{6}+\frac{1}{6}=\frac{2}{6}$. For the addition it is essential that only one of the two faces can lie on top;

otherwise this manner of calculating would be unjustified. The theorem is usually called the *theorem of addition*, and it must now be formulated as an axiom.

The condition of exclusion could be written in the form $(B \supset \bar{C})$, but it is sufficient to use the weaker statement

$$(A . B \supset \bar{C}) \tag{1a}$$

which can be derived from $(B \supset \bar{C})$, whereas the latter formula is not derivable from (1*a*). Although (1*a*) appears to be nonsymmetrical with respect to B and C, this is actually not so; for, because of (6*a* and 5*a*, § 4), formula (1*a*) is equivalent to

$$(A . C \supset \bar{B}) \tag{1b}$$

By the use of (1*a*) the theorem of addition may be written as follows:

III. Theorem of addition

$$(A \underset{p}{\Rightarrow} B) . (A \underset{q}{\Rightarrow} C) . (A . B \supset \bar{C}) \supset (\exists r)\, (A \underset{r}{\Rightarrow} B \vee C) . (r = p + q)$$

The addition theorem is a formula that expresses the transfer property of the calculus: it states a rule according to which the character of existence is transferred. It asserts the existence of the probability implication for the logical sum, if the individual probability implications are given. Nonetheless, we recognize the indispensability of the rule of existence (§ 11). For it is the existence rule that permits us to reverse the addition theorem; with its help we can derive the theorem

$$(A \underset{p}{\Rightarrow} B) . (A \underset{r}{\Rightarrow} B \vee C) . (A . B \supset \bar{C}) \supset (\exists q)\, (A \underset{q}{\Rightarrow} C) . (q = r - p) \tag{2}$$

This theorem cannot be obtained from axiom III alone, since the latter asserts existence only if the individual probabilities are given. The implicans of (2) differs from that of the axiom in that it contains only one individual probability and, moreover, the probability of the logical sum. Yet we recognize that the degree q of the probability implication, stated on the right side of (2), is determined by the addition theorem, provided this probability implication exists. Because of the univocality axiom I, the probability q, if it exists, must assume a value that, when added to p, furnishes the value r, that is, $q = r - p$. Now we can apply the existence rule, and the existence of the probability implication $(A \underset{q}{\Rightarrow} C)$ can be asserted.

The form of the relation (2) makes it clear that axiom III can be only partially reversed. The existence of the probability of the logical sum is not sufficient for the reversal; one of the two individual probabilities must also be given. Otherwise the degree of probability, q, would be undetermined, and the existence rule would not be applicable. The restricting condition is necessary because otherwise it would be possible to infer quite generally $(A \underset{q}{\ni} C)$, that is, the existence of a probability implication for any event. The *tertium non datur* (1*e*, § 4) and the formula $(\exists r)\,(A \underset{r}{\ni} C \vee \bar{C}) . (r = 1)$, which is obtained from it by the help of (8*c*, § 4) and axiom II,1, would give this result. The unwarranted generalization is made impossible by the existence rule, which demands that the probabilities under consideration be determined by those given.

The idea expressed in (2) is of great importance in the logical construction of the probability calculus. It is the validity of reversed formulas like theorem (2) and thus of the existence rule upon which rests the possibility of operating with numerical values of probabilities according to the rules of algebra. When we no longer incorporate the condition of exclusion into the formula, stating it only in the context, we may write, introducing the *P*-notation,

$$P(A,B \vee C) = P(A,B) + P(A,C) \tag{3}$$

With this way of writing we express the fact that the rules by which mathematical equations are manipulated can be applied to probability formulas. Thus it is permissible to proceed from (3) to the formula

$$P(A,C) = P(A,B \vee C) - P(A,B) \tag{4}$$

The admissibility of this step is expressed in theorem (2). We recognize that the mathematical symbolization of the probability calculus is made possible by a particular property of this calculus, a property that requires a special formulation. The property is expressed by the rule of existence in combination with the axiom of univocality.

Certain difficulties arise from the fact that we cannot incorporate into the mathematical symbolization the condition of exclusion, presupposed for (3) and (4), but must add it verbally. A formula that is not dependent on conditions to be added in the context will be developed later (see § 20).

A remark must be made concerning the univocality of the *P*-symbol. Since univocality of a probability $P(A,B)$ is restricted to the case that A is not empty, the *P*-symbol has only in this case the character of a *numerical functor*, a number variable determined by the argument in parentheses. In order to make equations like (3) hold also in the case of an empty class A, the convention is introduced that such equations then represent *existential statements* of the form, "There is a numerical value for the dependent probability that satisfies the equation when the independent probabilities are given". For instance, (3) states for an empty class A that, if for $P(A,B)$ and $P(A,C)$ any values are given, there is a probability value among those holding for $P(A,B \vee C)$ that satisfies (3). All equations, in this case, will represent trivial statements, because, if A is empty, a probability with A in the first term will have all real numbers as its values; the existential statement will therefore be trivially satisfied. The advantage of this convention is that it allows us to drop, for probability equations, the condition stating that A is not empty. The equations also hold in the contrary case, but then they say nothing. For the implicational mode of writing, no such convention is needed, since axiom III and formula (2) are existential statements and lead to univocal values of the probabilities only if A is not empty. The convention as to the P-symbol is therefore in agreement with the rule of translation (p. 49).

In the greater part of this book the mathematical notation will be employed. Except in this section and the next, the axioms formulated in the implicational notation will no longer be used as a basis for further derivations. Their place will be taken by theorems in the P-notation, derived from them. The transition to the P-notation restricts the logical operations to the inner part of the P-symbols. Supplementary remarks will be made in the context whenever other restricting conditions, on which the validity of the formulas depends, are added.

We now derive a few theorems that have been used in the preceding section. Because of the *tertium non datur*, the formula $(A \supset B \vee \bar{B})$ is always true, and we obtain the general formula

$$(\exists r)\,(A \underset{r}{\rightarrow} B \vee \bar{B}) \, . \, (r = 1) \tag{5}$$

or, in the P-notation.

$$P(A,B \vee \bar{B}) = 1 \tag{5'}$$

We may therefore add formula (5) to $(A \underset{p}{\rightarrow} B)$. The conditions of theorem (2) are satisfied if we substitute $\bar{B}$ for C, since $(A \, . \, B \supset \bar{\bar{B}})$ also is always valid. We thus obtain the theorem

$$(A \underset{p}{\rightarrow} B) \supset (\exists u)\,(A \underset{u}{\rightarrow} \bar{B}) \, . \, (u = 1 - p) \tag{6}$$

In the P-notation the theorem is written

$$P(A,B) + P(A,\bar{B}) = 1 \tag{7}$$

This formula is called the *rule of the complement.*

We can now demonstrate that the probability degree, for which we postulated in II,2 only the nonnegative character, can never become greater than 1. We can complement the term B by its negation to constitute a complete disjunction. Considering the fact expressed in II,2 that both probabilities occurring in (7) cannot be negative, we obtain from (7) the relation

$$0 \leqq P(A,B) \leqq 1 \tag{8}$$

Furthermore, we have from II,1 and (6) the theorem

$$(A \supset \bar{B}) \supset (\exists p)\,(A \underset{p}{\rightarrow} B) \, . \, (p = 0) \tag{9}$$

The mathematical symbolization of the calculus of probability may be illustrated by another problem. Given the three classes B_1, B_2, B_3, which are mutually exclusive but do not form a complete disjunction, and given the three probabilities

$$P(A,B_1 \vee B_2) \qquad P(A,B_2 \vee B_3) \qquad P(A,B_3 \vee B_1) \tag{10}$$

we wish to infer from them the existence of the three individual probabilities

$$P(A,B_1) \qquad P(A,B_2) \qquad P(A,B_3) \tag{11}$$

Theorem (2) is not applicable, because none of the individual probabilities is known to exist. However, we obtain from the addition theorem the equations

$$\begin{aligned} P(A,B_1) + P(A,B_2) &= P(A,B_1 \vee B_2) \\ P(A,B_2) + P(A,B_3) &= P(A,B_2 \vee B_3) \\ P(A,B_3) + P(A,B_1) &= P(A,B_3 \vee B_1) \end{aligned} \tag{12}$$

They can be solved for the individual probabilities:

$$\begin{aligned} P(A,B_1) &= \tfrac{1}{2}[P(A,B_1 \vee B_2) + P(A,B_3 \vee B_1) - P(A,B_2 \vee B_3)] \\ P(A,B_2) &= \tfrac{1}{2}[P(A,B_1 \vee B_2) + P(A,B_2 \vee B_3) - P(A,B_3 \vee B_1)] \\ P(A,B_3) &= \tfrac{1}{2}[P(A,B_3 \vee B_1) + P(A,B_2 \vee B_3) - P(A,B_1 \vee B_2)] \end{aligned} \tag{13}$$

The three individual probabilities (11) are therefore determined according to (13) by the or-probabilities (10); and it follows from the rule of existence that when (10) is given, the existence of (11) is also assertable. Owing to the rule of existence, we can apply, in the calculus of probabilities, the procedure of eliminating unknown quantities from a system of equations and use it to find new existing probabilities. Probability equations, therefore, *determine existence*, that is, the existence of any of the probabilities occurring in an equation is secured if all the other probabilities are known to exist.[1]

§ 14. The Theorem of Multiplication

The fourth and last group refers to an axiom that determines the probability of a combination of terms. It is a well-known theorem of the probability calculus that the probability of a combination—that is, the probability of a

[1] I am indebted to E. Tornier for having called my attention to the fact that the problem formulated in (10) and (11) cannot be solved by means of the formulas given in my paper on probability published in *Math. Zs.*, Vol. 34 (1932), p. 568. In that article I did not use the existence rule, but gave special *reversal axioms* that permitted the derivation of such theorems as (2) and, thereby, the application of the calculus of algebraic equations. But it turned out that, in this system, the existence-determining character is not always conserved when variables are eliminated. Equations (12) determine existence for my former system also, but equations (13) do not have this property. This fact led me to replace the reversal axioms by the rule of existence.

logical product—is represented by the arithmetical product of certain individual probabilities. This is the *multiplication theorem* of the probability calculus. The theorem is formulated by the following axiom:

IV. THEOREM OF MULTIPLICATION

$$(A \underset{p}{\ni} B).(A.B \underset{u}{\ni} C) \supset (\exists\, w)\, (A \underset{w}{\ni} B.C).(w = p \cdot u)$$

For the first time we deal with probability expressions in which the probability implication refers to three different classes, two of them occurring either in the first or in the second term. This does not cause any difficulty, because the translation rule (p. 49) determines the transition to the detailed notation for formulas of this kind also. In this case the domain of the probability implication is a triplet of sequences.

By a procedure of the kind used for the theorem of addition we can derive the converse of the multiplication theorem. We obtain two different conversions, since the three events A,B,C do not occur symmetrically in IV, whereas III is symmetrical with respect to B and C:

$$(A \underset{p}{\ni} B).(A \underset{w}{\ni} B.C) \supset (\exists\, u)\, (A.B \underset{u}{\ni} C).\left(u = \frac{w}{p}\right) \tag{1}$$

$$(A.B \underset{u}{\ni} C).(A \underset{w}{\ni} B.C) \supset (\exists\, p)\, (A \underset{p}{\ni} B).\left(p = \frac{w}{u}\right) \tag{2}$$

The proof of the theorems is based on the rule of existence, which applies because it can be demonstrated that the probability implications occurring on the right in (1) and (2) are determined by those on the left. Because of theorems (1) and (2), axiom IV can be replaced by the more comprehensive formula, written in the P-notation,

$$P(A,B.C) = P(A,B) \cdot P(A.B,C) \tag{3}$$

Theorems (1) and (2) mean that formula (3) can be solved according to the rules for mathematical equations for each of the individual probabilities occurring. Here again it is seen that the mathematical formalization of the probability calculus depends on the validity of the existence rule, as explained in §13.

Formula (3) is always true and does not require any restricting condition to be added verbally in the context, as was necessary for (3, § 13). Formula (3) will therefore be used in further discussion of the theorem of multiplication, without going back to axiom IV. The form selected here for theorem (3), characterized by the occurrence of three classes and of a term having two classes in the place of the reference class, has long been applied in the British and the American literature.[1] It has been used in the axiomatic construction in this work because only in this form is the axiom always correct. The probability from A to the logical product $B.C$ can be calculated only if the probability from A to B as well as that from $A.B$ to C is given.

In mathematical presentations the probability $P(A.B,C)$ is usually called "the relative probability of C with respect to B". This notation does not seem advisable because all probabilities are relative, and, furthermore, because the probability under consideration cannot be characterized by B and C alone but requires class A also.

For example, the probability that a person suffering from diphtheria subsequently contracts nephritis and dies is represented by a probability of the form $P(A,B.C)$, A denoting diphtheria; B, nephritis; and C, death. The probability is calculated as the product of the probability that a person suffering from diphtheria contracts nephritis, and the probability that a person dies who gets nephritis after having had diphtheria. The latter probability is different from the one that a person suffering from nephritis will die, since a patient who has had diphtheria is weakened and therefore is in greater peril of losing his life. This consideration shows why the last probability occurring in (3) must be characterized by three classes.

Another example is the probability that a thunderstorm follows a hot summer day with a subsequent change in the weather, which splits up into the product of two probabilities: the probability that a thunderstorm will follow a hot day and the probability that a change in the weather will follow a thunderstorm that was preceded by a hot day. The second probability is smaller than the probability that any thunderstorm brings with it a change in the weather, because the *convective* thunderstorms produced by local heat conditions usually do not result in a change in the weather, in contradistinction to *frontal* thunderstorms. The example illustrates once more the necessity of characterizing by three classes the probability that occurs in the last term of (3).

It must be regarded as a special case if two classes suffice for this term—a case arising when the actual three-class probability is equal to a certain two-class probability. Such specialization results if

$$P(A.B,C) = P(A,C) \tag{4}$$

Then (3) assumes the form of the *special theorem of multiplication*:

$$P(A,B.C) = P(A,B) \cdot P(A,C) \tag{5}$$

[1] In 1878 the form was used by C. S. Peirce. See his *Collected Papers* (Cambridge, Mass., 1932), Vol. II, p. 415. J. M. Keynes also employed the form in *A Treatise on Probability* (London, 1921), chap. XI, p. 6. The use of relative probabilities for the determination of dependent events is, of course, much older. P. S. Laplace gives a corresponding rule in his *Essai philosophique sur les probabilités* (Paris, 1814), chapter on "Principes généraux, quatrième principe." But he uses only two classes, my classes B and C, suppressing the general reference class A.

The condition (4) is paraphrased by the statement: *the events B and C are mutually independent with respect to A* (see also § 23). For example, the probability that a sudden gust of wind will capsize two sailboats is obtained as the product of the probability that the wind overturns one boat by the corresponding probability concerning the other boat. The two probabilities need not be the same, since the two sailboats may be of different construction. It is, however, necessary for (5) that the probability of the second boat's turning over be independent of whether the first boat turns over.

Another specialization of (3) is obtained if A can be represented as the product of two events A_1 and A_2 such that

$$P(A_1.A_2,B) = P(A_1,B) \qquad P(A_1.A_2.B,C) = P(A_2.B,C) \tag{6}$$

In this case (3) leads to

$$P(A_1.A_2,B.C) = P(A_1,B) \cdot P(A_2.B,C) \tag{7}$$

If we add the specialization analogous to (4)

$$P(A_2.B,C) = P(A_2,C) \tag{8}$$

we obtain

$$P(A_1.A_2,B.C) = P(A_1,B) \cdot P(A_2,C) \tag{9}$$

This case may be illustrated by the throwing of two dice: A_1 refers to the throwing of one die and A_2 to the throwing of the other. However, (9) would not be permissible without the conditions (6) and (8).

A third specialization results if

$$P(A.B,C) = P(B,C) \tag{10}$$

Then (3) becomes

$$P(A,B.C) = P(A,B) \cdot P(B,C) \tag{11}$$

Examples of this kind occur in certain causal chains: A may be represented by the occurrence of a storm; B, the falling of a tree; C, an accident caused by the falling tree. For the application of (11), however, we must inquire in each case whether (10) is satisfied.

The preceding discussion reveals that specializations of the multiplication theorem—some of which are used as axioms in representations of the probability calculus—do not provide formulas that are always true. They result from the general form (3) only for special cases. The latter are characterized by the equality of certain probabilities having different references classes, as stated in (4), (6), (8), (10). It follows that the question whether one of the special forms of the multiplication theorem can be applied is reduced to a question of the same type as that of how to determine the numerical value of a probability. It is always known whether two probabilities are equal when the probabilities themselves are known. Using the general form (3), or the form of axiom IV, for the theorem of multiplication eliminates certain logical difficulties that were connected with this theorem in the history of the calculus of probability.

§ 15. Reduction of the Multiplication Theorem to a Weaker Axiom

The theorem of multiplication is not independent of the other axioms; it can be reduced to a weaker assumption. In order to show this dependence I shall make use of the fact that the multiplication theorem can be split into two separate assertions. The first partial assertion states that the probability $P(A,B.C)$ *is determined* by $P(A,B)$ and by $P(A.B,C)$; the second assertion is that $P(A,B.C)$ is obtained, in particular, by the *arithmetical multiplication* of the two probabilities. The second assertion need not be stated explicitly as an axiom, but can be derived from the calculus with the use of the other axioms.

To prove this contention, multiplication theorem IV is replaced by the weaker axiom

IV*a*. $$(A \underset{p}{\ni} B).(A.B \underset{u}{\ni} C) \supset (\exists\, w)\, (A \underset{w}{\ni} B.C).[w = f(p,u)]$$

Here f stands for a mathematical function, temporarily undefined, that is to determine for any values p,u the corresponding w and, conversely, is required to be solvable unambiguously for p and u. Similarly to (1 and 2, § 14), it can be shown that the probability implication written at the right in these theorems assumes the degree of probability corresponding to the solution of $w = f(p,u)$ for p and u respectively; in these theorems the probability on the right side is replaced by

$$p = f'(w,u) \text{ and } u = f''(w,p), \text{ respectively,} \tag{1}$$

where f' and f'' represent the functions obtained by the solution. In this way it can be shown analogous to (3, § 14) that we may write

$$P(A,B.C) = f[P(A,B), P(A.B,C)] \tag{2}$$

The function f is the function occurring in IV*a*, and the comma between the probability symbols separates the two arguments of this function; that is, it serves as the comma between the arguments of a mathematical function.

In order to infer the form of f from (2), we substitute for C the disjunction of two mutually exclusive events C and D; then (2) becomes

$$P(A,B.[C \vee D]) = f[P(A,B), P(A.B,C \vee D)] \tag{3}$$

According to the first distributive law (4*a*, § 4), we dissolve

$$(B.[C \vee D] \equiv B.C \vee B.D) \tag{4}$$

and apply to both sides of equation (3) the addition theorem (3, § 13):

$$P(A,B.[C \vee D]) = P(A,B.C \vee B.D) = P(A,B.C) + P(A,B.D) \quad (5a)$$

$$P(A.B,C \vee D) = P(A.B,C) + P(A.B,D) \quad (5b)$$

The probabilities of the logical products occurring in (5a) are dissolved again according to (2):

$$P(A,B.C) = f[P(A,B), P(A.B,C)]$$

$$P(A,B.D) = f[P(A,B), P(A.B,D)] \quad (6)$$

Thus (3) is transformed into

$$f[P(A,B), P(A.B,C)] + f[P(A,B), P(A.B,D)]$$
$$= f[P(A,B), P(A.B,C) + P(A.B,D)] \quad (7)$$

Using the abbreviations

$$P(A,B) = p \qquad P(A.B,C) = u \qquad P(A.B,D) = v \quad (8)$$

we can write (7) as

$$f[p,u] + f[p,v] = f[p,u + v] \quad (9)$$

This is a functional equation for f; if it is to be valid for any values u and v the function f must have the form

$$f[p,u] = g(p) \cdot u \quad (10)$$

where $g(p)$ represents a function of p alone, which remains undetermined for the time being.[1]

In (2) we now substitute $[C \vee \bar{C}]$ for C; then (2) becomes

$$P(A,B.[C \vee \bar{C}]) = f[P(A,B),P(A.B,C \vee \bar{C})] \quad (11)$$

According to (5c, § 4), we have

$$(B.[C \vee \bar{C}] \equiv B) \quad (12)$$

and therefore

$$P(A,B.[C \vee \bar{C}]) = P(A,B) = p \qquad P(A.B,C \vee \bar{C}) = 1 \quad (13)$$

[1] I refer to a well-known theorem of mathematics. It may be proved as follows: we put $u = 0$; then we derive from (9) that $f(p,0) = 0$. Assuming v to be the differential increase du, we write (9):

$$f[p,0 + du] - f[p,0] = f[p,u + du] - f[p,u]$$

Dividing by du, we obtain for the limit $du = 0$ the differential equation

$$\left(\frac{\partial f[p,u]}{\partial u}\right)_0 = \left(\frac{\partial f[p,u]}{\partial u}\right)_u$$

The subscript marks the argument-place at which the differential quotient is to be formed. Since u can be chosen at random, the equation states the differential quotient for u to be constant; that is, the function f is linear with respect to u. It is even possible to drop the assumption that the function f is differentiable and continuous, but the proof will then be more complicated.

Using these results in combination with (10), we transform (11) into

$$p = f[p,1] = g(p) \cdot 1 = g(p) \tag{14}$$

With this determination of $g(p)$, the relation (10) assumes the form

$$f(p,u) = p \cdot u \tag{15}$$

Because of (2) and (8) this means

$$P(A,B.C) = P(A,B) \cdot P(A.B,C) \tag{16}$$

Thus we have proved the multiplication theorem (3, § 14).

It is seen from this demonstration that the theorem of multiplication represents a necessary formula within the frame of the calculus of probability. That the probability of the logical product is given by an arithmetical product is a consequence of the fact that the probability of a logical sum is given by an arithmetical sum, in combination with the first distributive law of logic.

The result enables us to introduce a new definition of the property of independence, defined in (4, § 14) or (5, § 14). Combining (4, § 14) with (2), we may define independence as follows.[2] Two events are independent with respect to A if the probability from A to their logical product is a function of their individual probabilities with respect to A alone, that is, if

$$P(A,B.C) = f[P(A,B),P(A,C)] \tag{17}$$

It then follows that f assumes the form of the arithmetical product. This characterization of independence is very instructive; it states that the probability of the combination of independent events is determined whenever the probabilities of the separate events are given. For instance, the probability $\frac{1}{6}$ for each of two dice determines the probability $\frac{1}{36}$ for the combination of any two faces.

§ 16. The Frequency Interpretation

Axioms I to IV suffice to derive all the theorems of the calculus in which probability sequences occur as wholes the structure of which is not considered. The totality of these theorems is called the *elementary calculus of probability.* With the given axioms we therefore control the *formal structure* of the elementary calculus of probability. But before developing the theorems of this calculus we wish to give the probability concept an interpretation over and above the characterization of its formal structure (see § 8).

This leads to a problem that has been under much discussion. The formal structure of the probability calculus that I have developed might be conceded

[2] I am indebted to Kurt Grelling for the suggestion that independence can be characterized in this manner; he thereby directed my attention to the foregoing proof for the product form of the function f.

by adherents of the most diverse theories about probability. But the question of the interpretation of the probability concept can be answered only on the basis of painstaking philosophical investigations, and different theories have answered it in different ways. It will be treated, therefore, in more detail later (see chap. 9).

The laws of the calculus of probability are difficult to understand, however, if one does not envisage a definite interpretation. Thus, for didactic reasons, an interpretation of the probability concept must be added, at this point, to the axiomatic construction. But this method will not prejudice later investigations of the problem. The interpretation is employed merely as a means of illustrating the system of formal laws of the probability concept, and it will always be possible to separate the conceptual system from the interpretation, because, for the derivation of theorems, the axioms will be used in the sense of merely formal statements, without reference to the interpretation.

This presentation follows a method applied in the teaching of geometry, where the conceptual formulation of geometrical axioms is always accompanied by spatial imagery. Although logical precision requires that the premises of the inferences be restricted to the meaning given in the conceptual formulation, the interpretation is used as a parallel meaning in order to make the conceptual part easier to understand. The method of teaching thus follows the historical path of the development of geometry, since, historically speaking, the separation of the conceptual system of geometry from its interpretation is a later discovery. The history of the calculus of probability has followed a similar path. The mathematicians who developed the laws of this calculus in the seventeenth and eighteenth centuries always had in mind an interpretation of probability, usually the frequency interpretation, though it was sometimes accompanied by other interpretations.

In order to develop the frequency interpretation, we define probability as the *limit of a frequency* within an infinite sequence. The definition follows a path that was pointed out by S. D. Poisson[1] in 1837. In 1854 it was used by George Boole,[2] and in recent times it was brought to the fore by Richard von Mises,[3] who defended it successfully against critical objections.

The following notation will be used for the formulation of the frequency interpretation. In order to secure sufficient generality for the definition, we shall not yet assume that all elements x_i of the sequence belong to the class A. We assume, therefore, that the sequence is *interspersed* with elements x_i of a different kind. For instance, the sequence of throws of a coin may be interspersed with throws of a second coin. In this case only certain elements x_i

[1] *Recherches sur la probabilité des jugements en matière criminelle et en matière civile* . . . (Paris, 1837).

[2] *The Laws of Thought* (London, 1854), p. 295.

[3] "Grundlagen der Wahrscheinlichkeitsrechnung," in *Math. Zs.*, Vol. V (1919), p. 52, and later publications.

will belong to the class A, if the class is defined as representing the throws of one of the coins only. Similarly, only some among the elements y_i will belong to the class B, which may signify the occurrence of tails lying up. It may happen that y_i represents a case of tails up, whereas the corresponding x_i does not belong to the class A, that is, the event of tails lying up is produced by the second coin. When the frequency is counted out in such a sequence pair, the result is expressed by the symbol

$$\overset{n}{\underset{i=1}{N}} (x_i \epsilon A) \tag{1a}$$

which means the number of such x_i between 1 and n that satisfy $x_i \epsilon A$. The symbol is extended correspondingly to apply to different variables and to different classes and also to a pair, a triplet, and so on, of variables. For instance, the expression

$$\overset{n}{\underset{i=1}{N}} (x_i \epsilon A).(y_i \epsilon B) \tag{1b}$$

represents the number of pairs x_i,y_i such that x_i belongs to A and simultaneously y_i belongs to B; it signifies the number of pairs x_i,y_i that are elements of the common class A and B. To abbreviate the notation, the following symbol is introduced:

$$N^n(A) =_{Df} \overset{n}{\underset{i=1}{N}} (x_i \epsilon A) \qquad N^n(A.B) =_{Df} \overset{n}{\underset{i=1}{N}} (x_i \epsilon A).(y_i \epsilon B) \tag{2}$$

Furthermore, the *relative frequency* $F^n(A,B)$ is defined by

$$F^n(A,B) = \frac{N^n(A.B)}{N^n(A)} \tag{3}$$

In the special case in which all elements x_i belong to the class A, that is, when the sequence x_i is *compact*, the denominator of the fraction is equal to n, whereas in the numerator the expression A may be dropped; then (3) assumes the simpler form

$$F^n(A,B) = \frac{1}{n} \cdot N^n(B) \tag{4}$$

With the help of the concept of relative frequency, the frequency interpretation of the concept of probability may be formulated:

If for a sequence pair x_iy_i the relative frequency $F^n(A,B)$ goes toward a limit p for $n \rightarrow \infty$, the limit p is called the probability from A to B within the sequence pair. In other words, the following coördinative definition is introduced:

$$P(A,B) = \lim_{n \rightarrow \infty} F^n(A,B) \tag{5}$$

No further statement is required concerning the properties of probability sequences. In particular, randomness (see § 30) need not be postulated.

§ 17. The Origin of Probability Statements

So long as we regard the probability calculus as a formal calculus by means of which formulas are manipulated, that is, so long as we do not speak of the meaning of the formulas, the origin of probability statements presents no problem. The question whether the individual probability statement is true or false, then, is not a problem of the calculus, as was explained above. The calculus deals solely with transformations of probability statements; and the statements of the mathematical calculus, therefore, represent exclusively tautological implications of the type, "If certain probability implications $a_1, \ldots a_n$ exist, then certain other probability implications $b_1, \ldots b_n$ exist also". I agree here with a conception emphasized by von Mises.

But it would be a shortsighted attitude if mathematicians were induced by this conception to regard the question of the origin of probability statements as unreasonable. With the given definition of the probability calculus, the question is merely shifted to another field. At the very moment at which an interpretation is assigned to the probability statement, there arises the question how to know whether, in a given instance, a probability statement holds. It follows from the nature of the interpretation that the question is equivalent to the question how to ascertain the existence of a limit of an infinite sequence.

Here an important distinction must be made. First, probability sequences may be regarded as mathematically given sequences, that is, as sequences that are defined by a rule. For instance, a probability sequence can be defined by means of an infinite decimal fraction in which every even number is regarded as the case B and every odd number as the case $\bar{B}$. Whether such a sequence has a frequency limit and what the limit is, is a question of purely mathematical nature to be answered by means of the usual mathematical methods. It is important that we have at our disposal such mathematically given sequences representing the frequency interpretation; on occasion they will be used as models (see §§ 30 and 66). In the practical application of the probability calculus, however, they do not play a part.

Second, sequences provided by events in nature may be considered. For such sequences, which include all practical applications of the calculus of probability, we do not know a definite law regarding the succession of their elements. Instead of a defining rule, we have a finite initial section of the sequence; therefore we cannot know, strictly speaking, toward what limit such a sequence will proceed. We assume, however, that the observed frequency will persist, within certain limits of exactness, for the infinite rest of the sequence. This inference, which is called *inductive inference*, leads to very difficult logical problems; and it will be one of the most important problems of this investigation to find a satisfactory explanation of the inference. For the present, however, the inference will not be questioned. Suffice it to say that the inference is actually used—sometimes under the name *a posteriori determination of a probability*—by statisticians as well as in everyday life. We shall therefore use it, too, in problems of the application of the formulas constructed.

It may sometimes be expedient, for mathematical reasons, to imagine a fictitious observer who can count out an infinite sequence and thus is able to determine its limit. But the picture serves only to illustrate certain logical relations and cannot replace the inductive inference where physical reality is concerned.

To summarize: for the present we shall regard as verifiable an assertion stating that there exists a probability sequence of a determinate degree of probability. The verification may be derived either mathematically, from the defining rule of the sequence, or by means of an inductive inference.

The given interpretation will now be used to elucidate some properties of the axiom system that so far, perhaps, have not been made sufficiently clear. First, we realize why the existence of an indeterminate probability implication has been regarded as a synthetic statement requiring empirical proof. The assertion that there exists a limit of the frequency, even without specification of the degree, represents a definite statement that is certainly not satisfied for every sequence pair $x_i y_i$. For this reason the rule of existence is necessary within our formal system; when interpreted, it expresses the assertion that a limit of the frequency exists in the cases concerned.

Second, we recognize that the indeterminate probability implication $(A \twoheadrightarrow B)$ states more than the existence of a mere possibility relation, which we write as $(\overline{A \supset \bar{B}})$.[1] The added meaning consists in the fact that the first statement asserts a certain regularity in the repetition of events. When a die is thrown upon a table, it is possible that a sudden thunderbolt may happen simultaneously; but such a statement of possibility does not mean that a probability implication exists between the two events. I do not wish to say that the probability is very small; I mean, rather, that it is not permissible to assert a definite regularity with respect to the occurrence of thunder when the die is thrown repeatedly. The illustration will make it clear that the existence of a probability cannot be inferred from the possibility of an event. But neither does the converse hold. From (1, § 12) it is seen that the possibility of an event cannot be inferred from the existence of a probability. The probability can be equal to zero, and the probability zero may or may not represent impossibility. In neither direction does an implication hold between the two statements $(A \twoheadrightarrow B)$ and $(\overline{A \supset \bar{B}})$. Probability and possibility are disparate concepts, that is, their extensions overlap.

If we were to assert that a frequency limit must exist for any two repetitive events observed for a sufficiently long time, we would commit ourselves to a

[1] This is the extensional possibility of § 80.

far-reaching hypothesis. On this assumption it would be possible to drop the existence rule; but, instead, we should have to introduce into the calculus an axiom of the form, "For all A and C, $(A \ni C)$ is valid". Obviously this addition would mean an extraordinary extension of the content of the calculus, with which we do not wish to burden the axiom system.

I therefore consider the assertion of a determinate as well as of an indeterminate probability implication to be a synthetic statement, the validity of which can be ascertained, when physical events are concerned, by means of statistics in combination with inductive inferences. This method of ascertainment will not be questioned throughout the mathematical part of the investigation, because the frequency interpretation does not enter into the content of the probability calculus to be developed. It constitutes only an illustrative addition and will not be used for the derivation of theorems.

§ 18. Derivation of the Axioms from the Frequency Interpretation

It will now be shown that all axioms of the calculus of probability can be derived from the frequency interpretation, that is, they are tautologies if the frequency definition of probability is assumed.

We start with the univocality axiom I. The case $(\bar{A})$, to which this axiom refers, signifies that the relative frequency F^n assumes the indeterminate form $\frac{0}{0}$, since the summation N^n in (3, § 16) leads to 0 for numerator as well as denominator. Therefore we also have $P(A,B) = \frac{0}{0}$, that is, the probability does not possess a determinate value. This result represents one assertion of the axiom. If the case $(\bar{A})$ does not hold, however, a definite limit exists; since there can be only one limit, the other assertion of the axiom is likewise satisfied. Notice that a limit exists even when only a finite number of elements x_i belong to A; the value of the frequency for the last element is then regarded as the limit. This trivial case is included in the interpretation and does not create any difficulty in the fulfillment of this or the following axioms.

Axiom II,1 concerns the case in which each element of the form $(x_i \epsilon A)$ is followed by an element $(y_i \epsilon B)$, since this is what the logical implication asserts. In this case all $F^n = 1$, a result following immediately from (3, § 16), so that II,1 is satisfied. The major implication in the axiom can be directed toward only one side, since the probability 1 can be obtained, also, if there are some cases in which $x_i \epsilon A$ is followed by $y_i \epsilon \bar{B}$. These cases, however, must be distributed so sparsely that the limit F^n becomes equal to 1, though every individual F^n may be smaller than 1. An example is given by a compact sequence A accompanied by a sequence B that has a $\bar{B}$ in all elements whose subscript i is the square of a whole number but which has a B in all other elements. Thus the frequency interpretation makes it clear why the probability 1 represents a wider concept than the logical implication.

This consideration shows also that the probability implication of the degree p represents a generalization of the general implication of symbolic logic. Whereas the general implication demands all elements $x_i \epsilon A$ to be followed by a $y_i \epsilon B$, the probability implication includes the case in which certain $x_i \epsilon A$ are followed by a $y_i \epsilon \bar{B}$, with the qualification, however, that between the numbers of the elements there must exist a frequency ratio that goes in the limit toward a determinate value. The probability implication, itself representing a general implication, therefore constitutes the generalization of the usual general implication for sequences in which the individual implication occurs only in a certain number of places. Instead of demanding the individual implication to be valid without exceptions, we require only a frequency ratio.

That II,2 is satisfied follows directly from the fact that the relative frequency F^n is a positive number (including 0). The condition, expressed in (8, § 13), that the probability degree cannot be greater than 1 likewise follows from the definition of the relative frequency.

We turn now to the addition theorem III. In order to prove this axiom, we form first

$$F^n(A,B \vee C) = \frac{N^n(A.[B \vee C])}{N^n(A)} \tag{1a}$$

If $(A.B \supset \bar{C})$ is valid, this is equal to

$$\frac{N^n(A.B)}{N^n(A)} + \frac{N^n(A.C)}{N^n(A)} \tag{1b}$$

and we obtain

$$F^n(A,B \vee C) = F^n(A,B) + F^n(A,C) \tag{2}$$

The equation remains unchanged in the transition to the limit, and for mutually exclusive events we have

$$P(A,B \vee C) = P(A,B) + P(A,C) \tag{3}$$

The exclusion condition suffices for the addition of probabilities having the same first term. We need not presuppose, in such a case, that the terms B and C belong to the same sequence; this represents a special case for which, of course, the theorem is also valid.

The given proof can be made clearer by the following consideration. We write the three sequences below one another, each in one row; however, we do not write the elements x_i, y_i, z_i, but only the classes A, B, C, to which the elements belong. For the sake of simplicity we shall assume that the sequence x_i consists only of the elements $x_i \epsilon A$ and thus is compact. We thereby arrive at the following arrangement:

$$\begin{array}{l} A\ A\ A\ A\ A\ A\ A\ A\ \ldots \\ B\ B\ \bar{B}\ B\ \bar{B}\ B\ B\ \bar{B}\ \ldots \\ \bar{C}\ \bar{C}\ C\ \bar{C}\ \bar{C}\ \bar{C}\ \bar{C}\ C\ \ldots \end{array} \tag{4}$$

The frequency $F^n(A,B \vee C)$ expresses the relative frequency of the A under which a B or a C is found. Because of the condition of exclusion, a B and a C can never stand simultaneously under the same A, and thus the relative frequencies of B and C add up to that of $B \vee C$.

The multiplication theorem IV, also, can be derived from the frequency interpretation. We obtain from (3, § 16)

$$\begin{aligned} F^n(A,B.C) &= \frac{N^n(A.B.C)}{N^n(A)} = \frac{N^n(A.B)}{N^n(A)} \cdot \frac{N^n(A.B.C)}{N^n(A.B)} \\ &= F^n(A,B) \cdot F^n(A.B,C) \end{aligned} \tag{5}$$

The equation remains valid for the transition to the limit, if the individual limits exist, and we have with the use of (5, § 16)

$$P(A,B.C) = P(A,B) \cdot P(A.B,C) \tag{6}$$

We thus arrive at the general theorem of multiplication (3, § 14). We now see why this form, which we used for the theorem, is always valid. Only in this form does the multiplication theorem represent a tautology in the frequency interpretation.

This proof, too, may be illustrated by a schema as used above:

$$\begin{array}{l} A\ A\ A\ A\ A\ A\ A\ A\ \ldots \\ B\ \bar{B}\ \bar{B}\ B\ \bar{B}\ B\ \bar{B}\ B\ \ldots \\ \underline{\underline{C}}\ C\ \bar{C}\ \underline{\underline{C}}\ C\ \underline{\underline{\bar{C}}}\ C\ \underline{\underline{\bar{C}}}\ \ldots \end{array} \tag{7}$$

The frequency $F^n(A,B.C)$ represents the frequency of the couples $B.C$; the first of the expressions standing on the right side of (5), $F^n(A,B)$, counts the frequency of B. Now B selects from the sequence of C's a subsequence, the elements of which are marked by a lower double bar in (7); this subsequence, of course, contains elements C as well as $\bar{C}$. The number of elements of this subsequence is given by $N^n(A.B)$; therefore $F^n(A.B,C)$ means the relative frequency of C in the subsequence. The consideration is always applicable: if a term is added before the comma within a probability expression, the frequency is counted within the subsequence that is selected by this term. Formula (5) states that the desired frequency of the pair $B.C$ can be represented as the product of the frequency of B by the frequency of C counted within the subsequence selected by B.

These considerations lead to an instructive interpretation of the independence relation defined in (4, § 14). The definition

$$P(A.B,C) = P(A,C) \tag{8}$$

states that, within the subsequence selected by B from the C-sequence, C has the same relative frequency as in the main sequence. This characterization reveals the meaning of the independence relation; that B does not influence C means that a selection by B from the C-sequence does not change the relative frequency.[1] For instance, when we throw with two dice and consider, within the sequence produced by the second die, only the subsequence of throws in which the first die simultaneously gives the result of face 6 lying up, we shall find, too, the relative frequency $\frac{1}{6}$ for any face of the second die.

Finally, it remains to prove that the rule of existence is derivable from the frequency interpretation. Since each of the axioms represents a tautological relation between frequencies, which holds strictly even before the transition to the limit, every probability formula derivable from the axioms will correspond also to a tautological relation between frequencies; and this relation will be strictly valid before the transition to the limit. Every such relation can be written in the form

$$f_m^n = r(f_1^n \ldots f_{m-1}^n) \tag{9}$$

In this formula the f_i^n stand for frequency expressions of the form

$$f_i^n = F^n(A_i,B_i) \tag{10}$$

The subscripts in (9) and (10) indicate the fact that we are dealing here with frequency quantities that belong to different events $A,B \ldots$. According to the existence rule, r is a single-valued function, free from singularities at this place. Passing to the limit $n \to \infty$, we derive from the laws governing the formation of a limit that, whenever the $f_1^n \ldots f_{m-1}^n$ go toward limits $p_1 \ldots p_{m-1}$, the f_m^n also must approach a limit p_m. In other words, the probability p_m must exist whenever the probabilities $p_1 \ldots p_{m-1}$ exist. This is the assertion made by the rule of existence.

At the same time we recognize why the existence of a probability is bound by the condition that it be determined by given probabilities. Assume that it is unknown in (9) for two quantities, say, f_m^n and f_{m-1}^n, whether they go toward a limit. Then we cannot infer, from the fact that the other quantities $f_1^n \ldots f_{m-2}^n$ approach certain limits, that the two residual quantities f_m^n and f_{m-1}^n go toward a limit. For instance, if the probability of a logical sum is given, the sum f_3^n of the two frequencies

$$f_1^n + f_2^n = f_3^n$$

approaches a limit p_3. Yet the individual frequencies f_1^n and f_2^n need not go toward a limit. A convergence can be inferred only when it is known that, apart from f_3^n, at least one of the other quantities, say f_2^n, approaches a limit.

This concludes the proof that all the axioms of the probability calculus follow logically from the frequency interpretation. The result holds not only for infinite but also for finite sequences, provided that in this case we regard the limit of the frequency as given by the value of $F^n(A,B)$ taken for the last

[1] R. von Mises has made this idea the starting point of his probability theory. See § 30.

element. All the axioms are satisfied tautologically, and are strictly, not only approximately, valid even before the transition to the limit.

The given proof guarantees that the frequency interpretation is an admissible interpretation of the theorems derivable from the axiom system. The interpretation will be applied in the examples used to illustrate the derived formulas.

§ 19. The Rule of Elimination

We may now proceed to the derivation of individual theorems of the probability calculus from the axiom system.

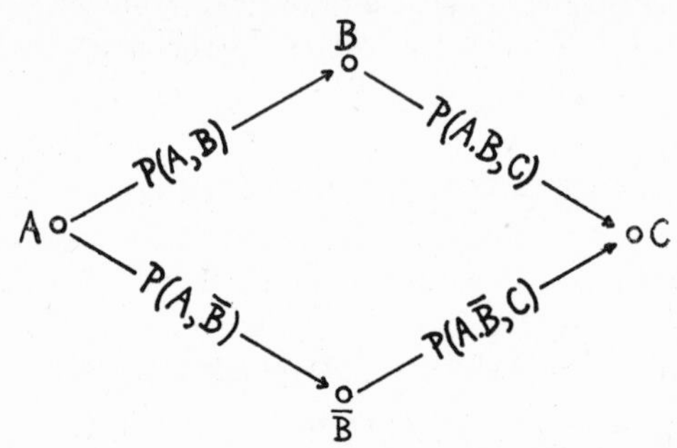

Fig. 4. Schema for rule of elimination, according to (2).

Many practical cases present the problem of calculating the probability from A to C, when C is linked to A by an intermediate term B and only the intermediary probabilities are given. Figure 4 may serve to illustrate the problem.

It represents the *divergent* probabilities $P(A,B)$ and $P(A,\bar{B})$, having the first term in common, and the *convergent* probabilities $P(A\,.\,B,C)$ and $P(A\,.\,\bar{B},C)$, which possess a common term after the comma. When the divergent and convergent probabilities are given, it is possible to calculate $P(A,C)$. For this purpose we use the logical equivalence

$$([B \vee \bar{B}]\,.\,C \equiv C) \tag{1}$$

and thus obtain the relations

$$P(A,C) = P(A,[B \vee \bar{B}]\,.\,C) = P(A,B\,.\,C \vee \bar{B}\,.\,C)$$
$$= P(A,B\,.\,C) + P(A,\bar{B}\,.\,C)$$

In the last equality the addition theorem has been applied because the terms are mutually exclusive. The use of the multiplication theorem gives the result

$$P(A,C) = P(A,B) \cdot P(A\,.\,B,C) + P(A,\bar{B}) \cdot P(A\,.\,\bar{B},C) \tag{2}$$

This formula is called the *rule of elimination.* It permits the elimination of a term B that is interpolated between the terms A and C, and the establishment of a direct probability from A to C. The rule of elimination performs with respect to probability implication the function that is performed for the logical implication by its transitivity (8*l*, § 4). But here the logical structure is much more complicated than it is for a transitivity. The elimination of B can be achieved, according to (2), only when $P(A.\bar{B},C)$ is known, apart from $P(A,B)$ and $P(A.B,C)$. The probability $P(A,\bar{B})$ is determined by $1-P(A,B)$, but $P(A.\bar{B},C)$ represents an independent probability that is not determined by the other quantities written at the right of (2). The convergent probabilities $P(A.B,C)$ and $P(A.\bar{B},C)$ will be called *nonbound probabilities,* since their sum can be greater or smaller than 1; the divergent probabilities $P(A,B)$ and $P(A,\bar{B})$ are *bound probabilities,* that is, they must add up to the value 1.

The theorem may be illustrated by an example previously used. Let A denote a hot summer day; B, the occurrence of a thunderstorm; C, a change in the weather. The probability of a change in weather occurring on a hot day can be calculated from the intermediary probability concerning the thunderstorm; but we must know the probability of the occurrence of a thunderstorm, the probability of a change in the weather on a hot day after a thunderstorm has occurred, and the probability of a change in the weather on a hot day on which no thunderstorm occurs.

In the frequency interpretation, (2) can easily be made clear: the number of C's to which a B is coördinated, and the number of C's to which a $\bar{B}$ is coördinated, add up to the total number of C's.

The rule of elimination contains some interesting special cases. First, we may have

$$P(A.B,C) = P(B,C) \tag{3a}$$

$$P(A.\bar{B},C) = P(\bar{B},C) \tag{3b}$$

Then (2) assumes the form

$$P(A,C) = P(A,B) \cdot P(B,C) + P(A,\bar{B}) \cdot P(\bar{B},C) \tag{4}$$

We can illustrate this form by choosing for B and $\bar{B}$ two bowls that contain black and white balls in different ratios, and for A another bowl containing, say, numerous tickets on which is written B or $\bar{B}$. The ticket drawn from A decides whether the second draw should be made from B or $\bar{B}$. By C we understand the event of a black ball being obtained.

A further specialization results for

$$P(A.\bar{B},C) = 0 \tag{5}$$

We then have

$$P(A,C) = P(A,B) \cdot P(A.B,C) \tag{6}$$

If the specialization (3a) is added, we obtain

$$P(A,C) = P(A,B) \cdot P(B,C) \tag{7}$$

Only in this very specialized case does the rule of elimination assume the form of a transitivity, in which the degrees of probability are simply multiplied. The case may be illustrated by the example above, with the qualification that the bowl $\bar{B}$ does not contain any black balls. Other examples are given in causal chains: for instance, when A means the presence of a hot day in summer; B, the occurrence of a thunderstorm; C, a flash of lightning hitting a house. In the special case where $P(A,B) = 1$ and $P(B,C) = 1$, the relation (7) determines also $P(A,C) = 1$; here the condition (5) is no longer required, since the second term in (2) drops out because of $P(A,\bar{B}) = 0$.[1] These relations are satisfied for logical implications of the form $(A \supset B)$ and $(B \supset C)$. The relation (3a), too, must hold in this case because with $(B \supset C)$ we have also $(A . B \supset C)$. This is why the logical implication follows a general rule of transitivity that is not restricted by any conditions. It is seen, further, that the transitivity (7), in general, produces a decrease in the degree of probability. If the intermediary probabilities written at the right in (7) are smaller than 1, the total probability at the left in (7) will be smaller than any of the intermediary probabilities. A corresponding statement cannot be made for the general case (2); here $P(A,C)$ represents a certain mean value between the other probabilities.

A third specialization results by the assumption

$$P(A.B,C) = P(A.\bar{B},C) \tag{8}$$

Introducing this condition into (2) and using the relation $P(A,B) + P(A,\bar{B}) = 1$, we obtain

$$P(A,C) = P(A.B,C) = P(A.\bar{B},C) \tag{9}$$

Comparison with (4, § 14) shows that this means the independence of B and C with respect to A. In the frequency interpretation, (9) means that if the subsequences selected from the C-sequence by B and $\bar{B}$, respectively, contain C with equal relative frequencies, this frequency is the same as in the main sequence.

It has been pointed out that $P(A.\bar{B},C)$ is not determined by $P(A,B)$ and $P(A.B,C)$; but (2) states that a determination results if $P(A,C)$ is added. This connection is expressed by the solution of (2) for $P(A.\bar{B},C)$:

$$P(A.\bar{B},C) = \frac{P(A,C) - P(A,B) \cdot P(A.B,C)}{1 - P(A,B)} \tag{10}$$

[1] If it is known that $P(B,A) > 0$, even the condition (3a) can be omitted, because this condition then follows from $P(B,C) = 1$. See (6, § 25).

The relation shows how a probability containing a negation in the first term is calculated from the probabilities of nonnegative reference. We must except the case $P(A,B) = 1$, since in this case the value of (10) is indeterminate; this condition is also understood for the relations (11), (12), and (14), to be derived presently.

As before, some important special cases must be considered. We see that with

$$P(A,C) = P(A.B,C) \tag{11a}$$

we also have

$$P(A.B,C) = P(A.\bar{B},C) \tag{11b}$$

in correspondence to (9); that is, the converse of the relation leading from (8) to (9) is valid. Furthermore, we infer that, if $P(A.B,C) > P(A,C)$, we have $P(A.\bar{B},C) < P(A,C)$, and, similarly, if $P(A.B,C) < P(A,C)$, we have $P(A.\bar{B},C) > P(A,C)$. This result follows because for $P(A.B,C) = P(A,C)$ the relation (10) supplies $P(A.\bar{B},C) = P(A,C)$, and this value is diminished or increased according as $P(A.B,C)$ is larger or smaller than $P(A,C)$.

For mutually exclusive events B and C, that is, $P(A.B,C) = 0$, relation (10) assumes the simple form

$$P(A.\bar{B},C) = \frac{P(A,C)}{1 - P(A,B)} \tag{12}$$

Another special case arises for

$$P(A,B) = P(A,C) \tag{13}$$

Then (10) is transformed into

$$\frac{P(A.\bar{B},C)}{P(A.B,\bar{C})} = \frac{P(A,B)}{P(A,\bar{B})} = \frac{P(A,C)}{P(A,\bar{C})} \tag{14}$$

From (10) we can derive two important inequalities that restrict the choice of the probabilities to be given. Since $P(A.\bar{B},C)$ is bound by the normalization (8, § 13), the expression on the right side of (10) must lie between 0 and 1 (with inclusion of the limits). This leads to the two inequalities

$$1 - \frac{1 - P(A,C)}{P(A,B)} \leqq P(A.B,C) \leqq \frac{P(A,C)}{P(A,B)} \tag{15}$$

The inequality on the left side results from transformation of the condition that (10) must not be greater than 1; the inequality on the right side arises from a transformation of the condition that the numerator of (10) must not be smaller than 0. The double inequality is not necessarily satisfied for given values $P(A,B)$ and $P(A,C)$, even if $P(A.B,C)$ is chosen according to the normalization (8, § 13). The relation (15) formulates an additional condition, which prescribes a narrower domain for $P(A.B,C)$ whenever we have

$1 - P(A,C) < P(A,B)$ or $P(A,C) < P(A,B)$. It can be shown that for independent events B and C, that is, for $P(A.B,C) = P(A,C)$, (15) is always fulfilled.[2] It is permissible, therefore, to give two events as independent, regardless of the values of their probabilities. But if two events are given as dependent, the degree of dependence must be kept within the limits defined by (15). The occurrence of such inequalities in regard to the choice of probabilities may be compared to the occurrence of similar inequalities in geometry. A triangle, for instance, can be constructed from three given determinations only when their values satisfy certain numerical restrictions. Notice that the inequalities (15) hold also for the case $P(A,B) = 1$, which had to be excepted for (10), since in this case the numerator of (10) must be $= 0$ in order to make possible a finite value of $P(A.\bar{B},C)$, and thus the conditions leading to (15) are satisfied. For mutually exclusive events B and C, that is, $P(A.B,C) = 0$, (15) leads to the trivial condition $P(A,B) + P(A,C) \leqq 1$.

We turn now to an extension of the rule of elimination to disjunctions of more than two terms. There are special kinds of such *many-term disjunctions* $B_1 \vee \ldots \vee B_r$ that play a particularly important role in the calculus of probabilities: disjunctions that are both *complete* and *exclusive*. A disjunction is called *complete* if it is true; it then follows that at least one of its terms is true. A disjunction is called *exclusive* if not more than one of its terms is true. These concepts, as applied to probability sequences, are used in an extended sense: the disjunction must have these properties for all elements of the sequence. Thus completeness, in this sense, is formulated by the statement

$$(B_1 \vee \ldots \vee B_r) \tag{16}$$

The parentheses express, according to the convention given in §§ 10, 12, the condition that the disjunction is true for all elements of the sequence; and it would be more correct to speak of completeness and exclusiveness *with respect to the sequence.* The latter qualification is always understood when the terms "complete" and "exclusive" are used in probability considerations.

The combination of the two conditions of completeness and exclusiveness is expressed by the following r formulas, which are all-statements:[3]

$$\begin{array}{c}(B_1 \equiv \overline{B_2} \,.\, \overline{B_3} \ldots \overline{B_r}) \\ (B_2 \equiv \overline{B_1} \,.\, \overline{B_3} \ldots \bar{B}_r) \\ \ldots\ldots\ldots \\ (B_r \equiv \overline{B_1} \,.\, \overline{B_2} \ldots \overline{B_{r-1}})\end{array} \tag{17}$$

The equivalence signs of the relations can be conceived as representing two mutual implications, according to (7*a*, § 4). The implication running from left to right expresses exclusiveness; the implication running from right to

[2] This is easily seen for the right-hand inequality. The proof for the left-hand inequality follows from the relation (5, § 23).

[3] The exclusive "or" cannot be used to express these conditions. See *ESL*, p. 45.

left expresses completeness. It can easily be shown that statement (16) is derivable from the relations (17).

For most of the following considerations it will be sufficient if the disjunctions are complete and exclusive with respect to A, that is, with respect to the subsequence selected by A. The symbolic expression is given by the formulas

$$A \supset \begin{cases} B_1 \equiv \overline{B_2} \,.\, \overline{B_3} \ldots \overline{B_r} \\ B_2 \equiv \overline{B_1} \,.\, \overline{B_3} \ldots \overline{B_r} \\ \quad \cdot \quad \cdot \quad \cdot \quad \cdot \\ B_r \equiv \overline{B_1} \,.\, \overline{B_2} \ldots \overline{B_{r-1}} \end{cases} \tag{18}$$

From these formulas the statement of completeness relative to A is derivable:

$$(A \supset B_1 \vee \ldots \vee B_r) \tag{19}$$

The condition (18) can be used to replace the stronger condition (17) in all cases in which only probabilities containing A in the first term are concerned. Thus when a die is thrown, the six possible results given by the six faces of the die constitute a disjunction that is complete and exclusive with respect to the sequence of events A represented by the throwing of the die. For the sake of simplicity, the condition (17) will always be used, leaving the reader to construct similar proofs on the basis of the weaker condition (18).

The introduction of many-term disjunctions in the rule of elimination is made in the same way as was used for the derivation of (2). Corresponding to (1), we have the relation

$$([B_1 \vee \ldots \vee B_r].C \equiv C) \tag{20}$$

Applying the inference leading to (2), we derive for many-term disjunctions the *extended rule of elimination*:

$$P(A,C) = \sum_{k=1}^{r} P(A,B_k) \cdot P(A\,.\,B_k,C) \tag{21}$$

Figure 5 (p. 82) may serve as an illustration. The divergent probabilities again are bound probabilities, so that

$$\sum_{k=1}^{r} P(A,B_k) = 1 \tag{22}$$

is valid; the convergent probabilities, however, are nonbound.

A schematized example for figure 5 is found in games of chance. Let $B_1 \ldots B_r$ represent bowls containing black and white balls, each in a different ratio. Let C be the drawing of a black ball, and A an auxiliary bowl containing numerous tickets, each carrying one of the numbers $1 \ldots r$. If there are more than r tickets in the bowl and each number occurs repeatedly, each number has a determinate probability of being drawn from the bowl. We draw first from the auxiliary bowl and determine from which of the other bowls we are to draw next. Repeating the two actions again and again, we obtain a statistical relation between A and C, the frequency of which is determined by $P(A,C)$ according to (21).

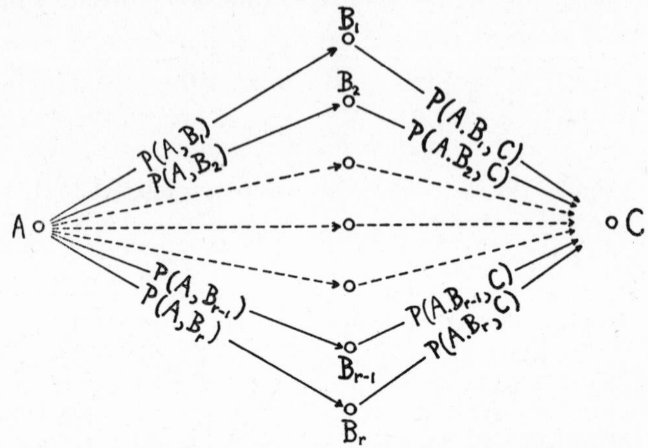

Fig. 5. Schema for extended rule of elimination, according to (21).

Another example results by taking for A the throwing of two dice, for C the occurrence of face 1 of the second die, for B_k the occurrence of face k of the first die. Then (21) means that the probability of obtaining 1 with the second die can be divided, additively, into the probabilities of the combinations in which this result is accompanied by one side k of the other die.

Both examples represent special cases of (21), namely, cases of such a kind that, for the first example, $P(A.B_k,C) = P(B_k,C)$ holds; for the second example, $P(A.B_k,C) = P(A,C)$. This corresponds to the causal conception of the problem, according to which, in the first example, B_k is the cause of C; in the second example, A is the cause of C. However, this is irrelevant to the treatment of the problem within probability theory; the lines in figure 5 represent probabilities, but not necessarily causal chains. The statement of the causal relationships requires specific investigation.

§ 20. The General Theorem of Addition

We shall now investigate the question how to calculate the probability of a disjunction if the terms of the disjunction do not mutually exclude one another, that is, if we are dealing with a nonexclusive disjunction. If, for example, two coins are thrown, what is the probability of obtaining tails with either coin, of obtaining *at least one* event of tails lying up? A simple addition would give $\frac{1}{2} + \frac{1}{2} = 1$—which obviously is a wrong result. But the conditions for applying the addition theorem are not satisfied, because it is possible to obtain tails simultaneously with both coins. In order to calculate the desired probability we must, therefore, transform the question into a form suitable for the application of the theorem of addition. Several such methods may be demonstrated.

We can start from the equivalence

$$(B \vee C \equiv B.C \vee B.\bar{C} \vee \bar{B}.C) \tag{1}$$

which leads to mutually exclusive terms and thus permits us to apply the theorem of addition:

$$\begin{aligned} P(A,B \vee C) &= P(A,B.C \vee B.\bar{C} \vee \bar{B}.C) \\ &= P(A,B.C) + P(A,B.\bar{C}) + P(A,\bar{B}.C) \end{aligned} \tag{2}$$

In the example of the two coins, the formula gives $P(A,B \vee C) = \frac{3}{4}$, because each of the probabilities of the combinations is equal to $\frac{1}{2} \cdot \frac{1}{2} = \frac{1}{4}$.

In practice, other methods may be used to solve the problem. Occasionally it is possible, using material thinking (see § 5), to contract certain steps that are made separately in the calculus. The following method may be used: (1) B occurs; then it is immaterial whether or not C also occurs. The probability for this case is $P(A,B)$. (2) B does not occur; then C must occur. The probability for this case is $P(A,\bar{B}.C)$. Since the cases (1) and (2) are mutually exclusive, the theorem of addition is applicable, and we obtain

$$P(A,B \vee C) = P(A,B) + P(A,\bar{B}.C) \tag{3}$$

a result that is identical with (2) because of $P(A,B) = P(A,B.C \vee B.\bar{C})$. This method differs from the former one in that the first two cases of the disjunction (1) are collected in one case by the help of material thinking. This thinking can also be formalized: in (5e, § 4) we have a formula that leads directly to (3).

A third method starts from the equivalence

$$(B \vee C \equiv \overline{\bar{B}.\bar{C}}) \tag{4}$$

which leads with (7, § 13) to the simple result:

$$P(A,B \vee C) = 1 - P(A,\bar{B}.\bar{C}) \tag{5}$$

Here the probability of the opposite case is calculated and then is subtracted from 1. For the example with the two coins, the probability of obtaining heads with both coins is equal to $\frac{1}{2} \cdot \frac{1}{2} = \frac{1}{4}$. Because in any other case at least one event of tails must happen, the desired probability is calculated to be $1 - \frac{1}{4} = \frac{3}{4}$.

We now establish for such probabilities a fourth formula that seems very convenient for technical reasons. It can be derived directly from the calculus without the aid of material thinking. Because of

$$\begin{aligned} P(A,B) &= P(A,B.C) + P(A,B.\bar{C}) \\ P(A,C) &= P(A,B.C) + P(A,\bar{B}.C) \end{aligned} \tag{6}$$

we can write, together with (2), the three formulas

$$\begin{aligned} P(A,B \vee C) &= P(A,B.C) + P(A,B.\bar{C}) + P(A,\bar{B}.C) \\ 0 &= -\,P(A,B.C) - P(A,B.\bar{C}) + P(A,B) \\ 0 &= -\,P(A,B.C) + P(A,C) - P(A,\bar{B}.C) \end{aligned} \tag{7}$$

Adding the three formulas, we obtain

$$P(A,B \vee C) = P(A,B) + P(A,C) - P(A,B.C) \tag{8}$$

This formula is called the *general theorem of addition*. It is a generalization of the addition theorem (3, § 13), applying to nonexclusive terms. In case $P(A,B.C) = 0$ it becomes identical with the *special theorem of addition* (3, § 13). In contradistinction to the latter, (8) represents an always-true formula because it is not contingent upon any conditions to be expressed in the context. The condition of exclusion, which had to be added verbally to the P-notation (3, § 13) as a logical condition, is *formalized mathematically* in (8); it is expressed by the case that a mathematical quantity assumes the value 0.

In the frequency interpretation, (8) can easily be made comprehensible. In dealing with the nonexclusive cases B and C, the couples $B.C$ will occur according to, say, the following schema:

$$\begin{matrix} A\ A\ A\ A\ A\ A\ \ldots . \\ B\ \bar{B}\ \bar{B}\ B\ \bar{B}\ B\ \ldots . \\ C\ C\ \bar{C}\ \bar{C}\ C\ C\ \ldots . \end{matrix} \tag{9}$$

Adding the frequencies B and C, we shall have counted the couples $B.C$ twice; therefore, to form $P(A,B \vee C)$, the frequency of the couples $B.C$ is to be subtracted once. This fact is expressed in (8).

It need not be expressed as a condition that the probability of the disjunction, as given by (8), satisfy the normalization of probabilities; this follows from the double inequality (15, § 19) previously established. After a simple transformation by means of the theorem of multiplication, the inequality on the left side of (15, § 19) leads to

$$P(A,B) + P(A,C) - P(A,B.C) \leqq 1 \tag{10}$$

Now the inequality on the right gives the result

$$P(A,B.C) \leqq P(A,C) \tag{11}$$

By interchanging B and C we obtain

$$P(A,B.C) \leqq P(A,B) \tag{12}$$

Therefore the following inequalities are satisfied:

$$P(A,B \vee C) \geqq P(A,B) \tag{13}$$

$$P(A,B \vee C) \geqq P(A,C)$$

The probability of a disjunction is never smaller and, in general, is even greater than the probability of its individual terms. Thereby the character of the disjunction as a logical sum is clearly expressed. Addition of a term connected by "or" signifies an increase in probability, and only in the limiting case does the probability remain the same.

Some examples may illustrate the general theorem of addition. The testing of a mechanical appliance reveals, on the average, 2% rejections because of material defects and 3% rejections because of defects in assembling the parts. What is the average rejection on the whole? Here the probabilities are given statistically, as is usual in practice. But we must not assume as total rejection 3% + 2% = 5%, since the two sources of defect are not mutually exclusive. An appliance that is faulty because of material defects may also show a defect owing to assembling. We know from experience that we are dealing here with independent probabilities; thus we can apply the special theorem of multiplication. The probability of both defects occurring simultaneously is given by the product 3% · 2% = 0.06%. Then (8) provides as average frequency for the total rejection 3% + 2% − 0.06% = 4.94%.

Another example is a firm that sells its products partly through traveling salesmen and partly through advertisements. The statistics on customers reveal that 80% of all products are sold by salesmen and 60% by advertisements. What is the percentage of customers won by advertisements as well as by salesmen? Since here $P(A,B \vee C) = 1$ (we assume that all products are sold only in these two ways), it follows that $P(A,B.C) = 80\% + 60\% - 100\% = 40\%$, that is, 40% of the customers are won by both means together.

Formula (8) permits a general calculation of the or-probability, but in applying it we must be sure that the case considered possesses the logical structure of the theorem of addition. Mistakes of this kind may be illustrated by two examples that were given by Richard von Mises[1] with the intention of showing that the addition must not be carried out uncritically, even for

[1] *Wahrscheinlichkeit, Statistik und Wahrheit* (Berlin and Vienna, 1928), p. 40.

mutually exclusive events. He wishes to restrict the theorem of addition to events belonging to the same "collective", that is, the same sequence. My formulation of the theorem is somewhat more general, since the theorem is not restricted to events belonging to the same sequence. Instead, another condition is used, specifying that the probabilities have the same reference class, or first term. I shall now show that my formulas are applicable to the examples given by von Mises, and permit the use of the "or" in a reasonable sense.

Assume that a tennis player has the probability 0.8 of winning in a tournament in Berlin; he may have the probability 0.7 of winning in a tournament played the same day at New York. The events are mutually exclusive; thus one might infer that the probability of the player winning in the one *or* in the other tournament was given by the addition of the probabilities, which would result in $0.8 + 0.7 = 1.5$. This is certainly a nonsensical result.

We are dealing here with a question of interpretation. A problem given in conversational language is to be translated into the strict language of the calculus; one cannot expect unambiguous rules to be available for such a translation. To assume that the special theorem of addition is applicable would be to interpret the problem in the form

$$P(A,B) = 0.8 \qquad P(A,C) = 0.7 \qquad P(A,B.C) = 0 \tag{14}$$

A representing the general situation before the tournaments; B, the victory in Berlin; C, in New York. It is obvious that the numerical values used in the interpretation violate the inequality (15, § 19), because $P(A,B.C) = 0$ implies $P(A.B,C) = 0$, whereas the expression on the left of the inequality assumes the value $\frac{5}{8}$. This illustrates the fact that the condition of exclusion represents a high degree of dependence and therefore can be combined only with suitable numerical values of the other given probabilities. It follows that (14) is not an admissible interpretation of the problem.

An interpretation that comes closer to what is intended by the formulation of the problem can be given. We consider the probability 0.8 of winning in Berlin as referring to the first term B_1, "if the player participates in Berlin"; and the probability 0.7 of winning in New York as referring to the first term B_2, "if the player participates in New York". If C represents "winning", we then can set down

$$P(B_1,C) = 0.8 \qquad P(B_2,C) = 0.7 \tag{15}$$

The two probabilities do not differ by their second term, as do the expressions (4), but by their first term. It is obvious that the probabilities do not permit the application of formula (8). The general condition A holding before the tournaments take place appears as a reference class in the sense of the theorem of elimination (fig. 5, p. 82), representing the fact that the player may decide to participate in one or the other of the tournaments; and the condition of exclusion must then be written

$$P(A,B_1.B_2) = 0 \tag{16}$$

When we wish to derive from these conditions the probability of winning, that is, $P(A,C)$, the two further probabilities

$$P(A,B_1) \qquad P(A,B_2) \tag{17}$$

must be given. This means that the probability of winning depends on the probabilities of the player deciding, respectively, to participate in New York or in Berlin.

In this interpretation the problem is solved, since $P(A,\overline{B_1 \vee B_2}.C) = 0$, by the equations

$$\begin{aligned} P(A,C) &= P(A,[B_1 \vee B_2 \vee \overline{B_1 \vee B_2}].C) \\ &= P(A,B_1.C) + P(A,B_2.C) \\ &= P(A,B_1) \cdot P(A.B_1,C) + P(A,B_2) \cdot P(A.B_2,C) \\ &= P(A,B_1) \cdot P(B_1,C) + P(A,B_2) \cdot P(B_2,C) \end{aligned} \tag{18}$$

because we may assume (10, § 14). That we cannot carry out the calculation numerically is due to the fact that the probabilities (17) are not given, but the failure to obtain a solution does not result from an inadmissible use of the "or". It is clear, furthermore, that in this interpretation the sum of $P(B_1,C)$ and $P(B_2,C)$ can be greater than 1, since these values represent nonbound probabilities (see § 19).

Von Mises presents another example that is supposed to demonstrate the use of an unreasonable "or". Let 0.011 be the probability that a man 40 years of age will die between his 40th and 41st birthdays; and let the probability that a man 41 years old marries in that year be 0.009. Both events are exclusive for one individual. If we now want to find the probability that a man 40 years of age either dies within the current year or marries in the following year, it may occur to us to add the given numbers, thus obtaining the result, $0.011 + 0.009 = 0.020$. Von Mises is right in asserting that this is a nonsensical result.

For the conception of the or-probability developed in this section, however, the problem is not meaningless. The probability of a man 40 years old dying this year or marrying next year can be interpreted to have a definite meaning. It may be expressed statistically: after a lapse of two years, we count among all the original quadragenarians those who died within the first year or married in the second year. These numbers may indeed be added, in agreement with (8). However, we must not add the numerical values given; the second value cannot be used because it states, not the probability that a man 40 years of age will marry in his 41st to 42d year, but the probability that a man 41 years of age will marry in that period. The probabilities are not the same, because some of the men will have died within the year. The value 0.009, therefore, is to be interpreted as the probability that a man 40 years old who reaches his 41st year will marry in his 41st to 42d year. This probability is represented by $P(A.\bar{B},C)$, if A stands for the class of quadragenarians, B for the class of deaths among them, and C for the class of men 41 years old who marry. We have, therefore,

$$P(A,B) = 0.011 \qquad P(A,B.C) = 0 \qquad P(A.\bar{B},C) = 0.009 \tag{19}$$

and obtain

$$\begin{aligned} P(A,B \vee C) &= P(A,B) + P(A,C) \\ &= P(A,B) + P(A,[B \vee \bar{B}].C) \\ &= P(A,B) + P(A,B.C) + P(A,\bar{B}.C) \\ &= P(A,B) + P(A,\bar{B}) \cdot P(A.\bar{B},C) \\ &= 0.011 + (1 - 0.011) \cdot 0.009 = 0.0199 \end{aligned} \tag{20}$$

This represents the probability that a man 40 years of age either will die in his 40th to 41st year or will marry in his 41st to 42d year.

In criticizing these examples I do not wish to deny that the probability calculus of von Mises supplies equally correct solutions. I intend merely to show that we can dispense with the relatively complicated operations of constructing new collectives, which von Mises has introduced, and that the desired probabilities can be conceived reasonably as or-probabilities.

We shall now derive from the general theorem of addition some consequences for later use. We can calculate a probability of the form $P(A,B \supset C)$ by resolving the implication into $\bar{B} \vee C$ according to (6*a*, § 4) and then applying the general theorem of addition. We obtain

$$\begin{aligned} P(A,B \supset C) &= P(A,\bar{B} \vee C) \\ &= P(A,\bar{B}) + P(A,C) - P(A,\bar{B}.C) \\ &= P(A,\bar{B}) + P(A,C) - P(A,\bar{B}) \cdot P(A.\bar{B},C) \end{aligned} \tag{21}$$

By the use of (10, § 19) we arrive at

$$P(A,B \supset C) = 1 - P(A,B) + P(A,B) \cdot P(A.B,C) \tag{22}$$

In a similar way we obtain for the equivalence, by the dissolution $([B \equiv C] \equiv [B.C \vee \bar{B}.\bar{C}])$, according to (7b, § 4), and with (10, § 19),

$$\begin{aligned} P(A,B \equiv C) &= P(A,B.C \vee \bar{B}.\bar{C}) \\ &= P(A,B.C) + P(A,\bar{B}.\bar{C}) \\ &= P(A,B) \cdot P(A.B,C) + P(A,\bar{B}) \cdot P(A.\bar{B},\bar{C}) \\ &= 1 - P(A,B) - P(A,C) + 2\,P(A,B) \cdot P(A.B,C) \\ &= 1 + P(A,B.C) - P(A,B \vee C) \end{aligned} \tag{23}$$

A formula containing an exclusive "or" will now be constructed. According to (1, § 4), this operation can be defined as

$$b \wedge c =_{Df} (b \vee c).\overline{b.c} \tag{24}$$

Because of the equivalence

$$(b \vee c).(\overline{b.c}) \equiv (b \vee c).(\bar{b} \vee \bar{c}) \equiv b.\bar{c} \vee \bar{b}.c \tag{25}$$

we can write, using (7b and 7c, § 4),

$$b \wedge c \equiv \overline{b \equiv c} \tag{26}$$

The symbol of the exclusive "or" can be used also in the class calculus. The class $B \wedge C$ represents, according to (24), the common class of $B \vee C$ and $\overline{B.C}$, that is, the part of the joint class of B and C that results by subtracting the common class of B and C. Because of the relation (26) we have

$$P(A,B \wedge C) = P(A,\overline{B \equiv C}) = 1 - P(A,B \equiv C) \tag{27}$$

With the use of (23) we obtain, applying (8),

$$P(A,B \wedge C) = P(A,B) + P(A,C) - 2\,P(A,B.C) \tag{28}$$

Although we have thus derived a formula dissolving an exclusive "or", the result shows that it is not possible, for the special theorem of addition, to eliminate the condition of exclusion by the use of a symbol for the exclusive "or". The formula

$$P(A,B \wedge C) = P(A,B) + P(A,C) \tag{29}$$

is false if it is conceived as holding for all B and C; it holds only if $P(A,B.C) = 0$, that is, if B and C are mutually exclusive. But if this condition must again be added, the introduction of the symbol of the exclusive "or" is useless. The aim of expressing the addition theorem completely in the mathematical notation is achieved, instead, in the general theorem of addition formulated in (8).

§ 21. The Rule of the Product

So far in this study, the multiplication theorem has been written in the form

$$P(A,B.C) = P(A,B) \cdot P(A.B,C) \tag{1}$$

Since the left side is symmetrical with respect to B and C, we may write the corresponding equation, dividing the product in a different way:

$$P(A,B.C) = P(A,C) \cdot P(A.C,B) \tag{2}$$

This is not a new axiom; it follows from (1) by substituting B for C and C for B, and in view of the fact that the "and" on the left is commutative. Because of the equality of the expressions written on the left in (1) and (2), we have

$$P(A,B) \cdot P(A.B,C) = P(A,C) \cdot P(A.C,B) \tag{3}$$

This equation is called the *rule of the product.*

For example, let $P(A,B.C)$ be the probability that a person, A, shows ability for physics, B, as well as for music, C.[1] Then (1) represents one form of splitting the probability of the product into two probabilities: the probability that a person has an ability for physics and the probability that a person endowed with an ability for physics also shows a talent for music. Formula (2) represents the opposite splitting of the probability of the product, namely, into the probability that a person has a talent for music and the probability that a musically gifted person also shows ability for physics. The two probabilities having two terms in their reference class are not equal; rather, according to (3), they have the ratio

$$\frac{P(A.B,C)}{P(A.C,B)} = \frac{P(A,C)}{P(A,B)} \tag{4}$$

The probability $P(A,C)$ of being musically gifted is, in general, much greater than the probability $P(A,B)$ of having any ability for physics. Therefore, according to (4), the probability that a person who is able in physics has a talent for music must be much greater than the probability that a musician shows an aptitude for physics. But we know from experience that some connection exists between ability in physics and in music, such that $P(A.B,C) > P(A,C)$; musical talent occurs more frequently among persons who are able physicists than corresponds to the general average. Therefore, because of (4) it must equally be the case that $P(A.C,B) > P(A,B)$, that is, among musicians also there must be a higher percentage of people with an ability in

[1] It does not matter that, in this example, according to the convention given, we should write for A,B,C the same capital letters but with different subscripts.

physics than corresponds to the average. The ratio must be the same in both cases, since (4) may be written

$$\frac{P(A.C,B)}{P(A,B)} = \frac{P(A.B,C)}{P(A,C)} \tag{5}$$

Furthermore, we derive from (5) that if $P(A.C,B) = P(A,B)$, then also $P(A.B,C) = P(A,C)$. This means that the independence of B and C with respect to A is a relation symmetrical in B and C (see p. 105). The condition of exclusion is also symmetrical, because if $P(A.B,C) = 0$, then $P(A.C,B) = 0$, according to (5), provided $P(A,B)$ and $P(A,C)$ are different from 0.

Solving (3) for $P(A.C,B)$, we obtain

$$P(A.C,B) = P(A.B,C) \cdot \frac{P(A,B)}{P(A,C)} \tag{6}$$

This relation shows that the probability $P(A.C,B)$ is determined by the three probabilities $P(A,B)$, $P(A,C)$, and $P(A.B,C)$. Since the latter three probabilities determine also the probability $P(A.\bar{B},C)$, as is shown in (10, § 19), it follows that all probabilities of B and C relative to A are determined by these three probabilities. Thus $P(A.\bar{C},B)$ is derivable by means of (6), when we substitute there $\bar{C}$ for C and use the rule of the complement (7, § 13).

$$P(A.\bar{C},B) = [1 - P(A.B,C)] \cdot \frac{P(A,B)}{1 - P(A,C)} \tag{7}$$

The three probabilities

$$P(A,B) \qquad P(A,C) \qquad P(A.B,C) \tag{7'}$$

will be called the *fundamental probabilities* of the three events A,B,C; they determine completely the *probability status of B and C with respect to A*. The choice of these values as fundamental must be regarded as a convention; any other three independent values might be chosen, for instance, the values $P(A,B)$, $P(A,C)$, $P(A,B.C)$. But the convention will be seen to be expedient.

The numerical values of the fundamental probabilities can be chosen arbitrarily when a problem is to be given; they are subject only to the restrictions of the inequalities (15, § 19). It has been shown in § 20 that these inequalities suffice to guarantee that the probabilities $P(A,B \vee C)$ and $P(A,B.C)$ are between 0 and 1, limits included. Formula (6) shows that then $P(A.C,B)$ also is bound to these limits, since we derive from the condition on the right of (15, § 19) that

$$P(A.B,C) \cdot \frac{P(A,B)}{P(A,C)} \leqq 1 \tag{8}$$

The inequalities (15, § 19) formulate, therefore, the necessary and sufficient restrictions to which the fundamental probabilities are subject.

The term "restrictions" is applied to an arbitrary choice of numerical values such as is made when fictitious problems are constructed. For all statistics that are empirically compiled, these restrictions are satisfied automatically. With respect to applications, the result can be stated as follows: when three events A,B,C, are concerned, it is sufficient to ascertain statistically the values of the three fundamental probabilities; these values, which will always satisfy the inequalities (15, § 19), are sufficient to derive all other probabilities of B and C that have the term A as reference class or as a factor of the reference class.

If B and C are events that stand to each other in the relation of cause to effect, (6) becomes of particular interest. For instance, let A represent the occurrence of a hot day in summer; B, the occurrence of a thunderstorm; C, the occurrence of a change in the weather. Then $P(A.C,B)$ represents the probability that a change of weather observed on a hot day has been preceded by a thunderstorm. In contradistinction to the example referring to talents in music and physics, in which the probabilities $P(A.B,C)$ and $P(A.C,B)$ express mere correlations, the probabilities refer here to causal relations: the thunderstorm is a possible cause of the change in the weather. The quantity $P(A.B,C)$ is therefore the probability that a certain cause will produce a particular effect, and the quantity $P(A.C,B)$ is the probability that an observed effect was produced by a specified cause. With respect to applications of this kind, (6) is also called the *rule for the probability of a cause*. In this interpretation (6) is usually given another form. Considering $\bar{B}$ as another possible cause of C, and expanding $P(A,C)$ according to the rule of elimination (2, § 19), we transform (6) into

$$P(A.C,B) = \frac{P(A,B) \cdot P(A.B,C)}{P(A,B) \cdot P(A.B,C) + P(A,\bar{B}) \cdot P(A.\bar{B},C)} \tag{9}$$

The expression obtains a more general form when the version (21, § 19) of the rule of elimination is used:

$$P(A.C,B_k) = \frac{P(A,B_k) \cdot P(A.B_k,C)}{\sum_{i=1}^{r} P(A,B_i) \cdot P(A.B_i,C)} \tag{10}$$

This formula carries the name of the English clergyman Thomas Bayes[2] and is called *the rule of Bayes*. The schema of figure 5 (p. 82) may serve again as illustration.

[2] Thomas Bayes' *Essay towards Solving a Problem in the Doctrine of Chances* was published after his death in *Philosophical Transactions of the Royal Society of London*, Vol. 53 (1763), p. 370. This paper gives only the simplified version (12) of the formula. The general formula (10) was introduced by Pierre Simon Laplace, *Théorie analytique des probabilités* (Paris, 1812), Vol. II, chap. 1 (3d ed.; Paris, 1820), p. 182. The major interest of both authors concerned the application of the rule to the derivation of convergence formulas of induction, as presented in § 62.

The quantities $P(A,B_i)$, which occur in Bayes's rule, have been named "*a priori* probabilities". The term is misleading because of its metaphysical connotations, and I prefer to call them *antecedent probabilities*. The name indicates that in these probabilities the event B_i is referred to certain general data A the acquisition of which precedes the observation of the specific data included in C. It goes without saying that antecedent probabilities are of the same type as all other probabilities.

The probabilities $P(A.C,B_k)$ are called *inverse probabilities*. Bayes's rule determines the *inverse probabilities* as functions of the *forward* probabilities, the latter term including both kinds of probabilities occurring on the right of (10). It is important to realize that such a determination is possible only if, among the forward probabilities, the antecedent probabilities are given; without a knowledge of the latter the problem would be indeterminate. Only when the antecedent probabilities are all equal, that is, when

$$P(A,B_1) = P(A,B_2) = \ldots = P(A,B_r) \tag{11}$$

do they disappear in the formula, since then (10) assumes the simplified form

$$P(A.C,B_k) = \frac{P(A.B_k,C)}{\sum_{i=1}^{r} P(A.B_i,C)} \tag{12}$$

But in order to apply (12) we must have the positive knowledge expressed in (11). It is by no means permissible to use (12) when the values of the antecedent probabilities are unknown. Absence of knowledge of numerical values is not equivalent to knowledge of their equality. The disregard of this simple logical fact has become the source of many erroneous interpretations of Bayes's rule.[3] When nothing about the antecedent probabilities is known, we must simply admit that the inverse probabilities cannot be determined.

The following example may serve as a numerical illustration of Bayes's rule. A factory A has three machines for the manufacture of a certain product; machine B_1 produces 10,000 pieces daily; machine B_2, 20,000 pieces; machine B_3, 30,000 pieces. All three machines occasionally produce faulty pieces, C; and, specifically, the first machine has on the average a rejection of 4%; the second, of 2%; the third, of 4%. A characteristic sample is found among the rejects, and we ask for the probability stating by which of the three machines it was produced. We have here

$$P(A,B_1) = \tfrac{10,000}{60,000} = \tfrac{1}{6} \qquad P(A,B_2) = \tfrac{20,000}{60,000} = \tfrac{1}{3}$$

$$P(A,B_3) = \tfrac{30,000}{60,000} = \tfrac{1}{2}$$

[3] These misinterpretations go back to Bayes and Laplace, who regarded it permissible to apply (12) when the antecedent probabilities are unknown; the name *a priori probabilities* was used with reference to such an "*a priori* reasoning". See the criticism of the principle of indifference in § 68.

$P(A.B_1,C) = 4\% = \frac{4}{100} \quad P(A.B_2,C) = 2\% = \frac{2}{100} \quad P(A.B_3,C) = 4\% = \frac{4}{100}$

$$P(A.C,B_1) = \frac{\frac{1}{6} \cdot \frac{4}{100}}{\frac{1}{6} \cdot \frac{4}{100} + \frac{1}{3} \cdot \frac{2}{100} + \frac{1}{2} \cdot \frac{4}{100}} = \frac{1}{5} = 20\%$$

$$P(A.C,B_2) = \frac{\frac{1}{3} \cdot \frac{2}{100}}{\frac{1}{6} \cdot \frac{4}{100} + \frac{1}{3} \cdot \frac{2}{100} + \frac{1}{2} \cdot \frac{4}{100}} = \frac{1}{5} = 20\%$$

$$P(A.C,B_3) = \frac{\frac{1}{2} \cdot \frac{4}{100}}{\frac{1}{6} \cdot \frac{4}{100} + \frac{1}{3} \cdot \frac{2}{100} + \frac{1}{2} \cdot \frac{4}{100}} = \frac{3}{5} = 60\%$$

We see clearly the influence of the antecedent probabilities $P(A,B_k)$, which are calculated in a simple way from the distribution of the total production over all the machines. Though the second machine works twice as well as the first, it is equally probable that the rejected piece originates from the second as from the first machine; this is due to the fact that the second machine produces twice as many pieces. The third machine, which supplies half of the total production, is to be assigned the probability $\frac{3}{5}$ of having produced the reject; this probability is greater than $\frac{1}{2}$ because one of the two other machines works more reliably. Therefore, of all rejects, 20% originate from the first, 20% from the second, and 60% from the third machine; this represents the statistical meaning of the inverse probabilities calculated. At the same time we recognize that without such antecedent probabilities the problem is not determined. Should we consider only the efficiency ratio of the machines, 4 : 2 : 4, and calculate the inverse probabilities as $\frac{4}{10}, \frac{2}{10}, \frac{4}{10}$, this would mean putting $P(A,B_1) = P(A,B_2) = P(A,B_3)$; but we must check whether the assumption is justified. The probability of causes cannot be calculated without a knowledge of the antecedent probabilities. (Further examples are given in the appendix to chap. 3, pp. 123–127.)

The range of application for Bayes's rule is extremely wide, because nearly all inquiries into the causes of observed facts are performed in terms of this rule. The *method of indirect evidence*, as this form of inquiry is called, consists of inferences that on closer analysis can be shown to follow the structure of the rule of Bayes. The physician's inferences, leading from the observed symptoms to the diagnosis of a specified disease, are of this type; so are the inferences of the historian determining the historical events that must be assumed for the explanation of recorded observations; and, likewise, the inferences of the detective concluding criminal actions from inconspicuous observable data. In many instances the use of probability relations is not manifest because the probabilities occurring have either very high or very low values. Thus, when a corpse is found, it is virtually certain that a murder has been committed; and a fingerprint on the handle of a pistol may be considered as strict evidence for the assumption that a certain person X has fired the pistol. That even in such cases the inference has the structure of Bayes's rule is often seen from the fact that appraisals of the antecedent probabilities are made. Thus an inquiry by the detective into the motives of a crime is an attempt to estimate the antecedent probabilities of the case, namely, the probability of a certain person committing a crime of this kind, irrespective of the observed incriminating data. Similarly, the general inductive inference from observational data to the validity of a given scientific theory must be regarded as an inference in terms of Bayes's rule.[4]

The theory of indirect evidence has been obscured by the assumption that there exists an inference leading from an implication $(B \supset C)$ to a probability implication $(C \underset{q}{\rightarrow} B)$, which would enable us to infer with probability from an observed effect C the presence of the cause B. This inference has been called an *inference by confirmation.*[5] The analysis of the calculus of probability shows that no such inference exists. The probability of a cause B can be inferred from the observation of the effect C only if all the probabilities occurring on the right-hand side of (9) are known. The relation $(B \supset C)$ supplies only $P(A.B,C) = 1$. There remain to be known, therefore, the antecedent probability $P(A,B)$ and the probability $P(A.\bar{B},C)$. These values are in no way restricted by the fact that $(B \supset C)$ holds and must be independently ascertained for this as well as for the general case $P(A.B,C) < 1$.

For the case $P(A.B,C) = 1$, a weaker inference can be made when it is known, at least, that the other probabilities on the right-hand side of (9) exist. Putting $P(A,B) = p$, $P(A.\bar{B},C) = v$, this side then assumes the form

$$\frac{p}{p + (1 - p)v} \tag{13}$$

If $v = 1$, the denominator will be $= 1$; if $v < 1$, it will be < 1. The fraction, therefore, will be $\geqq p$, and we have the inequality

$$P(A.C,B) \geqq P(A,B) \tag{14}$$

If we know that $v < 1$, we can say that the observation of C will increase the probability of B. But even the latter statement presupposes more than the observation of C; besides the knowledge that $P(A.\bar{B},C)$ exists and is < 1, it presupposes knowledge about the existence of the probability $P(A,B)$. It is obvious, furthermore, that when inferences of indirect evidence are made, the conclusion is not restricted to the assertion of a mere increase in prob-

[4] For a more elaborate discussion of this inference see §§ 84–85.

[5] R. Carnap, "Testability and Meaning," in *Philos. of Science*, Vol. III (1936), p. 420; Vol. IV (1937), p. 1. Instead of the relation of implication in $B \supset C$, other relations that make the inference even worse are sometimes used.

ability. We wish to assert more, namely, to arrive at an estimate of whether the probability $P(A.C,B)$ is a high value. This aim can be reached only when the values $P(A,B)$ and $P(A.\bar{B},C)$ are known to a certain degree of approximation.

The so-called inference by confirmation, therefore, represents an incomplete schematization of the inference actually made in such cases. When it seems that we sometimes do infer from an observed consequence that an assumption is probably true, as in the confirmation of a scientific theory by the observational test of its consequences, such a procedure is possible only because more is known than is explicitly stated in the inference, in other words, because we have estimates of the other necessary probabilities. This additional knowledge plays a part in the inferences actually made, as may be illustrated by the problems given in the exercises (see the appendix to chap. 3, pp. 123–124).

An inferential schema like the inference by confirmation, which omits this knowledge in its premises, must be regarded as an instance of the *fallacy of incomplete schematization*. Like other fallacies, it will sometimes lead to correct results; that will be the case when the additional premises are true. But it does not represent a valid inference, because it does not state all the premises required for the truth of the conclusion. Such mistaken interpretations of the method of indirect evidence make it clear that a satisfactory analysis of the method can be given only when it is construed as an inference that follows the rules of the calculus of probability.

§ 22. The Rule of Reduction

In connection with the schema of "bifurcation" as shown in figure 5 (p. 82), we shall now derive a theorem for a probability containing an "or" in its first term, that is, in the reference class. We can obtain such a probability by using theorem (1, § 14), which shows a way of bringing a symbol B from the second into the first term of a probability expression. It is convenient to use, instead, the mathematical notation (3, § 14), which, when we put D for C, may be written

$$P(A.B,D) = \frac{P(A,B.D)}{P(A,B)} \tag{1}$$

Let $B \vee C$ be an exclusive disjunction that is incomplete with respect to A. If we substitute $B \vee C$ for B and apply the distributive law and the special theorem of addition, we obtain

$$P(A.[B \vee C],D) = \frac{P(A,B.D) + P(A,C.D)}{P(A,B) + P(A,C)} \tag{2}$$

Solving the terms in the numerator by the theorem of multiplication, we arrive at the formula

$$P(A.[B \vee C],D) = \frac{P(A,B) \cdot P(A.B,D) + P(A,C) \cdot P(A.C,D)}{P(A,B) + P(A,C)} \tag{3}$$

This theorem, which is valid only when B and C are mutually exclusive, will be called the *special rule of reduction.* It solves a probability with a disjunction in the first place of the probability functor in terms of *individual* probabilities, that is, probabilities that do not contain a disjunction in the first place, but may contain a conjunction in that place. The theorem can easily be extended for exclusive disjunctions of the form $B_1 \vee \ldots \vee B_m$ that are incomplete relative to A:

$$P(A.[B_1 \vee \ldots \vee B_m],D) = \frac{\sum_{k=1}^{m} P(A,B_k) \cdot P(A.B_k,D)}{\sum_{k=1}^{m} P(A,B_k)} \tag{4}$$

The name *rule of reduction* is chosen in order to express the fact that the reference class on the left in (4) can be conceived as resulting from the general reference class A by a reduction. This is to be understood as follows. The reference class A is the same as the class $A.[B_1 \vee \ldots \vee B_r]$ when the latter disjunction is complete relative to A. The reference class $A.[B_1 \vee \ldots \vee B_m]$, containing a disjunction that is incomplete with respect to A, results from the former by the canceling of some of the B_i—a process that may be called a reduction. Such a reduction will be used when additional knowledge permits us to drop some of the B_i.

An example chosen from political elections may serve as an illustration. Let $B_1 \ldots B_r$ represent candidates of several political parties for a high office, say the presidency of a nation; let $P(A,B_k)$ be the probability with which the election of the candidate B_k may be expected in the situation A existing before the votes are cast; and let $B_1 \vee \ldots \vee B_r$ be a complete disjunction relative to A. This disjunction is also exclusive when the political office can be occupied by only one candidate. Let D be a certain action of economic importance; it may be expected with a probability $P(A.B_k,D)$ that the candidate B_k carries out this action successfully. For example, D may be the conclusion of a commercial pact with another country. [In this case, $P(A.B_k,D)$ would be equal to $P(B_k,D)$, which is, however, irrelevant to the example.] *Before* the beginning of the elections the probability $P(A,D)$ of the signing of the commercial treaty is calculated according to the rule of elimination (21, § 19). Now assume that the elections are under way, and that it is already known that certain candidates are not elected; so only a part $B_1 \vee \ldots \vee B_m$ $(m < r)$ of the candidates remain to be considered. The probability with which the signing of the commercial pact is to be expected is then obtained by a reduction of the reference class and is determined by (4).

Equation (4) expresses a characteristic asymmetry between the first and the second terms of a probability implication. An "or" in the second term leads to an *addition* of probabilities, but an "or" in the first term, as is recognized from (4), leads to an addition combined with a division, that is, to the formation of a *mean value*. This becomes obvious through consideration of the special case in which

$$P(A.B_1,D) = P(A.B_2,D) = \ldots = P(A.B_m,D) \tag{5}$$

Then we obtain from (4)

$$P(A.[B_1 \vee \ldots \vee B_m],D) = P(A.B_1,D) = \ldots = P(A.B_m,D) \tag{6}$$

Here the addition of or-terms in the first term does not change the probability. If we assign to each of the candidates who are not yet eliminated an equal probability that he will successfully carry out the signing of the pact, then it is immaterial which candidate is elected. The probability of the signing of the treaty does not depend on the further outcome of the election. Furthermore, it is of no importance for the relation (6) whether all the $P(A,B_k)$ are equal; the values of the $P(A.B_n,D)$ $(m < n < r)$ no longer matter, because the respective candidates are already eliminated from the election.

If the disjunction is complete relative to A, (4) represents a second form of the rule of elimination (21, § 19), since the denominator becomes equal to 1:

$$P(A,D) = P(A.[B_1 \vee \ldots \vee B_r],D) = \sum_{k=1}^{r} P(A,B_k) \cdot P(A.B_k,D) \tag{7}$$

When we write the disjunction in the form $B \vee \bar{B}$, we arrive at the equation

$$P(A,D) = P(A.[B \vee \bar{B}],D) = P(A,B) \cdot P(A.B,D) + P(A,\bar{B}) \cdot P(A.\bar{B},D) \tag{8}$$

From (7) and (8) we see that a B occurring in the first term can be eliminated like a B in the second term, whereas the right side of the equation assumes the same form as in (2, § 19) and (21, § 19).

The difference between the rule of reduction and the rule of elimination, however, is made clear when we consider disjunctions which, though exclusive, are incomplete with respect to A. The formula

$$P(A, [B_1 \vee \ldots \vee B_m].D) = \sum_{k=1}^{m} P(A,B_k) \cdot P(A.B_k,D) \tag{9}$$

which corresponds to the rule of elimination, is then always true, although this probability is not equal to $P(A,D)$. The corresponding formula with the disjunction in the first place is given by (4); here the sum in the denominator is added.

When we add the assumption (5) to (7) and extend (5) to hold for all terms of a disjunction $B_1 \vee \ldots \vee B_r$, which is complete relative to A, we obtain from (7), analogous to (6),

$$P(A,D) = P(A.[B_1 \vee \ldots \vee B_r],D) = P(A.B_1,D) = \ldots = P(A.B_r,D) \tag{10}$$

This represents the trivial assertion that, in this case, the probability from A to D *is* the same as the probability from A together with any B_i to D.

Formula (4) will now be presented in a different form in order to make its structure clearer. Taking into account the condition of exclusion, namely,

$$P(A.B_k,B_k) = 1 \qquad P(A.B_k,B_i) = 0 \text{ for } k \neq i \tag{11}$$

and substituting B_k for D, we obtain from (4)

$$P(A.[B_1 \vee \ldots \vee B_m],B_k) = \frac{P(A,B_k)}{\sum_{i=1}^{m} P(A,B_i)} \qquad 1 \leqq k \leqq m \tag{12}$$

These expressions may be called *reduced probabilities*; they represent the probability that B_k has with respect to A in combination with the terms of the incomplete disjunction. In the example given, they represent the probability of a candidate B_k being elected when we know that only the candidates $B_1 \ldots B_m$ remain. The reduced probabilities are bound probabilities because, according to (12),

$$\sum_{k=1}^{m} P(A.[B_1 \vee \ldots \vee B_m],B_k) = P(A.[B_1 \vee \ldots \vee B_m], [B_1 \vee \ldots \vee B_m]) = 1 \tag{13}$$

This also follows from axiom II,1; when the term in the square brackets in (13) is denoted by B, the second expression assumes the form $P(A.B,B)$, and this probability is $= 1$ because of $(A.B \supset B)$.

Using (12), we can write (4) in the form

$$P(A.[B_1 \vee \ldots \vee B_m],D) = \sum_{k=1}^{m} P(A.[B_1 \vee \ldots \vee B_m],B_k) \cdot P(A.B_k,D) \tag{14}$$

The desired probability containing the "or" in the first term is here determined by a summation of terms, each of which contains a probability $P(A.B_k,D)$ multiplied by the corresponding reduced probability. For the calculation of the probabilities having an "or" in the first term, the probabilities $P(A.B_k,D)$ do not suffice—a peculiarity reminiscent of Bayes's rule. They must first be multiplied by the reducéd probabilities $P(A.[B_1 \vee \ldots \vee B_m],B_k)$, which, in turn, are determined by the values $P(A,B_k)$ according to (12). Without these divergent or antecedent probabilities the problem remains indeterminate.

We turn now to the extension of these results to nonexclusive disjunctions. As before, we start from (1). However, when we substitute here for B the the nonexclusive disjunction $B \vee C$, we must use the general theorem of addition and thus obtain, instead of (2), the formula

$$P(A.[B \vee C],D) = \frac{P(A,B.D) + P(A,C.D) - P(A,B.C.D)}{P(A,B) + P(A,C) - P(A,B.C)} \tag{15}$$

Applying the general theorem of multiplication, we arrive at the relation

$$P(A.[B \vee C],D) = \tag{16}$$

$$\frac{P(A,B) \cdot P(A.B,D) + P(A,C) \cdot P(A.C,D) - P(A,B) \cdot P(A.B,C) \cdot P(A.B.C,D)}{P(A,B) + P(A,C) - P(A,B) \cdot P(A.B,C)}$$

This formula will be called the *general rule of reduction*. It contains the special rule, expressed in (3), as the special case resulting for $P(A.B,C) = 0$.

We can use formula (16) to determine generalized reduced probabilities, corresponding to (12); for this purpose we substitute B for D. Taking account of the fact that $P(A.B,B) = 1$ and using (3, § 21), we obtain

$$P(A.[B \vee C],B) = \frac{P(A,B)}{P(A,B) + P(A,C) - P(A,B) \cdot P(A.B,C)} \tag{17}$$

$$P(A.[B \vee C],C) = \frac{P(A,C)}{P(A,B) + P(A,C) - P(A,C) \cdot P(A.C,B)} \tag{18}$$

These formulas determine the value of the *reduced probabilities for nonexclusive disjunctions*. The denominators of (17) and (18) are equal because of (3, § 21).

Introducing the reduced probabilities into (16), we can give to the general rule of reduction a form corresponding to (14):

$$\begin{aligned} P(A.[B \vee C],D) &= P(A.[B \vee C],B) \cdot P(A.B,D) \\ &+ P(A.[B \vee C],C) \cdot P(A.C,D) - P(A.[B \vee C],B) \\ &\cdot P(A.B,C) \cdot P(A.B.C,D) \end{aligned} \tag{19}$$

For exclusive disjunctions the last term disappears because $P(A.B,C) = 0$, and the formula is thus transformed into (14) written for $m = 2$.

It is possible to extend the general rule of reduction to disjunctions of more than two events. The resulting formula, however, is cumbersome because of the complicated form that the general theorem of addition assumes for more than two events. Therefore it is not presented here.

Consider a schema that represents a combination of the rule of reduction with the rule of Bayes. Assume the observation of an event D that can be explained by several possible causes $B_1 \ldots B_r$. We do not know which of the causes exists, but we know their antecedent probabilities $P(A,B_k)$ relative

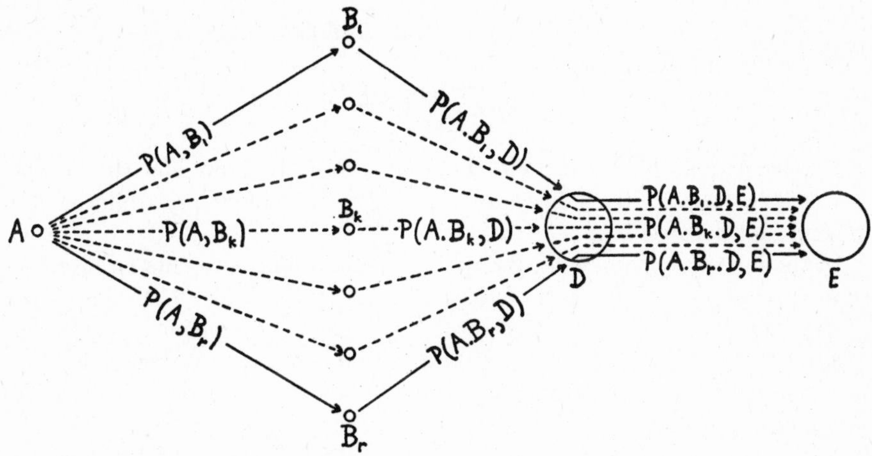

Fig. 6. Schema for rule of composition, according to (21).

to a common first term A and, furthermore, the probabilities $P(A.B_k,D)$ for the production of D by the individual causes B_k. The disjunction $B_1 \ldots B_r$ may be complete and exclusive. We ask for the probability $P(A.D,E)$ of an event E resulting from D. What are known, however, are only the individual probabilities $P(A.B_k.D,E)$, which confer upon E a probability relative to D and to the cause B_k that produced D. The schema is illustrated by figure 6.

For example, assume that D means a symptom of disease that may be explained by several possible causes, and let E mean the case of death. We know the probability of a lethal issue for each of the diseases B_k, and we ask for the probability of the death of the patient who shows the symptom D. The probability can be constructed as a mean in terms of the rule of reduction, after the probabilities of the individual causes B_k have been computed through the rule of Bayes. For this purpose, in turn, we must know the antecedent probabilities $P(A,B_k)$, in which A means the class of persons of a certain age and state of health; moreover, we must know the probabilities $P(A.B_k,D)$ for the production of the symptom D by the individual diseases B_k.

We have

$$P(A.D,E) = P(A.[B_1 \vee \ldots \vee B_r].D,E)$$
$$= \sum_{k=1}^{r} P(A.D,B_k) \cdot P(A.D.B_k,E) \tag{20}$$

according to (4), when we put in (4) $A.D$ for A and E for D, the denominator being $= 1$ because of the completeness of the disjunction. Putting for $P(A.D,B_k)$ its value resulting from the rule of Bayes (10, § 21), we obtain

$$P(A.D,E) = \frac{\sum_{k=1}^{r} P(A,B_k) \cdot P(A.B_k,D) \cdot P(A.B_k.D,E)}{\sum_{k=1}^{r} P(A,B_k) \cdot P(A.B_k,D)} \tag{21}$$

This formula may be called the *rule of composition*. It shows how the integral probability from D to E is composed of the individual probabilities that depend on the cause B_k.

In the interpretation given, the rule of composition represents an inference from the present (D) by way of the past (B_k) to the future (E). Such inferences occur in many kinds of scientific prognoses. As explained for the rule of Bayes, however, the temporal interpretation is not the only possible interpretation; formula (21) and figure 6*a* represent a logical structure capable of many interpretations.

Note that the rule of composition (21) becomes identical with the rule of elimination (21, § 19) when A is identical with D.

§ 23. The Relation of Independence

The independence of two events B and C was defined by the condition

$$P(A.B,C) = P(A,C) \tag{1}$$

We then derived for the theorem of multiplication the special form

$$P(A,B.C) = P(A,B) \cdot P(A,C) \tag{2}$$

It is also possible to consider (2) as the definition of independence and then to derive (1). This method has the disadvantage that it breaks down if $P(A,B) = 0$, since then the fraction $\frac{P(A,B.C)}{P(A,B)}$ assumes the indeterminate form $\frac{0}{0}$ and thus does not determine the value of $P(A.B,C)$. In this case, (1) is not derivable from (2), whereas (2) is always derivable from (1), even for the case $P(A,B) = 0$. This is why it is preferable to define independence by (1).

The general theorem of addition assumes for independent events the particular form

$$P(A,B \vee C) = P(A,B) + P(A,C) - P(A,B) \cdot P(A,C) \tag{3}$$

If the values $P(A,B)$ and $P(A,C)$ are small numbers, the value of their product is small within a lower order of magnitude; for such events the product term in (3) can be omitted—which means that in practice it is permissible, for low probabilities, to replace the general rule of addition by the special one. Thus, if $P(A,B) = P(A,C) = \frac{1}{1{,}000}$, their product is $\frac{1}{1{,}000{,}000}$; this value can be neglected in (3), and we have, with sufficient approximation, $P(A,B \vee C) = \frac{2}{1{,}000}$.

Although the general inequalities (10 and 13, § 20) show that the probability $P(A,B \vee C)$ cannot be greater than 1 or smaller than 0, a simple proof may be added to show that this condition is always satisfied for the form (3).[1] Let us put

$$P(A,B) = p \qquad P(A,C) = q \tag{4}$$

Then the condition under consideration requires that

$$0 \leqq p + q - pq \leqq 1 \tag{5}$$

Now this inequality holds for all numbers p and q between 0 and 1, limits included. To show this, we put

$$p = 1 - p' \qquad q = 1 - q' \tag{6}$$

Inserting these expressions in (5), we arrive at

$$0 \leqq 1 - p'q' \leqq 1 \tag{7}$$

This is indeed true if p' and q' are between 0 and 1, limits included. The case of equality with 0 in (7) can occur only when both p' and q' are $= 1$, that is, when both $p = 0$ and $q = 0$; and equality with 1 will occur only when $p' = 0$ or $q' = 0$, that is, when at least one of the two values p or $q = 1$. In all other cases the expression considered in (5) will differ from its lower and upper limit.

It is important to realize that the independence defined by (1) is a three-place relation, that is, a relation involving the terms B,C, and A. We must say, *B is independent of C with respect to A*. Without stating the reference term A we cannot speak of independence. This is contrary to linguistic usage, in which the reference term usually is not expressed.

It may be asked whether this usage can, perhaps, be justified by saying that certain events B and C are independent relative to all A as reference

[1] See also the remarks following (15, § 19).

classes, so that the three-place relation can be transformed into a two-place relation by generalization in A. However, it turns out that this assumption is erroneous, for there always exist events A in respect to which any two events B and C are mutually dependent.

This can be proved by choosing A as given by the disjunction $A.[B \vee C]$. We can then show that, relative to this reference class, B and C are not independent. Using formula (18, § 22) and applying (1), we obtain the following expression for a reduced probability, holding for any events B and C independent of each other with respect to A:

$$P(A.[B \vee C],C) = \frac{P(A,C)}{P(A,B) + P(A,C) - P(A,B) \cdot P(A,C)} \tag{8}$$

Applying the inequality (5) to the denominator, we see that, apart from the extreme cases $P(A,B) = 1$ or $P(A,C) = 1$, the expression (8) is $> P(A,C)$. Using (4*e*, § 4) and (1), we have

$$P(A.[B \vee C].B,C) = P(A.B,C) = P(A,C) \tag{9}$$

Therefore we have

$$P(A.[B \vee C].B,C) < P(A.[B \vee C],C) \tag{10}$$

Thus with respect to the reference class $A.[B \vee C]$ the two events B and C are not independent.

This result may be illustrated by an example concerning bets on two horses in different races. Let B be the case that the first horse wins; C, that the second horse wins. A is given by the general conditions before the races. Assume $P(A,B) = 50\%$, $P(A,C) = 80\%$. Relative to the general conditions A, the two results are independent and thus (1) is satisfied; if the first horse wins, the chances for the other are not changed. Assume that the races are over and we are told that one of the horses has won, but not which horse it is. The probability that the second horse has won, relative to what we know now, is given by (8); this formula furnishes the value 89%. So we now have a greater chance than before that the second horse has won. At this moment we learn that it was the first horse that won; the result as to the second horse is still unknown. Now the probability that the second horse has won is given by (9) and is the same as in the beginning, namely, 80%. This shows that relative to the situation $A.[B \vee C]$ the case that the second horse has won is not independent of whether the first horse has won. In this situation, additional knowledge as to the winning of the first horse will change the probability with which we may expect the second horse to have won.

This example shows that we must regard independence as a three-place relation, which is comparable, for instance, to the geometrical relation "between": the statement, "A lies between B and C", can be formulated only for three terms. Just as the between-relation is symmetrical with respect to the terms B and C, so is the independence relation symmetrical in B and C. For it has been shown in (5, § 21) that if (1) is valid, it is likewise true that

$$P(A.C,B) = P(A,B) \tag{11}$$

Furthermore, the independence relation is similar to the between-relation in that it is not transitive with respect to B and C. If B and C are mutually independent relative to A, and if C and D are mutually independent with respect to A, then B and D need not be mutually independent relative to A. In the case of the between-relation there even exists intransitivity for B and C, that is, if A is between B and C, and A between C and D, then A never lies between B and D. The independence relation is *only nontransitive*, that is to say, in the case considered B and D *may be* mutually independent with respect to A, but such is not necessarily the case. An instance of the nontransitive case is obtained when two dice are linked with a piece of string and, besides, a third, free die is used. The first two dice produce mutually dependent sequences, each of which, however, is independent of the third sequence.

A further property of the independence relation must now be presented. Let three events B,C,D be given, any pair of which is mutually independent with respect to A; then it does not follow that one of the events is *independent of the other two* with respect to A. We must understand this statement in the following way. From the relations

$$\begin{aligned} P(A.B,C) &= P(A.D,C) = P(A,C) \\ P(A.C,D) &= P(A.B,D) = P(A,D) \\ P(A.D,B) &= P(A.C,B) = P(A,B) \end{aligned} \tag{12}$$

it does not follow that the relations

$$\begin{aligned} P(A.B.C,D) &= P(A,D) \\ P(A.C.D,B) &= P(A,B) \\ P(A.D.B,C) &= P(A,C) \end{aligned} \tag{13}$$

also hold. This fact is expressed by saying that the independence relation is *not combinable.*

This is shown by the following considerations. If we add to the probabilities on the left side of (13) those obtained by negating B,C,D in the reference class, that is, if we regard probabilities of the kind $P(A.\bar{B}.C,D)$ or $P(A.B.\bar{C},D)$ or $P(A.\bar{B}.\bar{C},D)$, there are twelve probabilities that have a triple reference class. For these, according to the rule of elimination, there are only six independent equations of the form

$$P(A,D) = P(A.B,D) = P(A.B,C) \cdot P(A.B.C,D) + [1 - P(A.B,C)] \cdot P(A.B.\bar{C},D) \quad (14)$$

The probabilities having a triple reference class are, therefore, not determined by the probabilities having a single or a double reference class, and thus (13) does not follow from (12).

An example of a case for which (12) is valid, but not (13), is provided by the sequences

$$\begin{array}{l} A\,A\,A\,A\,A\,A\,A\,A\,\ldots \\ B\,\bar{B}\,B\,\bar{B}\,B\,\bar{B}\,B\,\bar{B}\,\ldots \\ C\,C\,\bar{C}\,\bar{C}\,C\,C\,\bar{C}\,\bar{C}\,\ldots \\ D\,\bar{D}\,\bar{D}\,D\,D\,\bar{D}\,\bar{D}\,D\,\ldots \end{array} \quad (15)$$

for which the first part written down is to be repeated periodically in the same order. Here all the probabilities (12) are equal to $\frac{1}{2}$. But $P(A.B.C,D)$ is equal to 1; so is $P(A.\bar{B}.\bar{C},D)$, and so on.

Sequences for which, apart from the relations (12), the relations (13) are fulfilled, are called *completely independent.* This notation applies similarly for a greater number of sequences.

§ 24. Complete Probability Systems

In § 16 the assumption of a compact sequence A was introduced and shown to be convenient for the frequency interpretation, because it leads to the simple formula (4, § 16). It is possible to introduce this assumption by a logical device that makes its truth analytic: by replacing the class A by the universal class $A \vee \bar{A}$. The condition $x_i \epsilon A \vee \bar{A}$ is then tautologically satisfied for every element x_i.

To simplify the notation we introduce the rule that the universal class may be omitted in the first term of a probability expression. This rule is expressed by the definition

$$P(B) =_{Df} P(A \vee \bar{A},B) \quad (1)$$

The probability $P(B)$ may be called an *absolute probability,* in contradistinction to the relative probabilities so far considered. An absolute probability can be regarded as a relative probability the reference class of which is the universal class.

If the statement $x_i \epsilon A$ is true for all x_i, though not analytic, the class A, for this sequence, is equivalent to the universal class $A \vee \bar{A}$. If a sequence is compact in A, the indication of the class A may therefore be omitted, and the probabilities may be treated as absolute probabilities.

The axioms and theorems of the calculus are transferred by the definition (1) to absolute probabilities. We find, for instance,

$$P(B) + P(\bar{B}) = 1 \tag{2}$$

$$P(B \vee C) = P(B) + P(C) - P(B.C) \tag{3}$$

$$P(B.C) = P(B) \cdot P(B,C) = P(C) \cdot P(C,B) \tag{4}$$

$$P(C) = P(B) \cdot P(B,C) + P(\bar{B}) \cdot P(\bar{B},C) \tag{5}$$

$$P(B \vee C,D) = \frac{P(B) \cdot P(B,D) + P(C) \cdot P(C,D) - P(B) \cdot P(B,C) \cdot P(B.C,D)}{P(B) + P(C) - P(B) \cdot P(B,C)} \tag{6}$$

and for exclusive events

$$P(B \vee C,D) = \frac{P(B) \cdot P(B,D) + P(C) \cdot P(C,D)}{P(B) + P(C)} \tag{7}$$

These special forms follow from the general forms (7, § 13), (8, § 20), (3, § 14), (2, § 19), (16, § 22), (3, § 22), when $A \vee \bar{A}$ is substituted for A. But it is not possible conversely to derive the general forms from the special forms; the latter hold only for the universal class as reference class, whereas the former hold for all reference classes. For the treatment of the relative probabilities occurring in the above formulas, therefore, formulas in terms of the general reference class A must be used.[1]

If two classes B and C are considered, the *complete system of probabilities* pertaining to them is given by the probabilities

$$\begin{array}{cccc} P(B) & P(\bar{B}) & P(C) & P(\bar{C}) \\ P(B,C) & P(\bar{B},C) & P(C,B) & P(\bar{C},B) \\ P(B,\bar{C}) & P(\bar{B},\bar{C}) & P(C,\bar{B}) & P(\bar{C},\bar{B}) \end{array} \tag{8}$$

The probabilities of the second and third line will be called *mutual probabilities*. The twelve values (8) are determined by the three fundamental probabilities

$$P(B) \quad P(C) \quad P(B,C) \tag{9}$$

which are the analogues of the three fundamental probabilities $P(A,B)$, $P(A,C)$, $P(A.B,C)$, introduced in (7′, § 21). The computation is made by

[1] If formulas containing the general reference class A are to be derivable from the corresponding formulas written in the notation by absolute probabilities, a particular rule of substitution must be introduced; see rule α, § 82.

the use of the relations (2)–(5); among these, (4) supplies the value $P(C,B)$; (5), the values with negated terms in the reference class.

It was mentioned above that the choice of these values as fundamental probabilities is a matter of convention, and that three other independent values could be used for the same purpose. From this point of view it is of some interest to select three values from the second line in (8) as fundamental probabilities. This line contains the two *affirmative* terms $P(B,C)$ and $P(C,B)$, which contain no negation signs, and the two terms of *negative reference* $P(\bar{B},C)$ and $P(\bar{C},B)$, which contain negation signs for reference classes. These four probabilities are not independent, but are connected by the relation

$$\frac{P(B,C)}{P(C,B)} \cdot \frac{P(\bar{C},B)}{P(\bar{B},C)} = \frac{1 - P(B,C) - P(\bar{C},B)}{1 - P(C,B) - P(\bar{B},C)} \tag{10}$$

This relation is derived as follows. We introduce the abbreviations

$$P(B) = b \qquad P(C) = c \qquad P(B,C) = c_1 \qquad P(\bar{B},C) = c_2$$
$$P(C,B) = b_1 \qquad P(\bar{C},B) = b_2 \tag{11}$$

Applying (4) to the three forms $P(B.C)$, $P(\bar{B}.C)$, $P(B.\bar{C})$, we construct the three relations

$$bc_1 = cb_1 \qquad c(1 - b_1) = c_2(1 - b) \qquad b(1 - c_1) = b_2(1 - c) \tag{12}$$

Solving the first two relations for b and c, we have

$$b = \frac{b_1c_2}{c_1(1 - b_1) + b_1c_2} \qquad c = \frac{c_1c_2}{c_1(1 - b_1) + b_1c_2} \tag{13}$$

Inserting these values in the third relation (12), we find

$$\frac{c_1b_2}{b_1c_2} = \frac{b_2 + c_1 - 1}{c_2 + b_1 - 1} \tag{14}$$

which is the relation (10).

Equations (13) determine the absolute probabilities as functions of the three mutual probabilities $P(B,C)$, $P(C,B)$, $P(\bar{B},C)$, and can be written in the form

$$\begin{aligned} P(B) &= \frac{P(C,B) \cdot P(\bar{B},C)}{P(B,C)[1 - P(C,B)] + P(C,B) \cdot P(\bar{B},C)} \\ P(C) &= \frac{P(B,C) \cdot P(\bar{B},C)}{P(B,C)[1 - P(C,B)] + P(C,B) \cdot P(\bar{B},C)} \end{aligned} \tag{15}$$

These results show that all the probabilities (8) are determined by the three mutual probabilities $P(B,C)$, $P(C,B)$, $P(\bar{B},C)$. Exception is to be made for the case that the denominator of (15) (which represents the determinant of the corresponding set of linear equations) vanishes. This is the case, in particular, for exclusive classes B and C, that is, for $P(B,C) = P(C,B) = 0$; (15) then gives the indeterminate form $\frac{0}{0}$.

It is possible to construct a formula that is not subject to this degeneration when the probability $P(\bar{C},B)$ is included in the arguments. For $P(B,C) = P(C,B) = 0$, formula (10) supplies the form $\frac{0}{0}$, and the four values of the second line of (8) are no longer connected by a restrictive condition; thus the last term of this line can be added as an independent parameter. For the derivation we use (5), first with B and C interchanged, then in the form given, thus arriving at the two equations

$$b = cb_1 + (1 - c)b_2 \qquad c = bc_1 + (1 - b)c_2 \tag{16}$$

which, solved for b and c, give the results

$$b = \frac{c_2 b_1 + b_2(1 - c_2)}{1 - (b_1 - b_2)(c_1 - c_2)} \qquad c = \frac{b_2 c_1 + c_2(1 - b_2)}{1 - (b_1 - b_2)(c_1 - c_2)} \tag{17}$$

Introducing the exclusion condition $b_1 = c_1 = 0$, we find

$$b = \frac{b_2(1 - c_2)}{1 - b_2 c_2} \qquad c = \frac{c_2(1 - b_2)}{1 - b_2 c_2} \tag{18}$$

These formulas can be written

$$P(B) = \frac{P(\bar{C},B)[1 - P(\bar{B},C)]}{1 - P(\bar{B},C) \cdot P(\bar{C},B)} \qquad P(C) = \frac{P(\bar{B},C)[1 - P(\bar{C},B)]}{1 - P(\bar{B},C) \cdot P(\bar{C},B)} \tag{19}$$

They determine the absolute probabilities in terms of the mutual probabilities for the exclusive classes B and C.

The two affirmative mutual probabilities $P(B,C)$ and $P(C,B)$ are connected by the relation

$$\frac{P(B,C)}{P(C,B)} = \frac{P(C)}{P(B)} \tag{20}$$

which follows from (4). Corresponding relations hold for negated mutual probabilities; they follow from (20) by the substitution of $\bar{B}$ for B and so on.

In many applications the two absolute probabilities $P(B)$ and $P(C)$ are unknown, and only the two affirmative mutual probabilities $P(B,C)$ and $P(C,B)$ are given. These values are subject to no restrictions other than that their values be between 0 and 1. The two probabilities are sometimes combined in a *mutual probability implication,* which is written in the implicative notation

$$(B \underset{q\;p}{\ominus} C) \tag{21}$$

This is equivalent to the conjunction

$$(B \underset{p}{\ni} C) \,.\, (C \underset{q}{\ni} \bar{B}) \tag{22}$$

As explained, the two values p and q are not subject to any connecting condition. It is even possible that $p = 1$ and $q = 0$ without B or C being empty. The corresponding general implications $(B \supset C)$ and $(C \supset \bar{B})$ are compatible only if B is empty, since we can derive from them, by the transitivity of the implication, the relation $(B \supset \bar{B})$. There is no such consequence for the probability implications because the probability 1 is not equivalent to certainty. Thus $P(C,B) = 0$ does not exclude the possibility that C is sometimes accompanied by B.

If the two mutual probabilities are given, the values of the absolute probabilities and those of the probabilities of negative reference are not determined. Only the ratio of the absolute probabilities is determined, according to (20). But it is possible to compute some other probabilities. First, we can replace the relation (20), which includes absolute probabilities, by a corresponding relation for relative probabilities. For this derivation, the relations (2)–(7) are not sufficient, and we must return to the notation in terms of the general reference class A. We apply (4, § 21), substitute for A the disjunction $B \vee C$, and use the tautological equivalences

$$([B \vee C].B \equiv B) \qquad ([B \vee C].C \equiv C) \tag{23}$$

We thus arrive at the formula

$$\frac{P(B,C)}{P(C,B)} = \frac{P(B \vee C,C)}{P(B \vee C,B)} \tag{24}$$

The expressions $P(B \vee C,B)$ and $P(B \vee C,C)$ may be called *disjunctive weights;* they determine the weight with which either of the terms B or C occurs in their mutual disjunction. Formula (24) states that the ratio of the mutual probabilities is equal to the ratio of the disjunctive weights.

There is a further relation, which makes it possible, in combination with (24), to determine the disjunctive weights in terms of the affirmative mutual probabilities. We have, with the general rule of addition in the A-notation,

$$P(B \vee C,B \vee C) = 1 = P(B \vee C,B) + P(B \vee C,C) - P(B \vee C,B.C) \tag{25}$$

The last term is transformed with (23) into

$$\begin{aligned} P(B \vee C,B.C) &= P(B \vee C,B) \cdot P([B \vee C].B,C) \\ &= P(B \vee C,B) \cdot P(B,C) \end{aligned} \tag{26}$$

Introducing this result in (25) and substituting for $P(B \vee C,C)$ the value resulting from (24), we arrive at an equation, which, solved for $P(B \vee C,B)$, gives the result

$$P(B \vee C,B) = \frac{P(C,B)}{P(B,C) + P(C,B) - P(B,C) \cdot P(C,B)} \tag{27}$$

This relation will be called the *general rule of the disjunctive weight.* It determines the disjunctive weight as a function of the affirmative mutual probabilities.

It is easily seen that the disjunctive weight of C is given by a similar expression, resulting from (27) when $P(B,C)$ is put for $P(C,B)$ in the numerator. The probability of the product results from (26) in the form

$$P(B \vee C,B.C) = \frac{P(B,C) \cdot P(C,B)}{P(B,C) + P(C,B) - P(B,C) \cdot P(C,B)} \tag{28}$$

As for (15), a qualification must be added. Formulas (27)–(28) depend on the condition that at least one of the two mutual probabilities is > 0. It follows that for exclusive classes the disjunctive weights are not determined by the affirmative mutual probabilities.

As before, a computation for exclusive classes is made possible by the use of mutual probabilities of negative reference. From (7) we derive, substituting B for D and putting $P(C,B) = 0$ because of the exclusion condition,

$$P(B \vee C,B) = \frac{P(B)}{P(B) + P(C)} \tag{29}$$

With the application of (19) we find

$$P(B \vee C,B) = \frac{P(\bar{C},B)[1 - P(\bar{B},C)]}{P(\bar{B},C) + P(\bar{C},B) - 2P(\bar{B},C) \cdot P(\bar{C},B)} \tag{30}$$

This formula, which holds only for exclusive disjunctions, will be called the *special rule of the disjunctive weight.* Since the mutual probabilities of negative reference used in the formula are sufficient to determine the absolute probabilities, a knowledge of the disjunctive weights, for exclusive disjunctions, is inseparable from a knowledge of the absolute probabilities.

We turn now to probability relations between three classes B_1,B_2,B_3. The complete probability system, written only for affirmative terms, is given here by the probabilities

$$\begin{array}{ccc} P(B_1) & P(B_2) & P(B_3) \\ P(B_1,B_2) & P(B_2,B_3) & P(B_3,B_1) \\ P(B_2,B_1) & P(B_3,B_2) & P(B_1,B_3) \\ P(B_1.B_2,B_3) & P(B_2.B_3,B_1) & P(B_3.B_1,B_2) \end{array} \tag{31}$$

The other forms result by substitution of $\bar{B}_1$ and so on, in these expressions. The probabilities of the last line will be called *compound mutual probabilities.* Those of the second and third lines are then called *simple mutual probabilities.* Note that the values (31) are not subject to restrictive conditions: it is not required that the three classes be exclusive or independent or that they form a complete disjunction.

For any two simple mutual probabilities, formula (4) leads to the relation

$$\frac{P(B_i,B_k)}{P(B_k,B_i)} = \frac{P(B_k)}{P(B_i)} \tag{32}$$

The probabilities of the third line of (31) are thus determined by those of the first and second lines. Probabilities having $\bar{B}_i$ or $\bar{B}_k$ in the reference class are derivable from the affirmative terms by means of (5).

The three compound mutual probabilities are connected by the relations, following from the rule of the product,

$$\frac{P(B_i.B_k,B_m)}{P(B_i.B_m,B_k)} = \frac{P(B_i,B_m)}{P(B_i,B_k)} \tag{33}$$

The three relations resulting for $m = 1$, $k = 2$; $m = 2$, $k = 3$; and $m = 3$, $k = 1$, are not independent, because the last is easily seen to be a consequence of the other two. Thus (33) represents two independent relations. If one of the affirmative compound mutual probabilities is given, the other two are thus determined. Probabilities with terms $\bar{B}_i$ in the reference class are computed from the affirmative terms by means of the relations (5) and (10, § 19).

The complete probability system for three classes is thus determined by the six values of the first two lines of (31) and, besides, one of the compound values of the last line of (31), that is, by seven independent probability values.

If the absolute probabilities $P(B_1)$, $P(B_2)$, $P(B_3)$ are unknown, the six values of the simple mutual probabilities in the second and third lines of (31) cannot be assumed arbitrarily, but are connected by the relation

$$\frac{P(B_1,B_2)}{P(B_2,B_1)} \cdot \frac{P(B_2,B_3)}{P(B_3,B_2)} \cdot \frac{P(B_3,B_1)}{P(B_1,B_3)} = 1 \tag{34}$$

which follows by the use of (32) and can be written in the form[2]

$$P(B_1,B_2) \cdot P(B_2,B_3) \cdot P(B_3,B_1) = P(B_1,B_3) \cdot P(B_3,B_2) \cdot P(B_2,B_1) \tag{35}$$

Only five of these values, therefore, are independent. Formula (35) will be called the *rule of the triangle.* It states that in a triangle $B_1B_2B_3$ the product of the three mutual probabilities is the same, whether we go clockwise or

[2] This relation was pointed out by Norman Dalkey, "The Plurality of Language Structures," doctoral dissertation, University of California at Los Angeles, 1942.

counterclockwise around the triangle. For two events there is no such dependence of mutual probabilities, because in this case there is only one direction for the "round trip". The distinction of two such directions begins with three events.

The relation (35) has a simple explanation in the frequency interpretation; it represents the identity

$$\frac{N^n(B_1 . B_2)}{N^n(B_1)} \cdot \frac{N^n(B_2 . B_3)}{N^n(B_2)} \cdot \frac{N^n(B_3 . B_1)}{N^n(B_3)}$$

$$= \frac{N^n(B_1 . B_3)}{N^n(B_1)} \cdot \frac{N^n(B_3 . B_2)}{N^n(B_3)} \cdot \frac{N^n(B_2 . B_1)}{N^n(B_2)} \qquad (36)$$

The rule of the triangle (35) is automatically satisfied if the absolute probabilities in combination with the second line of (31) are used for the determination of the values of the third line. But if the absolute probabilities are not used and, instead, five of the simple mutual probabilities are assumed arbitrarily, they are subject to the numerical restriction

$$P(B_1,B_2) \cdot P(B_2,B_3) \cdot P(B_3,B_1) \leqq P(B_3,B_2) \cdot P(B_2,B_1) \qquad (37)$$

which formulates the condition $P(B_1,B_3) \leqq 1$ for computation of this probability from (35). This inequality is to be added to the inequalities (15, § 19).

The following special conditions can be derived. If $P(B_1,B_2) = 1$ and $P(B_2,B_3) = 1$, it follows that $P(B_1,B_3) = 1$ if $P(B_2,B_1) > 0$. This transitivity is shown by the considerations added to (7, § 19). Another rule of transitivity is as follows: if transitivity holds in one direction of the triangle, it also holds in the other. This theorem, which applies, too, when the probabilities are < 1, is derivable from (35), because when we put there

$$P(B_1,B_3) = P(B_1,B_2) \cdot P(B_2,B_3)$$

we have

$$P(B_3,B_1) = P(B_3,B_2) \cdot P(B_2,B_1)$$

The values (37) determine the ratios of the absolute probabilities, according to (32). If a further condition is added, for instance, that the disjunction of the three classes be complete, the absolute probabilities are determinable. Instead of such a condition for the absolute probabilities, it is sufficient to give one simple mutual probability of negative reference. The computation of the absolute probabilities then follows the methods developed for two classes.

If, besides the values (37), one compound probability and one simple mutual probability of negative reference are given, all the other probabilities can be computed by the methods developed for two classes.

For three classes, the problem of a disjunctive reference class offers particular interest. The problem will first be treated for nonexclusive reference classes B_1 and B_2, for which case it can be solved in terms of affirmative mutual probabilities. When we insert the values (15) in (6), the denominator of (15) drops out and the term $P(\bar{B},C)$, which occurs in every term, can be canceled. Putting B_1, B_2, B_3, respectively, for B, C, D, we arrive at the formula

$$P(B_1 \vee B_2,B_3) = \tag{38}$$
$$\frac{P(B_1,B_2) \cdot P(B_2,B_3) + P(B_2,B_1) \cdot P(B_1,B_3) - P(B_1,B_2) \cdot P(B_2,B_1) \cdot P(B_1 . B_2,B_3)}{P(B_1,B_2) + P(B_2,B_1) - P(B_1,B_2) \cdot P(B_2,B_1)}$$

The formula differs from the general rule of reduction, in the forms (16, § 22) or (6), in that it includes neither absolute probabilities nor a term A common to all reference classes. Instead, the terms B_1 and B_2 of the disjunction are distributed into the first terms of the expressions on the right; we therefore call (38) the *general rule of distributive reference*. The occurrence of the term $P(B_1 . B_2,B_3)$ shows that the solution requires one compound mutual probability; but all the terms are affirmative mutual probabilities.

For exclusive classes, again, a different solution is necessary, because, for $P(B_1,B_2) = P(B_2,B_1) = 0$, formula (38) gives the form $\frac{0}{0}$. As before, the problem is solved by the use of probabilities of negative reference. Starting with (7), we insert the values of $P(B)$ and $P(C)$ from (19); we then substitute B_1, B_2, B_3, respectively, for B, C, D, and arrive at the result

$$P(B_1 \vee B_2,B_3) = \tag{39}$$
$$\frac{P(B_1,B_3) \cdot P(\bar{B}_2,B_1) \cdot [1 - P(\bar{B}_1,B_2)] + P(B_2,B_3) \cdot P(\bar{B}_1,B_2) \cdot [1 - P(\bar{B}_2,B_1)]}{P(\bar{B}_1,B_2) + P(\bar{B}_2,B_1) - 2P(\bar{B}_1,B_2) \cdot P(\bar{B}_2,B_1)}$$

This is the *special rule of distributed reference*, which holds only for exclusive disjunctions. It does not require the use of compound mutual probabilities, but presupposes terms of negative reference. Note that, in contradistinction to previous theorems to which similar names were assigned (§§ 14, 20, 22), the two special rules (30) and (39) do not follow from the two general rules (27) and (38) as special cases, but require separate derivations.

The considerations can be extended to n classes $B_1 \ldots B_n$. For every subset of m classes there exist compound mutual probabilities of the form $P(B_{k_1} \ldots B_{k_{m-1}},B_{k_m})$. They are connected with those of the next lower subset by the relations

$$\frac{P(B_{k_1} \ldots B_{k_{m-1}},B_{k_m})}{P(B_{k_1} \ldots B_{k_{m-2}} . B_{k_m},B_{k_{m-1}})} = \frac{P(B_{k_1} \ldots B_{k_{m-2}},B_{k_m})}{P(B_{k_1} \ldots B_{k_{m-2}},B_{k_{m-1}})} \tag{40}$$

which express the rule of the product. Given one probability of the subset, all the others are thus determined in terms of those of the next lower subset.

Probabilities having a term $\bar{B}_{k_r}$ in the reference class are determined by the use of the rule of elimination (10, § 19).

The total number μ of independent probabilities determining the complete probability system for n classes is computed as follows. First, the n absolute probabilities $P(B_i)$ can be given independently. Second, of each subset of m classes one probability must be given. This includes the case $m = 2$, for which we have the simple mutual probabilities; for every combination $P(B_i,B_k)$, the converse probability follows from (32) in terms of the absolute probabilities. The relation (32) is not restricted to three classes and is a special case of (40) resulting for $m = 2$. The number of subsets of m classes among n classes being $\binom{n}{m}$ we find, since $n = \binom{n}{1}$,

$$\mu = \sum_{m=1}^{n} \binom{n}{m} = 2^n - 1 \tag{41}$$

using the familiar theorem for binomial coefficients

$$\sum_{m=0}^{n} \binom{n}{m} = 2^n \tag{42}$$

For $n = 2$ we have $\mu = 3$, according to (9). For $n = 3$ we have $\mu = 7$, in correspondence with the above result.

The number ν of affirmative probabilities can be found as follows. For every subset of m classes, there are m affirmative probabilities, which result when, one after another, each of the classes is chosen as attribute class. This is true, too, for $m=2$ and $m=1$. There being $\binom{n}{m}$ subsets of m terms, we have

$$\nu = \sum_{m=1}^{n} \binom{n}{m} m = \sum_{m=1}^{n} \frac{n!}{m!\,(n-m)!} \cdot m$$

$$= n \sum_{m=1}^{n} \frac{(n-1)!}{(m-1)!(n-m)!} = n \sum_{m=1}^{n} \frac{(n-1)!}{(m-1)![(n-1)-(m-1)]!}$$

$$= n \sum_{m=1}^{n} \binom{n-1}{m-1} = n \sum_{m-1=0}^{n-1} \binom{n-1}{m-1} = n \cdot 2^{n-1} \tag{43}$$

For $n = 1$ we have $\nu = 1$; for $n = 2$, $\nu = 4$; for $n = 3$, $\nu = 12$, in correspondence with (9) and (31).

The number ρ of probabilities of the complete system, all being of the form occurring in (40), but including both affirmative and negated terms, is computed as follows. Each affirmative probability of m terms contributes 2^m into the total, since each term can be written once without, and once with, a negation sign. So we have, using the preceding transformation of the sum,

$$\rho = \sum_{m=1}^{n} \binom{n}{m} m \cdot 2^m = n \sum_{m-1=0}^{n-1} \binom{n-1}{m-1} \cdot 2^m = 2n \sum_{m-1=0}^{n-1} \binom{n-1}{m-1} \cdot 2^{m-1}$$
$$= 2n(2+1)^{n-1} = 2n \cdot 3^{n-1} \qquad (44)$$

For the transition to the second line we use the binomial theorem

$$(p+q)^n = \sum_{m=0}^{n} \binom{n}{m} p^m q^{n-m} \qquad (45)$$

choosing $p = 2$, $q = 1$, and putting $n - 1$ for n and $m - 1$ for m. For $n = 1$ we have $\rho = 2$; for $n = 2$, $\rho = 12$ (see 8); for $n = 3$, $\rho = 54$.

There are many applications for the relations developed for three classes. Let B_1 be a symptom of illness; B_2, a certain disease; B_3, the case of death. The simple mutual probabilities may be known from statistics; the relation (35) shows that only five are to be ascertained, the sixth being determinable. Furthermore, one of the compound probabilities must be ascertained, for instance, $P(B_1.B_2,B_3)$. When these values are known, all statistical questions referring to the three classes are answerable except those referring to absolute probabilities or probabilities of negative reference. A psychological application obtains when B_1 means a certain stimulus; B_2, a perception; B_3, a certain reaction of a person.

§ 25. Remarks Concerning the Mathematical Formalization of the Probability Calculus

Having carried through, to a large extent, the formalization of the calculus of probability, we are now free to discuss this procedure from a logical viewpoint. The "logification" by which this construction of the calculus was introduced has, in the meantime, been transformed into a "mathematization", a notation in which the logical operations are restricted to the inner part of the P-symbols. The resulting complexes of the P-symbols, into which these symbols enter as units, have the character of mathematical equations. Thus the probability calculus acquires a form that is convenient for the purpose of carrying out calculations.

This manner of writing—the mathematical notation—has the disadvantage that it cannot express certain relations of a nonmathematical kind that hold within the probability calculus. There are three different forms of such relations:

1. The dependence of a mathematical equation on the validity of another mathematical equation, that is, the implication between equations. An exam-

ple is given by the assertion that (4, § 14) is the condition of validity for (5, § 14).

2. The dependence of a mathematical equation on a nonmathematical condition. Of this kind is the condition $(A \supset B)$ in axiom II,1 or the condition of exclusion in axiom III.

3. Logical properties of the quantities occurring, such as are expressed in the statement of univocality formulated in axiom I.

The first case is irrelevant because the existence of a logical implication between equations is easily expressible by some connecting words in the context. This is the method usually applied in mathematics. The second case is serious because here the condition on which the validity of a mathematical equation depends is not expressible in mathematical notation. The third case is irrelevant again. It concerns only the assertion of univocality; this assertion, as is usual in mathematics, may be added in words.

It will now be shown that the second difficulty can be eliminated by the use of a method that translates the logical condition into a mathematical condition. The method may be illustrated by reference to the general theorem of addition. It was seen, in § 20, that the condition of exclusion, written for this axiom in the logical notation, could be replaced by the condition that the corresponding probability becomes 0. Since this assumption, according to (9, § 13), states less than the strict condition of exclusion, a certain generalization of the special theorem of addition has thus been constructed. It will now be shown that the same procedure is feasible in some other places, so that, by its use, relations of the form 2 can be reduced to those of the form 1. We are concerned here with the following theorems written in the implicational notation:

$$(A \supset B) \supset [(A \underset{q}{\twoheadrightarrow} C) \equiv (A \, . \, B \underset{q}{\twoheadrightarrow} C)] \tag{1}$$

$$(A \supset B) \supset (\exists p)(A \, . \, C \underset{p}{\twoheadrightarrow} B) \, . \, (p = 1) \tag{2}$$

$$(A \supset B) \supset [(A \underset{q}{\twoheadrightarrow} C) \equiv (A \underset{q}{\twoheadrightarrow} B \, . \, C)] \tag{3}$$

$$(C \supset B) \supset [(A \underset{q}{\twoheadrightarrow} C) \equiv (A \underset{q}{\twoheadrightarrow} B \, . \, C)] \tag{4}$$

The proof of the theorems is easily given. Let us prove immediately their generalization for the cases $P(A,B) = 1$ and $P(C,B) = 1$, respectively. From this result, of course, by the help of II,1, theorems (1)–(4) follow.

Instead of (1) we obtain: if $P(A,B) = 1$, then

$$P(A,C) = P(A \, . \, B,C) \tag{5}$$

This follows from the elimination theorem (2, § 19) because $P(A,\bar{B}) = 0$. Formula (5) states that, if $P(A,B) = 1$, B and any C are mutually independent with respect to A.

Instead of (2) we obtain: if $P(A,B) = 1$ and $P(A,C) > 0$, we have for any C

$$P(A.C,B) = 1 \tag{6}$$

The proof follows from the product rule (6, § 21) by the use of (5). If $P(A,C) = 0$, (6) is not derivable; in this case the value of $P(A.C,B)$ cannot be determined from $P(A,B)$, though it is possible that a determinate value $P(A.C,B)$ exists.

Instead of (3) we obtain: if $P(A,B) = 1$, then

$$P(A,C) = P(A,B.C) \tag{7}$$

The proof is given by the multiplication theorem (3, § 14) with the help of (6). This formula is also valid for the case $P(A,C) = 0$, since it then follows directly from the multiplication theorem without the use of (6).

Instead of (4) we obtain: if $P(C,B) = 1$ and $P(C,A) > 0$, then

$$P(A,C) = P(A,B.C) \tag{8}$$

The proof is given by the multiplication theorem, since, for the assumptions made, $P(A.C,B) = 1$, according to (6), if A and C are interchanged in (6).

With these proofs the mathematical formalization is carried through for the four theorems.[1] It will now be shown that in axiom II,1, too, the condition $(A \supset B)$ can be formally eliminated.

In this case we make use of the equivalence

$$([A \supset B] \equiv [B \equiv A \vee B]) \tag{9}$$

The formula is proved by solving the right side according to (7*b*, § 4), applying (4*e*, § 4) and (4*b*, § 4) and, finally, transforming the left side by (6*a*, § 4). Formula (9) is a tautology of the class calculus, that is, the expression inside the parentheses represents the universal class. The formula can be transcribed into the form

$$(A \supset B) \equiv (B \equiv A \vee B) \tag{9'}$$

This formula means that if A is a subclass of B, the joint class $A \vee B$ is identical with B.

Because of (9), axiom II,1 is equivalent to the expression

$$P(A,A \vee B) = 1 \tag{10}$$

For we have, on account of (9),

$$(A \supset B) \supset [P(A,B) = P(A,A \vee B)] \tag{11}$$

[1] The restrictions $P(A,C) > 0$ and $P(C,A) > 0$ added, respectively, to (6) and (8) are not required for the corresponding theorems (2) and (4). This is due to the fact that the logical implication represents a stronger assumption than the probability 1.

Therefore, II,1 follows from (10). That, conversely, (10) follows from II,1 can be shown by the use of (8a, § 4).

As a formula that cannot be formalized mathematically, there remains only axiom I, the axiom of univocality, apart from implications of the form 1. All other expressions can be formalized, and we may ask whether we can omit axiom I and construct a *mathematical axiom system* of the calculus of probability. By this term is understood a system in which the logical operations are restricted to the inner part of the P-symbols, whereas the symbols themselves enter into relations having the form of mathematical equations. Such an axiom system can be constructed; the condition of univocality is then added in words.

In order to set up this axiom system, we introduce the following changes from the axiom system written in the implicational notation. We replace II,1 by (10). Furthermore, we replace the addition theorem III by the general theorem of addition (8, § 20), so that we can free ourselves from the condition of exclusion. This requires, in the group of axioms of normalization, a further axiom, α,2, which defines the probability 0 in a way similar to that in which α,1 defines the probability 1. We thus obtain the following *mathematical axiom system of the calculus of probability:*

α) AXIOMS OF NORMALIZATION

$$1.\ P(A, A \vee B) = 1$$

$$2.\ P(A, B.\bar{B}) = 0$$

$$3.\ 0 \leqq P(A, B)$$

β) AXIOM OF ADDITION

$$P(A, B \vee C) = P(A,B) + P(A,C) - P(A, B.C)$$

γ) AXIOM OF MULTIPLICATION

$$P(A, B.C) = P(A,B) \cdot P(A.B, C)$$

Axiom α,3 needs no qualification demanding that A be nonempty, because, if A is empty, this inequality does not represent any restriction on the numerical values of probabilities. According to the convention concerning the use of the P-symbol for empty reference classes (see p. 59), the inequality α,3 expresses, in this case, merely a trivial existential statement. Thus the axiom does not depend on a special condition to be added in words. The only condition of this kind is the axiom of univocality. It may be convenient to formulate this axiom, together with the rule of existence and the convention about the use of P-symbol, as a rule given in the metalanguage.

It will be shown briefly how the rule of the complement (7, § 13) can be derived from these axioms. We substitute first, in α,1, $B \vee \bar{B}$ for B; because, according to (8c, § 4), the relation $(A \supset B \vee \bar{B})$ is always valid, we obtain, by the use of (11),

$$P(A,B \vee \bar{B}) = 1$$

Dissolving the term on the left side according to axiom β, we obtain, by the use of α,2,

$$P(A,B \vee \bar{B}) = P(A,B) + P(A,\bar{B}) = 1 \tag{12}$$

From this result we derive the special theorem of addition for the mutually exclusive events B and C, that is, for $(B \supset \bar{C})$. Since $(B \supset \bar{C}) \equiv (\overline{B.C})$, the relation $(A \supset \overline{B.C})$ follows from $(B \supset \bar{C})$ with the help of (8c, § 4); substituting in (11) $\overline{B.C}$ for B and using axiom α,1 we thus derive

$$P(A,\overline{B.C}) = P(A,A \vee \overline{B.C}) = 1 \tag{13}$$

With the help of (12) we now derive $P(A,B.C) = 0$ and thus obtain from axiom β the special theorem of addition.

Regarding the theorem of multiplication, the previous remarks hold good, according to which this theorem can be replaced by the weaker assumption of § 15. It is possible, furthermore, to replace the axioms β and γ by a compound axiom, as was shown by William Gustin. According to Gustin, the following mathematical axiom system is sufficient:

a) Normalization

1. $P(A.B,B) = 1$
2. $0 \leqq P(A,B)$

b) Axiom of the complement of the product

$$P(A,\overline{B.C}) = 1 - P(A,B) \cdot P(A.B,C)$$

The postulate of univocality must be added in words, as in the preceding system. The Gustin system shows that the axiom of addition can be replaced by the rule of the complement and that the latter can be combined with the axiom of multiplication in one axiom. In this system the rule of the complement is derivable as follows:

$$P(A,\bar{B}) = P(A,\overline{B.B}) = 1 - P(A,B) \cdot P(A.B,B) = 1 - P(A,B) \tag{14}$$

Using this result, we immediately derive from axiom b the general theorem of multiplication. The general theorem of addition is proved as follows:

$$\begin{aligned} P(A,B \vee C) &= P(A,\overline{\bar{B}.\bar{C}}) = 1 - P(A,\bar{B}) \cdot P(A.\bar{B},\bar{C}) \\ &= 1 - P(A,\bar{B}) \cdot [1 - P(A.\bar{B},C)] \\ &= P(A,B) + P(A,\bar{B}.C) \\ &= P(A,B) + P(A,C) \cdot P(A.C,\bar{B}) \\ &= P(A,B) + P(A,C) \cdot [1 - P(A.C,B)] \\ &= P(A,B) + P(A,C) - P(A,B.C) \end{aligned} \tag{15}$$

Axiom α,2 follows when $\bar{B}$ is substituted for C in the theorem of addition, and the rule of the complement is used. Axiom α,1 follows by substituting A for C in the theorem of addition. The Gustin system is thus proved to be equivalent to the system of axioms α–γ.

The mathematical axiom systems presented here are sufficient to prove all the theorems of the calculus of probability. They do not carry through the formalization completely; the condition of univocality and the implication between equations must be added verbally in the context. But the nonformalized residue is relatively small. It is true that my mathematical axiom systems require the use of symbolic logic for the inner part of the probability symbols, but I hope that the presentation shows that this feature only facilitates operations within the calculus. With the help of symbolic logic a probability calculus has been constructed that exhibits not only the mathematical but also the logical structure of its subject matter. I should be happy if the unification of mathematics and symbolic logic thus achieved would stimulate other authors to attempt similar constructions in other fields of research.

Historical remark concerning the axiomatic construction of the calculus of probability.—Axiomatic foundations of the calculus of probabilities have been given repeatedly within the last few decades. Corresponding to my division into a formal and an interpreted theory of probability, two groups may be distinguished. The interpreted form of axiomatic construction regards probability, from the beginning, as a frequency, and derives from this interpretation, by the possible inclusion of additional postulates, the rules of the theory. This group began with Richard von Mises' analyses[1] (1919) and was continued by the inquiries of Karl Dörge[2] (1930), Erhard Tornier[3] (1930), and Erich Kamke[4] (1932); it includes, also, the investigations by Arthur H. Copeland[5] (1928).

The formal conception introduces the concept of probability by the method of implicit definitions, and uses no properties of the concept other than those expressed in a set of formal relations placed as axioms at the beginning of the theory, leaving open various possibilities for its interpretation. The group includes the axiom system given in 1901 by Georg Bohlmann[6] and the analyses published by S. Bernstein[7] (1917) and Emile Borel[8] (1925). To it belongs also my own axiomatic presentation, which was first published in 1932.[9] It was followed by an axiomatic construction by A. N. Kolmogoroff[10] in 1933.

[1] "Grundlagen der Wahrscheinlichkeitsrechnung," in *Math. Zs.*, Vol. V (1919), pp. 52–99; *Vorlesungen aus dem Gebiete der angewandten Mathematik*, Vol. I: *Wahrscheinlichkeitsrechnung* . . . (Leipzig, 1931).

[2] "Zu der von R. v. Mises gegebenen Begründung der Wahrscheinlichkeitsrechnung," in *Math. Zs.*, Vol. 32 (1930), pp. 232–258.

[3] "Eine neue Grundlegung der Wahrscheinlichkeitsrechnung," in *Zs. f. Physik*, Vol. 63 (1930), p. 697; "Grundlagen der Wahrscheinlichkeitsrechnung," in *Acta Math.*, Vol. 60 (1933), pp. 239–380.

[4] *Einführung in die Wahrscheinlichkeitstheorie* (Leipzig, 1932).

[5] "Admissible Numbers in the Theory of Probability," in *Amer. Jour. Math.*, Vol. 50, No. 4 (1928), p. 535; and later papers.

[6] *Encykl. d. math. Wiss.*, Vol. I, Part 2 D 4*b* (1901), pp. 852–917.

[7] "Versuch einer axiomatischen Begründung der Wahrscheinlichkeitsrechnung," in *Mitt. d. math. Ges. Charkow* (1917), pp. 209–274.

[8] *Traité du calcul des probabilités* (Paris, 1924–), Vol. I, Part 1; *Principes et formules classiques du calcul des probabilités* (Paris, 1925).

[9] "Axiomatik der Wahrscheinlichkeitsrechnung," in *Math. Zs.*, Vol. 34 (1932), pp. 568–619.

[10] *Grundbegriffe der Wahrscheinlichkeitsrechnung* (Berlin, 1933).

Most inquiries of the formal group omit the development of a theory of the order of probability sequences. The problem was first attacked by von Mises, Dörge, and Copeland in articles applying the interpreted conception, whereas my presentation has shown that the problem can be dealt with even within the formal conception. The next chapter deals with the differences between my presentation and those of von Mises, Dörge, Tornier, and other authors. These differences result from the fact that my theory develops a system comprising all types of order, whereas the other systems are restricted to special types.

A third line of development, going much further back historically than the axiomatic inquiries, connects the treatment of probability with the methods of symbolic logic. This line can be traced to Leibniz,[11] whose program of a mathematical logic included that of a logic of probability. The idea of construing probability as a relation between statements, which includes logical implication as a special case, was proposed in 1837 by Bernard Bolzano.[12] British and American logicians have followed a similar course. In his fundamental work introducing the period of modern logic, George Boole[13] (1854) included a logic of probability; he was followed by John Venn[14] (1866), Charles S. Peirce[15] (1878), and John M. Keynes[16] (1921). The latter work, besides combining symbolic logic with the calculus of probability, contains a report on earlier attempts at constructing such a calculus. In this group belong also the publications of Harold Jeffreys.[17]

My own presentation undertakes to unite the axiomatic method with the construction of a logico-mathematical calculus, which I developed without a knowledge of the calculi published much earlier by the authors cited. My theory of probability implication originated within the context of inquiries into the nature of causality.[18] The table of rules of probability implication given there is to be replaced by my present formulation. A summary of my theory of probability was published in French.[19]

---

[11] See the presentation by Louis Couturat, *La Logique de Leibniz . . .* (Paris, 1901), pp. 239–250.

[12] *Wissenschaftslehre* (1837; Leipzig, 1929–), § 161; see also the remark by Walter Dubislav, in *Erkenntnis*, Vol. I (1930), p. 264.

[13] *The Laws of Thought* (London, 1854).

[14] *The Logic of Chance* (London and Cambridge, 1866).

[15] *The Doctrine of Chances* (1878), printed in *Collected Papers* (Cambridge, Mass., 1932), Vol. II, p. 395.

[16] *A Treatise on Probability* (London, 1921).

[17] *Theory of Probability* (Oxford, 1939).

[18] "Die Kausalstruktur der Welt und der Unterschied von Vergangenheit und Zukunft," in *Ber. d. bayer. Akad., Math.-phys. Kl.* (1925), p. 144.

[19] "Les Fondements logiques du calcul des probabilités," in *Ann. de l'Inst. Henri Poincaré*, Vol. VII, Part 5 (1937), pp. 267–348.

APPENDIX TO CHAPTER 3

Exercises

Problem 1

According to official statistics from 1937, published by the National Socialist government, Germany had 66,031,580 inhabitants (A); among them were 502,799 Jews (J) and 325,541 sentenced criminals (C). Among the latter category were 1,794 Jews. What is the probability

a) that a German Jew is a criminal?
b) that a non-Jewish German is a criminal?
c) that a German criminal is a Jew?
d) that a non-criminal German is a Jew?

For the solution use the frequency interpretation directly.

Problem 2

Out of 1,000 unmarried men who are 20 years old (A), 28.3 die in that year (D). Among these, 6.1 die from tuberculosis (T), and 6.6 die from accidents (C) (German statistics of 1937). What is the probability

a) that a man 20 years old dies from tuberculosis *or* accident?
b) that a reported case of death of a man 20 years old is due to tuberculosis?
c) that a reported case of death of a man 20 years old is due to tuberculosis *or* accident?

For the solution use the frequency interpretation directly.

Problem 3

Throwing (A) with two dice distinguished as B and C, what is the probability of getting a number smaller than 5 on die B *or* a number greater than 4 on die C?

Problem 4

Urn A contains 10 slips showing the number 1, 20 slips showing the number 2, 30 slips showing the number 3. Urn B_1 contains 30 black and 50 white balls; urn B_2 contains 50 black and 50 white balls; urn B_3 contains 60 black and 20 white balls. The drawing is made as follows. A slip is drawn from urn A. The number obtained determines with which urn B_i to continue, and a ball is then drawn from that urn.

a) Determine the probability of getting a white ball (C).
b) If a white ball has been drawn, it being unknown from which of the urns B_i it was obtained, what is the probability that it was drawn, respectively, from urns B_1, B_2, B_3?

Problem 5

Mr. Smith's gardener is not dependable; the chances are 2 to 1 that he will forget to water the rosebush during Smith's absence. The rosebush is in a questionable condition; it has even chances of recovery if it is watered, but only 25% chances of recovery if it is not watered. Upon returning, Smith finds that the rosebush has withered. What is the probability that the gardener did not water the rosebush?

Problem 6

Lady Catherine's poodle has been missing for five days. There are only three explanations. Either the poodle went to the town, in which case there is a 3% chance that he was run over by a car, and a 50% chance that he was taken to the dog pound; or he went astray in the woods and was accidentally hit by a hunter, the chance of such an accident being 1%; or he went to the village and was stolen by gypsies, who in the last year had stolen five out of ten stray dogs. On a dozen previous occasions the dog had been found four times in the town, twice in the village, and six times in the woods. An inquiry at the dog pound showed that he was not there. He had never been absent more than three days at a time.

a) What is the probability that the dog was stolen by gypsies?

b) If no inquiry at the dog pound had been made, what would be the probability that the dog was stolen by gypsies?

Problem 7

Under the present conditions (A) the chances for election as governor are $\frac{1}{4}$ for Brown (B), $\frac{5}{8}$ for Jones (J), $\frac{1}{8}$ for Robinson (R). Should Brown be elected, the chances for the construction of a certain highway (C) are 60%; the chances are 80% if Jones is elected; and 20% if Robinson is elected.

a) On the basis of the present conditions, what is the chance that the highway will be constructed?

b) On the evening of the election, before the counting of the votes, Jones dies of apoplexy because of excitement. A simple majority will decide the election. What is the chance now that the highway will be constructed?

Solutions

Problem 1

a) $P(A.J,C) = \dfrac{N^n(A.J.C)}{N^n(A.J)} = \dfrac{1{,}794}{502{,}799} = 3.57$ per thousand

b) $P(A.\bar{J},C) = \dfrac{N^n(A.\bar{J}.C)}{N^n(A.\bar{J})} = \dfrac{(325{,}541 - 1{,}794)}{66{,}031{,}580 - 502{,}799}$

$= 4.94$ per thousand

c) $P(A.C,J) = \dfrac{N^n(A.C.J)}{N^n(A.C)} = \dfrac{1{,}794}{325{,}541} = 5.52$ per thousand

d) $P(A.\bar{C},J) = \dfrac{N^n(A.\bar{C}.J)}{N^n(A.\bar{C})} = \dfrac{502{,}799 - 1{,}794}{66{,}031{,}580 - 325{,}541}$

$= 7.64$ per thousand

The figures show that criminality among Jews is smaller than among non-Jews.

Problem 2

a) $P(A,D.T \vee D.C) = \dfrac{N^n(A.[D.T \vee D.C])}{N^n(A)} = \dfrac{6.1 + 6.6}{10{,}000} = 0.00127$

b) $P(A.D,T) = \dfrac{N^n(A.D.T)}{N^n(A.D)} = \dfrac{6.1}{28.3} = 0.216$

c) $P(A.D,T \vee C) = \dfrac{N^n(A.D.[T \vee C])}{N^n(A.D)} = \dfrac{6.1 + 6.6}{28.3} = 0.45$

Problem 3

Notation: B_5 = number smaller than 5 on die B; C_4 = number greater than 4 on die C.

$$P(A,B_5 \vee C_4) = P(A,B_5) + P(A,C_4) - P(A,B_5.C_4)$$
$$= \tfrac{2}{3} + \tfrac{1}{3} - \tfrac{2}{3} \cdot \tfrac{1}{3} = \tfrac{7}{9}$$

Problem 4

$$P(A,C) = \sum_{i=1}^{3} P(A,B_i) \cdot P(A.B_i,C)$$
$$= \tfrac{1}{6} \cdot \tfrac{5}{8} + \tfrac{1}{3} \cdot \tfrac{1}{2} + \tfrac{1}{2} \cdot \tfrac{1}{4} = \tfrac{19}{48}$$

$$P(A.C,B_k) = \frac{P(A,B_k) \cdot P(A.B_k,C)}{\sum_{i=1}^{3} P(A,B_i) \cdot P(A.B_i,C)}$$

$$P(A.C,B_1) = \tfrac{5}{19} \quad P(A.C,B_2) = \tfrac{8}{19} \quad P(A.C,B_3) = \tfrac{6}{19}$$

Problem 5

This and problem 6 are examples of a formalization of an informal use of probability rules—in particular, of the rule of Bayes. The numerical values used should be considered as rough estimates of the probabilities concerned. The inference then leads to values that have some significance, at least qualitatively, and correspond to instinctive appraisals of probabilities made, for example, by detectives or other experts in indirect evidence, in situations of the kind described.

Notation: A = the situation before Smith's voyage: W = the watering of the rosebush; D = the withering of the rosebush.

$$P(A,W) = \tfrac{1}{3} \quad P(A,\bar{W}) = \tfrac{2}{3} \quad P(A.W,D) = \tfrac{1}{2} \quad P(A.\bar{W},D) = \tfrac{3}{4}$$

$$P(A.D,\bar{W}) = \frac{P(A,\bar{W}) \cdot P(A.\bar{W},D)}{P(A,W) \cdot P(A.W,D) + P(A,\bar{W}) \cdot P(A.\bar{W},D)} = \tfrac{3}{4}$$

Problem 6

Notation: A = general situation after the poodle's disappearance, but not yet including a statement that an accident has occurred; T = the poodle's going to the town; V = the poodle's going to the village; W = the poodle's going to the woods; D = the poodle's being in the dog pound; C = the poodle's having an accident of any kind, including the case of his being stolen by gypsies. The following values are given:

$$P(A,T) = \tfrac{4}{12} \qquad P(A,V) = \tfrac{2}{12} \qquad P(A,W) = \tfrac{6}{12}$$

$$P(A.T,C) = \tfrac{3}{100} \qquad P(A.T,D) = \tfrac{50}{100} \qquad P(A.V,C) = \tfrac{5}{10}$$

$$P(A.W,C) = \tfrac{1}{100}$$

Question a: The poodle is not in the dog pound. Because he has never been absent for more than three days but has now been missing for five days, we consider the assumption of an accident as true. The assumption that he was stolen by gypsies is equivalent to his having gone to the village *and* having an accident, that is to $V.C$. Therefore the probability sought for is given by

$$P(A.C,V) = \frac{P(A,V) \cdot P(A.V,C)}{P(A,C)} = 85\%$$

where $P(A,C) = P(A,T) \cdot P(A.T,C) + P(A,V) \cdot P(A.V,C) + P(A,W)$

$$\cdot P(A.W,C) = \tfrac{59}{600}$$

Question b: Here the situation is characterized by $C \vee D$, and the rule of reduction must be applied:

$$P(A.[C \vee D],V) = \frac{P(A,C) \cdot P(A.C,V) + P(A,D) \cdot P(A.D,V)}{P(A,C) + P(A,D)}$$

Now $P(A.D,V) = 0$ because the dog pound is not in the village but in the town. Furthermore, we have

$$P(A,D) = P(A,T) \cdot P(A.T,D) + P(A,\bar{T}) \cdot P(A.\bar{T},D)$$

Since the dog pound is in the town, $P(A.\bar{T},D) = 0$. Therefore

$$P(A,D) = P(A,T) \cdot P(A.T,D) = \tfrac{4}{12} \cdot \tfrac{50}{100} = \tfrac{1}{6}$$

and the probability asked for is given by

$$P(A.[C \vee D],V) = 31.5\%$$

This result shows that the probability of an accident is considerably smaller so long as there is a chance that the poodle is in the dog pound.

Problem 7

Question a

$$P(A,C)$$

$$= P(A,B) \cdot P(A.B,C) + P(A,J) \cdot P(A.J,C) + P(A,R) \cdot P(A.R,C)$$

$$= \tfrac{27}{40} = 67.5\%$$

Question b

$$P(A.[B \vee R],C) = \frac{P(A,B) \cdot P(A.B,C) + P(A,R) \cdot P(A.R,C)}{P(A,B) + P(A,R)}$$

$$= \tfrac{7}{15} = 46.6\%$$

Chapter 4

THEORY OF THE ORDER OF PROBABILITY SEQUENCES

THEORY OF THE ORDER OF PROBABILITY SEQUENCES

§ 26. The Task of the Theory of Order

The first part of the probability calculus, which thus far has been presented, is the *elementary calculus of probability.* It deals with the logical elements of the probability calculus, just as the elements of geometry comprise the logical foundations of geometry. In the elementary calculus, probability sequences are treated with respect to their *external connections.* Sequences to which degrees of probability are attached play in it the part of units the mutual relations of which are investigated, without regard to the *internal structure* of the units. The second part of the probability calculus, which we now begin, is concerned with the *internal order of probability sequences.* The elementary calculus is concerned with probability sequences only with respect to the one property of possessing a probability, or, in the frequency interpretation, of having a limit of the frequency; but no statements are made about the manner in which the elements of such sequences follow one another. As a consequence of this procedure, sequences of very different types of order are subsumed under the name of probability sequence. Thus we regard a strictly alternating sequence

$$B\,\bar{B}\,B\,\bar{B}\,B\,\bar{B}\,B\,\bar{B}\,B\,\ldots \tag{1}$$

as a probability sequence because the frequency of B approaches a limit, which in this case is $= \frac{1}{2}$. But statistical sequences of natural events show a different structure. For instance, they may have the form of the sequence

$$B\,B\,B\,\bar{B}\,\bar{B}\,B\,\bar{B}\,\bar{B}\,B\,B\,\bar{B}\,B\,B\,B\,\bar{B}\,\bar{B}\,B\,\bar{B}\,\bar{B}\,\bar{B}\,B\,\ldots \tag{2}$$

obtained by tossing a coin. Both the sequences (1) and (2) go toward the limit $\frac{1}{2}$ of the frequency; however, the sequence (1) shows a strict order, whereas the sequence (2) is irregular, that is, is a *random sequence.* Since randomness is an essential feature of a number of very important problems of probability, the structure of a sequence like (2) must now be described by logical means. Of this kind are the problems investigated by the theory of the order of probability sequences. We shall deal not only with types of extreme order as given by (1) and (2); we shall become acquainted also with a number of intermediate types which are not completely irregular like (2), but which do not possess the strict order of (1). Sequences of type (2) will also be called *normal sequences;* the other types are called, correspondingly, *nonnormal sequences.* Further distinctions among the latter will be specified subsequently.

To carry out this task we require, apart from the concepts already developed, some conceptual means with the help of which the internal structure of probability sequences can be characterized, and, furthermore, another group of axioms that is not used in the elementary probability calculus. Like the other axiom groups, the new group can be shown to follow logically from the frequency interpretation and thus to require of probability sequences only such characteristics as are expressed implicitly by the frequency interpretation.

An essential feature of my theory of order is that it deals with all possible forms of probability sequences and is not restricted to sequences of one type of order, such as normal sequences. In this respect my probability theory differs from others—in particular, from that developed by R. von Mises. Such theories regard randomness as an essential characteristic of the very concept of probability; and they contend that the meaning of probability cannot be exhaustively formulated without reference to randomness. But we shall find that such a restriction of the concept of probability is not expedient; it artificially creates sharp boundaries in a region in which more or less general types occur side by side in a natural order of types. Those who regard randomness as a necessary feature of all probability sequences forget that a sequence like (1) represents merely the limiting case of a partially ordered type of sequence that still shows, in a less rigid form, rather important characteristics of "disorder" (see § 33). It is possible, in the theory of order, to represent by the same conceptual means all the different types of probability sequences lying between the extreme cases (1) and (2). Therefore I believe my general probability calculus to be preferable to any special calculus that is restricted to a certain type of order. Linguistic usage, too, has decided in favor of a more general concept of probability. In physics an important role is played by sequences with a *probability aftereffect*, which represent a type intermediate between (1) and (2).

For these reasons we develop the theory of order by beginning with the characterization of normal sequences and proceeding thence, step by step, to more general types of sequences. The characterization of a particular sequence type is to be conceived as a matter of definition; we cannot ask, therefore, whether the "correct" characterization has been found. On the other hand, definitions will be constructed in such a manner that they provide types relevant to practical applications.

§ 27. Phase Probabilities

To define the different types of order, some *means of structural characterization* will be constructed by the use of a specific method: we derive certain *subsequences* from the sequence considered, the *major sequence*, and then investigate the probabilities of these subsequences. The method of representing the structure of order by statements about the probabilities in subsequences was first carried through by von Mises for the purpose of defining random sequences. Although I shall construct a somewhat different definition of such sequences, I shall follow the same method, in principle, for it turns out to be an excellent instrument of structural characterization. I shall, however, extend the use of the method to the construction of further types of sequences, following my plan of including, in the theory of probability sequences, the whole variety of types of structural order.

The first of the means of structural characterization is provided by *phase probabilities*. This concept will be developed with the help of the frequency interpretation; but it will be shown later that the method applies also in a purely formal conception of probability.

To begin with some practical examples, compare (2, § 26) with (1, § 26). An essential difference between the two sequences is expressed by the relation of each individual element to its predecessors. In (1, § 26) B is always followed by $\bar{B}$, and $\bar{B}$ by B; in (2, § 26), however, there is no such dependence on the predecessor: sometimes B is followed by B and sometimes by $\bar{B}$. The type of order can therefore be characterized, at least in one essential feature, by stating the probability of an element with respect to its predecessors. The normal sequence will then have the property that the predecessors of an element have no influence on whether the element is of the kind B or $\bar{B}$; in the normal sequence the probability of the occurrence of a specific element is independent of its predecessors.

The idea may be illustrated by an example from a more extended sequence. We reproduce in (1) a sequence obtained by throwing one die, writing B for an even number and $\bar{B}$ for an odd number. The sequence is written in such a way that the right end of a row is to be continued by the left end of the next row, like the lines of a printed page.

$$\begin{array}{l} B\ \underset{\cdot}{\bar{B}}\ B\ \underset{\cdot}{\bar{B}}\ B\ \underset{\cdot}{B}\ \underset{\cdot}{B}\ \underset{\cdot}{\bar{B}}\ B\ \underset{\cdot}{\bar{B}}\ \bar{B}\ \bar{B}\ B\ \underset{\cdot}{B}\ \underset{\cdot}{\bar{B}}\ \bar{B}\ \bar{B}\ \bar{B}\ \bar{B}\ B\ \underset{\cdot}{B}\ \underset{\cdot}{\bar{B}}\ B\ \underset{\cdot}{\bar{B}}\ B\ \underset{\cdot}{\bar{B}}\ \bar{B} \\ B\ \underset{\cdot}{\bar{B}}\ \bar{B}\ \bar{B}\ B\ \underset{\cdot}{B}\ \underset{\cdot}{\bar{B}}\ B\ \underset{\cdot}{\bar{B}}\ B\ \underset{\cdot}{B}\ \underset{\cdot}{B}\ \underset{\cdot}{\bar{B}}\ \bar{B}\ B\ \underset{\cdot}{B}\ \underset{\cdot}{\bar{B}}\ \bar{B}\ B\ \underset{\cdot}{\bar{B}}\ B\ \underset{\cdot}{\bar{B}}\ B\ \underset{\cdot}{B}\ \underset{\cdot}{\bar{B}}\ \bar{B}\ B \\ \underset{\cdot}{B}\ \underset{\cdot}{\bar{B}}\ B\ \underset{\cdot}{B}\ \underset{\cdot}{B}\ \underset{\cdot}{B}\ \underset{\cdot}{B}\ \underset{\cdot}{B}\ \underset{\cdot}{B}\ \underset{\cdot}{B}\ \underset{\cdot}{\bar{B}}\ B\ \underset{\cdot}{B}\ \underset{\cdot}{B}\ \underset{\cdot}{\bar{B}}\ \bar{B}\ B\ \underset{\cdot}{\bar{B}}\ \bar{B}\ \bar{B}\ \bar{B}\ B\ \underset{\cdot}{B}\ \underset{\cdot}{\bar{B}}\ \bar{B}\ \bar{B} \end{array} \qquad (1)$$

Among the 80 elements of this sequence are 41 B's, 39 $\bar{B}$'s, so that the probability $\frac{1}{2}$ is reasonably well satisfied. The 41 elements that are preceded by a B are marked underneath with a dot. They form a subsequence within the major sequence; and the probability, or the frequency, of B within this subsequence is the means of structural characterization to be used. If we count out the B-cases among the elements with the dot underneath, we obtain 20 as result, that is, the probability of a B-element in the subsequence is $\frac{20}{41}$, which again is approximately $= \frac{1}{2}$. In the same manner the probability of a $\bar{B}$-element in the subsequence is determined: it is $= \frac{21}{41} \sim \frac{1}{2}$. The probability that B follows $\bar{B}$ is calculated by counting B in the subsequence of elements without a dot (dropping the first element, which has no predecessor); it is $= \frac{20}{38}$. Similarly, the probability of $\bar{B}$ following $\bar{B}$ is found to be $\frac{18}{38}$. These two values also are nearly equal to $\frac{1}{2}$. The figures express the fact that within the sequence under consideration the predecessor does not influence the occurrence of a B-element. In the sequence (1, § 26) we would obtain, within a similarly selected subsequence, a probability of B different from that in the major sequence; here the probability of B would be equal to 0 in the subsequence selected by B as predecessor, because B is always followed by $\bar{B}$. This fact shows the nonnormal character of the sequence (1, § 26).

Correspondingly, the influence of a *combination of predecessors* can be enumerated. For instance, we can consider in (1) the subsequence of elements having two predecessors $B\,B$; we thus obtain the value $\frac{9}{20}$ for the probability that a combination $B\,B$ is followed by a B. Similarly, the probability that a combination $\bar{B}\,\bar{B}$ is followed by a B has the value $\frac{9}{17}$. That is nearly $\frac{1}{2}$ for both again. The latter phase probabilities are independent of the former, which refer to only one predecessor, because the last-mentioned figures cannot be derived from the values given first, but must be ascertained by a separate counting. It is possible that the first predecessor does not produce a selection deviating from the major sequence, whereas the second predecessor does so. The normal sequence, however, as shown in the example, is characterized by particularly simple relations; it satisfies the condition that a selection by means of any chosen combination of predecessors results in the same probability of B as exists in the major sequence.

The concept of probability was introduced for a sequence pair; a sequence such as (1) must, therefore, be coördinated to another sequence, so that a *normal sequence pair* results. It is possible to suppress the first sequence in (1), because the corresponding sequence x_i consists exclusively of elements $x_i \in A$, that is, it is *compact*—using a term defined in § 16. If this is not the case, and the sequence x_i is *interspersed*, we shall cross out the elements of the sequence y_i for which $x_i \in \bar{A}$ holds, and consider the structural order of the reduced sequence only, to which we assign subscripts by the rule

$$i' = \overset{i}{\underset{k=1}{N}} (x_k \in A) \tag{2}$$

This relation coördinates an i' to each value i of the elements x_i belonging to A. With the help of this reduction, the first sequence of a normal sequence pair assumes a very simple structure and can be identified with the sequence of the subscripts, so that the pair is replaced by one sequence. In this sense, reference may be made to a normal sequence without reference to a pair. The same simplification will be used in studying other types of sequences.

A detailed notation for phase probabilities will now be defined, to be followed by a corresponding extension of the abbreviated way of writing. The probability of a B following a B will be written in the detailed notation as

$$(i)\ (x_{i+1} \epsilon A \,.\, y_i \epsilon B \underset{p_1}{\ni} y_{i+1} \epsilon B) \tag{3}$$

Here a *phase* occurs in the subscript. The convention is now introduced that this phase, in the abbreviated notation, is added as a superscript to the letter specifying the class, so that (3) is abbreviated to the form

$$(A^1 . B \underset{p_1}{\ni} B^1) \tag{3'}$$

In the P-notation we write

$$P(A^1 . B, B^1) = p_1 \tag{4}$$

A further restriction for the sequences, by which certain mathematically irrelevant cases are eliminated, is expedient. Apart from the premise that the A-sequence is compact, it will be assumed that the class B, or the combinations of the classes $B . C . . . G$, always occur in infinitely many elements. In other words, it will be assumed that the probabilities occurring represent genuine limits in the frequency interpretation. It then follows from this interpretation that we can drop the phase in the term A of (4), since the element x_i as well as the element x_{i+1} belongs to A (see also § 28). We thus write for (4)

$$P(A . B, B^1) = p_1 \tag{5}$$

Such quantities as (5) in which superscripts occur are called *phase probabilities*. It must not be assumed, however, that the existence of the *major probability* of a sequence insures the existence of phase probabilities; it represents a certain specialization of the sequence type if we demand that all the phase probabilities of the sequence exist. The type of sequence is much more specialized if we demand that certain phase probabilities are equal. By means of such relations the characterization of the sequence type asked for will be given. For instance, the relation

$$P(A . B, B^1) = P(A, B) \tag{6}$$

expresses the property, illustrated by example (1), that the immediate predecessor has no influence on the probability.

Before going on with these considerations, I wish to comment on the principles underlying the construction of the theory of order.

§ 28. Axioms Concerning the Theory of Order

In the preceding section, the probabilities of subsequences occurring in the characterization of normal sequences were interpreted as frequencies. It must now be emphasized that this is not necessary; like any other probabilities, phase probabilities may be regarded from the point of view of the formal probability calculus, that is, may be conceived as undefined quantities of which it is known only that they satisfy the axioms of the probability calculus. Therefore it is possible to characterize the type of order of a sequence axiomatically, that is, to give a definition of the type of sequence even in the formal calculus.

To achieve this result it is essential to make a distinction between *order* and *frequency* of sequences. The requirement that the probability statement always concern sequences was incorporated in the logical foundation of the probability calculus; therefore, statements can be made about the order of the sequences within the formal system. These statements assert the existence of certain probabilities for subsequences. That the probabilities of subsequences represent limits of the frequency is an additional assertion, which need not be stated within the axiomatic formulation of the probability calculus; the assertion is dispensable, as well as for probabilities in general. We can therefore regard probability statements that contain terms with superscripts as uninterpreted statements, to which, however, the same kind of interpretation is to be given as is used for all other probability statements. For this purpose the assumption will be made that the letters having a phase sign designate objects for which the probability axioms are valid; so it is permissible to apply the usual operations and substitutions to letters carrying a phase sign. This assumption is not an addition to the axiom system, but it does extend the *field of application* of the axioms.

Some additions to the axiom system will be made, however. In the preceding section we performed an operation with the phase superscript, omitting this phase symbol on the letter A. That we are entitled to do so cannot be derived from the axioms given in the preceding chapter because those axioms do not refer to superscripts. Only from the frequency interpretation can the admissibility of this operation be derived. In order to be independent of this interpretation we shall introduce two new axioms that allow us to perform operations with superscripts in the formal system. These axioms, which refer only to infinite sequences, will be called *axioms of order*. Like the other axioms, they are derivable from the frequency interpretation.

To formulate these axioms we go back to the logical notation of the probability calculus. We must first formulate a premise of the axioms stating that we deal with sequences consisting of an infinite number of elements and referred to a compact sequence A. The latter property can be expressed immediately by

$$(A) \tag{1}$$

because, according to the translation rule, this expression, in the detailed notation, means

$$(i)(x_i \in A) \tag{1'}$$

The formulation includes the condition that the number of elements of the A-sequence is infinite, for we regard the range of the subscript i as given by the positive integers; if the number of A-elements were finite, the subsequent elements x_i would be $\bar{A}$-elements. To formulate the infinity of the B-sequence, the abbreviation $(\exists \infty B)$ is introduced by the following definition:

$$(\exists\ \infty\ B) =_{Df} \overline{(\exists m)(n)[\overset{n}{\underset{i=1}{N}}(y_i \in B) < m]} \tag{2}$$

The right side of the definition states verbally: there exists no number m such that, for increasing subscript n, the number of the elements B to the n-th place remains below m. In an analogous way the condition can be expressed for combinations:

$$(\exists\ \infty\ B^\alpha \ldots G^\tau) =_{Df} \overline{(\exists m)(n)[\overset{n}{\underset{i=1}{N}}(y_{i+\alpha} \in B) \ldots (g_{i+\tau} \in G) < m]} \tag{3}$$

With the help of this notation, the new group of axioms is written as follows:

V. Axioms of the Theory of Order

1. $(A).(\exists\ \infty\ C) \supset [(A^\alpha . C \underset{p}{\rightarrow} B^\beta) \equiv (A . C \underset{p}{\rightarrow} B^\beta)]$
2. $(A).(\exists\ \infty\ C^\alpha \ldots G^\sigma)$
   $\supset [(A . C^\alpha \ldots G^\sigma \underset{p}{\rightarrow} B^\tau) \equiv (A . C^{\alpha-\rho} \ldots G^{\sigma-\rho} \underset{p}{\rightarrow} B^{\tau-\rho})]$

These axioms, in the given order, have the same meaning as the following formulas in the P-notation:

$$P(A^\alpha . C, B^\beta) = P(A . C, B^\beta) \tag{4}$$

$$P(A . C^\alpha \ldots G^\sigma, B^\tau) = P(A . C^{\alpha-\rho} \ldots G^{\sigma-\rho}, B^{\tau-\rho}) \tag{5}$$

Whereas, however, the implicative formulation of the axioms states the premises for which the equalities hold, these conditions are omitted in the P-notation; a corresponding verbal qualification, therefore, must be understood for formulas (4)–(5). Apart from this restriction, it is possible to manipulate these formulas in the same way as all the other formulas of the system. For instance, in (4) we can substitute A^γ for C and thus derive, by applying (4) twice,

$$P(A^\alpha . A^\gamma, B^\beta) = P(A . A, B^\beta) = P(A, B^\beta) \tag{6}$$

Furthermore, in (5) it is possible to replace the terms $C^\alpha \ldots G^\sigma$ by the factor $C \vee \bar{C}$, which is always true and may be dropped. Thus we have, with (4),

$$P(A,B^\alpha) = P(A^\alpha,B^\alpha) = P(A,B) \tag{7}$$

when we choose the quantity ρ occurring in (5) equal to α.

In § 18, axioms I–IV were derived from the frequency interpretation. We shall construct a similar derivation for the v-th group of axioms. The derivation is greatly simplified by the use of F-symbols. The proof is given in detail, for we shall use this calculation again, later. First, we write for (4)

$$F^n(A^\alpha . C,B^\beta) = \frac{\overset{n}{\underset{i=1}{N}}(x_{i+\alpha} \epsilon A).(z_i \epsilon C).(y_{i+\beta} \epsilon B)}{\overset{n}{\underset{i=1}{N}}(x_{i+\alpha} \epsilon A).(z_i \epsilon C)} = \frac{a_n - a}{b_n - a}$$

$$a_n = \overset{n}{\underset{i=1}{N}}(x_i \epsilon A).(z_i \epsilon C).(y_{i+\beta} \epsilon B) \tag{8}$$

$$a = 0 \text{ for } \alpha \geqq 0$$
$$a = \alpha \text{ for } \alpha < 0$$

$$b_n = \overset{n}{\underset{i=1}{N}}(x_i \epsilon A).(z_i \epsilon C)$$

We now use the identity

$$\frac{a_n + \delta_n}{b_n + \eta_n} = \frac{a_n}{b_n} \cdot \frac{1}{1 + \frac{\eta_n}{b_n}} + \frac{\delta_n}{b_n + \eta_n} \tag{9}$$

If a_n and b_n (or only b_n) increase continually, while $|\delta_n|$ and $|\eta_n|$ remain below a finite limit for all n, it follows that

$$\lim_{n \to \infty} \frac{a_n + \delta_n}{b_n + \eta_n} = \lim_{n \to \infty} \frac{a_n}{b_n} \tag{10}$$

We assume that at least one of the two limits exists. Because $|\delta_n| = |\eta_n| = |a|$ is subject to the upper bound $|\alpha|$, or even equal to 0, and, also,

$$F^n(A . C,B^\beta) = \frac{a_n}{b_n} \tag{11}$$

it follows that (4) is valid.

To derive (5) we start from

$$F^n(A.C^\alpha \ldots G^\sigma,B^\tau)$$

$$= \frac{\overset{n}{\underset{i=1}{N}}(x_i \epsilon A).(z_{i+\alpha} \epsilon C) \ldots (g_{i+\sigma} \epsilon G).(y_{i+\tau} \epsilon B)}{\overset{n}{\underset{i=1}{N}}(x_i \epsilon A).(z_{i+\alpha} \epsilon C) \ldots (g_{i+\sigma} \epsilon G)} = \frac{a_n + \delta_n}{b_n + \eta_n} \tag{12}$$

$$a_n = \overset{n}{\underset{i=1}{N}}(x_i \epsilon A).(z_{i+\alpha-\rho} \epsilon C) \ldots (g_{i+\sigma-\rho} \epsilon G).(y_{i+\tau-\rho} \epsilon B)$$

$$b_n = \overset{n}{\underset{i=1}{N}}(x_i \epsilon A).(z_{i+\alpha-\rho} \epsilon C) \ldots (g_{i+\sigma-\rho} \epsilon G)$$

$$\delta_n \leqq \rho \qquad \eta_n \leqq \rho$$

Because of

$$F^n(A.C^{\alpha-\rho} \ldots G^{\sigma-\rho},B^{\tau-\rho}) = \frac{a_n}{b_n} \tag{13}$$

and by the use of (10), the derivation of equation (5) is thus achieved.

From this proof it follows that the axioms v, like the previous axioms, are valid for all probability sequences, for they could be derived from the frequency interpretation. They are therefore not restricted to normal sequences and will be used for calculations with both normal and nonnormal sequences.

Reference must be made to the application of the existence rule in combination with the two axioms v. That the rule of existence is valid for the frequency interpretation was inferred from the fact that probability equations correspond to tautological frequency equations that are valid even before the transition to the limit (§ 18). It could thus be derived that in the transition $n \to \infty$ the m-th probability quantity must approach a limit, if limits exist for the $m - 1$ other probabilities occurring. The situation is somewhat different for the axioms v, because they are not valid *before* the transition to the limit, but become tautologies only for the limits themselves. It follows that a probability equation derived by the use of the axioms v is not strictly valid for the coördinated frequencies; and at first it seems doubtful whether we can infer, for the transition to the limit, the existence of the limit of the m-th probability from the existence of the limits of the $m - 1$ other probability quantities. But it will now be shown that in this case, too, the existence of the m-th probability is determined; in other words, equations derived by the help of the axioms v also determine existence.

The problem under investigation is made clear by the following considerations. We take the probability equation *before* applying the axioms v; then there corresponds to it a tautologous frequency equation (symbols as in § 18),

$$r(f_1^n \ldots f_m^n, f_{m+1}^n \ldots f_{m+s}^n) = 0 \tag{14}$$

Here $f_{m+1}^n \ldots f_{m+s}^n$ are the frequency quantities that, provided they exist, will become equal to one another in the limit $n \to \infty$, according to one of the axioms v; $f_1^n \ldots f_m^n$ are other frequency quantities. From (14) we can go to the limit equation, constructed without the use of the axioms v:

$$r(p_1 \ldots p_m, p_{m+1} \ldots p_{m+s}) = 0 \tag{15}$$

This equation determines existence just as do other probability equations: from the existence of the limit for $m + s - 1$ quantities of the probabilities occurring, the existence of the limit for the $(m + s)$-th probability can be inferred. If the axioms v are now applied to (15), the equation

$$r(p_1 \ldots p_m, p_{m+1}) = 0 \tag{16}$$

obtains. We do not yet know whether this equation determines existence, because no strict frequency equation

$$r(f_1^n \ldots f_m^n, f_{m+1}^n) = 0 \tag{17}$$

corresponds to it. If, in (15), the existence of the limits is not known for all the $p_{m+1} \ldots p_{m+s}$, we cannot infer from (15) that the limits exist. However, assuming that (16) determines existence, we can conclude from (16) that p_{m+1} exists. Such an inference, therefore, must be justified by a separate proof.

For this purpose we derive a property of the frequency equations corresponding to the axioms v. The transition from (9) to (10), with the help of which we inferred the validity of v,1, was based on the assumption that at least one of the two limits exists. But we are able to obtain a relation free from this assumption when we form the difference of the two frequency quantities

$$f_1^n = \frac{a_n}{b_n} \qquad f_2^n = \frac{a_n + \delta_n}{b_n + \eta_n}$$

We then obtain

$$f_2^n - f_1^n = \frac{\delta_n - \eta_n \dfrac{a_n}{b_n}}{b_n + \eta_n}$$

It follows that if b_n increases toward infinite values with n, whereas $|\delta_n|$ and $|\eta_n|$ remain for all n below a definite bound, then

$$\lim_{n \to \infty} (f_2^n - f_1^n) = 0 \tag{18}$$

Because of $\frac{a_n}{b_n} \leqq 1$, it is irrelevant whether a_n also increases toward infinite values or whether it remains finite. The relation (18) is valid even if the separate limits for f_1^n and f_2^n do not exist. In an analogous way, the inference for the axiom v,2 is carried out by interpreting the quantities a_n and b_n as in (12).

We make use of this fact for a transformation of (14). We assume that the function r occurring in (14) is continuous with respect to all its arguments. Subtracting, now, from the arguments $f_{m+2}^n \ldots f_{m+s}^n$ the argument f_{m+1}^n we go from r to a function r':

$$r'(f_1^n \ldots f_m^n, f_{m+1}^n, g_{m+2}^n \ldots g_{m+s}^n) \tag{19}$$

$$g_{m+i}^n = f_{m+i}^n - f_{m+1}^n \tag{20}$$

which also is continuous for all its arguments. If we make the transition to the limit $n \rightarrow \infty$, the g_{m+i}^n go to the limit 0 because of (18); therefore (19) must be an equation that determines existence for the variables $f_1^n \ldots f_{m+1}^n$. It follows that, if limits exist for the $f_1^n \ldots f_m^n$, a limit must also exist for f_{m+1}^n; thus, because of (18), limits must exist for the values $f_{m+2}^n \ldots f_{m+s}^n$.

It is thereby demonstrated that it is compatible with the frequency interpretation to apply the rule of existence to probability equations constructed with the help of the axioms v. We shall therefore extend the existence rule to the wider probability calculus obtaining when the axioms v are included. In other words, all probability equations will be regarded as determining existence.

§ 29. Sequences without Aftereffect

Phase probabilities are closely connected with the probabilities of certain combinations of consecutive elements. Applying the general theorem of multiplication, we represent the probability of a combination $B\ B$ by

$$P(A,B.B^1 = P(A,B) \cdot P(A.B,B^1) \tag{1}$$

Correspondingly, the probability of a combination $B\ B\ B$ is written in the form

$$\begin{aligned} P(A,B.B^1.B^2) &= P(A,B.B^1) \cdot P(A.B.B^1,B^2) \\ &= P(A,B) \cdot P(A.B,B^1) \cdot P(A.B.B^1,B^2) \end{aligned} \tag{2}$$

Still higher phase probabilities occur for longer combinations. Only in those sequences for which all the phase probabilities exist do probabilities exist for all combinations of consecutive elements.

A particular type of sequence is characterized by the condition that all phase probabilities of the sequence be equal to the major probability. Such a sequence will be called *free from aftereffect*. This concept does not completely define the normal sequence, but establishes a more general type; however, we must study this type first because it is a necessary part of the definition of the normal sequence.

For the symbolic formulation it is useful to introduce, instead of the disjunction $B \vee \bar{B}$, the many-term disjunction $B_1 \vee \ldots \vee B_r$, which, however, must be complete and exclusive. Then we have as definition of the *sequence free from aftereffect* the relation

$$P(A.B^1_{i_1} \ldots B^{\nu-1}_{i_{\nu-1}}, B^{\nu}_{i_\nu}) = P(A, B_{i_\nu}) \tag{3}$$

The use of sub-subscripts is a convenient notational device; every total subscript i_ρ is a free variable for which any value may be chosen, whereas the sub-subscript ρ indicates the phase of the term to which the chosen value belongs. Thus i_ρ means: the subscript belonging to the term occurring with the phase ρ. That (3) is meant to hold for all subscripts and all phase lengths ν is expressed by the fact that every i_ρ as well as ν is a free variable.

The probability of combinations according to (1) obtains a very simple form by the use of (3); it is determined by the special theorem of multiplication. This is proved by using the general theorem of multiplication in combination with (3):

$$\begin{aligned} P(A, B^1_{i_1} \ldots B^{\nu}_{i_\nu}) &= P(A, B^1_{i_1} \ldots B^{\nu-1}_{i_{\nu-1}}) \cdot P(A.B^1_{i_1} \ldots B^{\nu-1}_{i_{\nu-1}}, B^{\nu}_{i_\nu}) \\ &= P(A, B^1_{i_1} \ldots B^{\nu-1}_{i_{\nu-1}}) \cdot P(A, B_{i_\nu}) \end{aligned} \tag{4}$$

Applying the same procedure to the first term on the right side, we arrive, after $\nu - 1$ steps, at the result

$$P(A, B^1_{i_1} \ldots B^{\nu}_{i_\nu}) = P(A, B_{i_1}) \ldots P(A, B_{i_\nu}) \tag{5}$$

The sequence without aftereffect, can, therefore, be characterized by the statement that it satisfies the special theorem of multiplication with respect to the succession of its elements. This is true, however, only for a certain kind of enumeration. Probabilities like those occurring on the left side of (1) or (5) refer, in the frequency interpretation, to a counting operation in which the subscript i always progresses by one unit, so that overlapping combinations are counted. A group $B\ B\ B$ then contributes two combinations, which may be indicated by the schema $B\ B\ B$; the sequence (1, § 27), for instance, would contain 21 combinations. This method of counting is called an *enumeration by overlapping segments*, or, to use a shorter name, an *enumeration with overlapping*. The sequence (1, § 27) satisfies the special multiplication theorem for this type of counting; this can be recognized for the combinations $B\ B$ from the fact that the probability of such a combination assumes the value $\frac{21}{79}$, which is nearly $= \frac{1}{2} \cdot \frac{1}{2}$.

The next section introduces a different method of counting.

§ 30. Normal Sequences

It was pointed out above that the property of being free from aftereffect is only a *necessary*, but not a *sufficient*, condition for the definition of the normal sequence that we wish to construct. In order to state the sufficient conditions for the normal sequence, another means of structural characterization must be added to the use of phase probabilities. The second method consists in the use of *selections*.

By a selection S we understand a rule that determines for every element y_i whether $y_i \epsilon S$ or $y_i \epsilon \bar{S}$ holds. Any probability sequence in which S is an attribute of the elements y_i can be regarded as a selection; it can be used to select from another sequence those elements x_i for which y_i belongs to S. But the concept of selection is somewhat more general, for we do not require a limit of the frequency for S. Thus the probability $P(A,S)$ need not exist. The selection may be given by an arithmetical rule—for instance, in the form, "every third element y_i"—or, as explained, by a probability sequence. In principle, however, we can imagine every selection to be given in the form of an infinite schema which states for every y_i whether it belongs to S.

The method of characterizing the structural order of probability sequences by reference to subsequences is thus extended to the use of subsequences defined by any form of selection. The probability within a subsequence selected by S will be written

$$P(A.S,B)$$

The use of general selections can be combined with the use of phase probabilities; thus in

$$P(A.S.B,B^1)$$

the probability of B is considered within the subsequence determined by the rule that the predecessor of the selected element belongs to S and has the attribute B.

There will always be selections that leave the probability of B invariant, and others that change the probability of B. Let S be the selection, "every third element y_i". This selection, applied to a sequence obtained by throwing a die, will leave the probability of B invariant; we shall have, for instance,

$$P(A.S,B) = P(A,B) \qquad P(A.S.B,B^1) = P(A.B,B^1)$$

$$P(A.S^1.B,B^1) = P(A.B,B^1)$$

A selection changing the probability of B is represented by the selection, "every element B of the major sequence and its successor"; in this subsequence, B has a frequency higher than in the major sequence. Using, instead of the disjunction $B \vee \bar{B}$, the complete and exclusive disjunction $B_1 \vee \ldots \vee B_r$, we introduce the definition:

A selection S belongs to the domain of invariance of the probability sequence B if it leaves the probability of B_i and, simultaneously, all the phase probabilities of B_i invariant for all i, whereas S may occur in any phase; that is, if

$$P(A.S,B_i) = P(A,B_i)$$
$$P(A.S^\alpha.B^1_{i_1}\ldots B^{\nu-1}_{i_{\nu-1}},B^\nu_{i_\nu}) = P(A.B^1_{i_1}\ldots B^{\nu-1}_{i_{\nu-1}},B^\nu_{i_\nu}) \tag{1}$$
$$1 \leqq \alpha \leqq \nu$$

A certain narrower class—the class of *regular divisions*—plays an important part among the selections. By a *regular division S* we understand a division of the major sequence into λ subsequences $S_{\lambda 1}, S_{\lambda 2}, \ldots S_{\lambda\lambda}$, such that to a specified $S_{\lambda\kappa}$ belong all the elements y_i for which

$$i = \kappa + (m-1)\cdot\lambda \qquad m = 1,2,3\ldots \qquad \kappa = 1, \text{ or } = 2, \text{ or } \ldots, \text{ or } = \lambda \tag{2}$$

Thus the sequence $y_2,y_6,y_{10},y_{14}, \ldots$ is determined by the selection $S_{\lambda\kappa}$ resulting for $\lambda = 4$ and $\kappa = 2$. In (2), m may be interpreted as the running subscript in the "self-numeration" of the subsequence $S_{\lambda k}$, that is, a numeration in which a series of consecutive numbers is assigned to the elements of the subsequence. If the regular divisions belong to the domain of invariance of a sequence, we speak of a *regular domain of invariance* and call the sequence *regular-invariant.* By means of these concepts the desired definition can now be given:

A sequence B is a normal sequence if it is free from aftereffect and if the regular divisions belong to its domain of invariance.

Therefore we have, for a normal sequence,

$$P(A.S_{\lambda\kappa},B_i) = P(A,B_i)$$
$$P(A.S^\alpha_{\lambda\kappa}.B^1_{i_1}\ldots B^{\nu-1}_{i_{\nu-1}},B^\nu_{i_\nu}) = P(A.B^1_{i_1}\ldots B^{\nu-1}_{i_{\nu-1}},B^\nu_{i_\nu}) = P(A,B_{i_\nu}) \tag{3}$$
$$1 \leqq \alpha \leqq \nu \qquad 1 \leqq \kappa \leqq \lambda$$

To show the consequences of this definition, we shall investigate the conclusions that can be drawn from (3). When, in the second equation, we use the special values $\nu = \lambda$, $\alpha = 1$, and $\kappa = 1$, and then apply the first of the equations with $\kappa = \lambda$, we obtain

$$P(A.S^1_{\lambda 1}.B^1_{i_1}\ldots B^{\lambda-1}_{i_{\lambda-1}},B^\lambda_{i_\lambda}) = P(A.S^\lambda_{\lambda\lambda},B^\lambda_{i_\lambda}) = p \tag{4}$$

The probability on the left would read, in the implicational mode of writing,

$$(i)([x_i \epsilon A]\cdot[y_{i+1}\epsilon S_{\lambda 1}]\cdot[y_{i+1}\epsilon B_{i_1}]\ldots[y_{i+\lambda-1}\epsilon B_{i_{\lambda-1}}] \underset{p}{\supset} [y_{i+\lambda}\epsilon B_{i_\lambda}]) \tag{5}$$

Now it is obvious that when the element y_{i+1} belongs to $S_{\lambda 1}$, the element y_{i+2} will belong to $S_{\lambda 2}$, and so on, so that the element $y_{i+\lambda}$ will belong to $S_{\lambda\lambda}$. The condition stated by the compound first term in (5) can therefore be satisfied only when the elements $y_{i+1}, \ldots y_{i+\lambda-1}, y_{i+\lambda}$, belong, respectively, to the first, ... $(\lambda - 1)$-th, λ-th subsequence selected by the regular division. This means that we count in (5) only the elements $y_{i+\lambda}$ that belong to the last of the subsequences. A further restriction is that we count only the elements $y_{i+\lambda}$ of this subsequence for which the corresponding elements of the other subsequences have, respectively, the attributes $B_{i_1} \ldots B_{i_{\lambda-1}}$. The relation (4), therefore, states that the probability of B_{i_λ} in the λ-th subsequence, given by the term on the right, is independent of the attributes occurring in the other subsequences. In other words, (4) expresses the condition that the subsequences selected by regular divisions are independent of one another.

This result can also be proved by the fact that (3) leads to the special theorem of multiplication in a form, however, that is somewhat different from the one used in (5, § 29). It was pointed out in § 29 that the latter equation is valid for enumeration with overlapping. But the combinations can be counted in a different way, by counting consecutive sections of the length λ that do not overlap. If we thus section off in (1, § 27) by separating strokes after every two elements, we obtain 40 sections of two elements each; among these we find 8 sections of the form $B\ B$. Here again the probability is close to the one we require, namely, is $= \frac{8}{40}$ and thus is nearly $= \frac{1}{2} \cdot \frac{1}{2}$. This is a different kind of counting, however, which is called *enumeration by sections.*

The enumeration by sections is formulated by means of the regular divisions. When we write

$$P(A.S^1_{\lambda 1}, B^1_{i_1} \ldots B^\lambda_{i_\lambda}) \tag{6}$$

this means that we count the combinations $B^1_{i_1} \ldots B^\lambda_{i_\lambda}$ only for the case that the element y_{i+1} belongs to $S_{\lambda 1}$. Since the denominator of this expression in the frequency interpretation has the form

$$\overset{n}{\underset{i=1}{N}}(x_i \epsilon A).(y_{i+1} \epsilon S_{\lambda 1}) \tag{7}$$

it counts directly the number of sections of length λ. Thus the combinations $B\ B$ in (1, § 27) will be counted by the expression

$$P(A.S^1_{21}, B^1.B^2) \tag{8}$$

Similarly, we shall count by the expression $P(A.S^1_{22}, B^1.B^2)$ the sections BB resulting after the cancellation of the first element of the sequence. By the use of the general term $S^1_{\lambda\kappa}$ in (6), instead of $S^1_{\lambda 1}$, we thus can express the possibility that the sectioning starts after cancellation of the first $\kappa - 1$ elements of the sequence, for any value κ.

It is now easy to show that, as a consequence of (3), a probability of the form (6) satisfies the special theorem of multiplication. Using the general theorem of multiplication, we have:

$$\begin{aligned} P(A.S^1_{\lambda\kappa},B^1_{i_1}\ldots B^\lambda_{i_\lambda}) &= P(A.S^1_{\lambda\kappa},B^1_{i_1}\ldots B^{\lambda-1}_{i_{\lambda-1}}) \\ &\quad\cdot P(A.S^1_{\lambda\kappa}.B^1_{i_1}\ldots B^{\lambda-1}_{i_{\lambda-1}},B^\lambda_{i_\lambda}) \\ &= P(A.S^1_{\lambda\kappa},B^1_{i_1}\ldots B^{\lambda-1}_{i_{\lambda-1}})\cdot P(A,B_{i_\lambda}) \end{aligned} \tag{9}$$

$$1 \leqq \kappa \leqq \lambda$$

Here we have applied (3) to the term on the second line. Repeating the procedure, we obtain

$$\begin{aligned} P(A.S^1_{\lambda\kappa},B^1_{i_1}\ldots B^{\lambda-1}_{i_{\lambda-1}}) &= P(A.S^1_{\lambda\kappa},B^1_{i_1}\ldots B^{\lambda-2}_{i_{\lambda-2}}) \\ &\quad\cdot P(A.S^1_{\lambda\kappa}.B^1_{i_1}\ldots B^{\lambda-2}_{i_{\lambda-2}},B^{\lambda-1}_{i_{\lambda-1}}) \\ &= P(A.S^1_{\lambda\kappa},B^1_{i_1}\ldots B^{\lambda-2}_{i_{\lambda-2}})\cdot P(A,B_{i_{\lambda-1}}) \end{aligned} \tag{10}$$

The last step, once more, is covered by (3). By further iteration of the procedure we arrive at the result

$$P(A.S^1_{\lambda\kappa},B^1_{i_1}\ldots B^\lambda_{i_\lambda}) = P(A,B_{i_1})\cdot\ldots\cdot P(A,B_{i_\lambda}) \tag{11}$$

Thus we have shown that normal sequences satisfy the special theorem of multiplication for counting by sections. This result goes beyond (5, § 29) because the latter relation refers only to counting with overlapping. That normal sequences satisfy also (5, § 29) follows because the definition of these sequences includes the condition that they be free from aftereffect.

The normal sequence was defined, not by specifying its total domain of invariance, but by stipulating only a certain condition for the domain of invariance. There is an important reason for such a procedure: it is virtually impossible to specify completely the domain of invariance of a sequence. The normal sequences dealt with in practical applications possess, in general, a domain of invariance wider than that stipulated here. But whether a certain selection S belongs to the domain of invariance of a sequence is a question that must be investigated for each selection separately. The answer to the question is often an important task of empirical science.

For example, we form the sequence of adult male inhabitants (case A) of a metropolis, say, according to the order in which their names are given in the city directory (the alphabet then defines the subscript sequence); and we note whether or not the person suffers from tuberculosis (case B). Then a normal sequence is obtained for B, since here a selection by predecessor or by regular division does not result in a different frequency. Should we make a selection, however, by considering only the inhabitants of certain quarters of poor housing conditions, we would obtain a higher frequency for B. In this case the major sequence contains a different selection; but it will correspond, not only to the definition (3), but to the ordinary linguistic usage if we nevertheless call it a normal sequence.

A sequence is completely and individually characterized only by stating its domain of invariance; but whether it is a normal sequence must be determined on the basis of a general definition. For this reason, the definition of the normal sequence is given in terms of a minimum condition for the domain of invariance. The regular divisions were chosen for this purpose because they are connected with the validity of the special theorem of multiplication in enumeration by sections. Moreover, the domain of regular divisions can be applied usefully to some nonnormal sequences also.

This definition of the normal sequence corresponds to one given by A. H. Copeland,[1] who defines the normal sequence by the postulate of the invariance of the limit of the frequency in subsequences selected by regular divisions, with the use of an additional postulate requiring the complete mutual independence of the subsequences.[2] It can easily be shown that Copeland's definition is equivalent to the one given above, for which the independence of the subsequences is a derivable theorem, whereas the invariance condition is extended to include phase probabilities. Furthermore, Copeland has been able to show that it is possible to construct normal sequences by means of mathematical rules—a highly satisfactory result, which gives the theory of normal sequences a secure position within the mathematics of infinite sequences.

The various forms of definitions of normal sequences were preceded by a definition of random sequences established by R. von Mises. His theory was the first to characterize the structural order of sequences by means of postulates concerning the limit of the frequency in subsequences. The specific postulate used, however, differs from the one used for normal sequences in that it establishes much stronger requirements.

The theory developed by von Mises can be summarized as follows. The notion of *place selection* is defined as a selection by any rule that does not make use of the attribute of the element selected, though reference may be made to attributes of other elements. The regular divisions are examples of

[1] "Admissible Numbers in the Theory of Probability," in *Amer. Jour. Math.*, Vol. 50, No. 4 (1928), p. 535. Instead of my term "normal sequences", Copeland uses the term "admissible numbers". He uses as attributes the numbers 1 and 0 instead of B and $\bar{B}$, and thus regards every sequence as a dyadic fraction defining a number. Copeland's publication precedes my own first publication of my definition of normal sequences (1932) by several years; but it was unknown to me at that time. I am glad to recognize his priority with respect to the definition of normal sequences.

[2] I used this definition in my first publication on this subject: *Math. Zs.*, Vol. 34 (1932), p. 603. I prefer the definition given in this book (and also in the German original of this book) because it can be extended to the case of certain nonnormal sequences (see § 33).

place selections. Other illustrations are given by the selections, "every element the subscript of which is a prime number", or "every element that has an element B as predecessor". But the selection, "every element that has the attribute B", is not a place selection, because it makes use of the attribute B of the element selected.

Von Mises now introduces the postulate that any place selection that is definable by a rule belongs to the domain of invariance. This postulate is called the *principle of randomness*; a sequence thus defined is called a *collective*. He further shows that a collective defined in this manner can never be given by a mathematical rule. He intends thus to formulate the peculiar feature of randomness that is exhibited by sequences produced by events in nature. In particular he wishes to express the fact that it is impossible to construct a *gambling system*, that is, a method of selection by which a gambler will, on the whole, gain money.[3]

The definition of randomness given by von Mises has been widely discussed. In particular, the objection was raised whether his principle of randomness is free from contradictions. It is questionable whether the phrase, "a place selection that can be defined by a rule", determines a unique class of selections. The class of selections so determined will depend on the language used; what is undefinable for one language may be definable for another.

In order to overcome this difficulty, two different ways of defining randomness can be used. The first approach is to restrict the class of selections to a well-defined class; this method leads to a *restricted randomness*. The plan of defining a *logical randomness*, that is, of identifying randomness with the impossibility of a linguistic formulation of deviating selections, is then abandoned. A restricted randomness is defined in terms of a restricted domain of invariance. Of this kind are the definitions of normal sequences given by A. H. Copeland and the author; the theories developed by K. Dörge,[4] A. Wald,[5] and Jean Ville[6] belong in the same group. The latter theories differ from the former in that the domain of invariance is defined differently. Thus Dörge

[3] The first publication by von Mises of his theory was made in *Math. Zs.*, Vol. V (1919), p. 57. In this publication he stipulated the further condition that other collectives—that is, other sequences produced by nature, if taken as selections—should belong to the domain of invariance of the sequence. This condition was dropped later—see his *Vorlesungen aus dem Gebiete der angewandten Mathematik*, Vol. I: *Wahrscheinlichkeitsrechnung* . . . (Leipzig, 1931), p. 12—and a differentiation between dependent and independent collectives (*ibid.*, p. 93) was carried through. My objections to the principle of randomness—"Axiomatik der Wahrscheinlichkeitsrechnung," in *Math. Zs.*, Vol. 34 (1932), p. 594—were, in part, directed against this earlier formulation, but they also expressed the criticism given here. It may be emphasized again that I consider the foundation of the probability calculus given by von Mises as a great advance in the construction of the calculus. Only the principle of randomness seems to me to be a condition that requires revision. (See the remarks below.)

[4] *Math Zs.*, Vol. 32 (1930), p. 232.

[5] *Die Widerspruchsfreiheit des Kollektivbegriffs, Ergebnisse eines mathematischen Kolloquiums* (Vienna), Vol. VIII (1937).

[6] "Etude critique de la notion de collectif," in *Monographies des probabilités* (ed. by Emile Borel; Paris, 1939).

used a definition that begins with a certain set of selections and then includes in the domain of invariance all iterations of these selections, that is, all selections that can be constructed by any combination of a number of these selections. Wald introduced the notion of a collective relative to a certain class of selections, and then investigated the class of sequences that can possibly satisfy such a definition. His results are particularly interesting when sequences with a continuous attribute space (see § 41) are used. The investigations of Ville concerned, in particular, the question whether it is possible to construct a gambling system for a collective as defined by Wald. Ville's findings were that the exclusion of a gambling system cannot be postulated universally, but must be formulated with respect to a class of systems of wagering.

The method developed by Copeland and the author differs from all the others in that it uses only certain minimum conditions for the domain of invariance and leaves it open whether further kinds of selections are to be included. This method has the advantage that, when a certain theorem is to be derived, no more presuppositions are used than are necessary for the derivation. Thus, when the validity of the special theorem of multiplication in enumeration by sections is to be proved, it is irrelevant whether the sequences possess more invariant selections than those specified for a regular domain of invariance. For instance, it is irrelevant whether the subsequence of elements the place number of which is a prime number has a different limit of the frequency.

The second approach takes up the challenge of defining a *logical randomness*, but attempts to avoid the difficulties of the original definition given by von Mises. In this group belong certain remarks made by Wald.[7] In particular, however, the definitions developed by Alonzo Church[8] must be mentioned. Using his notion of *effective calculability*, developed in continuation of Gödel's theorem about certain limitations of deducibility,[9] Church arrived at a definition of a random sequence that may perhaps be regarded as the solution of the problem, that is, as a definition of randomness in terms of pure logic. It is certainly a remarkable result that such a definition can be given. The complexity of the subject makes it impossible to explain here the methods used by Church, which refer to some of the most intricate developments of symbolic logic.

The significance of randomness definitions of the latter kind is chiefly in the realm of logic and mathematics. So far as the applied calculus of probability is concerned, all requirements can be satisfied by simpler methods. The methods used in the definition of normal sequence are sufficient as a basis for the mathematical treatment of randomness used in practical appli-

[7] A. Wald, *op. cit.*, p. 47.
[8] *Bull. Amer. Math. Soc.*, Vol. 46 (1940), p. 130.
[9] Compare the presentation of this theorem and of Church's methods in D. Hilbert and P. Bernays, *Grundlagen der Mathematik* (Berlin, 1939), Vol. II, § 5, and Supplement II.

cations. If the term *randomness* is to include further requirements, they can be stated by the use of *selections of physical reference*—selections not defined by mathematical rules, but by reference to physical (or psychological) occurrences. For practical statistics it is as important to know which physical selections belong to the domain of invariance as it is to know which mathematical selections are contained in this domain. If a sequence possesses randomness of the von Mises–Church type, there may still be physical selections that lead to a deviating frequency. This may be illustrated by the previous example of tuberculosis in a metropolis and a selection in terms of living quarters.

Now it is possible to formulate an equivalent of von Mises' condition of randomness in terms of physical reference, instead of the use of logical methods. Random sequences are characterized by the peculiarity that a person who does not know the attributes of the elements is unable to construct a mathematical selection by which he would, on an average, select more hits than would correspond to the frequency of the major sequence. In other words, such selections will be included in the domain of invariance. In this form, the impossibility of making a deviating selection is expressed by a psychological, not a logical, statement; it refers to acts performed by a human being. This may be called a *psychological randomness.*

For all practical purposes the psychological definition of randomness is sufficient. It has the advantage of being free from problems of consistency, and does not connect the calculus of probability with controversial problems of the theory of logical deduction. It is true that a formulation of this kind, instead of speaking of logical impossibilities, refers only to a limitation of the technical abilities of human observers. But such a psychological reference is indispensable, too, when selections in terms of physical observations are to be incorporated in the domain of invariance. We say, for instance, that observations of the initial velocity of the spinning roulette ball do not enable us to make a deviating selection of the results and that thus a selection based on such observations belongs in the domain of invariance of the sequence. But this is true only in view of the limited abilities of human observers, as far as both observation and mathematical computation are concerned. With precise observation of the initial velocity of the spinning ball, it should be possible to foretell with any degree of exactness where it will come to rest. It is only lack of technical ability that prevents us from equaling Laplace's superman. Once the velocity of the spinning ball has died down noticeably, such a computation comes into the scope of technical possibilities. This is well known to the owners of gambling places, who stop such attempts by the croupier's call, "*Rien ne va plus*".

The significance of the problem of the definition of random sequences should not be overestimated, however. Within the general calculus of probability, random sequences represent merely a special type; as in all other cases, the definition of special types is more or less arbitrary and is subject only to considerations of practical use. In actual applications, all kinds of probability sequences are encountered. Some show the features of randomness; others represent intermediate types between strictly ordered and random sequences. In the following sections, such nonnormal sequences will be considered. It would constitute a rather narrow conception of probability if the name of probability sequences were reserved for random sequences—a conception that certainly would contradict language usage. For instance, the sequence of the daily number of traffic accidents in a city will not represent a normal sequence because the seven-day period of Sundays will furnish a subsequence of a different frequency; nonetheless, this sequence will be regarded as a probability sequence. All types of probability sequences are found in nature. A mathematical theory of probability should not be restricted to the study of one specific type of sequence but should include suitable definitions of various types, chosen from the standpoint of practical use.

It is an important fact that all these special types can be dealt with by the same general system of axioms that was presented in chapter 3 and § 28. The specific conditions of a certain type can be formulated as equalities referring to certain probabilities of subsequences. All that we need, therefore, for the treatment of special types is a knowledge of whether certain probabilities are equal; the rest is derivable by the general rules of the calculus. If it is possible to ascertain the values of the probabilities of sequences, it will be possible also to know whether two probabilities are equal. The determination of the specific types of probability sequences occurring in practical applications can therefore be carried through by the same methods as those that allow us to determine whether a sequence is a probability sequence at all.

§ 31. Some Numerical Problems Referring to Normal Sequences

The characterization of normal sequences given above, though simple methods are used throughout, may appear rather complicated when regarded abstractly in its symbolic notation. A few additional considerations and examples will serve to elucidate the bearing of the special theorem of multiplication on the structure of the sequence.

We shall examine the question how often a die must be thrown so that the occurrence of face 6 in at least 1 of the throws can be expected with the probability $\frac{1}{2}$. We may use here the relation (5, § 20). We regard the occurrence of the "6" as the event B. Because of the logical equivalence

$$(B^1 \vee \ldots \vee B^n \equiv \overline{\overline{B^1} \ldots \overline{B^n}}) \tag{1}$$

and with the use of the special theorem of multiplication, the desired probability obtains in the form

$$P(A,B^1 \vee \ldots \vee B^n) = 1 - P(A,\overline{B^1} \ldots \overline{B^n}) = 1 - (1 - \tfrac{1}{6})^n \qquad (2)$$

If this probability is to be $= \frac{1}{2}$, we have the equation

$$\tfrac{1}{2} = 1 - (1 - \tfrac{1}{6})^n \qquad (3)$$

which, when solved for n, gives the value $n = 3.8$. Since n can be only an integer, this result means that the occurrence of face 6 cannot be expected for 3 throws with a probability $\frac{1}{2}$; but for 4 throws it can be expected with a probability greater than $\frac{1}{2}$. We can ask the same question for the case that, throwing with two dice, we obtain doublets, for example, the combination 6.6; we then substitute in (3) $\frac{1}{36}$ for $\frac{1}{6}$ and obtain, solving, for n, the value $n = 24.6$. Only after at least 25 throws, therefore, can we expect[1] the double "6" with a probability greater than $\frac{1}{2}$.

Jacob Bernoulli showed that these numerical values lead to a paradox when treated without due caution.[2] Imagine a series of 600 throws carried out with one die; divide the series into sections of 4 throws each. The value 3.8 can be replaced approximately by 4, and thus we may expect a "6" in every second group. There are 150 groups, and therefore we should find face 6 about 75 times. This contradicts another consideration: since the probability of obtaining face 6 is $\frac{1}{6}$, we should expect "6" 100 times within 600 throws. How can the contradiction be explained?

The error does not result from the approximative replacement of 3.8 by 4; the small inaccuracy cannot account for so great a deviation. It lies rather in the careless manipulation of the *inclusive* "or". What follows from (3) is that among 4 throws *at least* one "6" is to be expected with the probability $\frac{1}{2}$. Therefore, on the average, a "6" will occur in every second section; but some sections will have *several* values "6", so that altogether there are 100 results of "6". This makes clear a peculiar consequence to which the special theorem of multiplication leads, so far as the structural order of normal sequences is concerned. The theorem requires that a certain *clustering* of the results must occur. When the series is divided into sections of 4, there will be, on the whole, no "6" in every second section; but in the other sections the "6" will be found accumulated. Should we distribute the 100 values "6" artificially in an even density over the 600 throws, we would not obtain the normal frequency distribution. The random distribution differs from a distribution by artificial equalization; it gives not only the statistical frequency required for the *major probability*, but, simultaneously, those required for the

[1] This example played an important role in the history of the probability calculus. It represents the question that the Chevalier de Méré asked Pascal and that stimulated this mathematician to give the first scientific treatment of the probability calculus.

[2] *Ars conjectandi* (Basel, 1713), Part 1, chap. x, p. 29.

many *minor probabilities* that hold for the possible combinations of results. Persons not acquainted with mathematics often fail to understand this feature of probability sequences and are astonished at the clustering that occurs. The phrase, "Like attracts like", was coined in order to account for *runs* of pleasant or unpleasant happenings. If a person not trained in the theory of probability were asked to construct artificially a series of events that seems to him to be well shuffled, there would not be enough *runs* in it. (It would be a sequence with subnormal dispersion; see § 52.) Only if the sequence is not normal can the clustering be stronger than is compatible with the special theorem of multiplication; tests of the normal character of a sequence are, in fact, often based on the counting of clustering. Normal sequences are characterized by *normal clustering.* Here the clusters are not very long, and the clustering disappears when larger sections are considered.

In this connection should be mentioned another paradox to which careless inferences may lead. According to the multiplication theorem, long runs, that is, consecutive occurrences of the same event, are comparatively rare. The probability of a series of 5 results of "6" is equal to $\frac{1}{6^5} = \frac{1}{7{,}776}$, which is a very small number. If such a series once occurs, many people are inclined to believe that the occurrence of a further "6" is now more improbable than in other cases. They argue that the appearance of another "6" would produce the very improbable case of a series of 6 results of "6", to which only the probability $\frac{1}{6^6} = \frac{1}{46{,}656}$ can be assigned. But this statement contradicts another consideration. In every throw, the probability of the "6" must be $\frac{1}{6}$, because the preceding throws cannot influence the later ones. The paradox is resolved by recognizing that, for instance, the occurrence of a "5" in the sixth throw makes the present 5 values "6" into a sequence 6-6-6-6-6-5, the probability of which is likewise only $\frac{1}{6^6}$. Thus it is clear that after a series of 5 values "6" there are no different conditions for the "6" than for any other number. Therefore the statement is correct that even in this case the original probability $\frac{1}{6}$ will hold for the occurrence of the "6". Every series that actually occurs is extraordinarily improbable, since its probability is calculated as the product of factors smaller than 1. If a series of 6 values "6" is regarded as particularly improbable, this will be correct when it is compared, not to any *definite* different sequence, but to the case of *any other* sequence, that is, to an or-combination of other sequences. But if 5 values "6" once have occurred, a great number of possible cases is therewith excluded; and there remain only 6 cases given by the occurrence of a *1*, or *2*, . . . or *6* for the sixth throw. Thus the situation is exactly the same as in the first throw.

Apparently it is not easy to acquiesce in this conclusion. At the "green table" it can frequently be observed that, after the occurrence of a long run, the gambler is inclined to believe in the prospect of a change rather than in the continuation of the run. It requires some theoretical training not to submit to the suggestive power of this fallacy. The case may be compared to optical illusions, which we know to be misleading without being able to free ourselves from the compulsion of the optical image.

§ 32. Mutually Dependent Normal Sequences

The definition of the normal sequence in § 30 is given in terms of an internal property of the sequence; it states nothing about the relation of the normal sequence to other sequences. Thus two normal sequences may be mutually dependent. In this case the special theorem of multiplication holds with respect to the succession of elements within each sequence, but not for the combination of the two sequences. This makes clear that the special theorem need not hold true for all combinatory properties of sequences, but that there will be cases in which the special theorem is used for certain operations, whereas the general theorem of multiplication is used for others.

A trivial example for the case considered is given by two normal sequences that are exactly equal. If, in throwing with two dice, the same face should always appear on both dice, each of the sequences could be normal, nevertheless. Among the combinations, however, doublets would have the probability $\frac{1}{6}$, but the combination of unequal faces would have the probability 0, a result that obviously contradicts the special theorem of multiplication. This extreme case is of little interest because it does not occur in actual practice. A different example will therefore be supplied. The explanation of the physical mechanisms by which such sequences can be produced will be postponed to the end of the section. For the moment we consider only the structure of the sequences given:

$$
\begin{array}{l}
\bar{B}\,\bar{B}\,B\,B\,\bar{B}\,\bar{B}\,B\,\bar{B}\,B\,\bar{B}\,B\,\bar{B}\,B\,\bar{B}\,B\,\bar{B}\,\bar{B}\,B\,B\,B\,B\,B\,\bar{B}\,\bar{B}\,\bar{B} \\
\bar{C}\,\bar{C}\,C\,C\,C\,C\,C\,\bar{C}\,\bar{C}\,\bar{C}\,C\,\bar{C}\,C\,\bar{C}\,C\,C\,\bar{C}\,C\,C\,\bar{C}\,C\,C\,C\,\bar{C}\,C \\
\\
B\,B\,B\,\bar{B}\,\bar{B}\,B\,B\,\bar{B}\,B\,B\,B\,\bar{B}\,B\,B\,B\,\bar{B}\,\bar{B}\,B\,B\,\bar{B}\,B\,\bar{B}\,\bar{B}\,B\,B \\
C\,\bar{C}\,\bar{C}\,\bar{C}\,C\,C\,C\,\bar{C}\,C\,C\,C\,\bar{C}\,C\,\bar{C}\,\bar{C}\,\bar{C}\,\bar{C}\,\bar{C}\,C\,\bar{C}\,C\,C\,C\,C\,C \\
\\
B\,\bar{B}\,B\,B\,B\,\bar{B}\,\bar{B}\,\bar{B}\,\bar{B}\,\bar{B}\,\bar{B}\,B\,B\,B\,B\,\bar{B}\,\bar{B}\,B\,\bar{B}\,B\,B\,\bar{B}\,B\,B\,B \\
\bar{C}\,C\,C\,C\,C\,\bar{C}\,C\,\bar{C}\,C\,\bar{C}\,C\,\bar{C}\,\bar{C}\,\bar{C}\,\bar{C}\,C\,C\,\bar{C}\,\bar{C}\,C\,\bar{C}\,C\,C\,C\,\bar{C} \\
\\
\bar{B}\,\bar{B}\,\bar{B}\,B\,\bar{B}\,\bar{B}\,B\,\bar{B}\,B\,B\,B\,\bar{B}\,B\,B\,\bar{B}\,\bar{B}\,B\,\bar{B}\,\bar{B}\,\bar{B}\,B\,\bar{B}\,\bar{B}\,\bar{B}\,\bar{B} \\
\bar{C}\,\bar{C}\,\bar{C}\,\bar{C}\,\bar{C}\,\bar{C}\,C\,\bar{C}\,C\,\bar{C}\,C\,\bar{C}\,\bar{C}\,C\,\bar{C}\,\bar{C}\,C\,\bar{C}\,C\,\bar{C}\,C\,\bar{C}\,\bar{C}\,\bar{C}\,\bar{C}
\end{array}
\qquad (1)
$$

The two sequences, consisting of 100 elements each, are to be read continuously from left to right; the left end of a row of the B-sequence is to be connected to the right end of the preceding row of the B-sequence. The B-sequence contains 51 B's, 49 $\bar{B}$'s; the C-sequence has 49 C's, 51 $\bar{C}$'s; the probability, therefore, is virtually equal to $\frac{1}{2}$. That the sequences are normal may be shown by enumeration of the selection given by the first predecessor. Among the 51 elements of the B-sequence having the predecessor B are 26 B's, 25 $\bar{B}$'s, so that the probability $\frac{1}{2}$ is reasonably well satisfied. Among the 48 elements with the predecessor $\bar{B}$ are 25 B's, 23 $\bar{B}$'s, which also corresponds well to the probability $\frac{1}{2}$. The situation is similar for the C-sequence. Among the 49 elements with the predecessor C are 23 C's, 26 $\bar{C}$'s. Among the 50 elements with the predecessor $\bar{C}$ are 26 C's, 24 $\bar{C}$'s. Thus the probability $\frac{1}{2}$ obtains everywhere in good approximation. I do not wish to investigate the normal character further and am satisfied with the result that the sequences at least have the property of being normal with respect to the first predecessor.

However, we see at a glance that the frequency of the combination $B.C$ is not determined by the product $\frac{1}{2} \cdot \frac{1}{2} = \frac{1}{4}$; on the contrary, the combinations $B.C$ as well as $\bar{B}.\bar{C}$ are more frequent than the others. We count 33 $B.C$, 18 $B.\bar{C}$, 16 $\bar{B}.C$, 33 $\bar{B}.\bar{C}$. In comparison with the total number of 100 elements for each sequence, the frequency of the combinations $B.C$ and $\bar{B}.\bar{C}$ thus results as about $\frac{1}{3}$; the frequency of combinations $B.\bar{C}$ and $\bar{B}.C$ as about $\frac{1}{6}$. If we introduce the corresponding probabilities within the subsequences instead of using the probabilities of the combinations, we have as a result

$$
\begin{array}{llll}
P(A,B) = \frac{51}{100} \sim \frac{1}{2} & & P(A,C) = \frac{49}{100} \sim \frac{1}{2} & \\
P(A.B,B^1) = \frac{26}{51} \sim \frac{1}{2} & & P(A.\bar{B},B^1) = \frac{25}{48} \sim \frac{1}{2} & \\
P(A.C,C^1) = \frac{23}{49} \sim \frac{1}{2} & & P(A.\bar{C},C^1) = \frac{23}{50} \sim \frac{1}{2} & (2) \\
P(A.B,C) = \frac{33}{51} \sim \frac{2}{3} & & P(A.C,B) = \frac{33}{49} \sim \frac{2}{3} & \\
P(A.\bar{B},C) = \frac{16}{49} \sim \frac{1}{3} & & P(A.\bar{C},B) = \frac{18}{51} \sim \frac{1}{3} &
\end{array}
$$

A stands here for the first term, the reference class, which is compact for the two sequences and is not written down in (1). The first three lines of (2) correspond to the properties of normal sequences; the last two lines express a mutual dependence of the two sequences. Consider, for instance, a roulette wheel the inner part of which is covered so that the ball can come to rest only on the outer part of the sectors. If we use two balls connected by a piece of string that is much shorter than the width of the sectors, the balls will usually stop on fields of the same color. Only occasionally will they come to rest so that the string crosses the boundary of two sectors. The resulting sequences will be of type (1). The individual sequences will be normal, but if one ball lies on "black", the other ball will usually come to rest on "black" also, and the corresponding result will hold for "red".

In this example the probabilities have been chosen so that the disjunctions $B \vee \bar{B}$ and $C \vee \bar{C}$ possess equal probability for each of their terms. The preceding considerations are, of course, independent of this special case. But the case must be studied more closely because of its practical importance.

Consider the disjunctions $B_1 \vee \ldots \vee B_r$, $C_1 \vee \ldots \vee C_r$, which contain the B_i and C_k as equally probable cases. The fact that the special theorem of multiplication holds for combinations of these terms leads to an analogous equiprobability for all possible combinations. The mutual independence of the terms of such disjunctions is identical with the equiprobability of all combinations $B_i . C_k$. This can be derived immediately from[1]

$$P(A,B_1) = \ldots = P(A,B_r) = P(A,C_1) = \ldots = P(A,C_r) \tag{3}$$

$$P(A,B_i . C_k) = P(A,B_i) \cdot P(A,C_k) \tag{4}$$

It is of interest to note that another assumption can be used for the combination of terms, which can also be regarded as expressing an equiprobability of combinations, though in a different sense. In (4) the probability of a combination $B_i . C_i$ is assumed as equal to the probability of a combination $B_i . C_k$; the probability of a disjunction $B_i . C_k \vee B_k . C_i$ is then twice as large. If we throw with two dice, for instance, the probability of obtaining face 6 with both dice is equal to $\frac{1}{36}$; the probability of obtaining "5" with the first die and "6" with the second is likewise equal to $\frac{1}{36}$. If we speak of the combinations 6.6 and 5.6, respectively, without specifying on which die each individual face shows up, then the combination 5.6 is twice as probable as the combination 6.6; for 5.6 can result in two different ways, which may be symbolized by $B_5 . C_6 \vee B_6 . C_5$, whereas 6.6 has only the form $B_6 . C_6$. Combinations not specifying to which individual die the term appertains may be called *nonindividualized combinations.* When such combinations are used, no attention is paid to individual differences between the combinations $B_i . C_k$ and $B_k . C_i$. Now it is possible to carry through the assumption that nonindividualized combinations constitute equiprobable cases. This idea, which, of course, leads to a different kind of statistics, deserves some further investigation.

The assumption is satisfied by the conditions

$$\begin{aligned} P(A,B_i) &= \frac{1}{r} & P(A,C_k) &= \frac{1}{r} \\ P(A . B_i,C_i) &= \frac{2}{r+1} & P(A . C_i,B_i) &= \frac{2}{r+1} \\ P(A . B_i,C_k) &= \frac{1}{r+1} \quad i \neq k & P(A . C_k,B_i) &= \frac{1}{r+1} \quad i \neq k \end{aligned} \tag{5}$$

[1] The combinations will also be equally probable when the two disjunctions have different numbers of terms. I shall not discuss this case here, however.

These equations formulate the equiprobability of nonindividualized combinations, as can be seen from the relations

$$P(A,B_i.C_k \vee B_k.C_i) = P(A,B_i.C_k) + P(A,B_k.C_i)$$
$$= P(A,B_i) \cdot P(A.B_i,C_k) + P(A,B_k) \cdot P(A.B_k,C_i)$$
$$= \frac{1}{r} \cdot \frac{1}{r+1} + \frac{1}{r} \cdot \frac{1}{r+1} = \frac{2}{r(r+1)} \quad i \neq k \tag{6a}$$

$$P(A,B_i.C_i) = P(A,B_i) \cdot P(A.B_i,C_i) = \frac{1}{r} \cdot \frac{2}{r+1} = \frac{2}{r(r+1)} \tag{6b}$$

For the case of the die, we would then have the value $\frac{2}{7}$ for the probability of obtaining a "6" on the second die when the first shows a "6"; we would have the value $\frac{1}{7}$ for the probability of obtaining a "5" on the second die when the first shows a "6". The combinations $B_i.C_k$ are here only half as probable as the combinations $B_i.C_i$, so that equiprobability obtains only for nonindividualized combinations. We thus have the probability $\frac{1}{21}$ for every nonindividualized combination, that is, for a combination 6.6 as well as for 6.5.

It remains to demonstrate that equations (5) are free from contradictions. Such proof is necessary because the assumption that nonindividualized combinations are equally probable contains overdeterminations. Now the proof is easily given. We see that the relations

$$\sum_{k=1}^{r} P(A.B_i,C_k) = 1 \qquad \sum_{i=1}^{r} P(A.C_k,B_i) = 1 \tag{7}$$

follow from (5); this condition is required because of the completeness of the disjunction. Furthermore, we derive from (5), using the elimination theorem (21, § 19),

$$P(A,C_k) = \sum_{i=1}^{r} P(A,B_i) \cdot P(A.B_i,C_k)$$
$$= P(A,B_i) \cdot \sum_{i=1}^{r} P(A.B_i,C_k) = P(A,B_i) \tag{8}$$

Here we used the relation

$$\sum_{i=1}^{r} P(A.B_i,C_k) = 1 \tag{9}$$

which follows from (5). Although formula (9), in general, is not a necessary condition, the probabilities $P(A.B_i,C_k)$ being nonbound probabilities, it must be required in this case because otherwise (8) would lead to contradictions. The result (8) shows the admissibility of the conditions given in (5) for the $P(A,C_k)$. Finally, we derive from the rule of the product (6, § 12) that

$$P(A.C_k,B_i) = \frac{P(A,B_i)}{P(A,C_k)} \cdot P(A.B_i,C_k) = P(A.B_i,C_k) \qquad (10)$$

The last step follows because of $P(A,B_i) = P(A,C_k)$. Thus the remaining part of equations (5) is also proved to be free from contradictions.

This concludes the proof that it is logically admissible to assume that nonindividualized combinations are equally probable. The assumption, however, entails a mutual dependence of the terms B_i and C_k; for (5) contradicts (4, § 14). This is of interest, for, in recent times, quantum theory has made use of such an assumption in the *Einstein-Bose statistics*. The given analysis shows that the equiprobability of nonindividualized combinations, assumed in this statistics, must be interpreted as a physical dependence. It is seen, furthermore, that it is pointless to adduce *a priori* reasons with the intention of deciding whether the individualized or the nonindividualized combinations should be regarded as equiprobable. All that can be asked is whether the events considered are dependent or independent; and this is an empirical question.

The relations (5) enable us to construct physical mechanisms that supply sequences in which the nonindividualized combinations are equally probable. For the sake of simplicity we use a disjunction of only two terms. The conditions (5) can then be realized if we have a mechanism for the probability $\frac{1}{2}$, another for the probability $\frac{1}{3}$, and a third for the probability $\frac{2}{3}$. For the first mechanism a die may be chosen when we use only the results *even* and *odd*. The other two mechanisms can be constructed by means of colored dice without eyes and with the following properties: one die has two black faces and four white ones; the other, four black faces and two white ones. Call the first the *white die;* the other, the *black die.* For the white die the probability of obtaining white equals $\frac{2}{3}$, and the probability of black equals $\frac{1}{3}$; for the black die the situation is the reverse.

We proceed with the game as follows. Throw first with the numbered die and note the result as B if an even number appears, as $\bar{B}$ if an odd number appears. Then throw with one of the colored dice according to the following rule: if the result of the numbered die was B, throw with the black die; if the result of the numbered die was $\bar{B}$, use the white die. The appearance of a black face may be designated by C. After each set of two throws, consisting of one throw with the numbered die and one with a colored die, a new set is played according to the same rules. In this manner we coördinate to every B and $\bar{B}$ of the numbered die a corresponding C or $\bar{C}$. It follows at once from our knowledge of the properties of a die that the sequence of the B and $\bar{B}$ thus obtained is normal. That the sequence of the C and $\bar{C}$, obtained indirectly, is also normal, follows from (8); each individual element of this sequence is independently played for, with the probability $\frac{1}{2}$, when we regard as the starting point the situation before the throwing of the numbered die. In this manner example (1) was actually produced. It therefore illustrates the equiprobability of nonindividualized combinations. If we count from this viewpoint, we find in (1) 33 combinations $B.C$, 33 combinations $\bar{B}.\bar{C}$, and 34 mixed combinations $B.\bar{C} \vee \bar{B}.C$. We see that the equiprobability of the combinations is satisfied reasonably well.

These considerations show that the order of the individual sequence is independent of its relations to other sequences. The frequency within each individual sequence of (1), as well as its internal order, is not influenced by the fact that homogeneous combinations have a higher probability than mixed ones, taken individually. Dependence is expressed solely by a relation between the sequences, not by properties of individual sequences. The technique of the probability calculus, therefore, enables us to formulate dependence as a relation between sequences without submitting the individual sequence to any special conditions in regard to structure.

The example can be used to illustrate another feature exhibited by the relation of the dependence of sequences. It was shown above (§ 23) that the independence of sequences is not a transitive relation. It is now easy to show by an example that even the normal character of sequences does not change this result. If we add to the sequences B and C in (1) another die sequence D, obtained by playing independently with another die, noting down an even result as D and an odd result as $\bar{D}$, then B is independent of D, and D is independent of C. However, as shown in the example, B is not independent of C.

§ 33. Probability Transfer

We now turn to the analysis of nonnormal sequences and thus of cases where the probability of an element depends on its predecessors. The simplest and most important type is a sequence in which the immediate predecessor alone determines the probability of an element, whereas the other predecessors are irrelevant. Such sequences were first studied by Markoff and are often called *Markoff chains.* I shall use the name of *sequences with probability transfer.*

These sequences may be illustrated by an example, constructed with the help of the white and black dice described at the end of § 32. Starting with any of the two dice, we observe the following rule: if black occurs, the next throw is made with the black die; if white appears, the white die is used for the following throw. This is continued; black and white are taken as B and $\bar{B}$ within one sequence. In general, when a certain result has been obtained, the probability of the same result in the next throw equals $\frac{2}{3}$; that of a different result is equal to $\frac{1}{3}$. The probability transfer has here the character of a *drag;* there exists a tendency to stay, and the change in events is delayed.

The opposite case is obtained when we reverse the rule: if black occurs, we proceed to play with the white die; if white appears, we play the next throw with the black die. The probability transfer possesses here the character of a *compensation;* there exists a tendency to alternate, and the change in events is speeded up.

It is obvious that, on the average, the game will equal a game played with the probability $\frac{1}{2}$, such as obtaining for a die with three white and three black faces, or for heads and tails of a coin. The order of the sequence, however, will be different, because it is not determined by the special theorem of multiplication, but in a more complicated manner.

Two examples of sequences obtained by the procedure described are given. Example (1) is a sequence with drag, (2) a sequence with compensation.

$$\begin{array}{l}
B\,B\,B\,B\,\bar{B}\,B\,B\,B\,\bar{B}\,\bar{B}\,\bar{B}\,\bar{B}\,B\,\bar{B}\,\bar{B}\,\bar{B}\,\bar{B}\,\bar{B}\,\bar{B}\,B\,B\,B\,B\,\bar{B}\,\bar{B}\,\bar{B}\,\bar{B}\,\bar{B}\,B\,\bar{B} \\
B\,\bar{B}\,B\,B\,B\,B\,B\,B\,B\,B\,B\,B\,B\,\bar{B}\,\bar{B}\,\bar{B}\,\bar{B}\,\bar{B}\,B\,B\,\bar{B}\,B\,B\,\bar{B}\,\bar{B}\,\bar{B}\,\bar{B}\,\bar{B}\,B\,B \\
\bar{B}\,B\,B\,\bar{B}\,B\,B\,B\,\bar{B}\,B\,B\,B\,B\,B\,B\,B\,B\,B\,\bar{B}\,\bar{B}\,\bar{B}\,\bar{B}\,\bar{B}\,\bar{B}\,B\,\bar{B}\,\bar{B}\,\bar{B}\,\bar{B}\,B\,\bar{B} \\
\bar{B}\,B\,\bar{B}\,\bar{B}\,B\,B\,\bar{B}\,\bar{B}\,\bar{B}\,B\,B\,B\,B\,\bar{B}\,\bar{B}\,B\,B\,B\,B\,B\,B\,\bar{B}\,\bar{B}\,\bar{B}\,\bar{B}\,\bar{B}\,B\,B\,\bar{B}\,B
\end{array} \qquad (1)$$

$$\begin{array}{l}
B\,\bar{B}\,B\,B\,B\,\bar{B}\,\bar{B}\,\bar{B}\,\bar{B}\,B\,B\,\bar{B}\,B\,\bar{B}\,\bar{B}\,\bar{B}\,B\,B\,\bar{B}\,B\,\bar{B}\,\bar{B}\,B\,\bar{B}\,B\,B\,B\,B\,B\,B \\
B\,\bar{B}\,B\,B\,B\,\bar{B}\,B\,B\,\bar{B}\,B\,\bar{B}\,\bar{B}\,B\,B\,\bar{B}\,\bar{B}\,B\,\bar{B}\,\bar{B}\,B\,B\,B\,\bar{B}\,B\,\bar{B}\,B\,\bar{B}\,B\,\bar{B}\,B \\
\bar{B}\,B\,\bar{B}\,B\,\bar{B}\,B\,\bar{B}\,B\,\bar{B}\,\bar{B}\,B\,\bar{B}\,B\,\bar{B}\,\bar{B}\,\bar{B}\,B\,\bar{B}\,B\,\bar{B}\,B\,B\,\bar{B}\,\bar{B}\,B\,B\,\bar{B}\,B\,B\,\bar{B} \\
\bar{B}\,\bar{B}\,\bar{B}\,\bar{B}\,B\,\bar{B}\,B\,\bar{B}\,B\,\bar{B}\,\bar{B}\,B\,\bar{B}\,\bar{B}\,\bar{B}\,B\,B\,\bar{B}\,B\,\bar{B}\,B\,\bar{B}\,B\,B\,\bar{B}\,B\,B\,\bar{B}\,\bar{B}\,B
\end{array} \qquad (2)$$

It may be seen at a glance that probability drag, as represented by (1), is characterized by relatively long runs, that is, successions of equal results; and that probability compensation, as represented by (2), is characterized by comparatively frequent alternations, that is, transitions from one result to the other. This fact finds its expression in the following statistics. We have for (1)

$$\begin{array}{llll}
P(A,B) = \frac{63}{120} \sim \frac{1}{2} & & P(A,\bar{B}) = \frac{57}{120} \sim \frac{1}{2} & \\
P(A.B,B^1) = \frac{42}{62} \sim \frac{2}{3} & & P(A.\bar{B},B^1) = \frac{20}{57} \sim \frac{1}{3} & \quad (1') \\
P(A.B,\bar{B}^1) = \frac{20}{62} \sim \frac{1}{3} & & P(A.\bar{B},\bar{B}^1) = \frac{37}{57} \sim \frac{2}{3} &
\end{array}$$

In contradistinction to these results, we obtain for (2)

$$\begin{array}{llll}
P(A,B) = \frac{61}{120} \sim \frac{1}{2} & & P(A,\bar{B}) = \frac{59}{120} \sim \frac{1}{2} & \\
P(A.B,B^1) = \frac{22}{60} \sim \frac{1}{3} & & P(A.\bar{B},B^1) = \frac{38}{59} \sim \frac{2}{3} & \quad (2') \\
P(A.B,\bar{B}^1) = \frac{38}{60} \sim \frac{2}{3} & & P(A.\bar{B},\bar{B}^1) = \frac{21}{59} \sim \frac{1}{3} &
\end{array}$$

We turn now to the theoretical treatment of this sequence type. The elimination theorem (2, § 19) gives

$$\begin{aligned} P(A.A^1,B^1) &= P(A.A^1,B) \cdot P(A.A^1.B,B^1) \\ &\quad + P(A.A^1,\bar{B}) \cdot P(A.A^1.\bar{B},B^1) \end{aligned} \qquad (3)$$

Using (6, § 28), we find

$$P(A,B) = P(A,B) \cdot P(A.B,B^1) + [1 - P(A,B)] \cdot P(A.\bar{B},B^1) \tag{4}$$

We now make use of the existence rule, which was extended in § 28 to include phase symbols. According to this rule, (4) may be considered as determining existence even if we do not know whether the two quantities $P(A.A^1,B^1)$ and $P(A.A^1,B)$ in (3) exist. Now (4) contains as an unknown quantity only $P(A,B)$, since the two probabilities $P(A.B,B^1)$ and $P(A.\bar{B},B^1)$ are given as existing. It then follows from (4) that the mean probability $P(A,B)$ exists.

The value of $P(A,B)$ may be obtained from (4). We use the abbreviations

$$P(A,B) = p \qquad P(A.B,B^1) = p_1 \qquad P(A.\bar{B},B^1) = q_1 \tag{5}$$

Then (4) can be written in the form

$$p = pp_1 + (1 - p)q_1 = \frac{q_1}{1 - p_1 + q_1} \tag{6}$$

Hereby $P(A,B)$ is expressed as a function of $P(A.B,B^1)$ and $P(A.\bar{B},B^1)$. To make this relation clearer, we introduce a notation somewhat different from (5):

$$P(A,B) = p \qquad P(A.B,B^1) = p + \epsilon \qquad P(A.\bar{B},B^1) = p - \eta \tag{7}$$

Then (6) is transformed into the relation

$$\frac{p}{1-p} = \frac{\eta}{\epsilon} \tag{8}$$

It can be seen at once that ϵ and η must have the same sign. If both are positive, there is a drag; if both are negative, there is a compensation. In figure 7 the relation (8) is illustrated for both cases. We recognize that p, in general, does not lie exactly halfway between p_1 and q_1; for $p > 1 - p$, p lies closer to p_1, for $p < 1 - p$, p is closer to q_1. Only the case $p = \frac{1}{2}$ has symmetry, as is shown by examples (1) and (2).

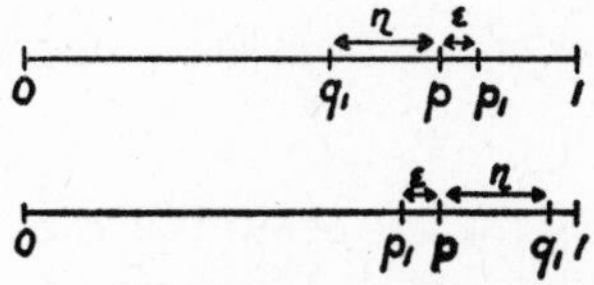

Fig. 7. Graphical representation of transfer probabilities, according to (6) and (8), for $p = \frac{3}{4}$. Upper diagram: $\epsilon = +\frac{1}{15}$, $\eta = +\frac{3}{15}$. Lower diagram: $\epsilon = -\frac{1}{15}$, $\eta = -\frac{3}{15}$.

We must investigate in which manner the probability depends on the second predecessor. It was pointed out above that for the case of probability transfer the immediate predecessor alone is relevant. The statement must now be made more precise. *Probability transfer* is defined by the condition

$$P(A.B^1_{i_1} \ldots B^{\nu-1}_{i_{\nu-1}},B^{\nu}_{i_\nu}) = P(A.B^{\nu-1}_{i_{\nu-1}},B^{\nu}_{i_\nu}) \tag{9}$$

This relation expresses the condition that the probability of an element is determined by the immediate predecessor alone. Since we have

$$P(A\,.\,B^{\nu-1}_{i_{\nu-1}},B^{\nu}_{i_\nu}) \neq P(A,B_{i_\nu}) \tag{10}$$

it follows that, for instance,

$$P(A\,.\,B^{\nu-2}_{i_{\nu-2}},B^{\nu}_{i_\nu}) \neq P(A,B_{i_\nu})$$

We find that an influence of the second and of further predecessors also exists; but this influence has its origin only in a *transfer from element to element.* For this reason the name of *probability transfer* has been chosen for these sequences. The transfer is formulated in (9) by the condition that, when the first predecessor is taken into consideration, the other predecessors no longer have any influence; they effect a deviating selection only in comparison to the major sequence. The influence upon the element immediately following is still to be felt for the succeeding elements. This is shown by the fact that in (1) the higher probability for a B as the first successor of a B results in a somewhat higher probability for a B as the second successor, though the latter probability is not so high as the former.

Analyzing the relation formally, we have, according to the elimination theorem,

$$\begin{aligned} P(A\,.\,B,B^2) &= P(A\,.\,B,B^1) \cdot P(A\,.\,B\,.\,B^1,B^2) \\ &\quad + P(A\,.\,B,\bar{B}^1) \cdot P(A\,.\,B\,.\,\bar{B}^1,B^2) \end{aligned} \tag{11}$$

Because of (9) we have the equalities

$$\begin{aligned} P(A\,.\,B\,.\,B^1,B^2) &= P(A\,.\,B^1,B^2) = P(A\,.\,B,B^1) \\ P(A\,.\,B\,.\,\bar{B}^1,B^2) &= P(A\,.\,\bar{B}^1,B^2) = P(A\,.\,\bar{B},B^1) \end{aligned} \tag{12}$$

Writing

$$P(A\,.\,B,B^2) = p_2 \tag{13}$$

we obtain from (11), with the use of the abbreviations (7),

$$p_2 = (p + \epsilon)^2 + (1 - p - \epsilon)(p - \eta) \tag{14}$$

By the use of (8) this can be transformed into

$$p_2 = p + \epsilon_1 \qquad \epsilon_1 = \epsilon \cdot \frac{\epsilon}{1 - p} \tag{15}$$

We now introduce the condition that

$$0 < p + \epsilon < 1 \qquad\qquad 0 < p - \eta < 1 \tag{16}$$

We thus exclude the degenerate case of a probability 1. Then it can be shown that $|\epsilon| < 1 - p$, and, consequently, $|\epsilon_1| < |\epsilon|$. For a positive value of ϵ it follows at once from $p + \epsilon < 1$ that $|\epsilon| < 1 - p$. For negative ϵ the inference is somewhat more complicated: either it is the case that $p \leqq 1 - p$, then we have, because of $p + \epsilon > 0$, also $p - |\epsilon| > 0$, that is, $|\epsilon| < p$ and $|\epsilon| < 1 - p$; or it is the case that $p > 1 - p$, then (8) gives $|\epsilon| < |\eta|$, and since, for a negative ϵ, η must also be negative, it follows from $p - \eta < 1$ that $p + |\eta| < 1$ and, therefore, that $|\eta| < 1 - p$ and finally $|\epsilon| < 1 - p$. So this relation is valid for any case. Combining the result with (15), we derive

$$|p_2 - p| < |p_1 - p| \tag{17}$$

This means that after the occurrence of an event B, the probability of obtaining another event B for the second successor lies closer to p than the probability of having an event B for the first successor.

Generalizing the result, we shall now prove that the quantities

$$P(A.B,B^{\nu}) = p_{\nu} \tag{18}$$

represent a sequence converging toward p. For this inference we use mathematical induction. Assuming that p_{ν} has the form

$$p_{\nu} = p + \epsilon_{\nu-1} \tag{19}$$

we derive that

$$p_{\nu+1} = p + \epsilon_{\nu} \qquad \epsilon_{\nu} = \epsilon_{\nu-1} \cdot \frac{\epsilon}{1-p} \tag{20}$$

For the proof we use the elimination theorem

$$\begin{aligned} P(A.B,B^{\nu+1}) &= P(A.B,B^{\nu}) \cdot P(A.B.B^{\nu},B^{\nu+1}) \\ &\quad + P(A.B,\bar{B}^{\nu}) \cdot P(A.B.\bar{B}^{\nu},B^{\nu+1}) \end{aligned} \tag{21}$$

Because of (9), and with the abbreviations (18) and (7), the equation can be written

$$p_{\nu+1} = p_{\nu} \cdot (p + \epsilon) + (1 - p_{\nu}) \cdot (p - \eta) \tag{21'}$$

With the help of (19) and (8) this can be transformed into (20), whereby the latter equation is proved to be valid.

Since p_2 possesses the form (19), it follows that (20) is valid for every ν. Substituting successively the value of ϵ_{ν}, we arrive at the result

$$p_{\nu} = p + \epsilon \cdot \left(\frac{\epsilon}{1-p}\right)^{\nu-1} \tag{22}$$

Since according to (16) we have $|\epsilon| < 1 - p$, the coefficient of ϵ converges toward 0 with increasing ν, so that p_{ν} converges toward p. For a positive ϵ, that is, for drag, the sequence converges from one side toward p, but for a negative ϵ, that is, for compensation, it converges alternatingly. The p_{ν} will then lie alternatively on both sides of p.

By corresponding considerations we can show that the probability of obtaining a B in the ν-th element after a $\bar{B}$ assumes the value

$$P(A.\bar{B},B^{\nu}) = q_{\nu} = p - \eta \cdot \left(\frac{\eta}{p}\right)^{\nu-1} \tag{23}$$

which likewise converges toward p.

The results may be summarized as follows. The probability of finding the attribute B in the element $y_{i+\nu}$, determined with respect to an arbitrary element y_i of the sequence (which may be either B or $\bar{B}$), converges toward p with increasing ν. The probability aftereffect dies down, therefore, and the influence of neighboring elements is obliterated. Seen from a distance, any element of the sequence has the probability p; only in the immediate environment can the influence of the predecessor be felt. Checking from this point of view, we count, for example (1), that

$$P(A.B,B^2) = \tfrac{36}{62} \qquad (24)$$

Since in this case $\epsilon = \frac{1}{6}$, we expect, theoretically, according to (15), the value $\frac{1}{2} + \frac{1}{18} = \frac{5}{9}$, which is in good agreement with (24). The number of elements in the examples is, of course, not large enough to admit an objective discrimination between such a value and the value $\frac{1}{2}$.

In spite of (24) and (15), it is the first predecessor only that determines the probability. This is seen by counting in (1) all the elements that are preceded by a group BB and comparing this number with the number of elements preceded by a group $\bar{B}B$. We find

$$P(A.B.B^1,B^2) = \tfrac{28}{42} \sim \tfrac{2}{3} \qquad P(A.\bar{B}.B^1,B^2) = \tfrac{13}{18} \sim \tfrac{2}{3} \qquad (25)$$

The probabilities are equal, and are the same as those resulting for a selection by the first predecessor alone, corresponding to (12). It should be emphasized again that these numerical values are given only as an illustration; the number of elements in the example is too small for numerical proof.

The quantity ϵ characterizes the degree of probability transfer, and may therefore be called the *degree of transfer*. A small value of ϵ represents a weak probability aftereffect; for $\epsilon = 0$ the sequence becomes free from aftereffect. In the case in which ϵ is positive and has a large value, there exists not only a high probability for B as immediate successor of B, but the high degree of transfer penetrates to further successors, so that the quantities p_ν become large, too. Because of (8), η likewise has a large and positive value. The maximum value in this case is $\epsilon = 1 - p$, that is, $p_1 = 1$. For $\eta \neq p$, that is, $q_1 \neq 0$, we then have $p = 1$ because of (6). This case may be realized, for instance, by a sequence in which only the y_i for which i is a square number belong to $\bar{B}$, all the other y_i belonging to B. Here we have $p_1 = 1$, $q_1 = 1$, $p = 1$. The sequence consisting of elements B alone, too, may be regarded as representing this case, but q_1 is indeterminate for such a sequence.

If, however, we have $q_1 = 0$ in addition to $p_1 = 1$ (that is, $\epsilon = 1 - p$ and $\eta = p$), then p is no longer determined by (6). This is a degenerate case, which would be realized by a sequence consisting of elements B only, as well as by a sequence of elements $\bar{B}$ alone. In this case (6) does not determine existence, and a limit p need no longer exist. An example may be constructed thus:

the sequence begins with 10 elements B; then follow 10^2 elements $\bar{B}$, 10^3 elements B, 10^4 elements $\bar{B}$, and so on, in an alternating fashion. Here $F^n(A,B)$ does not possess a limit, but fluctuates continuously between the limits 1 and 0; p_1 and q_1 exist, however, and we have $p_1 = 1$, $q_1 = 0$.

If ϵ is negative and has a large absolute value, we find a high probability for a change, which, according to (22), penetrates strongly to further successors, with a resulting high probability for B as second successor of B; that is, there will be a tendency toward double change. The extreme value here is $|\epsilon| = p$, to which the value $|\eta| = \dfrac{p^2}{1-p}$ corresponds; these values can be assumed, however, only if $p \leqq \frac{1}{2}$. This follows from the inequalities used to derive (17); they lead, for $p > 1 - p$ with $p - \eta \leqq 1$ (not only, as stated there, with $p - \eta < 1$), to the result $|\epsilon| < 1 - p$, which would represent a contradiction for $|\epsilon| = p$. But for $p = \frac{1}{2}$ this extreme case is possible, and the probabilities p_ν will converge also for $p < \frac{1}{2}$, according to (22). For the special case $p = \frac{1}{2}$, the strictly alternating sequence (1, § 26) represents a model; the probabilities p_ν assume here alternatively the values 0 and 1, corresponding to (22).

The example shows that there is a continuous transition from the unordered, or normal, sequences to the extremely ordered ones, corresponding to the continuous scale that is open to the degree of transfer. For the case $p = \frac{1}{2}$, $\epsilon = 0$ represents the normal sequence; and $\epsilon = -\frac{1}{2}$ supplies the strictly alternating sequence. The values of ϵ between these extremes, or between 0 and $+\frac{1}{2}$, represent intermediate types of sequences. For these reasons, a conception that excludes the extremely ordered sequences from the concept of probability sequence could hardly be regarded as consistent with the principles of a scientific terminology.

A comparison of a nondegenerate sequence that possesses probability transfer with a normal sequence of equal probability for B reveals that the sole difference is in the phase probabilities, a difference that diminishes with increasing phase length, so that the sequence gradually approaches the type of a normal sequence with respect to higher phase probabilities. It is possible to compare sequences having probability transfer with normal sequences in still another sense, when their domain of invariance is investigated. It turns out that we can even postulate a *regular domain of invariance* (see § 30) for sequences with probability transfer. We cannot, of course, tell from the definition of probability transfer as given, whether the sequences are regular-invariant. As in sequences without aftereffect, this is an additional property, which may be postulated by definition within the mathematical calculus of probability, or which must be tested by observation for empirical sequences. But there is no contradiction in combining the condition of a regular domain of invariance with that of probability transfer. This condition means, according to (1, § 30), that we have for the regular division $S_{\lambda\kappa}$ $(1 \leqq \kappa < \lambda)$ the relation[1]

$$P(A . S_{\lambda\kappa},B) = P(A,B) \tag{26a}$$

$$P(A . S^{\alpha}_{\lambda\kappa} . B^1_{i_1} \dots B^{\nu-1}_{i_{\nu-1}},B^{\nu}_{i_\nu}) = P(A . B^1_{i_1} \dots B^{\nu-1}_{i_{\nu-1}},B^{\nu}_{i_\nu}) \tag{26b}$$

Since the sequence is not free from aftereffect, the phase probability with selection $S_{\lambda\kappa}$ can be compared only with the same phase probability without selection $S_{\lambda\kappa}$; obviously the phase probability occurring in (26*b*) is not equal to $P(A,B_{i_\nu})$. This extension is made possible by the form of the definition of the domain of invariance given in (1, § 30).

Now it is possible to construct an important special case by assuming $\alpha = 1$, $\nu = \lambda + 1$, and dropping the terms $B^2_{i_2} \dots B^{\nu-1}_{i_{\nu-1}}$. Furthermore, put first B for B_{i_1} and B_{i_ν}, then $\bar{B}$ for B_{i_1}, and B for B_{i_ν}. Deducting the number 1 from the superscripts according to (5, § 28), we obtain

$$\left.\begin{aligned} P(A . S_{\lambda\kappa} . B,B^{\lambda}) &= P(A . B,B^{\lambda}) = p_\lambda \\ P(A . S_{\lambda\kappa} . \bar{B},B^{\lambda}) &= P(A . \bar{B},B^{\lambda}) = q_\lambda \end{aligned}\right\} \quad 1 \leqq \kappa \leqq \lambda \tag{27}$$

Since B and B^{λ} belong to the same subsequence selected by $S_{\lambda\kappa}$, (27) states that this subsequence is once more a sequence with probability transfer, except that here p_λ and q_λ replace the transfer probabilities p_1 and q_1 existing for the major sequence. Since p_λ and q_λ lie closer to p than do p_1 and q_1, (27) expresses the fact that the subsequences arising from a regular division are the more similar to a normal sequence, the greater λ. Even in these subsequences, only the immediate predecessor within the same subsequence is relevant to the aftereffect, in analogy to (9). This follows because with (9) and (26*b*) we have

$$\begin{aligned} P(A . S_{\lambda\kappa} . B_{i_0} . B^{\lambda}_{i_\lambda},B^{2\lambda}_{i_{2\lambda}}) &= P(A . B_{i_0} . B^{\lambda}_{i_\lambda},B^{2\lambda}_{i_{2\lambda}}) \\ &= P(A . B^{\lambda}_{i_\lambda},B^{2\lambda}_{i_{2\lambda}}) \end{aligned} \tag{28}$$

The latter equality is easily derived as follows. Using the decomposition (21), we can apply (9) and find

$$\begin{aligned} &P(A . B^1_{i_1} \dots B^{\nu}_{i_\nu} . B^{\nu+1}_{i_\nu+1},B^{\nu+3}_{i_\nu+3}) \\ &\qquad = P(A . B^{\nu+1}_{i_\nu+1},B^{\nu+3}_{i_\nu+3}) = P(A . B_{i_\nu+1},B^2_{i_\nu+3}) \end{aligned} \tag{29}$$

We then extend the relation recursively, using the decomposition (21), for any chosen phase distance.

Whether a given sequence with probability transfer is regular-invariant is a condition to be tested in every single case, as was pointed out above.

[1] We write here $B_{i_1} \dots B_{i_\nu}$, although we deal only with an alternative B and $\bar{B}$. The subscripts $i_1 \dots i_\nu$ admit, then, only of the two values 1,2: B_1 signifies B; B_2 stands for $\bar{B}$.

This additional specification represents a part of our knowledge about the sequence under consideration—a knowledge that is not included in the statement of probability transfer alone. Thus the sequences with probability transfer, (1) and (2), produced by casting dice according to a certain rule, are regular-invariant.

An example of a sequence with transfer that is not regular-invariant can be obtained as follows. At a seaside town we record continuously the height of the tides; these numbers possess probability transfer, because a very high tide is always followed by a reasonably high tide. If we now carry out a regular division, the sequence remains invariant, in general. An exception, however, is provided by the division $\lambda = 14$. This value of λ corresponds to the 14-day half-period of the moon, which results in particularly high tides for the full moon and the new moon, so that we have $P(A \, . S_{\lambda\kappa}, B) \neq P(A,B)$.

The type of nonnormal sequence produced by probability transfer, according to (9), is not only important for purely statistical applications but has great significance for physics in general. For it enables us to characterize the case of *causal connection* as it occurs in causal chains. Physical action is always *action by contact.* When causal connection assumes the form of a probability connection, we shall find sequences in which the immediately preceding event alone determines the probability of the succeeding event. We can therefore interpret (9) as the mathematical expression of the physical phenomenon of *action by contact.* In particular, it is the case of a positive sign for ϵ and η, that is, probability drag, which occurs in nature.

To give an idealized example, observe the path of a gas molecule within a closed vessel, dividing the vessel into r cells corresponding to the disjunction $B_1 \vee \ldots \vee B_r$. Note regularly at short intervals Δt in which cell the molecule happens to be. Because of collisions, the molecule will follow a zigzag path with varying velocity. If the molecule is at the time Δt_i in the cell B_k, it is most likely to be found in the neighboring cells at the time Δt_{i+1}. Over a longer period, however, this probability will die down; eventually, every cell will possess equal probability, so that the influence of the initial position is obliterated.

§ 34. The Probability Lattice

It is a consequence of the definition of probability given in § 9 that the degree of probability is considered as a property of a sequence in its entirety. In many applications, however, we deal with sequences for which we want to express the fact that a definite probability exists for each individual element, that is, the probability is constant from element to element. In order that such a statement will not contradict the general logical structure of the probability concept, we must investigate how the statement about an individual element of a probability sequence can be translated into a statement about a whole sequence—that is, in the language of the frequency interpretation, how it can be expressed as a statistical statement.

The definition of the normal sequence as well as of probability transfer represents a step in this direction. For instance, we say that in a normal sequence the probability of an element does not depend upon its predecessor; this is formulated so that we speak about the probability within a new sequence, namely, in a subsequence. In the case of probability transfer, we likewise translate the statement that the probability of an element does depend on its predecessor into a statement about a new sequence. The procedure indicates, in principle, how to transform statements about the probability of individual elements into statements about the properties of an entire sequence.

The methods given so far, however, are still insufficient. The reason is that a selection collects an infinite number of elements of the original sequence into a new sequence; therefore it supplies a statement, not about an individual element, but about a subsequence. The formulations so far given thus do not permit us to express certain assertions that are actually made in practical statistics. When we produce a probability sequence by throwing a die, we demand that each throw be played with the same probability, that is, there should not be occasional exceptions where a loaded die is used or the die is insufficiently shaken. The requirement signifies a well-defined condition for the physical production of the sequence, but the condition is not incorporated in the foregoing definition of a normal sequence. For example, if we do not make the fourth throw properly, but produce it by deliberately placing face 6 up, the incorrectness will not show in the limit of the frequency, because a single throw does not alter the limit properties of the whole sequence. The definition of the normal sequence would exclude only the possibility of producing every fourth throw artificially.

We can supply even farther-reaching examples in which all the elements of a sequence are played with different probabilities, whereas we are not able to detect this mode of playing when we use only the means of structural characterization so far developed. For instance, a roulette wheel can be constructed with a variable width of its sectors, so that a suitable adjustment produces any chosen degree of probability. When we now play, with the probabilities p_i changing from one throw to the next so that they converge to a limit p, the sequence will possess all the properties of a normal sequence. It follows that by means of the methods so far used we should not be able to discover from the observational results that the sequence was nonnormal; we should call it a normal sequence.[1]

[1] This statement holds not only for my definition but also for all other definitions discussed above, including the definitions of logical randomness given by von Mises and Church. See also p. 280.

In order to develop methods that characterize a sequence from this new point of view, we must analyze what we mean by saying that the probability varies with the width of the roulette sectors. If we keep to a sector width once adjusted and continue playing, the result for the sequence will be a certain limit that differs from the limit of the sequence mentioned first. The idea can be formulated statistically when we refer the probability statement, not to a single sequence, but to an infinite class of sequences, which may be written

$$\begin{matrix} y_{11} & y_{12} & y_{13} & y_{14} & \dots\dots y_{1i} & \dots \\ y_{21} & y_{22} & y_{23} & y_{24} & \dots\dots y_{2i} & \dots \\ \dots & \dots & \dots & \dots & \dots & \dots \\ y_{k1} & y_{k2} & \dots & \dots & \dots\dots y_{ki} & \dots \\ \dots & \dots & \dots & \dots & \dots & \dots \end{matrix} \qquad (1)$$

Such an arrangement may be called a *sequence lattice*, or a *probability lattice*. Assume that the horizontal rows represent sequences of equal probabilities p for B and are *normal sequences in the wider sense* (as we shall now call the normal sequences defined above). It does not follow from this definition, however, that the vertical sequences also possess the limit p of the frequency or that they are normal; this could not be derived even if the mutual independence of the horizontal sequences were assumed. Imagine, for instance, that all the horizontal sequences contain, in the fourth throw, one result obtained by placing face 6 up; this would show in the lattice by the fact that the fourth vertical sequence possesses for the "6" the frequency 1 as its limit. Or we could produce all horizontal sequences by playing with the adjustable roulette wheel according to the procedure mentioned; the probabilities p_i would then change from element to element and approach a limit p. This procedure would show up in the limits of the vertical sequences, since these limits would be equal to p_i.

Reversing the inference, we can express the assumption that the probabilities within the horizontal sequences are the same from element to element by postulating that each vertical sequence should likewise possess the limit p for its frequency. This presents a definition of normal sequence that is narrower than the one previously given. We shall speak, therefore, of *normal sequences in the narrower sense*. The definition makes use of a new means of structural characterization: the *probability lattice*.

The narrower definition of normal sequences differs essentially from the previous one. The first definition was concerned with an individual sequence, and reference was made to the properties of this sequence alone. The second definition concerns an infinite sequence of sequences, or a class of sequences, and thus specifies not a property of the single sequence but, rather, a property of the totality of sequences. Whether an individual sequence conforms to this definition depends not only on the sequence itself but also on all the other sequences belonging to the class under consideration.[2] This peculiarity of the definition, however, has a great advantage from the logical point of view. In many applications we deal with sequences of sequences, which are defined by a single physical rule. For instance, the rule about throwing dice, in a general formulation, defines not only a single sequence but a class of sequences, the properties of which are characterized by the use of the lattice. To say that the probability of the horizontal sequences is constant from element to element has a meaning because every individual element is imbedded in another sequence, the vertical sequence; and again we have expressed the individual property by means of a class property.

But this procedure also is limited. If in the lattice (1) the individual element y_{23} were played with a deviating probability, the fact would not be recognizable in the sequence lattice. This consequence does not indicate a mistake in the conception of probability; it merely proves that we can speak of a specific probability of an element only when the element is imbedded in a specific sequence. When, in the example of the adjustable roulette wheel, we speak of a varying probability in the horizontal sequence, the statement has meaning only because the varying probability can be represented once more as a property of sequences. Every adjustment of the roulette wheel determines a sequence, namely, the sequence of throws obtainable by the adjustment. And only because of the interpretation in terms of a sequence can we claim meaning for statements correlating a probability to every adjustment of the wheel.

The lattice makes it possible to express such statements formally. Even the assertion that, for instance, the individual term y_{23} is played with a different probability can be formulated in a similar manner, when we use a lattice of three dimensions, in which the sequence correlated to y_{23} in the third dimension indicates a different probability. It must be possible, in principle, to translate statements about the variability or the constancy of certain probabilities produced by physical devices into statements concerning sequences of an n-dimensional lattice, if the statements are to have any meaning. The lattice is a conceptual means of characterizing certain important properties of sequences that cannot be formulated by the use of selections from the sequence under consideration. Hence we shall turn now to the definition of various types of sequences in terms of a lattice.

Let us first give a precise definition of normal sequences in the narrower sense. Apart from the condition concerning the limits in the vertical direction, we wish to incorporate in this concept also the condition of the mutual independence of the sequences. It was explained in § 23 that the independence

[2] This type of definition, which is suitably called a *relative definition*, is well known in the theory of implicit definitions. For instance, it is used in the definition of geometrical elements such as point, line, and plane. See H. Reichenbach, *Philosophie der Raum-Zeit-Lehre* (Berlin and Leipzig, 1928), p. 118.

of every two sequences does not suffice to insure the independence of three and more sequences; the independence relation is not combinable. Consequently, we should demand the mutual independence of all sequences; but such a notion would lead to difficulties for infinitely many sequences, since it would refer to probability expressions containing infinitely many class symbols in the first term. We can avoid this consequence by introducing a corresponding condition in a finitized form: we lay down the condition that for every combination of any ν sequences there exists complete independence in the sense of § 23. Sequences that satisfy this condition may be called *independent by combination.* We now set up the definition:

A sequence of sequences constitutes a system of normal sequences in the narrower sense if all the horizontal and vertical sequences are normal in the wider sense and possess the same probability p, and if the horizontal as well as the vertical sequences are independent by combination.

An extension of the abbreviated notation will be used to represent the definition within the calculus. Instead of the expression

$$(i)(x_{ki} \epsilon A \underset{p}{\ni} y_{ki} \epsilon B) \tag{2}$$

we write, as an abbreviation,

$$(A \underset{p}{\ni} B^{ki})^i \tag{3}$$

or, in the P-notation,

$$P(A,B^{ki})^i = p \tag{3'}$$

We use here the superscripts of class symbols, as in B^{ki}, for the expression of the subscripts that are added, in the detailed notation, to the symbols x,y of the elements. The phases that were formerly expressed by the superscripts can now be added by the use of additive terms, for instance, in the form

$$P(A \, . \, B^{ki},B^{k,i+\lambda})^i \tag{4}$$

The superscript of A may be dropped on account of (4, § 28), since the lattice A is assumed to be compact. The general rule of translation (p. 49) may then be supplemented as follows:

The superscripts of B stand, in general, for subscripts belonging in the detailed notation to the corresponding y; one superscript signifies only the phase of the subscript in y; two superscripts represent the subscripts of y themselves. The superscript added to the parentheses of the P-symbol indicates the variable bound by the all-operator, that is, the running superscript.

Besides bound, or running, superscripts, free superscripts, such as the superscript k in (3) and (3′), will occur. Such a superscript constitutes a free variable. When a formula contains a free variable, this means that the formula is valid for all values of the variable; thus the relation (3′) is meant to hold for all values k. In this manner we can express in the P-notation that a formula is generally valid, without being compelled to use an all-operator.

The new symbolism will now be used for some necessary definitions:

A lattice is homogeneous if all the horizontal and vertical sequences possess equal probabilities, that is, if

$$P(A,B^{ki})^i = P(A,B^{ki})^k = p \tag{5}$$

Furthermore, we can express in the calculus the concept of being *independent by combination.* It seems advisable to write this expression for a many-term disjunction $B_1 \vee \ldots \vee B_r$, which, as usual, may be complete and exclusive. Then the phrase "independent by combination" is defined by the following relations, which are supposed to hold for any chosen values $\rho_1 \ldots \rho_{\nu-1}$:

$$P(A.B^{k+\rho_1,\, i}_{m_1} \ldots B^{k+\rho_{\nu-1},\, i}_{m_{\nu-1}}, B^{ki}_{m_0})^i = P(A,B^{ki}_{m_0})^i \tag{6a}$$

$$P(A.B^{k,\, i+\rho_1}_{m_1} \ldots B^{k,\, i+\rho_{\nu-1}}_{m_{\nu-1}}, B^{ki}_{m_0})^k = P(A,B^{ki}_{m_0})^k \tag{6b}$$

The property is formulated for horizontal sequences in (6*a*), for vertical sequences in (6*b*). When a lattice of normal sequences in the wider sense satisfies the conditions (5) and (6)—in other words, when in a homogeneous lattice all the horizontal and vertical sequences are normal in the wider sense and independent by combinations—the lattice constitutes a *system of normal sequences in the narrower sense.*

The homogeneous lattice, then, represents a wider concept than the normal sequences in the narrower sense; it defines a more general type of sequence. A still more general sequence type is obtained by the condition that, whereas the horizontal sequences all have the same probability p, the vertical sequences possess the probabilities p_i converging toward the limit p. The system may then be said to form a *convergent lattice.*

The behavior of sequences with respect to regular divisions was characterized above by the use of the concept *regular-invariant.* In a similar manner the concept *lattice-invariant,* which concerns the behavior of sequences with respect to lattice enumeration, can be defined as the property that every phase probability of each individual sequence has the same value as the corresponding probability in lattice enumeration.[3] When, once more, a complete and exclusive disjunction $B_1 \vee \ldots \vee B_r$ is used and the phases are added, according to (4), the definition of *lattice-invariance* may be written in the form

$$P(A.B^{ki}_{m_0} \ldots B^{k,\, i+\nu-1}_{m_{\nu-1}}, B^{k,\, i+\nu}_{m_\nu})^k = P(A.B^{ki}_{m_0} \ldots B^{k,\, i+\nu-1}_{m_{\nu-1}}, B^{k,\, i+\nu}_{m_\nu})^i \tag{7}$$

$$\nu > 0$$

The normal sequences in the narrower sense are lattice-invariant, since all the probabilities (7) are equal to $P(A,B_{m_\nu})$, according to (5), (6), and (3, § 29).

[3] I do not include in the concept "lattice-invariant" the requirement that also the major probabilities in the horizontal and the vertical direction should possess the same value, as would correspond to the homogeneous lattice; a nonhomogeneous lattice may equally be lattice-invariant.

But sequences with probability transfer can also be lattice-invariant, just as these sequences can be regular-invariant. Consider a lattice in which all horizontal sequences possess for the alternative B or $\bar{B}$ the same probability transfer, statable in terms of p, ϵ, η. It then follows from the property of the lattice invariance (7) that, using (18 and 23, § 33), we have

$$\begin{aligned} P(A.B^{ki}, B^{k,i+\nu})^k &= P(A.B^{ki}, B^{k,i+\nu})^i = P(A.B,B^\nu) = p_\nu \\ P(A.\bar{B}^{ki}, B^{k,i+\nu})^k &= P(A.\bar{B}^{ki}, B^{k,i+\nu})^i = P(A.\bar{B},B^\nu) = q_\nu \end{aligned} \tag{8}$$

By the help of these relations a particularly simple case can be constructed, when all horizontal sequences are assumed to begin with the term B. Because of (8) we then have

$$P(A,B^{ki})^k = P(A.B^{k1},B^{ki})^k = P(A.B,B^{i-1}) = p_{i-1} \tag{9}$$

The sequences form a convergent lattice the vertical sequences of which represent the phase probabilities p_ν with $\nu = i - 1$. Such a lattice may be called a *lattice of mixture.*

We cannot prove that every lattice of sequences with probability transfer has the property of lattice invariance formulated by (8). We must rather make this requirement the definition of a lattice type. The derivation in § 33 shows only that the values p_ν hold for an enumeration in the horizontal direction; it does not concern an enumeration in the vertical direction. When the assumption is extended to vertical sequences, this means, physically speaking, that every individual element of the sequences is played, respectively, with the probability $p + \epsilon$ or $p - \eta$ pertaining to it. This property, in fact, is found in many applications. When we construct a lattice by playing with dice according to the rules for probability transfer given in § 33, the vertical sequences fulfill equation (9) for p_ν, with $\nu = i - 1$. We see that, as before, certain properties of the physical setup can be formulated only by the use of a lattice; they cannot be expressed for a single sequence.

Examples of the lattice of mixture are physical processes that represent a mixing of substances. Imagine two liquids in a vessel that is divided by a wall into two compartments, with one liquid in each half. The first half of the vessel may be designated by B, the other by $\bar{B}$. We then withdraw the partition wall and permit the liquids to mix. Imagine that during the process we can note regularly, at frequent intervals, for each molecule of the first liquid whether it happens to be in B or $\bar{B}$. We then obtain for each molecule a sequence with the probability $\frac{1}{2}$ for B; but it has probability transfer, because a molecule that happens to be in B will, soon afterward, still be found in B (similar to the example mentioned at the end of § 33). We now write the sequences for the molecules of the first liquid one under another, disregarding the molecules of the second liquid. The sequences will represent a lattice of mixture in which the first element of every horizontal sequence is given by B. In the vertical sequences next to the first one (which, incidentally, do not have probability transfer in this case), the element B will still be prevalent corresponding to the probabilities p_ν. As we go to the right, the limits of the frequency of B in the vertical sequences will gradually approach the value $\frac{1}{2}$, that is, the liquid will spread over the whole space of the vessel. In this manner is expressed the diminishing influence of the initial position, which is precisely what characterizes the peculiarity of all mixing processes. Without the use of a probability lattice the mixing process cannot be formulated in the theory of probability.

It should be emphasized that the fact of mixing cannot be derived from the general principles of the probability theory.[4] This is as impossible as to derive, from these general principles, that a sequence of throws of a die is a normal sequence. Like the latter case, the phenomenon of mixture must be regarded, in the general theory of probability, as a special case, the occurrence of which can be asserted only on empirical grounds. The physical phenomenon of mixture is not a consequence of the general laws of probability. That natural processes will represent a process of mixture can be mathematically derived only when we know that the processes constitute sequences with probability transfer that are lattice-invariant. Then, however, the result is a tautological statement. The theory of probability cannot prove that the process of mixture will occur; it can only supply the logical schema by which the process is to be interpreted. This is done by the use of the lattice of mixture, which formulates the conditions from which the occurrence of the process is derivable.

The kinetic theory of gases includes the *inference from the time totality to the space totality*.[5] It is based on the assumption that the time sequence given by the states of one molecule (or of one system) exhibits the same statistical relations as the space sequence given by the states of different molecules (or systems) at the same time. This inference, too, is a lattice inference; it must be interpreted as the assumption that the sequences form a homogeneous lattice. The theory of probability cannot decide whether this assumption is valid. Such an assumption, rather, represents a physical hypothesis the validity of which must be ascertained in the same manner as for any other hypothesis. The hypothesis is tested by the observational examination of its consequences, applying the usual inference discussed in §§ 21 and 85. But to suppose a hidden mathematical secret in the inference from the time totality to the space totality is to misunderstand the theory of probability. Like all other forms of deductive inference, the calculus of probability cannot bring to light more results than were invested in the premises by suitable assumptions.

---

[4] The opinion has often been expressed that Bernoulli's theorem leads to a proof of the mixing process. This, however, is not correct. See p. 280.

[5] See P. Hertz in Weber-Gans, *Repertorium der Physik* (Leipzig, 1916), Vol. I, Part 2, Sec. 242.

Chapter 5

PROBABILITY SEQUENCES WITH COÖRDINATED AMOUNTS

PROBABILITY SEQUENCES WITH COÖRDINATED AMOUNTS

§ 35. The Average of a Sequence of Quantities

Leaving the investigations concerning the theory of order, we turn to the treatment of concepts that were developed for the application of the theory of probability to statistical problems. Many such problems involve classes B_m to which numerical *amounts* u_m are coördinated (amounts of money, lengths of distances measured, etc.). For the treatment of such classes the concepts of *average* and *dispersion* were introduced. We begin with the concept of average. For its definition, we first define the analogous concept with respect to a sequence of a finite number of quantities, which is called the *mean value;* the average is then the extension of this concept to a sequence of infinitely many quantities.

The amount u_m coördinated to the class B_m represents one of the *possible values* of the amounts occurring. From it we distinguish the *individual value* u^i that is found combined with the element y_i of the probability sequence. This distinction may be indicated by the use of subscripts and superscripts. If y_i belongs to the class B_m, we have $u^i = u_m$. The subscript of u runs through the r values of the disjunction $B_1 \vee \ldots \vee B_r$; the superscript of u assumes all numerical values from 1 to ∞. It will be obvious from the context that the superscripts do not express arithmetical powers.

In the simpler case in which the sequence A is compact, that is, where all x_i belong to A, all u^i fall into the sequence considered. If we cut off this sequence at an element u^n, the *mean value* of this finite sequence is defined by

$$M^n(u^i)^i = \frac{1}{n}[u^1 + \ldots + u^n] = \frac{1}{n}\sum_{i=1}^{n} u^i \qquad (1)$$

The repetition of the superscript i outside the parentheses in the term on the left side is to indicate its character as a superscript of summation. With increasing number n we can coördinate to every element u^n the mean value $M^n(u^i)^i$ taken at this place; and the question arises whether the quantities $M^n(u^i)^i$ approach a limit. If this is the case, we define as the *average* of the quantities u the value

$$M(u^i)^i = \lim_{n \to \infty} M^n(u^i)^i = \lim_{n \to \infty} \frac{1}{n}\sum_{i=1}^{n} u^i \qquad (2)$$

The average is thus the limit of the mean values. We call (1) and (2) the *statistical definition,* respectively, of the mean value and of the average; this term expresses the fact that the definition characterizes these quantities through direct enumeration.

Consider, for example, the sequence produced by the throwing of a die. The numbers 1 to 6 written on the faces of the die supply the values u_m; u^i is the number of the face of the die that lies up in the i-th throw. We determine $M^n(u^i)^i$ by adding all the numbers that have appeared up to the n-th throw and dividing them by n. The average of this value, that is, its limiting value for $n \to \infty$, is $M(u^i)^i$.

We can extend the definitions (1) and (2) to the case in which the sequence A is not compact and for which we do not take into account all the elements u^i, but only those that are coördinated to the elements y_i that correspond to an x_i belonging to A. To express this definition in the symbolism we use the symbol $V(a)$, introduced in § 4, meaning, "truth value of the statement a". However, we deviate from our previous notation by assigning to $V(A)$ the values 1 or 0 according as a is true or false. With the help of this convention we can replace the symbol $\overset{n}{\underset{i=1}{N}}$, which was introduced in § 16, by a summation:

$$\overset{n}{\underset{i=1}{N}} (x_i \epsilon A) = \sum_{i=1}^{n} V(x_i \epsilon A) \tag{3a}$$

Abbreviating the V-symbol in the form

$$V(A^i) =_{Df} V(x_i \epsilon A) \qquad V(A^i . B^i) =_{Df} V[(x_i \epsilon A).(y_i \epsilon B)] \tag{3b}$$

we can write for (3a), using the abbreviation (2, § 16) for the N-symbol,

$$N^n(A) = \sum_{i=1}^{n} V(A^i) \tag{3c}$$

The frequency interpretation of probability is then written

$$F^n(A,B) = \frac{\sum_{i=1}^{n} V(A^i . B^i)}{\sum_{i=1}^{n} V(A^i)} \tag{4}$$

$$P(A,B) = \lim_{n \to \infty} F^n(A,B)$$

The statistical definition of the mean value and of the average can now be symbolized by

$$M^n(A;u^i)^i = \frac{\sum_{i=1}^{n} u^i \cdot V(A^i)}{\sum_{i=1}^{n} V(A^i)} \tag{5}$$

$$M(A;u^i)^i = \lim_{n \to \infty} M^n(A;u^i)^i$$

The symbols written on the left side represent generalizations of the symbols occurring in (1) and (2); their definitions are identical with (1) and (2) if the sequence A is compact.

This presentation of the concepts of mean value and of average makes it clear that the sequences of quantities u^i involve a certain generalization of the concept of probability sequence. Whereas in a probability sequence each element y_i is referred to a class B_m and thus possesses a qualitative property, the sequences introduced in this section are composed of elements y_i that possess quantitative properties u_m. It is often convenient to forget about the sequence of the events y_i and to deal directly with the sequence of the numbers u^i. This number sequence has the character of a probability sequence if B_m is regarded, not as a class of elements y_i, but of elements u^i selected by the condition $u^i = u_m$. Instead of the *enumeration* of elements, the *addition* of quantities can then be employed. The enumeration may be regarded as a special case of addition, for which the u^i are represented, in particular, by the truth values 1 and 0. For this particular case the concept of average is identical with the concept of probability, since if in (5) u^i is replaced by $V(B^i)$, formula (4) obtains.

Conversely, replacing addition by enumeration, we can reduce the concept of average to the concept of probability. For this purpose we use the classification of amounts as given by u_m, counting how often the amount u_m occurs and multiplying u_m by this number. The addition of these values gives the same result as the direct addition of the individual values u^i. Employing immediately the more general definition (5), we are able to formulate this idea within the calculus when we add the always-true disjunction $B_1^i \vee \ldots \vee B_r^i$ and carry out the multiplication

$$M^n(A;u^i)^i = \frac{\sum_{i=1}^{n} u^i \cdot V(A^i.[B_1^i \vee \ldots \vee B_r^i])}{\sum_{i=1}^{n} V(A^i)}$$

$$= \frac{\sum_{i=1}^{n} u^i \cdot V(A^i.B_1^i) + \ldots + \sum_{i=1}^{n} u^i \cdot V(A^i.B_r^i)}{\sum_{i=1}^{n} V(A^i)} \tag{6}$$

With (4) we have

$$\frac{\sum_{i=1}^{n} u^i \cdot V(A^i.B_m^i)}{\sum_{i=1}^{n} V(A^i)} = u_m \cdot \frac{\sum_{i=1}^{n} V(A^i.B_m^i)}{\sum_{i=1}^{n} V(A^i)}$$

$$= u_m \cdot F^n(A,B_m) \tag{7}$$

Substituting this value in (6) and introducing the new symbols for mean value and average written on the left side of the following formulas, we obtain

$$M^n(A,B_m;u_m)_m = \sum_{m=1}^{r} F^n(A,B_m) \cdot u_m \tag{8a}$$

$$M(A,B_m;u_m)_m = \lim_{n \to \infty} M^n(A,B_m;u_m)_m = \sum_{m=1}^{r} P(A,B_m) \cdot u_m \tag{8b}$$

Because of the completeness of the disjunction the relation holds

$$\sum_{m=1}^{r} P(A,B_m) = 1 \tag{9}$$

The expressions (8) are called the *theoretical definitions*, respectively, of the mean value and of the average. This term expresses the fact that the definition is achieved by means of a theoretical transformation of the statistical definition, and thus is constructed in an indirect manner. The definition reduces the concept of average to the concept of probability. The equivalence of both kinds of definition follows from the fact that the right side of (8a) is a transformation of (6); this equivalence is formulated by the equations

$$M^n(A;u^i)^i = M^n(A,B_m;u_m)_m \tag{10a}$$

$$M(A;u^i)^i = M(A,B_m;u_m)_m \tag{10b}$$

To simplify the notation, the symbols occurring in (10) are abbreviated by omitting the classes A and B. This omission is, in general, without danger, since it is usually obvious to which probability classes the mean value is referred; in doubtful cases we shall return to the exact notation (10). We then obtain from (10) the equations

$$M^n(u^i)^i = M^n(u_m)_m = M^n(u) \tag{11a}$$

$$M(u^i)^i = M(u_m)_m = M(u) \tag{11b}$$

The last notation is further simplified by dropping the subscripts, or superscripts, upon which the expression does not depend. This simplification is possible because of the equivalence of statistical and theoretical definitions, which makes superfluous a distinction between the two definitions.

The importance of formulas (8) consists in the reduction of the calculation of an average to the knowledge of certain probabilities, so that the direct calculation of the average according to (5) is avoidable. Thus in the example of the die we may calculate the average for all m from (8) with $P(A,B_m) = \frac{1}{6}$ as

$$M(u) = \tfrac{1}{6} \cdot [1 + 2 + \ldots + 6] = 3\tfrac{1}{2}$$

Furthermore, the theoretical definition enables us to define the concept of average in the formal calculus of probability, for which (5) and the transformations resting upon the frequency interpretation are not applicable. For this purpose we write the expression (8*b*) by the use of the abbreviations (11*b*), omitting the relation (8*a*) concerning the mean value M^n. Thus we have

$$M(u) = \sum_{m=1}^{r} P(A,B_m) \cdot u_m \tag{12}$$

This relation, which is a transcription of (8*b*), is called the *formal definition* of the average. The expression $u_m \cdot P(A,B_m)$ is called *mathematical expectation;* correspondingly, the average is also called *expectation value.*

A relation following from (12) for a constant κ may be noted:

$$M(\kappa \cdot u) = \kappa \cdot M(u) \tag{13}$$

If the disjunction $B_1 \vee \ldots \vee B_r$ has a finite number of terms, the existence of the average according to (12) is insured if the probabilities $P(A,B_m)$ exist and (9) holds. These conditions guarantee simultaneously, for the statistical definition, that there exists the limit $M(u^i)^i$ of the sequence $M^n(u^i)^i$. The existence of this limit is thus reduced to the existence of the limit of the frequencies $F^n(A,B_m)$. However, if $r \rightarrow \infty$, that is, if the disjunction $B_1 \vee B_2 \vee \ldots$ has infinitely many terms, $M(u)$, according to (12), is the limit of a sum. Even if all the probabilities $P(A,B_m)$ exist and (9) holds, the existence of the limit $M(u)$ is then not warranted by (12), but is linked to certain conditions for the u_m. Such examples will be given in § 36.

It should be realized that the concept of average constitutes an arbitrary definition, or convention, by which a number of quantities are combined in a single value so as to achieve an abbreviation. Of course, this single value does not represent the totality in a completely adequate manner; it depends upon the special case whether the average is at least a suitable substitute for the totality. There are cases for which it is more important to know an upper limit for the u_m than their average. For instance, the stability of a bridge is to be adjusted not to the average, but to the maximum, load. The literature on probability includes such definitions as *central value*, *mode*, and *quartile*, which are used in a manner similar to "average", but which are not very useful for technical reasons. The arbitrary character of the concept of average is made obvious in applications where the average cannot be interpreted directly, as, for instance, in the statement that in a certain country the average number of children to a marriage is 2.35. In some cases the average is of great practical importance because, by the nature of the amounts considered, the addition of amounts supplies relevant information. For instance, if a merchant averages a profit of $M(u)$ dollars in his business, this means that for n transactions his income is greater by $n \cdot M(u)$ dollars than his expenditures; and only this total sum is of relevance to him.

Two further examples may be discussed, in which the probability structure is more clearly seen. The first is provided by insurance companies. They make contracts with a large number of clients; their policies stipulate that an accidental damage B_m occurring with low probability—for instance, damage by fire—is made good by payment of a large sum u_m. For the case B_l, by which we understand $\bar{B}_m$, and which occurs with a high probability, insurance companies take in a small amount u_l, the premium. If the number of clients is great, the relative frequency may be replaced, in practice, by the probability; and the average amount of the payments made is given by $M(u)$. For $M(u) = 0$ the premium would be exactly adequate to the risk; but insurance companies adjust the premium in their favor, so that $M(u)$ is somewhat greater than zero. With this surplus amount they not only cover the expenses of their organization, but accumulate reserve capital for the eventuality that there should once occur more accidents than are expected.

A second example is provided by games of chance, in which certain amounts u_m of money, depending on the probability, represent winnings or losses. A game is called fair if $M(u) = 0$; then the mathematical expectations of the gamblers are equal. If, in a game with one die, a gambler bets \$5 that the "6" will show up, the other gambler can demand \$1 as winnings if every occurrence in which the "6" fails to appear is counted in his favor. The calculation of the chances in complicated games, which played an important role in the writings of eighteenth-century mathematicians, is usually linked not to the concept of probability but to that of expectation; but this formulation obviously represents only a different mode of speech.

Games played in gambling clubs or casinos are never fair in the sense of this definition, since for them $M(u) > 0$ holds in favor of the bank. Thus in a game with two dice they offer five times the stake for a bet on "any 7", whereas the chances are only one in six to win. Now it is true that the word "fair" does not have a moral meaning in the mathematical definition, but merely defines the condition of an average balancing of the winnings. But it is strange that so many persons participate in gambling that guarantees them a loss if it is continued long enough. The flourishing of gambling places should be proof of the hopelessness of successful gambling. However, the suggestive power of the amounts that *may* be won seems to rule out a sound reasoning about what *will* be won.

The corruption of reasoning is one of the foremost dangers of gambling. Patrons of gambling clubs try all sorts of systems in the hope that they can outwit the bank, disregarding the fact that gambling machines furnish random sequences and thus are immune to attack by a gambler's skill. Various superstitions have arisen from gambling; certain numbers are regarded as "lucky", others as "unlucky". And the attempt to "try his luck" has been dearly paid for by many a man.

If the state, or a welfare organization, owns the bank, as in the institution of public lotteries, the odds in favor of the bank may seem to be excusable. There remains, however, the mischief created by false hopes in the mind of the lottery participant. The probability of winning in a lottery is usually seriously overestimated. If someone buys a ticket for $5 with the chance of winning $20,000, he considers only this pleasant ratio of numbers, forgetting that because of this very ratio the winning probability is smaller than $\frac{5}{20{,}000} = \frac{1}{4{,}000}$, a probability so small that it is regarded as zero in all other cases. For instance, the probability of being killed within a year by an automobile amounts to about the same value, but no one is concerned in this case about such a probability.[1]

It is true that mathematics cannot supply us with value judgments and so cannot determine whether gambling and lotteries are morally good or bad. But in pointing out the discrepancies between the gambler's behavior and his mathematical chances the mathematician can contribute his share to public education.

§ 36. Formation of an Average When the Summation Is Extended to Infinitely Many Terms

The summation (12, § 35), when extended over infinitely many terms, need not necessarily be convergent even if the condition (9, § 35) remains fulfilled. But it is, of course, possible that the summation does converge. These conditions will be illustrated by examples for convergence and opposite examples.

For instance, we play for a certain result B with the probability p, repeating the play until B happens for the first time; this set of plays is called a group. If the groups are repeated, what is the average length of a group? In other words, after how many elements on the average will B occur? We shall assume that $0 < p < 1$.

The different groups, each of which extends to the occurrence of B, may be collected in a single sequence of normal character, in which B occurs with the probability $P(A,B) = p$. Every group is of the form $\bar{B}^1 \ldots \bar{B}^{\lambda-1}.B^\lambda$, so that the whole sequence is divided into such groups of different length λ; the predecessor of each group is always an element B. We wish to find the average length $M(\lambda)$ of a group.

[1] In 1947 there were 32,300 fatal automobile accidents in the United States, in a population of 142,673,000. This is one death case through automobile accident in about 4,400 persons within a year.

The probability of a group of this kind is written symbolically as

$$P(A.B,\bar{B}^1 \ldots \bar{B}^{\lambda-1}.B^\lambda) \tag{1}$$

That the enumeration by elements is replaced by the counting of groups is expressed by the occurrence of B in the first term of the probability expression. For in the frequency interpretation the denominator of (1) is given by $N(A.B)$, that is, by the frequency of B, since the sequence A is compact. The amount u_λ, the average of which is to be found, is the group length λ, that is, we have

$$u_\lambda = \lambda \tag{2}$$

According to (12, § 35), the desired average is given by

$$M(\lambda) = M(u_\lambda)_\lambda = \sum_{\lambda=1}^{\infty} P(A.B,\bar{B}^1 \ldots \bar{B}^{\lambda-1}.B^\lambda) \cdot u_\lambda \tag{3}$$

Since the normal sequence is free from aftereffect, the probability (1) can be represented by the product of the individual probabilities within the major sequence. Thus we have

$$P(A.B,\bar{B}^1 \ldots \bar{B}^{\lambda-1}.B^\lambda) = (1-p)^{\lambda-1} \cdot p \tag{4}$$

With (2) and (3) we derive

$$M(\lambda) = \sum_{\lambda=1}^{\infty} \lambda \cdot (1-p)^{\lambda-1} \cdot p \tag{5}$$

Formula (5) determines an average by the summation of an infinite number of terms; therefore the existence of the average $M(\lambda)$ must first be proved. Proof of convergence, however, can be given. With the abbreviation q for $(1-p)$, we have

$$M(\lambda) = p \cdot \sum_{\lambda=1}^{\infty} \lambda \cdot q^{\lambda-1} \tag{6}$$

In the infinite sequence occurring in this expression, the quotient of two successive terms is equal to

$$\frac{(\lambda+1) \cdot q^\lambda}{\lambda \cdot q^{\lambda-1}} = \frac{\lambda+1}{\lambda} \cdot q \tag{7}$$

The expression (7) has, from a certain λ onward, a value < 1, because $q < 1$; and its limit is also < 1, namely, equal to q. According to a well-known mathematical theorem the results provide a sufficient condition of convergence for (6).

The expression (6) can easily be evaluated directly. We have

$$\sum_{\lambda=1}^{\infty} \lambda \cdot q^{\lambda-1} = \frac{d}{dq}\left(\sum_{\lambda=1}^{\infty} q^\lambda\right) \tag{8}$$

Since we are dealing with power series, we can calculate the sum written on the left side by calculating the sum on the right side and then differentiating this value. According to the summation formula of a geometrical series, we have

$$\sum_{\lambda=1}^{\infty} q^{\lambda} = \lim_{\lambda \to \infty} \frac{q^{\lambda+1} - q}{q - 1} = \frac{q}{1 - q} \quad \text{since } 0 < q < 1 \tag{9}$$

Substituting this value in (8), we arrive at

$$\sum_{\lambda=1}^{\infty} \lambda \cdot q^{\lambda-1} = \frac{d}{dq}\left(\frac{q}{1 - q}\right) = \frac{1}{(1 - q)^2} \tag{10}$$

Replacing $(1 - q)$ by p and introducing (10) in (6), we derive

$$M(\lambda) = \frac{1}{p} \tag{11}$$

This very simple expression answers the question about the average group length. For instance, when we throw a die we can expect a certain face to turn up on the average after six throws.

This result, which was derived purposely by a procedure that, though cumbersome, exhibits the general method, is made clear immediately by the following considerations. If we throw a die repeatedly until face 6 appears, and write the separate groups one after another, a normal die sequence obtains. When we denote by λ^i the length of the i-th group and by m the number of groups, the expression

$$n = \sum_{i=1}^{m} \lambda^i \tag{12}$$

measures the number of all the throws made. But the number m of the groups equals the number m of all the events B within the sequence, since every group has one and only one B, which stands as its final element. The frequency interpretation supplies

$$\lim_{n \to \infty} \frac{m}{n} = p \tag{13}$$

The average group length, therefore, is given in the statistical definition by

$$M(\lambda) = M(\lambda^i)^i = \lim_{m \to \infty} \frac{1}{m} \sum_{i=1}^{m} \lambda^i = \lim_{n \to \infty} \frac{n}{m} = \frac{1}{p} \tag{14}$$

This relation represents the statistical meaning of (11). For instance, in 600 throws the "6" will appear about 100 times; the average distance between two occurrences of a "6" is therefore $\frac{600}{100} = 6 = \frac{1}{p}$.

It is owing to the value u_λ given in (2) that the summation occurring in (3) is convergent. An example in which no convergence occurs can easily be constructed. If we put

$$u_\lambda = 2^{\lambda-1} \tag{15}$$

the sum (3) will not converge for certain values p. For instance, for $p = \frac{1}{2}$ we obtain from (3) with (4) and (15)

$$M(u_\lambda)_\lambda = \sum_{\lambda=1}^{\infty} \frac{1}{2^\lambda} \cdot 2^{\lambda-1} = \infty \tag{16}$$

This example, for which no finite mean value exists, is spoken of in the literature as the *Petersburg problem* because it was originally communicated in the publications of the Academy of Petersburg (now Leningrad). It is usually presented as follows. Peter flips a coin repeatedly until tails show up. If tails appear at the first throw, Peter receives \$1 from Paul; if tails occur for the first time in the second throw, he receives \$2; if tails occur at the third throw, he gets \$4. Generally speaking, Peter receives $2^{\lambda-1}$ dollars if tails occur for the first time in the λ-th throw. Each game is continued until tails show up. The question is: How large a sum may Paul ask Peter to pay as his stake in every game?

The "fair" stake would be given by Peter's average winnings. Since every separate game is identical with the group $\bar{B}^1 \ldots \bar{B}^{\lambda-1}.B^\lambda$, the stake is to be calculated by (3) together with (4) and (15), and thus is represented by (16). Therefore Paul may ask Peter to bet an infinitely high amount against him, to be paid for each individual game. The result may at first seem to be paradoxical; no one would feel inclined to risk an infinite amount of money if he were offered Peter's winning chances. We would rather be tempted to infer that there must be a finite λ for which tails will occur, so that even in the most favorable case Peter's winnings can only be finite, and it would be excluded that he could win back his stake in the first game. But the stake should be chosen in such a way that even for the first game there should exist at least the possibility of winning more than the amount placed.

Eighteenth-century mathematicians were greatly concerned about this paradox. Daniel Bernoulli tried to solve it by revising the concept of mathematical expectation. According to him the so-called moral expectation increases more slowly than the mathematical expectation and remains finite if the latter becomes infinite. He was guided by the thought that an increase in money is the less valuable for a person, the more money he has—an idea that was taken over by the economists as the *law of diminishing utility*. It is a mistake, however, to believe that such a consideration can solve the paradox. When we translate the probability statements into frequency statements, the solution is easily given, and the problem loses its paradoxical character.

Assume that a maximum length ν of the game is introduced. Only ν throws are made; if tails have not occurred up to this time, Paul has won the game. By cutting off the summation at the value ν, we calculate from (16) $M(u_\lambda)_\lambda = \frac{\nu}{2}$ and thus have a game free from any paradox. Peter's average winnings in each game are given by $\frac{\nu}{2}$; therefore he must pay Paul, for each game, the amount $\frac{\nu}{2}$, which seems to be a fair stake. For instance, for $\nu = 4$, Peter cannot win more than \$8, and his stake amounts to \$2. The result seems fair in view of the greater probability of smaller winnings. If ν is greater, $\frac{\nu}{2}$ again represents the fair stake; this means that the average winnings of Peter are larger. That the expression (16) goes to infinite values means that the average winnings of Peter increase continuously with ν, so that they are always equal to $\frac{\nu}{2}$ and become infinite with $\nu \to \infty$. If the game is played without any limit to ν, the average gain of Peter will indeed be infinite, and the infinitely large stake seems justified.

A clear picture of the situation is obtained when Peter's winnings are reconstructed statistically. When we denote Peter's gain in the i-th game by u^i, Peter's average winnings up to the m-th game are given by

$$M^m(u^i)^i = \frac{1}{m} \sum_{i=1}^{m} u^i \tag{17}$$

That the average supplied by the theoretical definition (16) goes toward infinity means that the statistically defined average is not convergent either, that is, we have, contrary to (14),

$$M(u^i)^i = \lim_{m \to \infty} M^m(u^i)^i = \infty \tag{18}$$

The result has the following simple meaning: although large winnings rarely occur for Peter, they influence the average so strongly that, until the next large winnings occur, it is not essentially lowered by the many small winnings accumulating. Owing to the strong increase of the powers, the amount of the winnings increases more strongly than the corresponding probability decreases.

If we use this consideration for a translation of the statement about the infinity of the mathematical expectation into a statement about finite quantities, the solution of the Petersburg paradox may be expressed as follows. *If Paul accepts the game for any finite amount to be paid by Peter in each game as his stake, Paul will lose in the long run when the game is continuously repeated.* With this formulation the paradox disappears.

The Petersburg "paradox" is related to a familiar gambling system. A person bets $1 on a certain result; if he does not win, he places $2; if he loses again, he bets $4, and so on. Thus after every loss he bets twice the amount. By this system he must win eventually, because his result must once occur. But the gain is not very high; it is equal to the stake of the first bet—$1. This follows from the relation

$$\sum_{i=0}^{n} 2^i = 2^{n+1} - 1 \tag{19}$$

Such a gambling system is not very lucrative. However, it carries extraordinarily high risks; since every gambler possesses only a finite amount of money, he may be unable to continue the game and must suffer a complete loss. The system is safe only for a gambler with an infinite bank account. If the Petersburg game is played with a finite stake for Peter, he can be certain of winning only if he possesses infinite capital; otherwise he may be forced to stop after he has lost all his money. Such an eventuality is not improbable, even for a millionaire; so this game, too, offers a high risk.

I cannot share the opinion that the Petersburg problem creates logical difficulties for the theory of probability. That special conditions are required for the convergence of an infinite summation, such as occurs in (12, § 35), seems quite natural. Strangely enough, we find in the literature peculiar ideas about the Petersburg problem and faulty attempts to solve it. Thus it is argued that it is not permissible to extend each single game until the event B occurs, and that the game must be limited by an upper limit ν for the group length λ, since otherwise the individual game would "possibly never end". This conception is erroneous; as soon as a probability is interpreted as the limit of a frequency, there must always exist a finite λ such that B occurs. Furthermore, it is not the infinity of the summation that makes the sum in (16) become infinite; the nonconvergence results rather from the specification of the amounts u_λ given in (15). This may be seen from the fact that different specifications of u_λ—for instance, the specification (2)—lead, even for infinite summation, to a finite mean value, as shown in (11). Thus if Peter always receives λ dollars when B occurs in the λ-th throw, his fair stake would be $= \frac{1}{p} =$ $2, according to (11), and the game would not differ in principle from other games, though neither the length of a game nor Peter's winnings are subject to an upper limit.

§ 37. The Dispersion

The average represents only a first comparatively rough characterization of the amounts u^i. The characterization can be made more precise by stating also the dispersion.

Every single amount u^i deviates from the average $M(u)$ by the amount

$$\delta u^i = u^i - M(u) \tag{1}$$

The δu^i are called *deviations*. If the δu^i are small, the average is a relatively good substitute for the totality of values; if they are great, the average characterizes the totality insufficiently. Since it is cumbersome to state the totality of the δu^i, we replace the δu^i by a mean value.

The average of the δu^i themselves is not a suitable characteristic, however, because it vanishes. If we define the possible deviations analogous to (1)

$$\delta u_m = u_m - M(u) \tag{2}$$

we obtain, with (9, § 35) and (12, § 35), for the average of the deviations the relation

$$M(\delta u^i)^i = M(\delta u_m)_m = \sum_{m=1}^{r} P(A,B_m) \cdot \delta u_m$$

$$= \sum_{m=1}^{r} P(A,B_m) \cdot u_m - M(u) \cdot \sum_{m=1}^{r} P(A,B_m) = 0 \tag{3}$$

The corresponding statement can be asserted, by the way, for the mean value $M^n(\delta^* u^i)^i$ of the deviations $\delta^* u^i$ from the mean value $M^n(u^i)^i$. We write

$$\delta^* u^i = u^i - M^n(u^i)^i \tag{4}$$

With (8*a* and 11*a*, § 35) we arrive at the result

$$M^n(\delta^* u^i)^i = M^n(\delta^* u_m)_m = \sum_{m=1}^{r} F^n(A,B_m) \cdot \delta^* u_m$$

$$= \sum_{m=1}^{r} F^n(A,B_m) \cdot u_m - M^n(u_m)_m \cdot \sum_{m=1}^{r} F^n(A,B_m) = 0 \tag{5}$$

That the average of the δu^i is $= 0$ results from the fact that the δu^i are partly positive and partly negative, and the two sums cancel each other. To free ourselves from this consequence we must form a mean value that does not depend on the sign of the individual elements δu^i. The absolute amount $|\delta u^i|$ might be used for such a mean value, but since this quantity is not easy to handle, the value $\delta^2 u^i$, which is likewise independent of the sign, seems preferable. The average of these quantities, for which the symbol Δ^2 is introduced, obtains, according to (12, § 35), as

$$\begin{aligned} \Delta^2(u) &= M(\delta^2 u) = \sum_{m=1}^{r} P(A,B_m) \cdot \delta^2 u_m \\ \Delta(u) &= \sqrt{M(\delta^2 u)} = \sqrt{\sum_{m=1}^{r} P(A,B_m) \cdot \delta^2 u_m} \end{aligned} \tag{6}$$

This quantity is called the *dispersion*. More specifically, Δ^2 is called the *quadratic dispersion;* Δ the *linear dispersion.* The usual name for Δ is *standard deviation;* for Δ^2, *variance.* In mathematical calculations Δ^2 is more useful, but for all numerical evaluations Δ is employed since it is of the same order of magnitude as the δu_m. The sign of the root in the expression for $\Delta(u)$ is always taken as positive. The choice of the dispersion as defined in (6) for the abbreviated characterization of the deviations of the u_m must be regarded, from a logical standpoint, as a convention, comparable to the choice of the average. The dispersion, of course, cannot characterize the deviations exhaustively.

A statistical definition may be added to the theoretical definition (6) of the dispersion. Employing the definition (4), we define the *mean square of deviations* $\Delta^{n2}(u^i)^i$, restricting ourselves for reasons of simplicity to the case of a compact sequence A, in the form

$$\Delta^{n2}(u^i)^i = \frac{1}{n} \sum_{i=1}^{n} \delta^{*2} u^i$$

$$= \frac{1}{n} \sum_{m=1}^{r} \delta^{*2} u_m \cdot \sum_{i=1}^{n} V(B_m^i)$$

$$= \sum_{m=1}^{r} F^n(A,B_m) \cdot [u_m - M^n(u^i)^i]^2 \tag{7}$$

The dispersion is defined as the limiting value of the mean square of deviations. According to the familiar rules for limits, this limit is constructed when we introduce in (7) the limits for F^n and M^n:

$$\Delta^2(u) = \lim_{n \to \infty} \Delta^{n2}(u^i)^i = \sum_{m=1}^{r} P(A,B_m) \cdot [u_m - M(u)]^2 \tag{8}$$

The statistical definition, therefore, leads in (8) to the same value as the theoretical definition in (6).

A few simple mathematical relations concerning the dispersion may be noted. First we have, from the definition (6),

$$\Delta^2(\kappa \cdot u) = \kappa^2 \cdot \Delta^2(u) \tag{9a}$$

$$\Delta(\kappa \cdot u) = |\kappa| \cdot \Delta(u) \tag{9b}$$

By κ we denote a constant factor. Second, we compute the change in the value of the dispersion arising when the deviations δu_m and δu^i, respectively, are not referred to the average $M(u)$ but to any other constant value u_0, that is, when we employ the deviations $\delta_0 u_m$ and $\delta_0 u^i$, which are defined by

$$\delta_0 u_m = u_m - u_0 \qquad\qquad \delta_0 u^i = u^i - u_0 \tag{10}$$

Using the relation

$$\delta_0 u_m = [u_m - M(u)] + [M(u) - u_0] = \delta u_m - \delta u_0 \qquad (11)$$

we derive for the dispersion[1] calculated in terms of the deviations (10)

$$\Delta_0^2(u) = M(\delta_0^2 u_m)_m = M(\delta u_m - \delta u_0)^2$$

$$= \sum_{m=1}^{r} P(A,B_m) \cdot [\delta^2 u_m - 2\delta u_m \delta u_0 + \delta^2 u_0]$$

$$= M(\delta^2 u_m)_m - 2\delta u_0 \cdot M(\delta u_m)_m + \delta^2 u_0$$

With (3) and (6) the equation assumes the form

$$\Delta_0^2(u) = \Delta^2(u) + [M(u) - u_0]^2 \qquad (12)$$

This simple relation may be called the *theorem of the shift of the reference point*, or *shift theorem*, in a notation introduced by von Mises. The dispersion $\Delta^2(u)$ referred to the average is a minimum when compared to dispersions referred to other values u_0. For in (12) there is added to $\Delta^2(u)$ the always-positive term $[M(u) - u_0]^2$, which vanishes only for $u_0 = M(u)$.

If we put in (10) $u_0 = 0$, we have $\Delta_0^2(u) = M(u^2)$; and solving (12) for $\Delta^2(u)$ we obtain the relation

$$\Delta^2(u) = M(u^2) - M^2(u) \qquad (13)$$

The relations (12) and (13) hold even for the quantities Δ^{n2} and M^n, since the derivation given remains valid when we introduce these quantities in place of Δ^2 and M and replace $P(A,B_m)$ by $F^n(A,B_m)$. Thus we have

$$\Delta_0^{n2}(u) = \Delta^{n2}(u) + [M^n(u) - u_0]^2 \qquad (14)$$

$$\Delta^{n2}(u) = M^n(u^2) - M^{n2}(u) \qquad (15)$$

There exists a certain probability that plays a characteristic part in the distribution of the amounts u^i. If we envisage a new element of the sequence, its value u^i, which is not yet known, depends on which of the B_m will be realized. Now we can ask for the probability w_δ that the amount u^i does not deviate by more than δ from the average $M(u)$.

For the probability w_δ it is possible to derive a characteristic inequality. For this purpose the amounts u_m are divided into two classes: to class I belong all the amounts for which $|\delta u_m| \leqq \delta$; to class II belong all the other u_m. Formulas (6) can be divided in two partial sums by extending the summation in the first partial sum over the amounts of class I, and in the second partial sum over the amounts of class II:

$$\Delta^2(u) = \sum_{\substack{m=1 \\ \text{I}}}^{r} P(A,B_m) \cdot \delta^2 u_m + \sum_{\substack{m=1 \\ \text{II}}}^{r} P(A,B_m) \cdot \delta^2 u_m \qquad (16)$$

[1] We write $M(u + v)^2$ instead of $M([u + v]^2)$, and $M^2(u + v)$ for $[M(u + v)]^2$.

Since none of the terms is negative, the right side of the equation will either diminish or remain the same when we drop the partial sum I; the partial sum II is further decreased (or left unchanged in the limiting case) when we write in it everywhere δ^2 instead of $\delta^2 u_m$, for in this sum we always have $|\delta u_m| > \delta$. Therefore we have

$$\Delta^2(u) \geqq \delta^2 \cdot \sum_{\substack{m=1 \\ \text{II}}}^{r} P(A,B_m) \tag{17}$$

But the sum of the probabilities is equal to the probability of $|\delta u_m|$ being greater than δ, that is, it equals $1 - w_\delta$. By solving the inequality (17) for w_δ we thus derive

$$w_\delta \geqq 1 - \frac{\Delta^2(u)}{\delta^2} \tag{18}$$

This important inequality is called *Tchebychev's inequality* in honor of the mathematician who first formulated it. It establishes a relation between the desired probability w_δ and the dispersion $\Delta^2(u)$. Conversely, the dispersion $\Delta^2(u)$ acquires by this inequality a special importance, as it determines a lower limit for the probability w_δ that an amount u^i does not deviate from the average by more than δ. But the inequality is of practical interest only if the limit δ considered is greater than the dispersion, since for $\delta \leqq \Delta(u)$ the inequality states merely the trivial fact that $w_\delta \geqq 0$. Nevertheless, the inequality will be used later for an important purpose.

§ 38. Average and Dispersion for a Combination of Events

We extend the concepts developed to cover the case of several sequences that lead to combinations $B_m . C_l$ and thus to the combination of amounts. Assume we have the two disjunctions $B_1 \vee \ldots \vee B_r$ and $C_1 \vee \ldots \vee C_t$, which belong to different sequences and to which the amounts $u_1 \ldots u_r$ and $v_1 \ldots v_t$ are coördinated; the amount w_{ml} may be coördinated to a combination $B_m . C_l$. For this amount w_{ml} we make the special assumption that it is composed additively of the amounts u_m and v_l, that is,

$$w_{ml} = u_m + v_l \tag{1}$$

We calculate first the average $M(w)$, which must be written in the detailed notation according to (8*b*, § 35):

$$M(w) = M(A,B_m . C_l;w_{ml})_{ml} = \sum_{m=1}^{r} \sum_{l=1}^{t} P(A,B_m . C_l) \cdot w_{ml} \tag{2}$$

or with (1), when we divide the whole expression into two terms and factorize the probability of the logical product according to the two forms of the rule of the product:

$$M(u+v) = \sum_{m=1}^{r} \sum_{l=1}^{t} P(A,B_m . C_l) \cdot (u_m + v_l)$$

$$= \sum_{m=1}^{r} P(A,B_m) \cdot u_m \cdot \sum_{l=1}^{t} P(A . B_m,C_l)$$

$$+ \sum_{l=1}^{t} P(A,C_l) \cdot v_l \cdot \sum_{m=1}^{r} P(A . C_l,B_m)$$

$$= M(u) + M(v) \qquad (3)$$

Here we have used the relation holding on account of the completeness of the disjunction analogous to (9, § 35):

$$\sum_{l=1}^{t} P(A . B_m,C_l) = 1 \qquad \sum_{m=1}^{r} P(A . C_l,B_m) = 1 \qquad (4)$$

The average is therefore additive when the amounts are additive.

For the sake of clarity, (3) may be written with the statement of the subscripts or superscripts of summation. We have two ways of writing:

$$M(u^i + v^i)^i = M(u^i)^i + M(v^i)^i \qquad (3a)$$

$$M(u_m + v_l)_{ml} = M(u_m)_m + M(v_l)_l \qquad (3b)$$

The first notation results from the statistical, the second from the theoretical, definition of the average. The plus-sign in the argument of the M-symbol on the left side of (3) has only a symbolic significance. If we wish to transform its meaning into an arithmetical significance, as on the left sides in (3a) and (3b), the two different subscripts m and l must be introduced when we use subscripts, as in (3b).

It is easily seen that additivity corresponding to (3) holds even for the mean values M^n; a proof results by writing everywhere F^n instead of P in the derivation given for (3). We have, therefore,

$$M^n(u+v) = M^n(u) + M^n(v) \qquad (3c)$$

Equation (3) was derived by general methods without any assumption concerning the dependence or independence of B_m and C_l. If we wish to derive an analogous result for the dispersion, however, we must make the assumption that the events B_m and C_l are mutually independent. We write with (3)

$$\delta w_{ml} = \delta(u_m + v_l) = u_m + v_l - M(u+v)$$

$$= u_m - M(u) + v_l - M(v) = \delta u_m + \delta v_l \qquad (5)$$

and thus we obtain

$$\Delta^2(u+v) = \sum_{m=1}^{r}\sum_{l=1}^{t} P(A,B_m.C_l)\,(\delta u_m + \delta v_l)^2 \tag{6}$$

Assuming the independence of B_m and C_l, we can factorize the probability according to the special theorem of multiplication. We therefore have

$$\Delta^2(u+v) = \sum_{m=1}^{r} P(A,B_m)\cdot\delta^2 u_m \cdot \sum_{l=1}^{t} P(A,C_l)$$
$$+ \sum_{l=1}^{t} P(A,C_l)\cdot\delta^2 v_l\cdot\sum_{m=1}^{r} P(A,B_m)$$
$$+ 2\sum_{m=1}^{r} P(A,B_m)\cdot\delta u_m\cdot\sum_{l=1}^{t} P(A,C_l)\cdot\delta v_l \tag{7}$$

The last term vanishes because of (3, § 37); with (9, § 35) and (6, § 37) we thus derive

$$\Delta^2(u+v) = \Delta^2(u) + \Delta^2(v) \tag{8a}$$

$$\Delta(u+v) = \sqrt{\Delta^2(u) + \Delta^2(v)} \tag{8b}$$

The quadratic dispersion is additive for independent quantities when the amounts are additive.

For the sake of clarity we may reintroduce the notation specifying the subscripts or superscripts of summation:

$$\Delta^2(u^i + v^i)^i = \Delta^2(u^i)^i + \Delta^2(v^i)^i \tag{8c}$$

$$\Delta^2(u_m + v_l)_{ml} = \Delta^2(u_m)_m + \Delta^2(v_l)_l \tag{8d}$$

Formula (8*c*) corresponds to the statistical, (8*d*) to the theoretical, definition of the dispersion.

The relation (8*b*) expresses the important *law of the spreading of the disper-sion*. Because of the inequality

$$\sqrt{\Delta^2(u) + \Delta^2(v)} < \Delta(u) + \Delta(v) \tag{9}$$

which is easily derived on the condition that the dispersions do not vanish we have

$$\Delta(u+v) < \Delta(u) + \Delta(v) \tag{10}$$

For nonvanishing dispersions, the linear dispersion of the sum is smaller than the sum of the linear dispersions; thus the dispersion does not increase at the same rate as the amount. The result can be explained as follows. The fluctuations of the amounts u^i and v^i were assumed to be mutually inde-pendent; therefore an extreme value of u^i will rarely coincide with an extrem value of v^i. The extreme deviations of one amount will thus be smoothed out or even compensated by the behavior of the other amount.

If, for instance, u^i and v^i represent each the number of passengers in a streetcar, counted every day at a specific time in a certain street, fluctuations in the number will occur. The total number of persons in both streetcars will then fluctuate a little more than in each car separately; the deviations, however, will not be twice as strong, but will grow according to (8*b*). Thus, when the dispersions $\Delta(u)$ and $\Delta(v)$ are assumed to be equal, the total number will merely fluctuate $\sqrt{2}$ times as much as the number in one car. If only a few passengers happen to be in one car, then, in general, the same will not be the case in the other car; and so the fluctuations are compensated. Formula (8*b*) in combination with (10) is therefore often called the *law of the compensation of the dispersion*. The independence of the fluctuations is essential for this result. If a large deviation in the number of persons in one car were always linked to a large deviation in the other car, no compensation would result, and the fluctuations of the total number of persons would be twice as great as those of the individual number. This simple consideration may explain why the relations (8) can be derived only if the amounts are assumed to be independent.

The case of extreme dependence can be illustrated through the preceding example by reference to the amount of fares received. If each passenger pays 10 cents, the number of cents paid is 10 times the number of persons; the fluctuations in the amounts of money will then be 10 times as strong as the fluctuations in the number of persons. No compensation results here, because the amounts of money are linked strictly to the number of persons present. The example corresponds to the relation formulated in (9*b*, § 37).

§ 39. Average and Dispersion in the Lattice

The principle of the compensation of the dispersion, in contradistinction to the multiplication of the dispersion as expressed in (9*b*, § 37), can be illustrated even better when the relations (8, § 38) are transferred from two to a greater number of amounts, which belong to further sequences. In order to have a suitable notation, we no longer denote the amounts by u_m and v_l, or u^i and v^i, but write u^k_m and u^{ki}, thus indicating by the first superscript that the amount belongs to the sequence k. The second superscript refers to the element i of the sequence k; the subscript indicates the value m of the amount belonging to the sequence k. As before, the second superscript can be interchanged with the subscript in summation expressions. The number of sequences may be n. The sequences constitute a lattice, which is finite in respect to the dimension of the first superscript and infinite for the dimension of the second superscript.

The dispersion of the mean value of the u_m^k may be calculated, starting with the statistical definition of the mean value, which is illustrated with the help of the schema

$$\begin{array}{cccccc|c|c} u^{11} & u^{12} & u^{13} & . & . & . & \hat{u}^1 & \Delta^2(u^1) \\ u^{21} & u^{22} & u^{23} & . & . & . & \hat{u}^2 & \Delta^2(u^2) \\ . & . & . & . & . & . & . & . \\ u^{k1} & u^{k2} & u^{k3} & . & . & . & \hat{u}^k & \Delta^2(u^k) \\ . & . & . & . & . & . & . & . \\ u^{n1} & u^{n2} & u^{n3} & . & . & . & \hat{u}^n & \Delta^2(u^n) \\ \hline \beta^1 & \beta^2 & \beta^3 & . & . & . & \hat{\beta} & \Delta^2(\beta) \end{array} \tag{1}$$

The horizontal rows represent the individual sequences. In the vertical direction the mean value β^i of the u^{ki} is indicated below each column; in the horizontal direction each individual sequence has a mean value $\hat{u}^k$. We thus have the notation

$$\hat{u}^k = M(u^{ki})^i \qquad \Delta^2(u^k) = M((u^{ki} - \hat{u}^k)^2)^i \tag{2a}$$

$$\beta^i = M^n(u^{ki})^k = \frac{1}{n} \cdot \sum_{k=1}^{n} u^{ki} \tag{2b}$$

$$\hat{\beta} = M(\beta^i)^i = M(M^n(u^{ki})^k)^i \qquad \Delta^2(\beta) = M((\beta^i - \hat{\beta})^2)^i \tag{2c}$$

We begin by calculating the average $\hat{\beta}$ of the mean values β^i. For this purpose we make use of the relation

$$M(\sum_{k=1}^{n} u^{ki})^i = \sum_{k=1}^{n} M(u^{ki})^i \tag{3}$$

which is easily derived from (3, § 38) by generalization. With (13, § 35) we then obtain

$$M(M^n(u^{ki})^k)^i = \frac{1}{n} \cdot M(\sum_{k=1}^{n} u^{ki})^i$$

$$= \frac{1}{n} \cdot \sum_{k=1}^{n} M(u^{ki})^i = M^n(M(u^{ki})^i)^k \tag{4}$$

or, by dropping the superscripts,

$$M(M^n(u)) = M^n(M(u)) \tag{4a}$$

Average and mean are therefore commutative operations. Using the notation employed in schema (1), we write for (4)

$$\hat{\beta} = M(\beta^i)^i = M^n(\hat{u}^k)^k \tag{4b}$$

This means that the quantity $\hat{\beta}$, which represents the average of the last row, can simultaneously be constructed as the mean value of the averages $\hat{u}^1 \ldots \hat{u}^n$ presented in the column ending with $\hat{\beta}$.

In order to calculate the dispersion of the β^i, we regard the mean value M^n as resulting from an addition of amounts. Now we can easily see that the law of the addition of the quadratic dispersion (8*a*, § 38) can be extended to apply to further amounts if they are additive and if the sequences are completely independent (§ 23). We thus derive the generalization of (8*a*, § 38):[1]

$$\Delta^2\left(\sum_{k=1}^{n} u^{ki}\right)^i = \sum_{k=1}^{n} \Delta^2(u^{ki})^i \tag{5}$$

obtaining, with the help of (9*a*, § 37) and (2*b*),

$$\Delta^2(M^n(u^{ki})^k)^i = \frac{1}{n^2} \cdot \Delta^2\left(\sum_{k=1}^{n} u^{ki}\right)^i = \frac{1}{n^2} \cdot \sum_{k=1}^{n} \Delta^2(u^{ki})^i$$

$$= \frac{1}{n} \cdot M^n(\Delta^2(u^{ki})^i)^k \tag{6}$$

By dropping again the superscripts of summation, we write (6) in the form

$$\Delta^2(M^n(u)) = \frac{1}{n} \cdot M^n(\Delta^2(u)) \tag{6a}$$

The quadratic dispersion of the n-fold mean value is only the n-th part of the n-fold mean value of the quadratic dispersion. Average and dispersion are noncommutative operations; if the operations are interchanged, the factor $\frac{1}{n}$ is to be added. In the presentation of schema (1) this result means: the quantity $\Delta^2(\beta)$ does not constitute the mean value of the amounts $\Delta^2(u^k)$ in the corresponding column, but represents only the n-th part of this mean value. In this respect the dispersion differs from the average. With the notation

$$s^2 = M^n(\Delta^2(u)) \qquad \sigma^2 = \Delta^2(M^n(u)) \tag{6b}$$

[1] I should like to draw attention to a possible fallacy. If the amounts u_m^k are equal for the different values of k, that is, if corresponding amounts are the same in the individual sequences, we might be inclined to infer from equation (5) the relation $\Delta^2(n \cdot u) = n \cdot \Delta^2(u)$, in contradiction to (9*a*, § 37). The inference is fallacious, however. From the notation (8*d*, § 38) we see that amounts with different subscripts are to be added; the inference, therefore, is impossible. Furthermore, we cannot derive $u^{ki} = u^{si}$ from $u_m^k = u_m^s$, so that it is also impossible to make the inference when the notation of the superscripts is used.

TABLE 2

Number of Passengers at a Railway Station

A. NUMBERS IN THE ORDER OBSERVED

Train No.	1	2	3	4	5	6	7	8	9	10	11	12	13	14	15	16	17	18	19	20	21	22	23	24	$\hat{u}^k$	$\Delta^2(u^k)$	$\Delta(u^k)$
Car No. I	41	16	13	28	25	33	32	21	26	37	38	36	37	30	57	26	9	44	40	38	19	33	27	40	31.08	95.06	9.75
" " II	25	11	14	22	20	27	17	13	25	16	24	27	29	22	30	16	8	49	44	32	11	18	10	24	22.25	84.64	9.20
" " III	21	7	2	23	15	30	11	5	14	17	29	21	27	14	19	7	3	18	23	29	15	22	11	25	17.00	67.40	8.21
" " IV	24	5	13	16	21	22	33	6	25	24	30	30	54	15	41	13	4	27	40	42	14	33	13	21	23.58	151.04	12.29
" " V	24	22	5	19	25	36	18	8	32	25	33	19	31	19	23	15	3	28	24	33	16	31	21	27	22.37	97.22	9.86
" " VI	39	22	10	42	43	31	33	15	59	47	51	43	37	19	47	14	10	33	45	53	19	22	12	46	33.00	219.20	14.57
																									$\hat{\beta}$	$\Delta^2(\beta)$	$\Delta(\beta)$
β^i	29.0	13.9	9.5	24.8	24.8	29.8	24	11.3	30.1	27.7	34.2	29.3	35.8	19.8	36.2	15.2	6.2	33.2	36.0	37.8	15.7	26.5	15.7	30.5	24.88	83.41	9.13

$s = 10.91$ $\sigma = 9.13$

B. NUMBERS AFTER SHUFFLING EACH HORIZONTAL ROW

Col. No.	1	2	3	4	5	6	7	8	9	10	11	12	13	14	15	16	17	18	19	20	21	22	23	24	$\hat{u}^k$	$\Delta^2(u^k)$	$\Delta(u^k)$
Car No. I	19	21	33	25	38	37	44	9	16	37	40	36	41	26	40	27	57	26	30	33	28	32	13	38	31.08	95.06	9.75
" " II	14	22	25	30	32	16	27	24	18	44	24	25	22	16	8	49	17	11	10	27	20	29	11	13	22.25	84.64	9.20
" " III	14	11	15	5	29	22	17	25	14	23	29	2	11	19	27	21	7	30	21	23	15	18	7	3	17.00	67.40	8.21
" " IV	14	13	33	15	4	13	42	33	40	24	41	30	24	13	16	30	5	21	21	54	25	22	6	27	23.58	151.04	12.29
" " V	25	27	31	33	18	23	5	19	32	16	24	21	3	24	19	19	36	28	33	31	25	22	8	15	22.37	97.22	9.86
" " VI	47	39	51	33	12	45	43	10	46	37	31	22	59	10	42	43	19	33	53	14	15	19	22	47	33.00	219.20	14.57
																									$\hat{\beta}$	$\Delta^2(\beta)$	$\Delta(\beta)$
β^i	22.1	22.1	31.3	23.5	22.1	26.0	29.7	20.0	27.7	30.2	31.5	22.7	26.7	18.0	25.3	31.5	23.5	24.8	28.0	30.3	21.3	23.7	11.2	23.8	24.88	21.89	4.68

$s = 10.91$ $\sigma_{\text{theor.}} = \frac{s}{\sqrt{6}} = 4.45$ $\dot{\sigma}_{\text{obs.}} = 4.68$

we may write for (6*a*)

$$\sigma^2 = \frac{1}{n} \cdot s^2 \qquad \sigma = \frac{1}{\sqrt{n}} \cdot s \qquad (6c)$$

The quantity s^2 can be regarded as a characterization of the dispersion of the whole lattice of the u^{ki}. The quantity σ^2 characterizes only the dispersion of the mean values β^i; the fluctuations of the individual u^{ki} within the vertical columns do not find an expression in the quantity σ^2. But for independent sequences both quantities are connected by the relation that the quantity σ^2 is the n-th part of the quantity s^2. Correspondingly, the linear dispersion σ of the mean value is the $\sqrt{n}$-th part of the linear dispersion s of the whole lattice, and therefore we speak of the *$\sqrt{n}$-law of the dispersion.*

This law, which is a generalization of (8*b*, § 38), is an impressive manifestation of the compensation of the dispersion. It formulates the assertion that the mean value of a number of independent quantities fluctuates the less, the more quantities are concerned. The law possesses an extraordinary importance in the physical world. It entails the consequence that the phenomena of fluctuation exhibited by microscopical and submicroscopical particles are not observable for macroscopic objects. Since the elementary processes are mutually independent, or at least very nearly so, their fluctuations compensate one another to a high degree. That we can speak, for instance, of the temperature of a macroscopic body and determine the quantity objectively by exact measurements derives from this law: for 1 cc. of a gas the number n of molecules is about 10^{18}, and even if we assume a dispersion s of 100% for the kinetic energy of the individual molecule, the dispersion of the mean value of all these kinetic energies, which corresponds to the temperature, levels off to $\sigma = \frac{100}{\sqrt{10^{18}}}\%$, that is, to one ten-millionths of 1%. *The probability law of the compensation of the dispersion formulates the very law of nature to which we owe the uniformity of the macroscopic world.*

This law may be illustrated by an example from everyday life, which at the same time makes it evident that the condition of independence is indispensable. At a railway station the number of passengers occupying each car was counted for every passing train.[2] Most of the trains had six cars; only three of the twenty-four trains counted had eight cars, in which case, instead of the fifth and sixth cars, the seventh and eighth were substituted for the statistics. The number of passengers is given in table 2*A* (p. 198). Each vertical column corresponds to one train; each horizontal row represents a certain car defined in terms of the same location in different trains. The first and the last car carry more passengers than the others, owing to the fact that they

[2] These observations were compiled in 1933 at a station of the Berlin Metropolitan Railway, which connects the various parts of the city.

stop nearer the exits of the station. The cars in the middle have fewer passengers because they contain second-class compartments, for which the fare is higher. The trains exhibit a *characteristic occupancy,* that is, some trains are fully occupied and others are almost empty; the differences result from the fact that the trains come from different directions or branch off later.

For this table the quantities $\hat{\alpha}^k$ are constructed by taking the mean values horizontally; the β^i are constructed as the vertical mean values. By the use of these values the quantities s and σ are calculated. We thus find the value $\sigma = 9.13$, which is only a little lower than the value $s = 10.91$. The smallness of the difference is explained by the fact that the relation (6*c*) is not applicable, because, owing to the characteristic occupancy of the trains, the horizontal rows are not mutually independent. Therefore we find only a weak compensation of the dispersion.

In order to obtain independent horizontal rows, the numbers of each horizontal row were written on slips of paper and the slips were thoroughly shuffled for each row separately. The result of the shuffling is given in table 2*B*. The horizontal mean values are the same as in table 2*A*, since every horizontal row as a whole remains unchanged. The vertical mean values are changed essentially, however; it can be seen at a glance that they fluctuate much less than the values of the individual rows. Thus the extreme values are given by 31.5 and 11.2, whereas the extreme values of the total lattice are 59 and 2. Correspondingly, the calculation of σ leads to a smaller value than before; we find $\sigma = 4.68$. Since s is the same as in table 2*A*, we would obtain from (6*c*) for $n = 6$ the value $\sigma = 4.45$, with which the observed value $\sigma = 4.68$ is in good agreement. This example[3] confirms the law of the compensation of the dispersion.

[3] The German edition of the book contains at this place a presentation of three kinds of dispersion definable for a lattice, a distinction which clarifies a puzzle connected with the dispersion of the mean value, namely, the problem of whether the observed dispersion should be divided by n or by n—1. I refer the reader who is interested in these problems to the German edition.

Chapter 6

CONTINUOUS EXTENSIONS OF THE CONCEPT OF PROBABILITY SEQUENCE

CONTINUOUS EXTENSIONS OF THE CONCEPT OF PROBABILITY SEQUENCE

§ 40. The Geometrical Interpretation of the Axiom System

The axiom system of the theory of probability was constructed in such a manner that it could be treated as a formal system of implicit definitions without the use of any interpretation of the concept of probability. An interpretation was introduced later in the form of the frequency interpretation. Since all formal systems of axioms have various admissible interpretations, we can coördinate interpretations of different kinds to the axiom system of probability. An admissible interpretation, which has no connection with the frequency interpretation, will now be developed.

As before, we understand, by A,B,C, classes, or sets, but we do not relate them to sequences; the x_i and y_i are elements of classes for which no definite order is assumed. However, a one-one correspondence of the x_i, y_i, and z_i, expressed by the subscript, is assumed as before. As explained in § 7, we understand by the logical sum of two classes their joint class, or narrower couple disjunct; by the logical product, their common class, or narrower couple conjunct, depending on whether we are concerned with classes of the same or separate domains of elements. The symbols "or" and "and" thus assume the meaning that they have in the calculus of classes.

We simplify the investigation by assuming the classes to be classes of geometrical points in a plane, thus arriving at a *geometrical interpretation* of the axiom system. The classes are then given by geometrical areas. This interpretation also facilitates the introduction of a measure, which we need for this interpretation, in that it permits us to understand by the measure $M(B)$ of a class B the size of the area B. The presentation will be further simplified by assuming the elements x_i, y_i, z_i, carrying the same subscript, to be identical. Such a restriction to what was called "internal probability implication" in (5 and 6, § 9) actually does not impair the generality of the following considerations, since the extension to pairs of nonidentical elements can easily be carried through. The classes are illustrated in figure 8; the measure $M(A)$ is represented by the size of the area denoted by A, and, correspondingly, the measure $M(B)$ is represented by the size of the area denoted by B. The common class $A.B$ is indicated by shading; the joint class is given by the area covered by A and B together, for which the shaded area is counted but once.

By means of the measure of the classes, the probability coördinated to both classes is now defined in the form

$$P(A,B) = \frac{M(A.B)}{M(A)} \qquad (1)$$

The probability from A to B is defined as the ratio of the measure of the common part of both classes to the measure of class A. This coördinative definition is used instead of the frequency interpretation for the concept "probability" occurring in the axiom system.

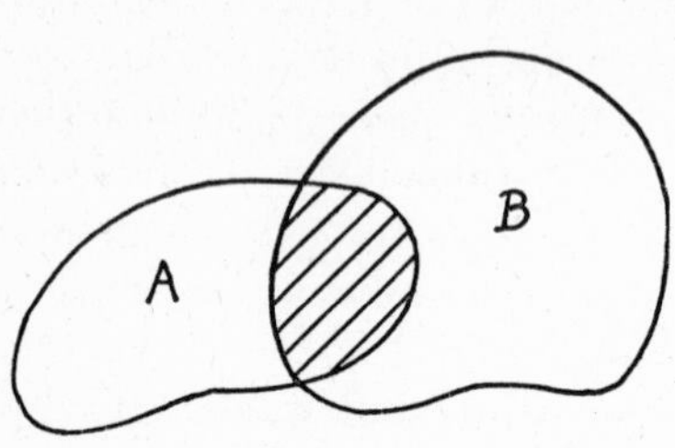

Fig. 8. Geometrical interpretation of probability concept by measure of areas.

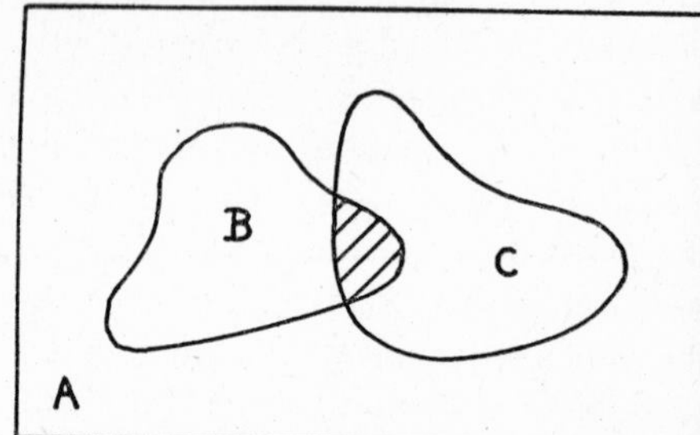

Fig. 9. Simplified geometrical interpretation of probability concept: $M(A) = 1$, B and C lie within A.

For formulas all containing A in the first term it is advisable to choose a measure such that $M(A)$ becomes equal to 1; besides, the conditions are simplified by assuming the classes B and C as lying completely within A. This simplification is permissible because we use only the parts of B and C that are situated inside A. The resulting relations are illustrated in figure 9, in which the class A is symbolized by the rectangle. We thus arrive at the following interpretation of symbol combinations occurring in the formulas:

$$\begin{aligned} P(A,B) &= M(B) \\ P(A,C) &= M(C) \\ P(A,B.C) &= M(B.C) \\ P(A,B \vee C) &= M(B \vee C) \\ P(A.B,C) &= \frac{M(B.C)}{M(B)} \end{aligned} \qquad (2)$$

The geometrical interpretation also illustrates a certain peculiarity of the calculus of probability that was mentioned above (§ 21): that the probability relation existing between two classes B and C relative to a third class A are determined by three fundamental probabilities, namely, the probabilities $P(A,B)$, $P(A,C)$, and $P(A.B,C)$, the latter being replaceable by the probability $P(A,B.C)$. That the two probabilities $P(A,B)$ and $P(A,C)$ are not sufficient is seen from figure 9; the size of the areas B and C does not determine the size of the area of their mutual overlapping. The areas can be moved individually, and only when their *degree of coupling* is given, for instance, by the size of their common area, is their mutual position determined as far as necessary for the computation of all their probabilities. Furthermore, the geometrical interpretation supplies an instructive illustration of the general theorem of addition. The area of the joint class $B \vee C$ is equal to the sum of the areas of B and C diminished by their common area; otherwise the latter would be counted twice.

It is easily seen that all the axioms I–IV are satisfied by the geometrical interpretation. The logical implication $A \supset B$, occurring in axiom II,1, denotes the relation of class inclusion according to (21, § 7), which in the geometrical interpretation means that the area A lies completely within the area B. For this case the definition (1) leads to the result $P(A,B) = 1$, in agreement with axiom II, 1. Axiom I asserts that the ratios of the areas become indeterminate only if the area A is zero. That the other axioms II,2–IV are fulfilled can easily be verified by similar considerations. Only the axioms V do not find an interpretation, since in the geometrical interpretation it is impossible to find an equivalent of the phase symbols. This impossibility results from the fact that the elements of the geometrical point sets are arranged in a continuous order, whereas the elements of probability sequences constitute discrete sequences. In the geometrical model, therefore, the theory of the order of probability sequences cannot be represented. All other probability relations, however, find a geometrical interpretation in the model given.

§ 41. Definition of Probability Sequences with Continuous Attribute

The geometrical interpretation of the concept of probability is connected with an extension of the concept of probability sequence now to be explained.

The extension may be illustrated by the example of shooting at a target. The individual hit, that is, the element x_i of the probability sequence, is characterized by a point on the target; the classes A and B are then areas of the target. For instance, A may represent the total area of the target; B may be a certain inner circle around the center of the target (as shown in fig. 10, p. 206). If we determine statistically the probability $P(A,B)$—that is, the probability that a shot fired at the target A hits the inner circle B—it is sufficient to know about each hit whether it lies within the areas A or B. But if we specify the exact location of the hit on the target we know more about the element x_i than merely whether it belongs to A or to B. From this more detailed statement we can see whether the hit is contained in A or in B; and, moreover, we can determine whether the hit belongs to any other area of the target that we may wish to consider. The precise location at which the shot has hit the target is called the *attribute* of the event x_i. This location is characterized by the use of two coördinates u,v on the target. The manifoldness scaffolded by the coördinates u,v is called the *attribute space;* in the illustration it is given by the target. If we know for each x_i its attribute, or its point in the attribute space, we can determine the probability of a hit for any arbitrarily chosen area of the target.

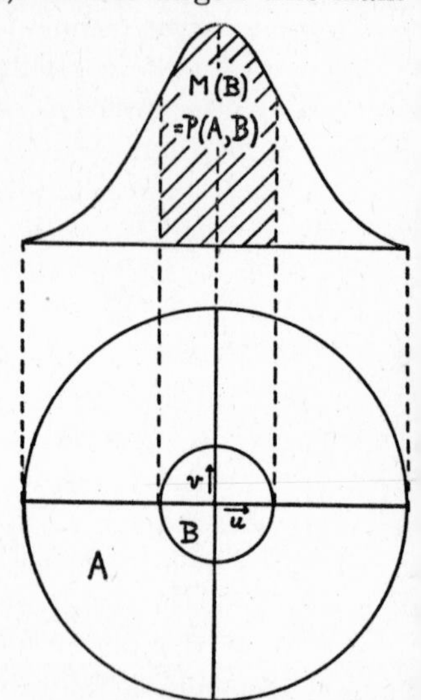

Fig. 10. Probabilities referring to a target. Top view shows target; front view indicates measure function corresponding to the probability, according to (1) and (4).

The probability for such areas A and B is determined through the frequency interpretation; we count the number of hits according to (3, § 16), that is, we take the ratio of the number of hits for the common class $A.B$ to the number of hits for A. The ratio will usually not be equal to the ratio of the corresponding geometrical areas. But the considerations of § 40 show how to proceed. It was explained there that the axioms of the probability theory can be interpreted by ratios of areas, and we must therefore be able to introduce on the target a measure function $M(G)$ for areas G such that the measures coördinated to the areas $A,B,A.B$ determine the probability $P(A,B)$ in the sense of (1, § 40).

It is convenient to assume that the measure function $M(G)$ can be expressed as the integral of a scalar function $\varphi(u,v)$ taken over the domain G:

$$M(G) = \iint_G \varphi(u,v)dudv \tag{1}$$

This way of writing the measure function is possible when the function $M(G)$ possesses certain limit properties for $G \to 0$. These conditions can be expressed in various ways. I shall use the following form: assume that for every series of *boxed-in* areas $G_1 \ldots G_n \ldots$, which contract toward a point, the limit

$$\lim_{n\to\infty} \frac{M(G_n)}{G_n} \tag{2}$$

exists and is a continuous function of the point. By "boxed-in" is meant that every G_n includes the following area G_{n+1}; by "contraction toward a point", that $\lim_{n\to\infty} G_n$ is a point. The limit (2) defines the value of the function $\varphi(u,v)$ at the point considered.

Since the measure function $M(G)$ is to supply the probability of the area, a condition corresponding to (2) must be satisfied by the probability values of the sequence, that is, we must require that for every series of boxed-in areas $B_1 \ldots B_n \ldots$, which contract toward a point, the limit

$$\lim_{n\to\infty} \frac{P(A,B_n)}{B_n} \tag{3}$$

exists and is a continuous function of the point. The condition (3) may be regarded as an axiom that probability sequences with a continuous attribute must satisfy, in addition to the axioms I–V. This axiom, which may be called the *axiom of continuity*, has no analogue for sequences with a finite number of attribute classes, $B_1 \ldots B_n$, and therefore does not belong in the elementary calculus of probability.

If the condition (3) is satisfied, the statistical probability may be represented by the expression

$$P(A,B) = \frac{\iint_{A.B} \varphi(u,v)dudv}{\iint_{A} \varphi(u,v)dudv} \tag{4}$$

For reasons of simplicity, the measure function M is chosen so that $M(A) = 1$; then (2, § 40) gives

$$P(A,B) = \iint_{B} \varphi(u,v)dudv \tag{5}$$

If φ is introduced as the third coördinate perpendicular to the plane u,v, the measure function φ of the target, according to general experience, assumes the form of a bell-shaped surface corresponding to the front view of figure 10 (p. 206) and resulting from rotation of the curve around its vertical axis. According to (1), the measure $M(G)$ coördinated to an area G of the target is presented by the volume of the cylinder erected above G, the upper boundary of which is given by the bell-shaped surface. For instance, the measure coördinated to the area B in the top view is the cylinder that is produced by a rotation of the shaded area in the front view.

A probability represented by an integral over an area, as symbolized in (4) and (5), is called *geometrical probability*. The possibility of a geometrical representation of probabilities results from the considerations given in § 40. By showing that both the frequency interpretation and the geometrical interpretation satisfy the axioms of the formal system of probability, that is, are interpretations of this system, we have demonstrated the *isomorphism*, or structural identity, of the two interpretations. Every operation carried out in terms of probability formulas entails analogous operations in the frequency interpretation and the geometrical interpretation. Any derived probability relation is, therefore, symbolized in the geometrical interpretation by those geometrical relations that have been specified above for the geometrical interpretation of the probability concept. For instance, the probability $P(A,B \vee C)$ is determined by the measure of the joint class of B and C when we use the same measure function φ that was introduced in the plane for B and C separately.

The foregoing considerations can easily be generalized for an attribute space of more than two dimensions. The attribute coördinated to the element x_i of the probability sequence is then a point in a multidimensional attribute space $u,v, \ldots$

Probability sequences of a continuous attribute are more general than the sequences discussed in the preceding chapters, so far as the classes A,B,C are not given constants of the sequences, but constitute variables. I call such a sequence a *primitive probability sequence;* I wish to express by this name that the sequence represents the root of a number of different probability sequences of the usual kind, or *classified probability sequences*, each of which results for some division of the attribute space. The transition from the primitive to the classified probability sequence, which is determined by the statement of the areas A and B, may be called *classification*. Classification is an operation by which the statement of the precise attribute coördinated to every element x_i of the sequence is replaced by the weaker statements of whether the element belongs to A or to B, respectively.

The possibility of extending the concept of probability sequence to that of primitive probability sequence, that is, of a sequence with variable classes, derives from the isomorphism existing between the geometrical interpretation of probability and the frequency interpretation. To this isomorphism we owe the existence of geometrical probabilities. The function φ determining the metric of the attribute space is called a *probability function;* it may also be called the *distribution* (a term introduced by von Mises) of the primitive probability sequence, since it determines the distribution of the attributes of the sequence.

As for sequences with coördinated steps of amounts, it is often convenient for primitive probability sequences to forget about the event sequences, or thing sequences, to which they refer, and to consider directly the sequence of number combinations, that is, of attribute points, given by the attribute sequence. This point sequence has the character of a probability sequence in which the classes A and B, the members of which are attribute points, replace the classes of events, or things.

The incorporation of primitive probability sequences into the axiomatic calculus of probability offers no difficulties. We can transfer the axioms I–V of the probability calculus to primitive probability sequences. Such a transfer is achieved by the condition that *these axioms are to be valid for every classified sequence that can be derived from the primitive probability sequence.* In this manner we can incorporate the theory of primitive sequences into the formal calculus of probability without employing the frequency interpretation. The theory of order can be transferred in the same way; for instance, we define the normal primitive probability sequence by the condition that any classified sequence derivable from it be a normal sequence in the sense of § 30.

The only addition to the axiom system that is required for primitive probability sequences is the postulate of continuity (3). An important consequence of this axiom, which greatly simplifies operations with an infinite number of attribute classes, may be explained: the satisfaction of the limit postulate (3) leads to the commutativity of the limit operation and the probability operator, that is, to the relation

$$\lim_{n \to \infty} P(A,B_n) = P(A, \lim_{n \to \infty} B_n) \tag{6}$$

The proof of this relation follows from the theorem, derived in the theory of integration, according to which the integral is a continuous function of its boundaries. This theorem holds even for certain places where the function φ is discontinuous; in fact, we are often concerned with probability functions that jump discontinuously from one value to another at individual points. For such points the postulate (3) is to be formulated somewhat differently by the distinction of an approach from one side or the other.

Although the postulate of continuity (3) is not derivable from the frequency interpretation, if offers no difficulties so far as the application of the theory of geometrical probabilities to physical reality is concerned. The postulate occupies a place similar to the logical position of the specializing conditions of the theory of order, such as the condition of absence of aftereffect. Whether the postulate (3) holds can be verified by empirical observation. How such verification is achieved will be shown in § 42, in the course of a study of the application of geometrical probabilities to practical problems. In this study of a chapter of the calculus of probability, which is of great practical importance, the frequency interpretation will always be used.

§ 42. Empirical Determination of a Probability Function

It will now be explained how a probability function is found by statistical observations. Retaining the example of the target, imagine its area to be covered by a rectangular system of coördinates with the distances du and dv, respectively (fig. 11); each rectangular area may be determined by the statement of the coördinates u_m, v_l, belonging to its lower left corner. By counting the number of hits for each rectangle we determine statistically the probability w_{ml} of hitting the rectangular area $u_m v_l$. If x_i is an individual shot, then x_i belongs to the class B_{ml} if x_i makes a hit inside the rectangle $u_m v_l$; and x_i belongs to the class A if x_i hits the target at all. We have then (for the N-notation see § 16)

$$w_{ml} = P(A,B_{ml}) = \lim_{n\to\infty} \frac{N^n(A\,.\,B_{ml})}{N^n(A)} \tag{1}$$

In practice we cannot reach the limit, of course, but must break off after some large number of shots. Theoretically, however, the frequency interpretation requires the existence of the limit (1) as the condition that enables us to speak of the probability w_{ml}. The transition from the observed number of hits to the limit represents the usual inductive inference, without which practical applications of the calculus of probability cannot be constructed (see § 17). After the existence and the value of the limit have been inductively ascertained for one rectangle, corresponding statements are made for the other rectangles. For a none-too-large number of rectangles this statement can be tested as for the first rectangle; the extension to further rectangles represents another inductive inference.

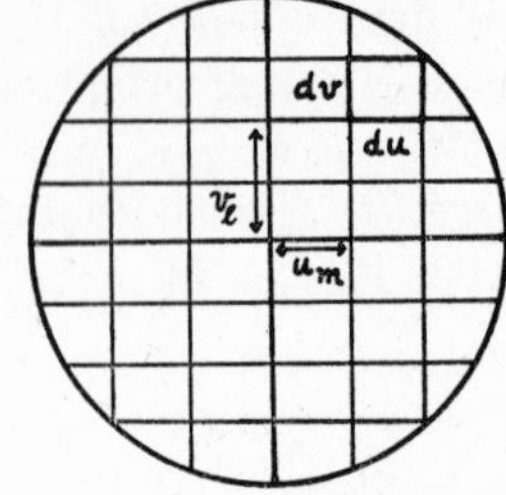

Fig. 11. Coördinates on a target.

Above the rectangle $u_m v_l$ we now draw as the third, or vertical, coördinate a quantity φ^*_{ml} defined by

$$\varphi^*_{ml} \cdot du_m \cdot dv_l = w_{ml} \tag{2}$$

Thus we erect on top of the area $u_m v_l$ a rectangular column with a volume equal to w_{ml}. The totality of the columns, which has a staircase-shaped surface, is called a *histogram.*

If the same construction is repeated for smaller rectangles $du_m dv_l$, the probabilities w_{ml} become smaller, because a smaller area is hit more rarely; but the ordinates φ^*_{ml} need not change much thereby, since the smaller frequency is expressed by the decrease of the width of the column. Assume that the height of the column remains virtually constant, or, more precisely, that when the construction is repeated for smaller and smaller rectangles, there exists at every place the limit

$$\lim_{\substack{du_m\to 0\\ dv_l\to 0}} \frac{w_{ml}}{du_m dv_l} = \varphi(u_m,v_l) \tag{3}$$

and that this limit is a continuous function of the point u_m,v_l. Of course, this assumption, including the existence of the limit (3), is not capable of a strict proof; but it can be inferred on the basis of the observational material by means of an inductive inference, an extrapolation of the observed regularities, corresponding to the inductive determination of the limit of the frequency in an infinite sequence (see § 17). We thus arrive at an empirical verification of the postulate (3, § 41) that was introduced for primitive probability sequences. This postulate does not impose general conditions on the physical world, but is treated like all other specializing conditions: we restrict its application to empirical material that satisfies the postulate.

The introduction of the condition (3) allows us to replace the staircase-shaped surface by a smooth surface $\varphi(u,v)$; for finite areas B the condition means that the probability is determined by the double integral

$$P(A,B) = \iint_G \varphi(u,v)\, dudv \tag{4}$$

This means that the probability relations are characterized by a probability function φ in the sense previously specified. In the example of the target, the function $\varphi(u,v)$ is represented by the bell-shaped surface drawn in the front view of figure 10 (p. 206).

Although the relation (3) is not strictly verifiable, because the limit of an infinite sequence is not accessible to observation, the practical procedure reflects the double transition to a limit expressed in the two relations (1) and (3). The transition to smaller rectangles cannot be continued until the number of observed hits in each rectangle is sufficiently large; otherwise we would arrive at a wrong picture of the distribution. Theoretically speaking, this means that the two transitions to a limit expressed in (1) and (3) are not commutative.

From (3) and (4) we see that φ does not possess the character of probability, which, rather, can be ascribed only to the integral over φ; the function φ is therefore called a *probability density*. In particular, $\varphi(u,v)$ is called a *two-dimensional probability density*, since only a double integration leads to a probability. Correspondingly, we speak of a *one-dimensional probability density* for a one-dimensional attribute space and, similarly, of *multidimensional probability densities* for attribute spaces of more than two dimensions. For small areas $dudv$ we can regard the value

$$\varphi(u,v)\, dudv \tag{5}$$

as the approximate value of the probability. This notation becomes strictly correct whenever we proceed from such expressions to integrations, and we shall therefore use it for the sake of brevity in the following discussions. The notation clearly illustrates the fact that the probability goes toward zero if the area $dudv$ goes to zero; the probability of hitting a precisely prescribed point, therefore, is $= 0$. It is this very fact that makes φ a density, as distinct from a probability.

It is a general property of all probability functions that they satisfy the relation

$$\int_{-\infty}^{+\infty}\int_{-\infty}^{+\infty} \varphi(u,v)\, dudv = 1 \tag{6}$$

This integral represents the probability that the attribute point is located at some point within the attribute space, a probability which is $= 1$. We call (6) the *condition of normalization* for probability functions.

The empirical determination of a probability function may be illustrated by an example. Table 3 (p. 213) presents the result of measurements of the height of American recruits.[1] A total of 25,878 individuals were measured. A division by intervals of one inch was made, and the number of persons

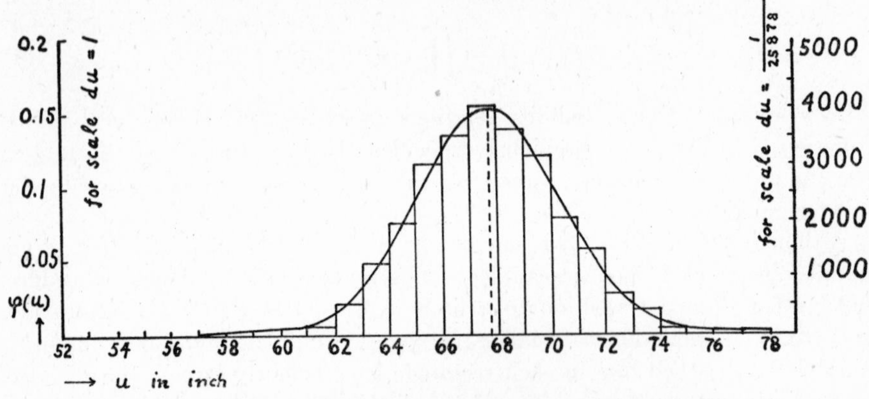

Fig. 12. Frequency distribution found in measurements of height of 25,878 American recruits.

falling into each interval was recorded. The last column of the table will be discussed later.

The table is represented graphically in figure 12. We are dealing with a one-dimensional attribute space; so only two axes are required. The values u of the height are given by the abscissa, and as the interval du the value of one inch is chosen, in correspondence with the table. Above every interval I have drawn a rectangle the height of which is chosen so that the area of the rectangle equals the probability that a certain person falls into this interval. The probability is calculated by dividing the corresponding number in the table by the total number 25,878. If the scale of the abscissa is chosen so that $du = 1$, the height of each rectangle is equal to this quotient because $\varphi(u)du$ represents the considered probability. I have used this scale for the ordinates inscribed on the left side of the diagram. One could also select for the ordinate a scale that makes the height of each rectangle directly equal to the corresponding number of individuals as given in the table. Then the length $\frac{1}{25{,}878}$ must be ascribed to the interval du. This scale is employed for the notation of the ordinates given on the right side of the diagram.

[1] This example is taken from Karl Pearson, *The Chances of Death* . . . (London, 1897), p. 276.

In this manner the histogram of figure 12 is obtained, which has the shape of a staircase. If the construction is repeated for a finer division by intervals, the steps become narrower; and the procedure would define a continuous

TABLE 3
HEIGHT OF AMERICAN RECRUITS

Height (inches)	Observed number of recruits	Theoretical number
Below 55	4	1
55–56	1	0
56–57	3	1
57–58	7	2
58–59	6	7
59–60	10	29
60–61	15	85
61–62	50	224
62–63	526	535
63–64	1237	1065
64–65	1947	1854
65–66	3019	2788
66–67	3475	3582
67–68	4054	3980
68–69	3631	3818
69–70	3133	3126
70–71	2075	2221
71–72	1485	1350
72–73	680	703
73–74	343	325
74–75	118	125
75–76	42	42
76–77	9	12
77–78	6	3
78–79	2	1

curve for the limit $du \to 0$ if the number of individuals measured were sufficiently increased with the decrease of the interval. For practical reasons, of course, the limit cannot be reached, but it can be constructed by extrapolation. The continuous limiting curve in the diagram has been constructed thus. We can imagine the curve as being drawn "intuitively"; another procedure will be explained later.

The question arises whether a simple analytic expression can be found for such a symmetrical bell-shaped curve. Gauss has answered this question affirmatively by showing that the expression

$$\varphi(u) = \frac{h}{\sqrt{\pi}} e^{-h^2u^2} \tag{7}$$

represents a curve of the type explained; it is usually called the *normal curve.* The choice of the parameter h is left open; thus a more or less steep form can be given to the curve. In figure 13 several such curves are drawn for different values of h; a larger value of h means a steeper curve. Therefore h is called the *measure of precision;* a larger value of h means that all values are crowded closer together around the mean value $u = 0$. For $u = 0$ we have $\varphi = \frac{h}{\sqrt{\pi}}$;

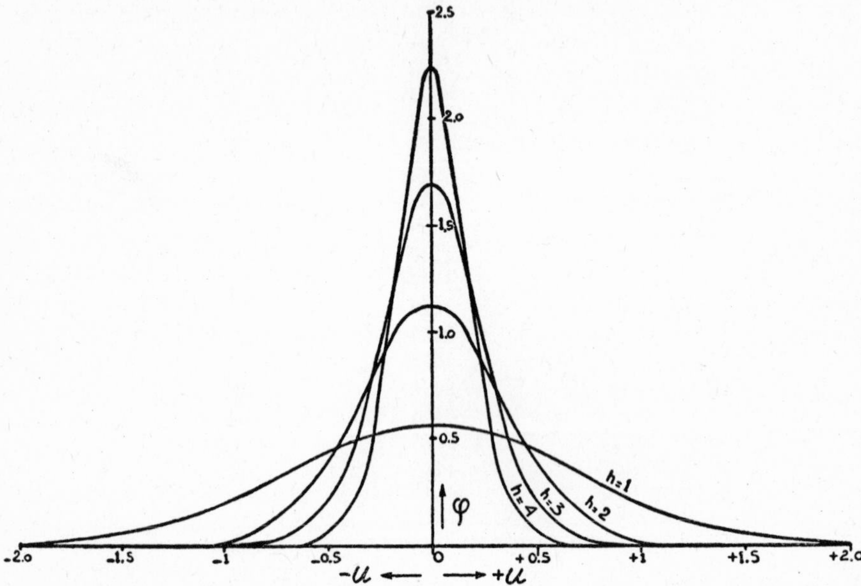

Fig. 13. Normal curves for different values of h, according to (7).

this is the maximum value of φ, whereas for positive or negative u the amount of φ decreases and finally goes asymptotically toward 0. The factor of normalization $\frac{h}{\sqrt{\pi}}$ is put before the exponential expression in order to satisfy the condition of normalization (6)

$$\int_{-\infty}^{+\infty} \varphi(u)du = \frac{h}{\sqrt{\pi}} \int_{-\infty}^{+\infty} e^{-h^2u^2}du = 1 \tag{8}$$

This result follows from the relation known from the theory of the exponential function.

$$\int_{-\infty}^{+\infty} e^{-h^2u^2}\,du = \frac{\sqrt{\pi}}{h} \tag{9}$$

This definite integral possesses a simple value, but the integration cannot be carried through explicitly for arbitrarily chosen limits. Tables have been constructed, however, for numerical calculations.

If the peak of the normal curve is not situated at $u = 0$ but at $u = u_0$, we obtain, instead of (7), the expression

$$\varphi(u) = \frac{h}{\sqrt{\pi}} e^{-h^2(u-u_0)^2} \tag{10}$$

For this function, too, (8) is fulfilled. In (10) the Gauss function has a form that can be made to fit an observed distribution. We choose for u_0 the mean value of all measurements; in the example about the recruits the mean value is equal to 67.701 in. The parameter h is to be selected so that the curve follows the observed distribution as smoothly as possible. A procedure achieving this purpose will be explained in § 43. For the measurement of recruits, h is $= 0.274$ (in the scale given for the ordinate on the left side of fig. 12). With the help of the values u_0 and h we can calculate from (10), conversely, the value $\varphi(u)$ belonging to each interval du; by comparing the calculated values with the observed values we are able to judge how well the function (10) fits the problem. The calculated values of $\varphi(u)$ are listed under the heading "Theoretical number" in the last column of table 3. (They refer to the scale indicated for the ordinate on the right side of fig. 12, p. 212.) We recognize the excellent correspondence, which proves that the function (10) fits the observed curve very well.

That it is possible to represent a distribution, as given in the example, by an expression like (10) must be regarded as an empirical fact. Unfortunately, some authors have believed all sorts of secrets to be hidden in the normal distribution; they have even regarded the normal curve as a mysterious law of all natural phenomena. But the normal curve is certainly not a law for *all* objects; there are distributions that follow the Gauss law and others that do not. It has to be tested for every set of observational data whether it satisfies the normal distribution.

The great practical importance of the normal distribution consists in the fact that we find many applications for it; we are thus in a position to express various sets of statistical material in terms of the same simple analytic expression, the individual form of each set being characterized by only two parameters, namely, the quantities h and u_0 (for this point see p. 221).

For the two-dimensional attribute space, also, normal distributions can be defined. They are characterized[2] by the analytic expression

$$\varphi(u,v) = \frac{h_1 h_2}{\pi} e^{-(h_1^2 u^2 + h_2^2 v^2)} \tag{11}$$

[2] In the general case the exponent contains a positive-definite quadratic form; but it can always be changed into (11) by transforming the main axes.

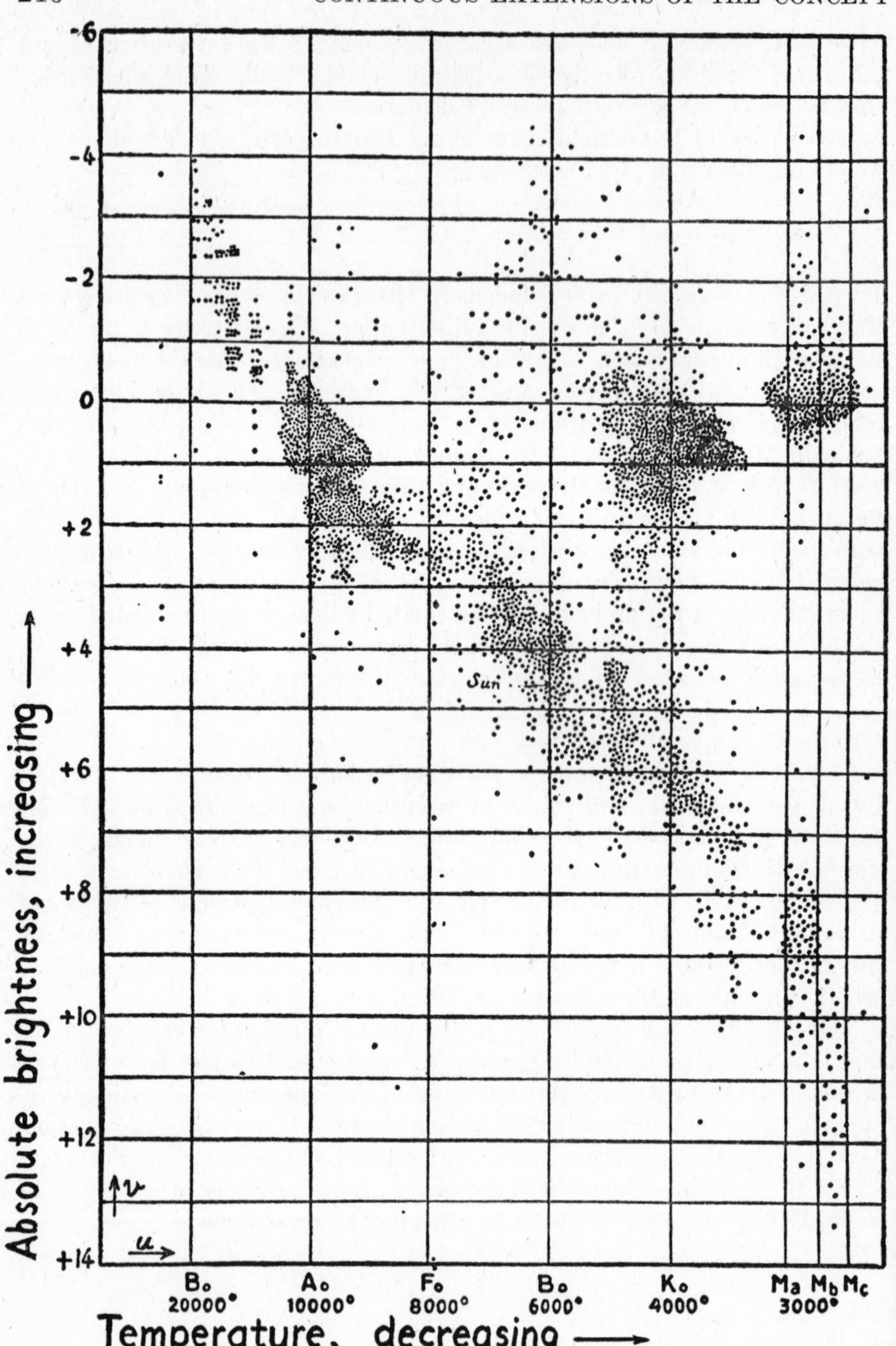

Fig. 14. Hertzsprung-Russell diagram of fixed stars.

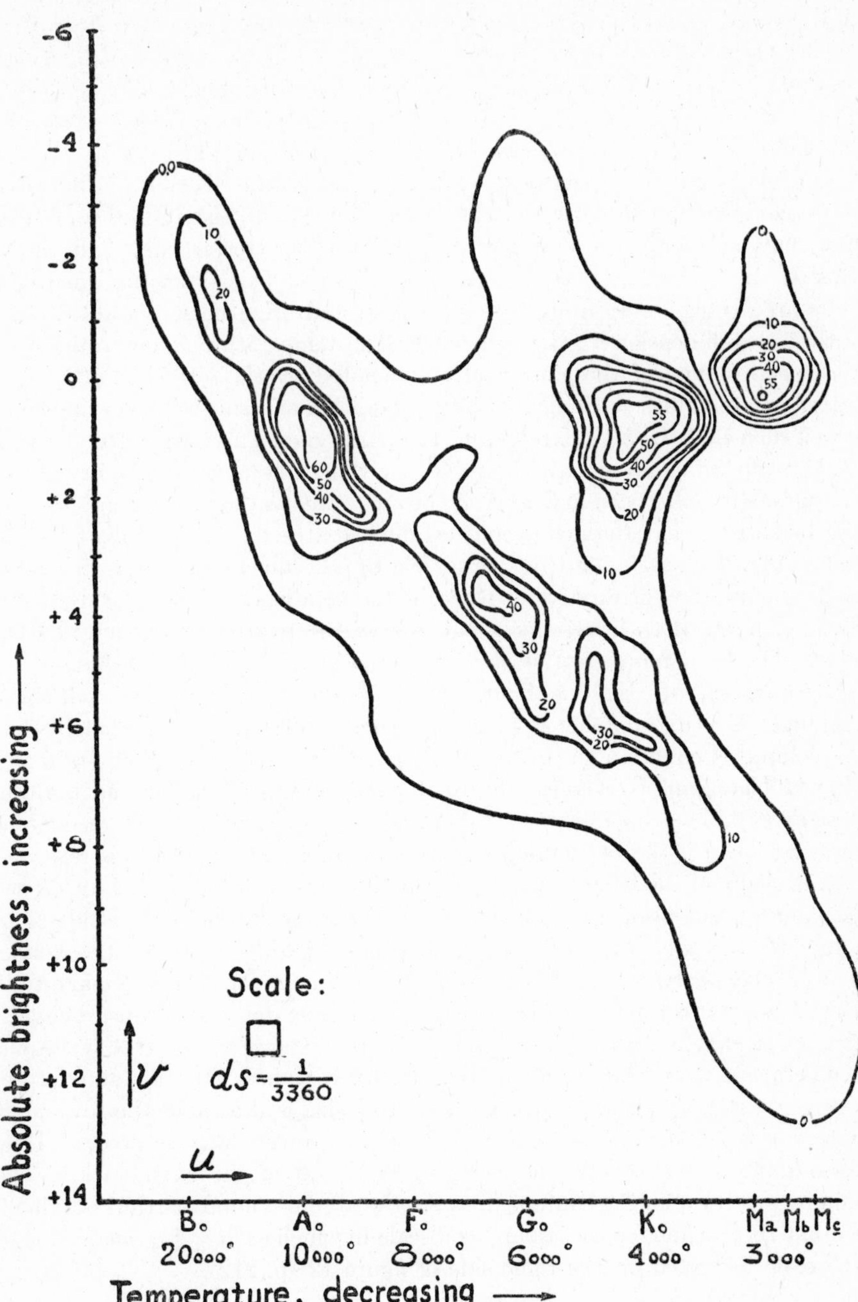

Fig. 15. Hertzsprung-Russell diagram drawn by means of contour lines.

For the case of rotational symmetry, in particular, we have $h_1 = h_2$; such a "Gauss bell" is produced by the rotation of one of the curves of figure 13 around the vertical axis. The distribution of shots hitting a target (according to fig. 10, p. 206) is of this type. The extension of these considerations to attribute spaces of more than two dimensions offers no difficulty.

A second example concerns an instance in which the distribution does not have the character of a normal curve. In recent years the statistics of stars have assumed major importance in astronomy. E. Hertzsprung and H. N. Russell, in particular, have compiled statistics of fixed stars in which the stars are arranged according to temperature and brightness. We are dealing with a two-dimensional attribute space. The temperature is determined by the spectral type of the star, which is denoted by the letters A, B, . . . ; intermediate values are indicated by the addition of a subscript varying from 0 to 9 such that A_0, B_0, . . . represent the pure types. This parameter, denoted by u, is drawn as abscissa in figure 14, for which the temperature in degrees centigrade has also been indicated. As ordinate we use the absolute brightness, the brightness that the star would exhibit for a certain normal distance. In figure 14, which is taken from an article by H. Mineur,[3] a dot is made for each star at the corresponding place in the coördinate system, so that the density of the dots represents a measure of the relative frequency at each place. The total number of stars represented in the diagram is 3,360.

By analogy with the one-dimensional case we can erect above each small rectangle $ds = dudv$ of the u,v-plane a prismatic column of such a height that its volume is equal to the corresponding relative frequency. The staircase-shaped histogram converges toward a continuous surface for the limiting case $ds = 0$. The shape of these "probability mountains" may be recognized from figure 14 if the blackening given by the density of the dots is regarded as a measure of the height. A presentation in contour lines, as used for mountain maps, can also be employed; a presentation of this kind is given in figure 15. For this purpose figure 14 was covered with a net of little squares ds. The side of the square, compared to the ordinate v, was $dv = \frac{1}{2}$ magnitude (of the stars); compared to the abscissa u, it was $du = \frac{1}{4}$ distance between two spectral types. Since the relation between distance of spectral types and temperature is not linear, du represents different intervals of temperature, varying with the place. A square ds of this size is drawn in the lower left corner of figure 15. The number of stars for each square was counted, and the contour lines were drawn according to the result of the enumeration. The numbers written on the contour lines are the absolute numbers thus obtained, so that they define for the square ds a scale in which $ds = \frac{1}{3{,}360}$, analogous to the scale indicated on the right side of figure 12 (p. 212).

[3] *Bull. Soc. Astron. de France*, Vol. 45 (1931), pp. 4–20.

In figure 15 the shape of the probability mountains is clearly visible; from the upper left corner to the lower right corner runs a continuous ridge, and at the upper right corner lie two further crests, relatively isolated. The shape of the probability surface expresses a peculiar law of the stellar system, the deeper significance of which has not yet been completely understood. Astronomers, however, have ventured to make the inference that the line of the ridge running from the isolated crests at the upper right to the upper left and thence to the lower right corner represents a picture of the life history of an individual star. This is an inference in terms of a probability lattice; it is assumed that the life lines of the individual stars, conceived as probability sequences, form a homogeneous lattice, so that a vertical cross section at a time t = const. is identical with a horizontal cross section in the t-direction.

§ 43. The One-Dimensional Attribute Space

The special case of an attribute space having only one dimension may be regarded as a generalization of the probability sequence with coördinated amounts. The generalization consists in that the amounts u^i are no longer restricted to certain fixed steps of amounts u_m, but may assume any values u located on a continuous scale, for which there need not exist an upper or a lower limit. Because of this relation it is possible to transfer to the one-dimensional attribute space the formulas developed previously for average and dispersion.

The *statistical definitions* of average and dispersion are the same as in the case treated above, since the expressions

$$M^n(u^i)^i = \frac{1}{n} \sum_{i=1}^{n} u^i \tag{1a}$$

$$M(u^i)^i = \lim_{n\to\infty} M^n(u^i)^i \tag{1b}$$

$$\Delta^{n2}(u^i)^i = M^n(u^i - M(u))^2 \tag{2a}$$

$$\Delta^2(u^i)^i = \lim_{n\to\infty} \Delta^{n2}(u^i)^i \tag{2b}$$

can be applied to a continuous variable u as well as to quantities u_m varying by steps. Such an application is possible because the summation by the superscript occurring in these expressions refers to the elements y_i of the sequence and is therefore not affected by the transition to a continuous variability of the u. It is different, however, with the *theoretical definition.* Since in this definition we take the sum of steps of amounts, the summation must be replaced in the continuous case by an integration. In a manner similar to that used for the probability, we assume a division by steps of amounts du_1, du_2,

. . . du_m, . . . of a small finite width; the probability of hitting such an interval is then approximately equal to $\varphi(u_m)du_m$, and instead of (12, § 35) we write

$$M(u) = \lim_{du_m \to 0} \sum_{m=-\infty}^{+\infty} \varphi(u_m)du_m \cdot u_m = \int_{-\infty}^{+\infty} u \cdot \varphi(u)du \qquad (3)$$

The proof that this expression is equal to (1*b*) is easily given by the use of the relation derived above for amounts varying by steps in combination with certain familiar methods of the differential calculus. This proof is based upon the simple inequality

$$\sum_{m=0}^{\infty} \varphi^*(u_m)du_m \cdot u_m + \sum_{m=-\infty}^{0} \varphi^*(u_m)du_m \cdot u_{m+1} \leqq M(u^i)^i$$

$$\leqq \sum_{m=0}^{\infty} \varphi^*(u_m)du_m \cdot u_{m+1} + \sum_{m=-\infty}^{0} \varphi^*(u_m)du_m \cdot u_m \qquad (3')$$

in which we have put $m = 0$ for $u_m = 0$, that is, $u_0 = 0$; φ^* has the meaning introduced in (2, § 42). Equation (3) follows from (3′) because the expressions written on the left and right sides assume the same value for the limit $du_m \to 0$ and become equal to the integral on the right side of (3). According to a previous remark (p. 181), the convergence of the expressions (3) and (3′) must be regarded as an additional assumption, which is not guaranteed by (6, § 42). Corresponding assumptions are to be made for the definition of similar quantities given in the following considerations.

Similarly, we obtain the theoretical definition of the dispersion by analogy with (6, § 37):

$$\Delta^2(u) = M(\delta^2 u) = \int_{-\infty}^{+\infty} \delta^2 u \cdot \varphi(u)du \qquad \delta u = u - M(u) \qquad (4)$$

Furthermore, all relations derived previously remain valid in a similar translation. Thus we have, in analogy to (3, § 37),

$$M(\delta u) = \int_{-\infty}^{+\infty} \delta u \cdot \varphi(u)du = \int_{-\infty}^{+\infty} u \cdot \varphi(u)du - M(u) \cdot \int_{-\infty}^{+\infty} \varphi(u)du = 0 \qquad (5)$$

By means of this relation we can easily derive the theorem of the shift of the reference point, analogous to (12, § 37):

$$\delta_0 u = u - u_0 \qquad \delta u_0 = M(u) - u_0$$

$$\Delta_0^2(u) = M(\delta_0^2 u) = \int_{-\infty}^{+\infty} [\delta^2 u - 2\delta u \delta u_0 + \delta^2 u_0]\, \varphi(u)du$$

$$= \Delta^2(u) + [M(u) - u_0]^2 \qquad (6)$$

Choosing the zero point $u = 0$ as point of reference u_0 and solving for $\Delta^2(u)$, we obtain the relation analogous to (13, § 37):

$$\Delta^2(u) = M(u^2) - M^2(u) = \int_{-\infty}^{+\infty} u^2 \cdot \varphi(u) du - M^2(u) \tag{7}$$

The last form is a convenient transformation of the definition (4) of the dispersion.

It is of special interest to determine the mean value and the dispersion for the normal distribution. We have with (9, § 42) and the substitution $x = h(u - u_o)$,

$$\begin{aligned} M(u) &= \frac{h}{\sqrt{\pi}} \cdot \int_{-\infty}^{+\infty} u \cdot e^{-h^2(u-u_0)^2} du \\ &= \frac{1}{\sqrt{\pi}} \cdot u_0 \int_{-\infty}^{+\infty} e^{-x^2} dx + \frac{1}{h \cdot \sqrt{\pi}} \cdot \int_{-\infty}^{+\infty} x \cdot e^{-x^2} dx = u_0 \end{aligned} \tag{8}$$

$$\begin{aligned} \Delta^2(u) &= \frac{h}{\sqrt{\pi}} \cdot \int_{-\infty}^{+\infty} (u - u_0)^2 \cdot e^{-h^2(u-u_0)^2} du \\ &= \frac{1}{h^2 \cdot \sqrt{\pi}} \int_{-\infty}^{+\infty} x^2 \cdot e^{-x^2} dx = \frac{1}{2h^2} \end{aligned} \tag{9a}$$

$$\Delta(u) = \frac{1}{h \cdot \sqrt{2}} \tag{9b}$$

For these equations we have used the relations, known from the theory of the exponential function,

$$\int_{-\infty}^{+\infty} x \cdot e^{-x^2} dx = 0 \qquad \int_{-\infty}^{+\infty} x^2 \cdot e^{-x^2} dx = \tfrac{1}{2} \cdot \sqrt{\pi} \tag{10}$$

The value u_0 of u at which the peak of the Gaussian bell curve is situated represents the average; and the most probable value of u coincides with the average. If it is known that a given distribution is of the normal type, it is permitted to treat the mean value of a number of measurements as the most probable value. The same result holds for distributions of a different character, provided they are symmetrical with respect to their peak value; but the theorem does not hold for all distributions, and it is not possible, therefore, unless specific reasons are known, to equate the mean value with the most probable value.

We see from (9*b*) that the dispersion, or standard deviation, of the normal curve is connected by a simple relation to the measure of precision. Since the normal distribution contains only the two parameters u_0 and h, it is completely determined by the average and the dispersion. The normal curve drawn in the diagram of the recruits in § 42, figure 12 (p. 212), was calculated by these relations. Its peak was drawn at the place of the average $u_0 = 67.701$ in., and the parameter $h = 0.274$ was calculated from the statistical dispersion $\Delta(u) = 2.5848$ by means of (9*b*). The numbers given in the last column of table 3 (p. 213) supply a test, which shows that the observed distribution is well represented by the normal curve thus computed. Usually, whenever

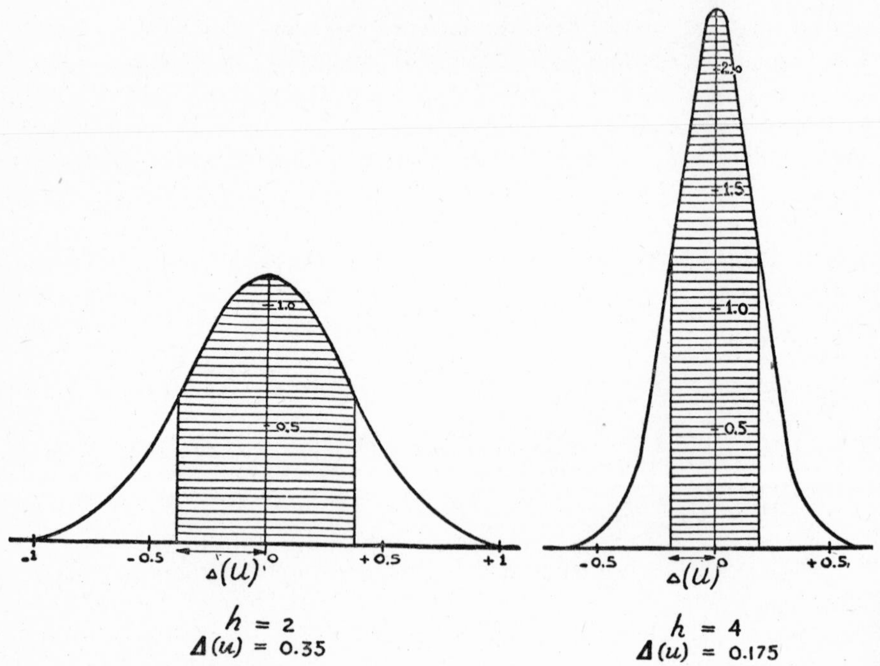

Fig. 16. Standard deviation for two different normal curves.

we have good reason to assume that the distribution is of the normal type, it will be unnecessary to draw the step curve and to make the test; the normal curve is then calculated directly from the values $M(u)$ and $\Delta(u)$ of the statistics.

It is of interest to calculate for the normal distribution the probability w_Δ belonging to the linear dispersion, that is, the probability that a value u^i deviates from the average only within the limits $\pm\,\Delta(u)$, given by the standard deviation. In Tchebychev's inequality (18, § 37) we found a more general probability w_δ of that kind; but the inequality merely demonstrates the trivial fact that w_Δ must be $\geqq o$, so that in the case of distributions of a general type no definite statement about w_Δ can be made. But for normal distributions w_Δ can be determined. If the average is assumed to be situated at the zero point (a simplification that does not change the result), the probability obtains the form, with the substitution $x = h \cdot u$,

$$w_\Delta = \frac{h}{\sqrt{\pi}} \cdot \int_{-\frac{1}{h \cdot \sqrt{2}}}^{+\frac{1}{h \cdot \sqrt{2}}} e^{-h^2 u^2}\, du = \frac{1}{\sqrt{\pi}} \cdot \int_{-\frac{1}{\sqrt{2}}}^{+\frac{1}{\sqrt{2}}} e^{-x^2}\, dx = 0.68260 \ldots \sim \tfrac{2}{3} \tag{11}$$

The last integration is carried out by means of a table of the Gauss function.

It is particularly important that the probability w_Δ is independent of the measure of precision h. This result makes it possible—if the Gaussian character of the distribution under consideration is known—to interpret the dispersion found statistically as follows. We may expect with the probability $\frac{2}{3}$ that an amount u^i deviates from the average only within the limits $\pm\ \Delta(u)$. This relation justifies the usual procedure of stating the result of a series of measurements in the form of the mean value by adding the limits $\pm\ \Delta(u)$. In dealing with normal distributions, it is always the same probability, that is, approximately the probability $\frac{2}{3}$, with which the result may be regarded as lying within these limits of error.

The dispersion, or standard deviation, $\Delta(u)$ for two different normal curves is presented in figure 16. The shaded areas represent the probability w_Δ; these areas, for both curves, are equal to about $\frac{2}{3}$ of the total area. For the steeper curve the shaded strip must, therefore, be narrower.

§ 44. Many-Dimensional Attribute Spaces

For many-dimensional attribute spaces the concepts of average and dispersion can be defined in a generalized meaning. The resulting concepts are mathematically analogous to certain concepts of mechanics: the probability function may be compared to the mass density, the average to the center of gravity, the dispersion to the momentum of inertia. Therefore the average is frequently called the momentum of the first order, the dispersion the momentum of the second order. This analogy makes it clear that it is possible to introduce momenta of higher order into the probability calculus, corresponding to those of mechanics (this holds also, of course, for the one-dimensional attribute space); and the interesting theorem has been derived that a distribution is determined by the totality of its infinitely many momenta. But these problems will not be discussed, since they belong in the purely mathematical parts of the probability calculus.

However, the many-dimensional attribute spaces will be considered from another point of view. We can regard the different dimensions of the attribute space as representing different attributes, combinations of which are coördinated to the elements x_i of the sequence; and thus every attribute pair, attribute triplet, and so on may be regarded from the standpoint of the probability of combinations. Call B_m the case, for which u lies within an interval from u_m to u_{m+1}, and C_l the case for which v lies within the interval from v_l to v_{l+1}; then the probability that the attribute point falls into the rectangle u_m to u_{m+1}, v_l to v_{l+1} is given by the expression

$$P(A,B_m . C_l) = \int_{u_m}^{u_m+1} \int_{v_l}^{v_l+1} \varphi(u,v)dudv \tag{1}$$

The probability function of many variables may therefore be regarded as the density of a probability of combinations. In order to arrive at the probability of the attribute B_m, taken alone, we construct the always true disjunction

$$(C_l \vee \bar{C}_l \equiv C_0 \vee C_1 \vee C_{-1} \vee C_2 \vee C_{-2} \vee \ldots) \tag{2}$$

and then apply the theorem of addition, which means the same as extending the integration over the whole domain of the variables. Thus we have

$$\begin{aligned} P(A,B_m) &= P(A,B_m . [C_0 \vee C_1 \vee C_{-1} \vee C_2 \vee C_{-2} \vee \ldots]) \\ &= \int_{u_m}^{u_m+1} \int_{-\infty}^{+\infty} \varphi(u,v)dudv \end{aligned} \tag{3}$$

We put

$$\varphi_1(u) = \int_{-\infty}^{+\infty} \varphi(u,v)dv \tag{4}$$

The one-place function $\varphi_1(u)$ is the probability function that controls u taken alone, that is, it measures the probability of a u-value irrespective of the values v. Analogously, the one-place function

$$\varphi_2(v) = \int_{-\infty}^{+\infty} \varphi(u,v)du \tag{5}$$

is the probability function controlling v taken alone. Therefore, we have

$$P(A,B_m) = \int_{u_m}^{u_m+1} \varphi_1(u)du \qquad P(A,C_l) = \int_{v_l}^{v_l+1} \varphi_2(v)dv \tag{6}$$

Equations (4) and (5) represent a continuous generalization of the rule of elimination.[1] In conjunction with (6) and (1) they permit us to derive the relative probabilities $P(A . B_m,C_l)$ and $P(A . C_l,B_m)$ when we employ the general theorem of multiplication:

$$P(A . B_m,C_l) = \frac{\int_{u_m}^{u_m+1} \int_{v_l}^{v_l+1} \varphi(u,v)dudv}{\int_{u_m}^{u_m+1} \varphi_1(u)du} \tag{7a}$$

[1] For a form corresponding more closely to (21, § 19) see (7, § 45).

$$P(A.C_l,B_m) = \frac{\int_{u_m}^{u_m+1}\int_{v_l}^{v_l+1} \varphi(u,v)\, dudv}{\int_{v_l}^{v_l+1} \varphi_2(v)dv} \tag{7b}$$

The relative probability appears, as in (2, § 41), as the quotient of two integrals. The special case in which the two dimensions u and v are mutually independent is characterized by the equations

$$P(A.B_m,C_l) = P(A,C_l) \tag{8}$$
$$P(A.C_l,B_m) = P(A,B_m)$$

which lead, with (7) and (6), to the relation

$$\int_{u_m}^{u_m+1}\int_{v_l}^{v_l+1} \varphi(u,v)dudv = \int_{u_m}^{u_m+1} \varphi_1(u)du \cdot \int_{v_l}^{v_l+1} \varphi_2(v)dv \tag{9}$$

If these equations are to be satisfied for any choice of the coördinates u_m and v_l and of the magnitude of the intervals $u_m - u_{m+1}$, $v_l - v_{l+1}$, the relation

$$\varphi(u,v) = \varphi_1(u) \cdot \varphi_2(v) \tag{10}$$

must hold. For the attribute space of independent dimensions the probability function splits up into a product of one-place probability functions. The function $\varphi(u,v)$, when considered as a surface in a three-dimensional space, thus represents a surface of the following properties: the two sets of planes u = const. and v = const. intersect with this surface in such a way that any two intersection figures of the same set result from each other by a multiplication of their ordinates by a factor. This factor is given for the first set by the ratio $\frac{\varphi_1(u_m)}{\varphi_2(v_l)}$, if u_m = const. and v_l = const. represent the intersecting planes; a corresponding relation holds for the other set. The process of expansion (or shrinking) of a figure in only one dimension is usually called a *dilatation;* we may say, therefore, that any two intersection figures of the same set result from each other by a dilatation. Thus we call a surface $\varphi(u,v)$ of the property (10) a *dilatation surface.*[2]

An example of a dilatation surface is presented by the Gauss bell erected over the target, as shown in figure 10 (p. 206); its vertical plane sections are

[2] Since the splitting (10) results only for a certain choice of the coördinates, the surface of dilatation must be defined more precisely by the condition that there exist rectangular coördinates for which the equation of the surface splits up according to (10); or that there exist two sets of planes at right angles to each other, intersecting with the surface in such a way that any two intersection figures of the same set result from each other by the operation of dilatation.

normal curves of different scales. Analytically, this property follows from the relation

$$e^{-h^2(u^2+v^2)} = e^{-h^2u^2} \cdot e^{-h^2v^2} \tag{11}$$

However, dilatation surfaces are not necessarily rotation surfaces; and, vice versa, rotation surfaces are not always dilatation surfaces.

These considerations may be extended with respect to attribute spaces of several dimensions. The coördinates of an n-dimensional attribute space are denoted by $u^{(1)} \ldots u^{(n)}$; the parentheses around the superscript are to indicate that this superscript stands for the dimension of the attribute space and not for the individual value of u coördinated to the n-th element (see the notation introduced in § 35). This role is then assigned, as in § 39, to a second superscript. When we write the second superscript, the parentheses around the first superscript are no longer required; for instance, u^{ki} would denote the attribute value of the i-th element of the probability sequence in respect to the k-th dimension of the attribute space. The rule of elimination assumes the form of an $(n - 1)$-fold integration:

$$\varphi_1(u^{(1)}) = \int_{-\infty}^{+\infty} \ldots \int_{-\infty}^{+\infty} \varphi(u^{(1)} \ldots u^{(n)})\, du^{(2)} \ldots du^{(n)} \tag{12}$$

The other one-place probability functions, $\varphi_2\,(u^{(2)})$ and so on, are introduced in a corresponding manner. The relative probabilities can be defined in analogy to (7); this definition can be given in different ways, depending on how many dimensions are used in the first term, that is, the reference class. But we consider here only the case of independent dimensions, which is characterized by the relation

$$\varphi(u^{(1)} \ldots u^{(n)}) = \varphi_1(u^{(1)}) \cdot \ldots \cdot \varphi_n(u^{(n)}) \tag{13}$$

Consequently, the function $\varphi\,(u^{(1)} \ldots u^{(n)})$ represents a dilatation surface in an $(n + 1) =$ dimensional space.

The attribute space of several dimensions can be applied to the treatment of event combinations. For this purpose the attribute spaces of separate sequences are combined in a single higher-dimensional attribute space. For instance, the physical state x_i of the wind as observed daily is represented in a two-dimensional attribute space $u^{(1)}u^{(2)}$, when we characterize the wind by its intensity and direction; the daily temperature y_i and the rainfall z_i may each be characterized, respectively, by the one-dimensional attribute $u^{(3)}$ or $u^{(4)}$. The data may be combined so that we can interpret them as an attribute point in a four-dimensional space $u^{(1)} \ldots u^{(4)}$; this attribute point is coördinated to the triplet x_i, y_i, z_i, of events. By regarding such combinations of events as one event, we can extend the considerations developed for the attribute space of many dimensions to the treatment of the combination of different sequences.

For the sake of simplicity, assume that the individual sequence has a one-dimensional attribute space; the n-sequences then determine an n-dimensional attribute space. If the sequences are completely independent of each other (§ 23), the probability function of the n-dimensional attribute space can be factorized according to (13) as the product of the one-place probability functions. Such a case may be discussed in a manner similar to the treatment of steps of amounts, as explained in § 38.

Beginning with the derivation of the results for average and dispersion of additive amounts in two attribute dimensions, analogous to (3, § 38), we obtain

$$\begin{aligned} M(u+v) &= \int_{-\infty}^{+\infty}\int_{-\infty}^{+\infty} (u+v)\cdot\varphi(u,v)dudv \\ &= \int_{-\infty}^{+\infty} u\cdot\left[\int_{-\infty}^{+\infty}\varphi(u,v)dv\right]du + \int_{-\infty}^{+\infty} v\cdot\left[\int_{-\infty}^{+\infty}\varphi(u,v)du\right]dv \\ &= \int_{-\infty}^{+\infty} u\cdot\varphi_1(u)du + \int_{-\infty}^{+\infty} v\cdot\varphi_2(v)dv = M(u)+M(v) \quad (14) \end{aligned}$$

It is not necessary to presuppose the independence of the dimensions. The corresponding result for the dispersion, however, is valid, as before, only for independent attribute dimensions, in analogy to (8, § 38). We obtain, with the help of (14),

$$\delta(u+v) = u+v-M(u+v) = u-M(u)+v-M(v) = \delta u + \delta v$$

$$\begin{aligned} \Delta^2(u+v) &= M(\delta^2(u+v)) \\ &= \int_{-\infty}^{+\infty}\int_{-\infty}^{+\infty} (\delta^2 u + 2\delta u\delta v + \delta^2 v)\cdot\varphi(u,v)dudv \quad (15) \end{aligned}$$

Using (10) we derive

$$\begin{aligned} \Delta^2(u+v) = &\int_{-\infty}^{+\infty}\delta^2 u\cdot\varphi_1(u)du\cdot\int_{-\infty}^{+\infty}\varphi_2(v)dv \\ &+ 2\int_{-\infty}^{+\infty}\delta u\cdot\varphi_1(u)du\cdot\int_{-\infty}^{+\infty}\delta v\cdot\varphi_2(v)dv \\ &+ \int_{-\infty}^{+\infty}\delta^2 v\cdot\varphi_2(v)dv\cdot\int_{-\infty}^{+\infty}\varphi_1(u)du \\ = &\ \Delta^2(u)+\Delta^2(v) \quad (16) \end{aligned}$$

since the middle term vanishes because of (5, § 43).

The extension of these formulas to n attribute dimensions need not be explained, since it is identical with the considerations of § 39. We can take over all the formulas of § 39. There will be no change in the formulation of the derivations, since we have always employed only the relations (3 and 8*a*, § 38) and (13 and 15, § 37), which are written in a symbolic notation and which we have now derived once more in (14) and (16) and in (7, § 43). The summations occurring in the formulas of § 39 remain summations for the continuous attribute space, since the sums refer to the individual dimensions.

From the viewpoint of the continuous distribution another problem may be considered. We can inquire after the distribution

$$\varphi(u) \qquad u = \frac{1}{n} \cdot \sum_{k=1}^{n} u^{(k)} \tag{17}$$

that is, after the probability function that controls the mean value of the n one-dimensional attributes $u^{(k)}$; u then corresponds to the quantity that we would denote by β_m in the notation of § 39. It can be shown that the function $\varphi(u)$ converges with increasing n toward a normal distribution if the attribute dimensions are completely independent of each other and if certain other properties are fulfilled. The proof cannot be derived without a mathematical apparatus more elaborate than falls within the scope of this book; the mathematical literature may be consulted for this purpose.[3] For practical applications the theorem is of great importance, because it explains the fact that the normal distribution finds many applications. The fluctuations of observable quantities can frequently be reduced to additive fluctuations of a great number of elementary quantities; in all such instances the observable result corresponds with good approximation to the normal distribution. An example is the theory of observational errors, since the error of each single measurement is to be conceived as the superposition of numerous elementary errors; or the kinetic theory of matter, in which a normal distribution is derived for the velocities of the molecules (the Maxwell distribution).

§ 45. Relative Probability Functions

In the attribute space of independent dimensions the relative probability (7, § 44) becomes identical with the one-place probability (6, § 44) because of (8, § 44); but in the attribute space of dependent dimensions the relative probability for intervals is to be expressed by the quotient of two integrals (7, § 44). For this case, too, we can derive a few general properties of the relative probability. If in (7*a*, § 44) the value u_m is made to approach u_{m+1}, while u_l and u_{l+1} remain fixed, the quotient does not go toward zero but to the expression

$$\int_{v_l}^{v_{l+1}} \frac{\varphi(u_m,v)}{\varphi_1(u_m)}\, dv \tag{1}$$

[3] See, for instance, Richard von Mises, *Vorlesungen aus dem Gebiete der angewandten Mathematik*, Vol. I: *Wahrscheinlichkeitsrechnung* (Leipzig, 1931), p. 216.

The corresponding statement holds for (7*b*, § 44). When we put

$$\psi_1(u;v) = \frac{\varphi(u,v)}{\varphi_1(u)} \qquad \psi_2(v;u) = \frac{\varphi(u,v)}{\varphi_2(v)} \tag{2}$$

the two functions $\psi_1(u;v)$ and $\psi_2(v;u)$ represent *one-dimensional probability densities;* integrated over the second variable they lead to probabilities. I shall call them *relative probability functions.* Integrated over v, the function $\psi_1(u;v)$ supplies the probability that v lies within a certain interval if u possesses a precisely given value; and, correspondingly, integrated over u, the function $\psi_2(v;u)$ supplies the probability that u lies within a certain interval if v has a precisely given value. These probabilities are not zero, since the precise value occurs in the first term, or reference class. For instance, if a certain precise value of u is given, the probability that v assumes any value within the whole domain between $-\infty$ and $+\infty$ is $= 1$. An integration over the first variable is not permissible, that is, it has no meaning in the theory of probability.[1] In order to indicate that the two variables are of a different kind, a semicolon is placed between the variables. If we wish to obtain the relative probability for an expression with a finite interval in the first term, we must not integrate over the first variable but have to go back to the integral quotient (7*a*, § 44) or (7*b*, § 44). The reason is that the probability of an "or" in the first term, according to (4, § 22), does not represent an addition, but indicates the formation of a mean value; (7*a*, § 44) and (7*b*, § 44) are analogous to (4, § 22). However, for small intervals ϵ (the interval v_l to v_{l+1} may be large) we have the approximate equality

$$\frac{\int_u^{u+\epsilon}\int_{v_l}^{v_{l+1}} \varphi(u,v)\,du\,dv}{\int_u^{u+\epsilon} \varphi_1(u)\,du} \sim \int_{v_l}^{v_{l+1}} \frac{\varphi(u,v)}{\varphi_1(u)}\,dv \tag{3}$$

so that it is possible, in practice, for small intervals in the first term, to construct the relative probability functions (2) by methods of enumeration similar to those used for absolute probability functions.

As examples of relative probability functions, figures 17 and 18 present certain transformations of the diagram of the Hertzsprung-Russell statistics. In figure 17 the function $\varphi(u,v)$ was integrated in the v-direction according to (4, § 44), and thus the function $\psi_1(u;v)$ was constructed by the help of (2). In figure 18 the function $\varphi(u,v)$ was integrated in the u-direction according to (5, § 44), and then the function $\psi_2(v;u)$ was constructed according to (2). The calculation was carried out as follows: the diagram of figure 14 (p. 216) was covered by the net of squares ds, and the numbers of points per square

[1] See, however, formula (7).

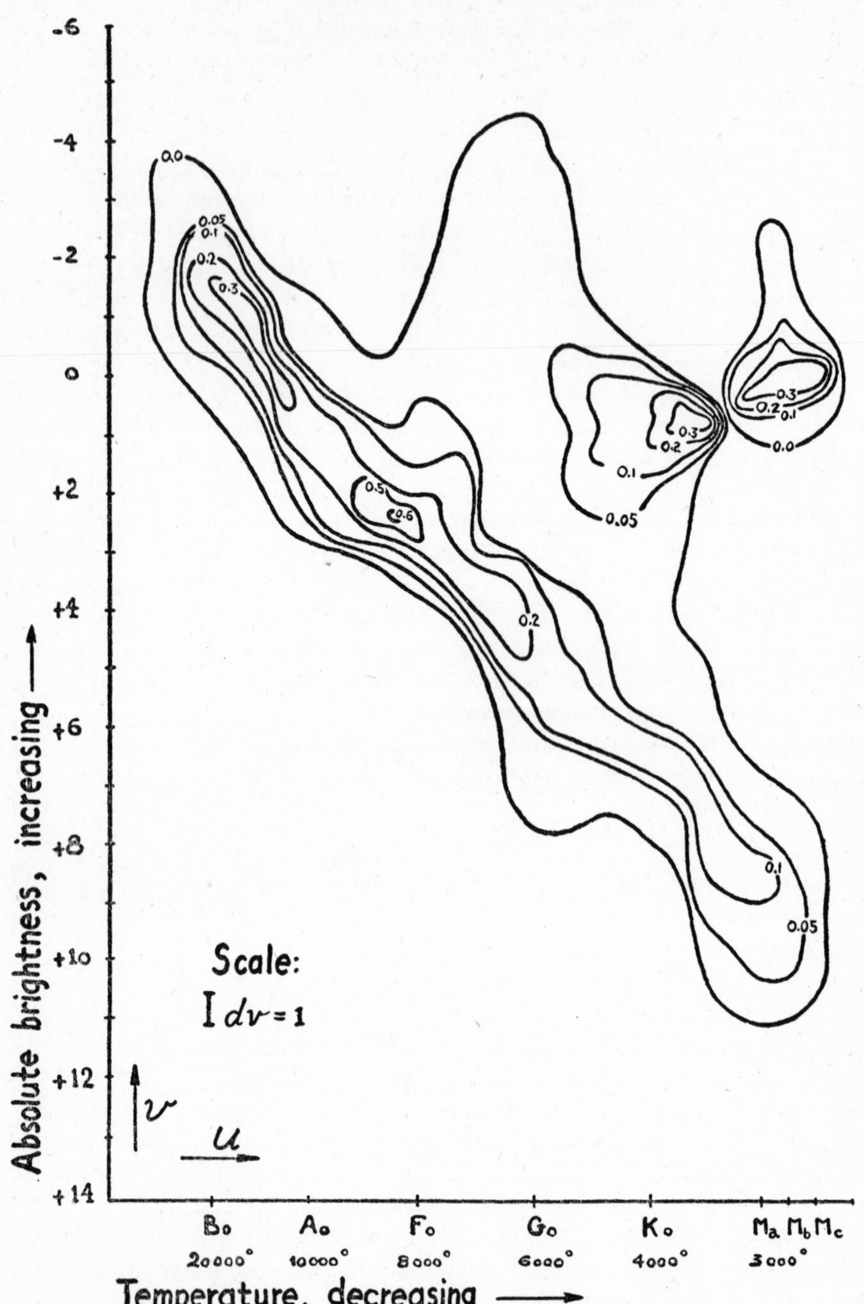

Fig. 17. Hertzsprung-Russell diagram redrawn as relative probability function: integration of fig. 15 in vertical direction; function $\psi_1(u;v)$ of (2) represented by contour lines.

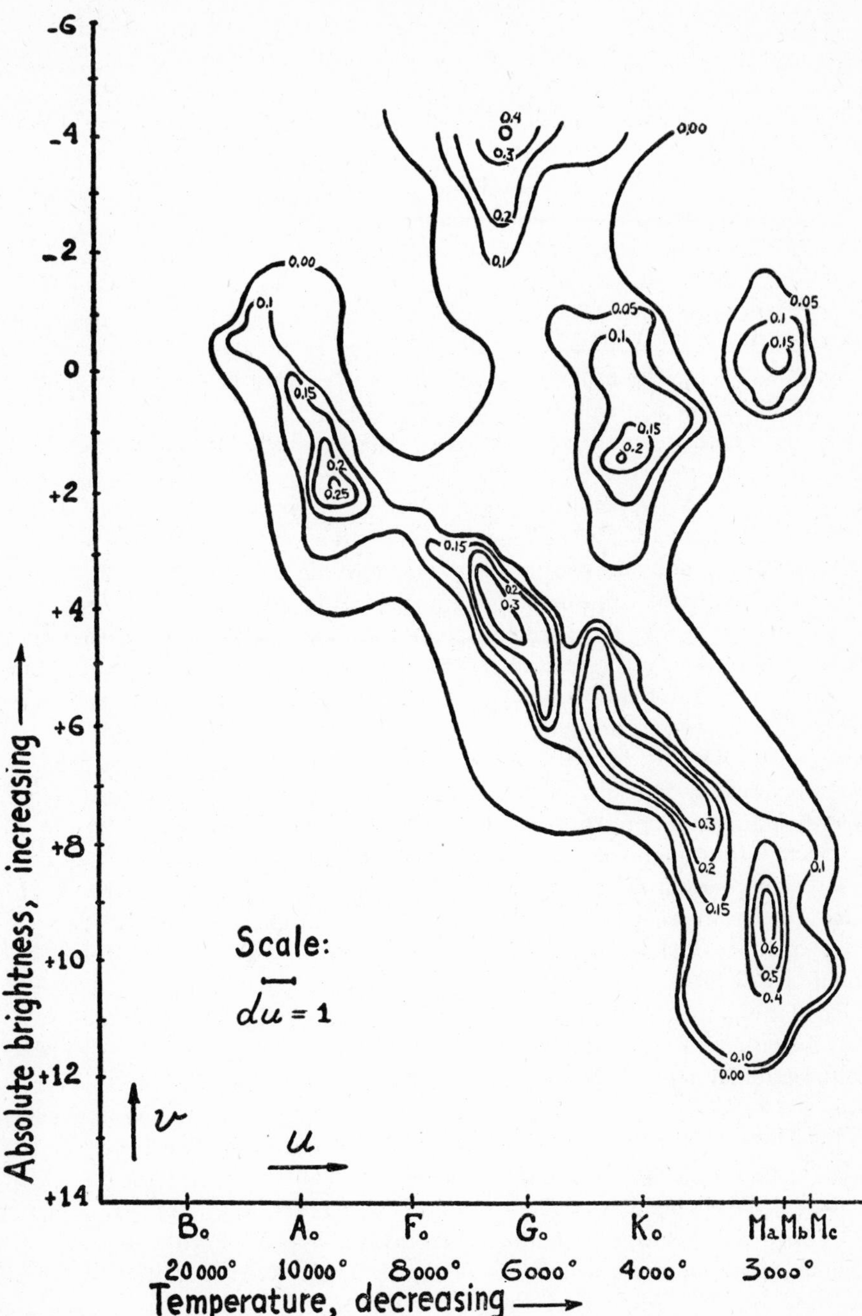

Fig. 18. Hertzsprung-Russell diagram redrawn as relative probability function: integration of fig. 15 in horizontal direction; function $\psi_2(v;u)$ of (2) represented by contour lines.

were added for each horizontal row and vertical column, respectively; then the number of points in each square was divided by the amount of its vertical column or horizontal row. Thus the contour lines in figures 17 and 18 were drawn. The numbers indicated for the contour lines correspond to a scale $ds = 1$. Thus they cannot be compared directly to the corresponding numbers of figure 15 (p. 217); a scale like that of figure 15 cannot be employed for these diagrams, since the denominators of the fractions obtained are different for each column or row. The following examples, which can be read directly from the diagrams, may explain the meaning of the diagrams: for a star of spectral type M_b, the probability that its absolute brightness lies between $+ 9$ and $+ 10$ has the value 0.05 (fig. 17); for a star of the absolute brightness $+ 10$, the probability that its temperature lies between 3250° K and 2750° K has the value 0.5 (fig. 18).

Both diagrams exhibit contour lines similar in shape to those of figure 15, but some characteristic differences are noticeable. In figure 18 the ridge running from the upper left to the lower right corner is ascending, but it descends in figure 15. Furthermore, in figure 18 a new crest appears, meaning that, although not many stars have a magnitude as large as $- 2$ to $- 4$, there is a high relative probability of a spectral type lying close to G_0 *if* such a star occurs at all. The significance of the diagrams consists in that they show more strongly the relation between brightness and temperature, eliminating from this relation certain accidental features and even systematic errors. For instance, in figure 15 the close relation between brightness and the spectral type G_0, which is shown in figure 18 by the new crest mentioned, is suppressed by the accidental feature that few of these bright stars exist. A systematic misrepresentation, however, is produced in figures 14 and 15 in the following manner. Since it is more difficult to find faint stars than to find bright ones, relatively few stars will appear in the lower part of the diagram. Or, more precisely speaking, since a bright star can be seen for great distances, where a dark star can no longer be seen, the bright stars as counted will fill a greater part of the cosmic space than the dark ones. The original diagrams of figures 14 and 15 thus are richer in bright stars, so that for larger values of the ordinate v (that is, the upper part of the diagram) the probability mountains become too high. Figure 18 is free from this error, as is shown by the upward slope of the ridge extending from the upper left to the lower right corner of the diagram, whereas the downward slope of the ridge in figure 15 must be regarded as an effect of the systematic misrepresentation mentioned.

By means of relative probability functions it is possible to construct continuous extensions of the schema of figure 5 (p. 82) and of the theorems referring to it. Assume that the disjunction $B_1 \vee \ldots \vee B_r$ is replaced by intervals du of a continuous variable u. We then replace B_k by the expression $B_{u,du}$, meaning "a value in the interval from u to $u + du$", and thus put

$$P(A,B_k) = P(A,B_{u,du}) = \varphi(u)du \tag{4}$$

For a precise definition of these symbols, we would first give the definition for finite intervals du and then proceed to the limit $du = 0$. Our way of writing means that the expression (4) leads to correct results if transitions to the limit $du = 0$ are derived.

Assume further that a second quantity v is related to u by a relative probability function $\psi_1(u;v)$, so that we have

$$P(A.B_k,C) = P(A.B_{u,du},C_{v,dv}) = P(A.B_u,C_{v,dv}) = \psi_1(u;v)dv \tag{5}$$

The transition from the second to the third term is possible because a relative probability has a precise value in the first variable; for small intervals du any value u of the interval may be chosen, according to (3). Generally speaking, a term $B_{u,du}$ can be replaced by B_u if it stands for the reference class of probability expressions; but if it stands for the attribute class, the corresponding probability function is to be multiplied by du. Like (4), formula (5) is correct for transitions to the limit $du = 0$, $dv = 0$.

Finally, we put

$$P(A,C) = P(A,C_{v,dv}) = \chi(v)dv \tag{6}$$

This formulation is subject to the same qualifications as the preceding ones.

We can now write the *continuous form of the rule of elimination* (21, § 19):

$$\chi(v) = \int_{-\infty}^{+\infty} \varphi(u) \cdot \psi_1(u;v)du \tag{7}$$

The value dv is canceled on the two sides, and thus (7) presents the probability density $\chi(v)$ as a function of the two probability densities $\varphi(u)$ and $\psi_1(u;v)$. Formula (7) shows that a relative probability function can be meaningfully integrated over the first variable if it is multiplied by another probability function.

For an illustration, let u be a man's height, whereas v is his weight. Height and weight are connected by the relative probability function $\psi_1(u;v)$. This means that the height of a man does not determine his weight; but if his height u is known, we know with a certain probability that his weight is $= v$ within the interval dv. Note that this probability is practically the same whether his height is given as a precise value or within a certain small interval du; but the interval dv, of course, has an influence on the numerical value of the probability. The distributions $\varphi(u)$ of the height and $\chi(v)$ of the weight are connected by the relation (7).

The *continuous form of the rule of Bayes* (10, § 21) results as follows:

$$\psi_2(v;u) = \frac{\varphi(u) \cdot \psi_1(u;v)}{\int_{-\infty}^{+\infty} \varphi(u) \cdot \psi_1(u;v)du} \tag{8}$$

The value dv, by which the expressions in numerator and denominator are multiplied if they are to represent probabilities, drops out. Like (7), formula (8) is thus a relation between probability densities. In the illustration used, $\psi_2(v;u)$ is the probability density that a man of weight v has the height u.

Let $\psi'_1(u_1,u_2;v)$ be the probability density that, if u is in the interval from u_1 to u_2, the other quantity has the value v. This probability is determined by the *continuous form of the special rule of reduction* (4, § 22):

$$\psi'_1(u_1,u_2;v) = \frac{\int_{u_1}^{u_2} \varphi(u) \cdot \psi_1(u;v)du}{\int_{u_1}^{u_2} \varphi(u)du} \tag{9}$$

In the illustration, $\psi'_1(u_1,u_2;v)$ is the probability density that a man whose height is between u_1 and u_2 has the weight v. Like (7) and (8), formula (9) is a relation between probability densities.

Formula (9) shows that the transition from $\psi_1(u;v)$ to a finite interval in the reference class leads to a somewhat involved expression, which requires knowledge of the antecedent probability density $\varphi(u)$. The transition from $\psi_1(u;v)$ to a finite interval in the attribute class, however, is achieved by a simple integration:

$$\psi''_1(u;v_1,v_2) = \int_{v_1}^{v_2} \psi_1(u;v)dv \tag{10}$$

Formulas (9) and (10) exhibit the intrinsic difference between the two variables of relative probability functions.

Relative probabilities have great practical importance, but this fact has rarely been recognized clearly. They always occur when physical laws are found or tested by experiment. A physical law presents a relation between two quantities such that *if* one quantity has a certain value, the other quantity also assumes a certain value. But in the application of physical laws it is never possible to take into consideration all the effective factors. Thus, if the first quantity has a definite value, the value of the other quantity can be determined only with a certain probability. If we regard the first quantity as a variable, all the mathematical functions occurring in the laws of nature have the character of relative probability functions. The assertion of the redrawn Russell diagram—If a star belongs to the spectral type M_b (that is, if it has a surface temperature of about 3000° K), then the probability is

0.05 that its absolute brightness lies between + 9 and + 10, and the probability is 0.3 that its brightness lies between − 0.5 and + 0.5—represents the prototype of a physical law. The fact of the probability character of physical laws is overlooked because usually the most probable value alone is taken into account, and the relation is then treated as a logical implication. Thus we say, If the temperature of a steam boiler is 121° C, the pressure

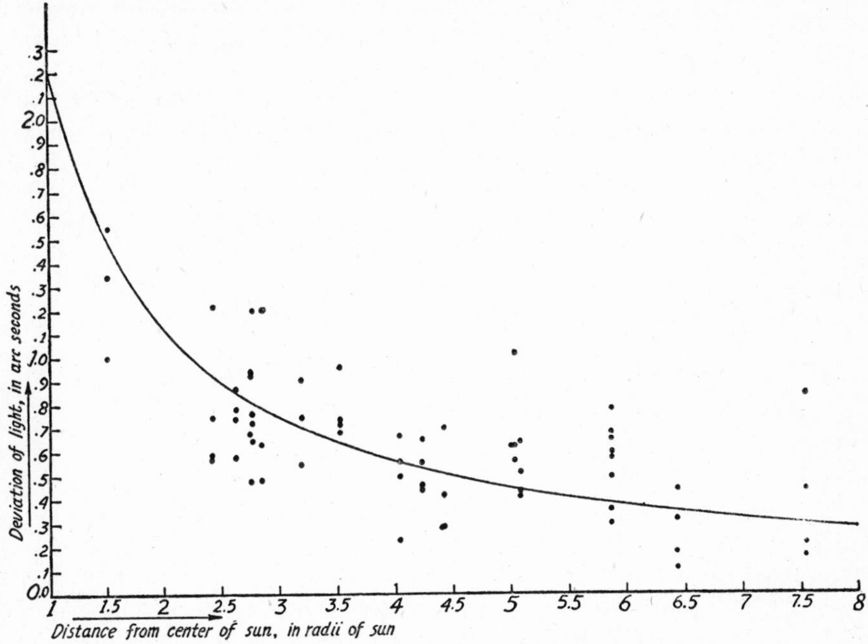

Fig. 19. Deviation of stellar light in gravitational field of sun, according to results of E. Freundlich's eclipse expedition in Sumatra, 1929.

amounts to 2 atmospheres. That this assertion does not hold with absolute certainty, but merely with a very high probability, becomes clear when we refer the second clause to the pressure indication of the manometer, which may fluctuate because of inhomogeneous conditions in the boiler.

The probability character of physical laws becomes obvious whenever an exact measurement is made, because the required relations are never strictly satisfied, but are disturbed by errors in observation. The rather irregular picture obtained by representing the results of measurements in a diagram can be understood only when it is interpreted by means of the theory of probability. For an illustration I refer to the result of Freundlich's eclipse expedition in Sumatra,[2] which was undertaken in order to measure the deviation of

[2] E. Freundlich, H. von Klüber, and A. von Brunn, "Ueber die Ablenkung des Lichtes im Schwerefeld der Sonne," in *Abhandl. d. Berliner Akad., math.-phys. Kl.*, 1931, p. 35.

a light ray of a star under the influence of gravitation when the ray passes close to the sun, according to Einstein's theory. In figure 19 the distance of the light ray from the sun (that is, the angular distance of the star from the sun) is indicated on the abscissa in multiples of the sun radius; the ordinates represent the observed deviation of the ray in angular seconds. Each dot of the diagram corresponds to an observed star.

The theory demands that to every distance there corresponds a certain deviation of the light, increasing with smaller distance such that the product of deviation by distance remains constant. This relation, which specifies a hyperbola, is fulfilled qualitatively by the curve as drawn, which collectively represents the measurements. The stars, however, show a large dispersion and the shape of the curve cannot be directly recognized; at some places many ordinate values lie above each other for the same value of the abscissa. The picture, strictly speaking, would have to be characterized by a relative probability function $\psi\ (u;v)$, which determines for every distance u a number of values v of the deviation, each with a certain probability. But the number of measured points does not suffice for the determination of this probability function, that is, of the probability mountains to be constructed above the plane of the drawing.

In such cases we usually ask for the *simplest curve* that can be traced through the points measured. In the interpretation according to the probability theory, the procedure means that we are satisfied if we can construct the *ridge line of the probability mountains.* The ascertainment of the curve is achieved by the *method of least squares,* which is too complicated mathematically to be dealt with in this book. Since the number of measurements usually is not large enough for a precise definition of the curve inquired, the balancing of observational results is carried through with the help of conditions concerning the shape of the curve, which results from the physical theory. In the instance considered, the balancing is based on the condition that the curve must possess the character of a hyperbola. In spite of this adaptation to the theory, the procedure supplies a quantitative test of the theory. The astronomic measurements of the illustration gave the result that the observed deviation is larger than the one demanded by the theory, because the theoretical curve would lie somewhat lower than the curve of the diagram.

The large dispersion of the measurements in the illustration may surprise those who believe that the laws of nature should exhibit "absolutely certain truth"; but the example does not represent an exceptional case. Other experimental results of great import present a like appearance, since precision measurements are always carried out close to the limits of experimental exactness. Freundlich's measurement of the deviation of light, obtained as it was by the help of precise and well-constructed apparatus, is a good example: only when we use the concepts and methods of the theory of probability are we able to uncover all the laws and regularities that are hidden in the observational material.

§ 46. Continuous Probability Sequences

The extension of the concept of probability sequence so far considered was constructed by means of a transition from discrete to continuous attributes. A disjunction of a finite number of terms was replaced by a disjunction of infinitely many terms, which could be interpreted as members of a continuous manifold, the attribute space. The possibility of this generalization is based on the existence of a geometrical interpretation of the axiom system; the duality of interpretation provides for an isomorphism between the probability interpreted as a frequency and the geometrical relations in the attribute space. The transition from the elements of the attribute space to finite domains entails, with respect to the probability, an infinite summation of infinitesimal amounts. The transition may therefore be regarded, in correspondence to the theorem of addition, as a transition to an or-probability; and thus the measure of the probability of finite domains of the attribute space assumes the form of an integral expression.

It is possible to introduce another generalization of the concept of probability sequence for which the transition to continuous variability concerns not the attribute but the element of the sequence itself. To simplify the exposition, the sequences of the elements x_i and y_i will be identified, so that we deal with an internal probability implication (see § 9). In the transition to be explained, a continuous sequence of the elements x_t takes the place of the sequence of the discrete elements x_i. The subscript may thereby be assumed to represent the continuous time-variable t, so that to each time point t a certain element x_t corresponds. The transition may be illustrated by the following example. If we shoot with a machine gun at a target, the continuous attribute space—the target—is filled with a discrete number of elements; each individual hit x_i represents an element of the sequence. But one can also shoot at the target with a continuous beam of water, ejected, say, from a hose moving irregularly and thus tracing an irregular zigzag line on the target. Then the hits x_t form a continuous event-sequence, in which a definite event is defined for each time t. Of this type is the continuous sequence of events now to be discussed.

It will be clear from the illustration that this generalization of the concept of probability sequence is possible only if we start with a sequence that has a probability aftereffect. If the event x_t is located at a certain point u,v of the attribute space, then an event x_{t+dt} occurring a short time dt later will still lie close to the attribute point u,v, since the event-sequence is continuous. This means, however, that there exists a probability aftereffect, manifesting itself as a *tendency to remain.* It will therefore be possible to apply to the sequences considered, in particular, the conditions of a probability transfer, in which the probability aftereffect depends on the immediate predecessor. This case will be given when the degree of the probability aftereffect depends only on the *position* of the attribute point u,v but not on its *velocity*. In this case the continuous probability sequence corresponds to the generalization of the sequence type classified in § 33 as probability drag. The peculiarity of continuous probability sequences, in contradistinction to discrete sequences, consists in the feature that the aftereffect is to be treated by a transition to infinitesimal distances. The probability of remaining, which was denoted in the case of probability drag by $p + \epsilon$, will go by infinitesimal steps toward 1, that is, the probability of the event x_{t+dt} lying within the attribute region $u + du$, $v + dv$ approaches 1 if dt goes toward zero. The convergence toward 1 results from the fact that for $dt = 0$ the events x_t and x_{t+dt} coincide. It would therefore be impossible to generalize the normal sequence so as to form a continuous event-sequence; only the sequence with a probability aftereffect lends itself to such a generalization, the probability of the aftereffect approaching certainty by infinitesimal steps.

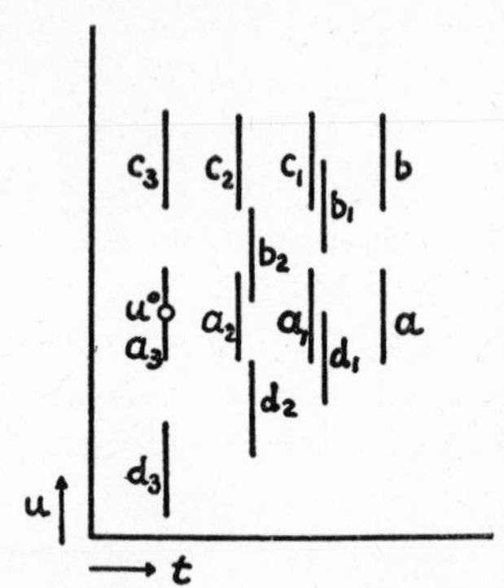

Fig. 20. Probabilities for sequences of intervals in migration space of continuous probability sequences.

These conditions may be illustrated by a translation into probability properties of sequences of discrete intervals, represented in figure 20 for the one-dimensional attribute space u. As abscissa the time t is used; as ordinate, the attribute u. Let u^o be the position of the attribute point at time t_o, and w the probability that a specified interval is reached from u^o. For this interval a there exists a certain probability w, lying between 0 and 1. If the interval a is shifted to a_3 by way of the positions a_1 and a_2, w goes toward 1. The same holds for the interval b if it is shifted by way of b_1 and b_2 to a_3. The convergence toward 1 does not hold, however, if b is shifted by way of c_1 and c_2 to c_3, since this w goes toward 0; the same applies to w if a is shifted by way of d_1 and d_2 to d_3. The probability state of a point environment is therefore represented by sequences the probability of which converges either to 1 or to 0.

The path of the attribute point of a continuous probability sequence is represented for the one-dimensional attribute space by a zigzag line. For an illustration I refer to the *Brownian motion* of a rotating mirror, drawn after a photograph taken by E. Kappler.[1] (See fig. 21.) In the experiment of Kappler a small mirror, 1–2 sq. mm. in size, was suspended by a quartz fiber

[1] *Ann. d. Physik.*, Vol. 11 (1931), p. 242.

of a length of some centimeters; the thickness of the quartz fiber amounted to only a few ten-thousandths of a millimeter. The mirror and the suspension were placed in a glass vessel, which was evacuated to low pressure. Because of the small diameter of the fiber and the weak damping by the air, such an arrangement is very sensitive to rotation; when the surrounding air molecules impinge upon the mirror, the differences in impact become noticeable and the mirror describes an irregular rotation.

It is characteristic of the Brownian motion, and also of other fluctuation phenomena, that the number n of the air molecules participating is not so

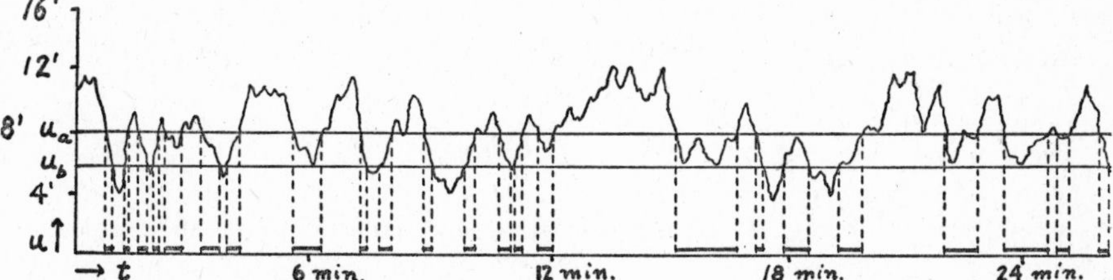

Fig. 21. Brownian motion of rotating mirror drawn after photograph by E. Kappler. u = angular movement of mirror in minutes of arc (zero point arbitrarily chosen); t = time in minutes. Reproduced in approximately same size as original photograph. Curve constitutes example of a continuous probability sequence.

great as is required if the $\sqrt{n}$-law (6c, § 39) is to provide for a virtually complete uniformity of the average air pressure. The apparatus is so sensitive that minute fluctuations in the average air pressure can be noticed. For the observation a special registering apparatus is employed. A beam of light from a lamp is reflected by the mirror and projected by means of a lens system on a strip of photographic film; while the light beam moves from side to side, the film passes simultaneously from top to bottom, producing the "jerky curve" shown in figure 21. However, the curve cannot be subsumed under the case of probability transfer because the probability of the aftereffect depends not only on the position but also on the velocity of the attribute point (because of the inertia of the mirror). Here we have a more general case of probability aftereffect.

Figure 21 indicates how the transition to the probability of finite regions is carried through. For discrete event-sequences the probability that the attribute point lies within the interval u_a to u_b is given by the number of hits falling into this region; for continuous event-sequences the probability is measured by the sum of the time intervals during which the attribute point remains in the interval from u_a to u_b. In figure 21 we thus determine the probability by adding the distances marked by a heavy line and dividing them by the total length of the t-axis. The enumeration of frequencies is here replaced by the measurement of time intervals. In analogy to the N-symbol used previously (1a, § 16), we introduce the symbol

$$\overset{T}{\underset{t=0}{L}}(u \epsilon B) \tag{1a}$$

for the length of time t during which the point u remains in the region B of the attribute space within the time interval from 0 to t. Instead of speaking of the sequence x_t, it is convenient to regard directly the sequence of attributes u^t coördinated to it, u^t meaning the value of the variable u at the time t. Instead of regarding B as a class of events, we then conceive B as a class, or an interval, of numbers (see § 41). The symbol (1a) may be abbreviated, analogously to the abbreviation (2, § 16), by means of the definition

$$L^T(B) =_{Df} \overset{T}{\underset{t=0}{L}}(u \epsilon B) \tag{1b}$$

Let A be another region of the attribute space. In analogy to the frequency interpretation (5, § 16), the probability for continuous event-sequences can be interpreted as

$$P(A,B) = \lim_{T \to \infty} \frac{L^T(A.B)}{L^T(A)} \tag{2}$$

If the sequence A is compact (as assumed for fig. 21), (2) can be written in the simpler form

$$P(A,B) = \lim_{T \to \infty} \frac{1}{T} L^T(B) \tag{3}$$

Like the geometrical probability (§ 41), this generalization of the concept of probability sequence depends on the possibility that the axiom system can be given a geometrical interpretation. Instead of the enumeration of elements, we have here the measurement of lengths of time; it is a measurement along the t-dimension of a space that is formed by the attribute dimensions together with the t-dimension. I call this space the *migration space.* The proof that the interpretation given by (2) for the probability concept is admissible is therefore included in the general proof of the possibility of a geometrical interpretation.

No difficulties arise for the incorporation of continuous probability sequences in the formal calculus of probability. It is irrelevant for the application of the axiom system I–IV whether the elements x_i, y_i form discrete or continuous sequences. Continuous probability sequences, which, according to the definition, include the properties of probability sequences with a continuous attribute, can be incorporated in the formal calculus of probability exactly in the same manner as was adopted for primitive probability sequences (see § 41). How to determine the probability coördinated to the sequences is not a problem of the formal calculus but of the interpretation. It is only with respect to the frequency interpretation that continuous probability sequences differ from discrete sequences: continuous probability sequences provide a model of the formal probability calculus in which an enumeration is replaced by a measurement. But all the concepts of the formal calculus are applicable. For instance, $P(A.B,C)$ means the probability of C within the subsequence selected by the condition that u lies in B. In the interpretation given by (2) we have, therefore,

$$P(A.B,C) = \lim_{\tau \to \infty} \frac{L^{\tau}(A.B.C)}{L^{\tau}(A.B)} \tag{4}$$

With the existence of probabilities of this kind, the concept of mutual dependence or independence is applicable to continuous probability sequences.

Only the axioms v (§ 28) require special treatment. The condition that the sequences contain an infinite number of elements is always fulfilled for continuous probability sequences. This condition must be strengthened by the requirement that the length of the sequence along the time axis [for the classes written in the first term of (4 and 5, § 28)] is infinite. With this addition the axioms v can be transferred to continuous probability sequences. In phase probabilities a superscript like α then denotes a time of finite but otherwise arbitrary length α. With this qualification the formulas of the theory of order are made applicable; thus the concepts of invariant selection, lattice formation, and so on can be defined. The only difference is that no normal sequences exist, since all sequences have an aftereffect.

The mathematical treatment of continuous probability sequences will be explained only for the case where the attribute space is divided into two regions B and $\bar{B}$; so reference will be made only to alternating sequences.[2] Consideration will therefore be restricted to the projection of the curve on the t-axis in figure 21 (p. 239). The distances in heavy print along the t-axis correspond to the case where $u \in B$; the other distances along the t-axis correspond to the case $u \in \bar{B}$. As a further simplification (apart from the infinite length of the sequence in B and $\bar{B}$), the three conditions are introduced:

1. The sequence satisfies the condition of probability transfer according to § 33, that is, the probability $f(t)$ that u, if it is in B at the time 0, remains in B during the time interval t does not depend on whether u was in B or $\bar{B}$ before the time 0.[3]
2. $f(t)$ is a continuous and differentiable function of t.
3. $f(0) = 1$.

[2] The mathematical treatment of the continuous attribute space may be found in the appendix of the German edition of this book (§ 81), and in H. Reichenbach, "Stetige Wahrscheinlichkeitsfolgen," in *Zs. f. Physik*, Vol. 53 (1929), p. 274.

[3] Notice that $f(t)$ is not a probability function, since $f(t)$ represents a probability and not a probability density.

Corresponding assumptions are made for the probability $g(t)$ that u, if it is in $\bar{B}$ at the time 0, remains in $\bar{B}$ during the time interval t. The third condition is required because of the continuity of the probability sequence.

We derive from the third condition, together with the second condition, if we write the Taylor series only for the first two terms (other terms are not needed, since we shall later go to the limit $dt = 0$),

$$\begin{aligned} f(0 + dt) &= f(0) + f'(0)dt \ldots \\ &= 1 - \alpha dt \qquad \alpha = -f'(0) \end{aligned} \tag{5}$$

With the notation of § 33 we obtain, putting $\beta = -g'(0)$,

$$\begin{aligned} P(A.B,B^{dt}) &= p_1 = 1 - \alpha dt \\ P(A.B,\bar{B}^{dt}) &= 1 - p_1 = \alpha dt \\ P(A.\bar{B},B^{dt}) &= q_1 = \beta dt \\ P(A.\bar{B},\bar{B}^{dt}) &= 1 - q_1 = 1 - \beta dt \end{aligned} \tag{6}$$

α and β will be called *coefficients of transfer*. We can now apply (6, § 33) and thus have

$$P(A,B) = p = \frac{\beta dt}{\alpha dt + \beta dt} = \frac{\beta}{\alpha + \beta} \tag{7}$$

The existence of the mean probability p according to (3) is thus made sure. in order to determine the probability $f(t)$, we imagine the t-axis to be divided Into finite small intervals dt; if the interval dt lies completely within B, we regard it as belonging to B, otherwise to $\bar{B}$. We then have

$$f(t) = \lim_{\substack{dt \to 0 \\ ndt = t}} P(A.B, B^{dt}.B^{2dt} \ldots B^{ndt}) \tag{8a}$$

The expression written on the right can be factorized as a product of n factors of the form $P(A.B,B^{dt})$; this follows from the property of probability transfer expressed in (9, § 33). Therefore (8a) assumes the form

$$f(t) = \lim_{\substack{dt \to 0 \\ ndt = t}} (1 - \alpha dt)^n \tag{8b}$$

With $\alpha dt = \dfrac{\alpha t}{n} = \dfrac{1}{m}$, that is, $n = m\alpha t$, we have

$$\begin{aligned} f(t) &= \lim_{m \to \infty} \left(1 - \frac{1}{m}\right)^{m\alpha t} \\ &= \left[\lim_{m \to \infty} \left(1 - \frac{1}{m}\right)^m\right]^{\alpha t} = \left[\lim_{m \to \infty} \left(1 + \frac{1}{m}\right)^m\right]^{-\alpha t} \\ &= e^{-\alpha t} \end{aligned} \tag{9}$$

In the last line the familiar definition of the number e is used. We see that the probability $f(t)$ of u remaining in B during the time t decreases exponentially. The form of this function is presented in figure 22*A*. In a similar manner we derive with (1, § 31) for the probability $\bar{f}(t)$ that u, if we have $u \epsilon B$ at the time 0, changes over to $\bar{B}$ during the following time interval t,

$$\bar{f}(t) = \lim_{\substack{dt \to \infty \\ ndt = t}} P(A.B,\bar{B}^{dt} \vee \bar{B}^{2dt} \vee \ldots \vee \bar{B}^{ndt})$$

$$= 1 - f(t) = 1 - e^{-\alpha t} \qquad (10)$$

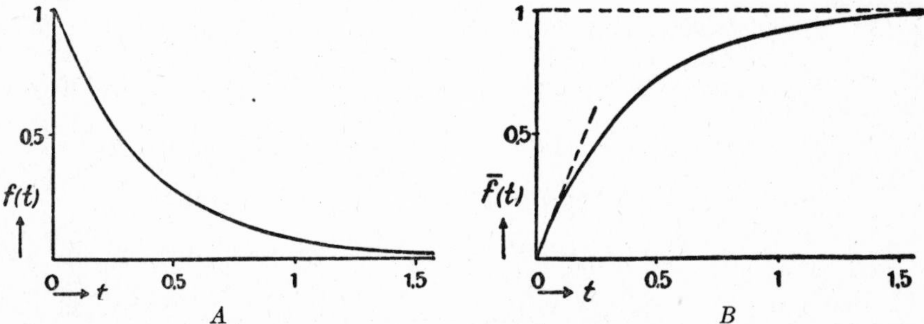

Fig. 22. *A*. Exponential decrease $f(t) = e^{-\alpha t}$, according to (9), with $\alpha = 2.5$. *B*. Exponential increase $\bar{f}(t) = 1 - e^{-\alpha t}$, according to (10), with $\alpha = 2.5$.

This equation determines an exponential increase for the probability of a change to $\bar{B}$, as drawn in figure 22*B*. When we apply the Taylor series to (10), we obtain for small t, according to (6),

$$\bar{f}(t) \sim \alpha t \qquad (11)$$

This relation corresponds to the tangent shown as a dotted line in figure 22*B*; for large t, however, the function $\bar{f}(t)$ increases more slowly.

These relations have many illustrations. For instance, the continuous probability sequence may be given by the path of a gas molecule (its world line); B may be a certain region inside the vessel in which the molecule moves. If the molecule happens to be in the region B at the time $t = 0$, then the probability that during the time t (that is, at any time point within the interval t) it changes over to $\bar{B}$ at least once is given by (10). Or the continuous probability sequence may be the life line of a human being; $\bar{B}$ may signify that the person is ill. Then (10) represents the probability that the individual, if in good health at the time $t = 0$, falls ill during time t. According to (11), this probability increases proportionally to time only for small t, otherwise more slowly. Formula (9) states the probability that the person remains healthy during the time t.

We calculate now the average length of time during which the attribute point remains constantly in B. Such a period is characterized by the condition that it starts with a change from $\bar{B}$ to B and stops with a change from B to $\bar{B}$. [Notice that we did not demand for the definition of $f(t)$ according to (8a) that the stay in B should cease; thus in $f(t)$ cases are counted where the attribute point remains in B during a period longer than t.] We inquire after the probability $\varphi(t)dt$ of a period of the kind described, the length of which is between t and $t + dt$, counted among all periods of an uninterrupted stay in B. The number of the latter periods being equal to the number of changes from $\bar{B}$ to B, we can represent this probability, before going to the limit, by

$$\varphi^*(t)dt = P(A.\bar{B}.B^{dt},B^{2dt} \ldots B^{(n+1)dt}.\bar{B}^{(n+2)dt}) \tag{12a}$$

Because of the property (9, § 33) of the probability transfer this expression is equal to

$$\begin{aligned} \varphi^*(t)dt &= P(A.B^{dt},B^{2dt} \ldots B^{(n+1)dt}.\bar{B}^{(n+2)dt}) \\ &= P(A.B,B^{dt} \ldots B^{ndt}) \cdot P(A.B,\bar{B}^{dt}) \\ &= (1 - \alpha dt)^n \cdot \alpha dt \end{aligned} \tag{12b}$$

When we make the transition to the limit $dt = 0$ for the first expression only, employing (9), we have

$$\varphi(t)dt = e^{-\alpha t} \cdot \alpha dt \tag{13}$$

For the desired average we thus find, according to (3, § 43) with $x = \alpha t$,

$$\begin{aligned} M(t) &= \int_0^\infty t \cdot \varphi(t)dt = \int_0^\infty t \cdot e^{-\alpha t} \cdot \alpha dt \\ &= \frac{1}{\alpha}\int_0^\infty x \cdot e^{-x}\, dx = \frac{1}{\alpha}\,\Gamma(2) = \frac{1}{\alpha} \end{aligned} \tag{14}$$

Γ is Euler's gamma function, which satisfies for integers z the relation

$$\Gamma(z) = \int_0^\infty x^{z-1} \cdot e^{-x}\, dx = (z - 1)! \tag{15}$$

The average length of an uninterrupted stay in B is therefore given by $\frac{1}{\alpha}$.

Similarly, the dispersion connected with this average can be calculated. From (7, § 43) we have

$$\Delta^2(t) = M(t^2) - M^2(t)$$

With (15) we find

$$\begin{aligned} M(t^2) &= \int_0^\infty t^2 e^{-\alpha t}\alpha dt = \frac{1}{\alpha^2}\int_0^\infty x^2 e^{-x} dx \\ &= \frac{1}{\alpha^2}\,\Gamma(3) = \frac{2}{\alpha^2} \end{aligned}$$

We thus derive

$$\Delta^2(t) = \frac{2}{\alpha^2} - \frac{1}{\alpha^2} = \frac{1}{\alpha^2} = M^2(t) \qquad (16)$$

$$\Delta(t) = \frac{1}{\alpha} = M(t)$$

We meet here with the special case that the linear dispersion is equal to the average, that is, the dispersion amounts to 100%.

The meaning of the quantity α can be made clear as follows. When we introduce a division by intervals dt that are so small that each dt at most contains one change, the number ν_τ of changes from B to $\bar{B}$ during the time τ is given by

$$\nu_\tau = \overset{\tau}{\underset{t=0}{N}}(u^t \epsilon B).(u^{t+dt} \epsilon \bar{B})$$

if t assumes the discrete values 0, dt, $2dt$, . . . 1. The value ν_τ is independent of the length of dt. Now we have

$$\alpha dt = P(A.B,\bar{B}^{dt}) = \lim_{\tau\to\infty} \frac{\left[\overset{\tau}{\underset{t=0}{N}}(u^t \epsilon B).(u^{t+dt} \epsilon \bar{B})\right] dt}{\left[\overset{\tau}{\underset{t=0}{N}}(u^t \epsilon B)\right] dt}$$

$$= \lim_{\tau\to\infty} \frac{\nu_\tau dt}{l_\tau} \qquad l_\tau = \overset{\tau}{\underset{t=0}{L}}(u^t \epsilon B) \qquad (17a)$$

If we drop dt on both sides and reverse the fractions, we have

$$\frac{1}{\alpha} = \lim_{\tau\to\infty} \frac{l_\tau}{\nu_\tau} \qquad (17b)$$

On the right side we have the quotient of the total length of all distances in B divided by the number of these distances, that is, the mean value of the distances; its limit corresponds to the average. This is the statistical meaning of (14).

Let us introduce the notation

$$\nu = \lim_{\tau\to\infty} \frac{1}{\tau}\nu_\tau \qquad (18)$$

ν means the average number of changes from B to $\bar{B}$ during the unit of time, or the average number of *one-sided* changes during the unit of time; we call ν the *alternation frequency*. The corresponding number of *two-sided* changes, or the *total alternation frequency*, is twice as large. Multiplying numerator and denominator of the expression on the right side of (17*b*) by $\frac{1}{\tau}$ and reversing the fractions, we obtain

$$\alpha = \lim_{\tau \to \infty} \frac{\frac{1}{\tau} \nu_\tau}{\frac{1}{\tau} l_\tau} = \frac{\nu}{p}$$

$$\alpha p = \nu \tag{19}$$

This is a very simple relation holding between the alternation frequency ν the mean probability p, and the transfer coefficient α; the alternation frequency is, for given p, determined by α. Therefore we call α the *relative measure of alternation;* it measures the alternation with respect to p. For statistical calculations (19) is used, conversely, to determine α, since p can be found empirically by a simple measurement of time intervals and ν by enumeration. This method represents a very simple statistical determination of the transfer coefficient α and thereby of the exponent in the exponential laws (9) and (10).

The corresponding statement is valid for β, which likewise may be called the *relative measure of alternation,* since the relation

$$\beta(1 - p) = \nu \tag{20}$$

can be derived in analogy to (19). The quantity β thus measures the alternation with respect to $1 - p$. The identity of (19) and (20) is warranted by (7). When we denote by $\bar{t}$ the time of the stay in $\bar{B}$, we have, correspondingly,

$$M(\bar{t}) = \frac{1}{\beta} \tag{21}$$

In the example of the life line of a person, for which $\bar{B}$ denotes illness, $\frac{1}{\beta}$ represents the average duration of an attack of illness; ν is the number of attacks of illness that the person suffers during the unit of time. The total duration of all attacks of illness is measured, in comparison to the length of life, by $1 - p$.

The insurance company that pays financial subsidies during the period of illness takes $1 - p$ into account as the quantity determining the insurance premium; but with respect to the physical constitution of the person the quantity β is also a characteristic, since β measures the probability that the person falls ill. If the coefficient β is large for a given value of p, the frequency of attacks of illness is high and the average duration of the attack is short.

The example concerning the theory of gases supplies an illustrative interpretation of ν if we understand by $\bar{B}$ the collision of one molecule with another. Then ν represents the number of collisions in the time unit. If we assume the duration of a collision $\bar{B}$ to be very short, compared to the time period during which no collisions occur, we derive from (3) that p is nearly equal to 1. Therefore β must be very large, and according to (19) we have approximately $\alpha = \nu$. This relation, well known in the theory of gases, likewise provides a good illustration of the measure of alternation α. According to (14), the quantity $\frac{1}{\alpha}$ represents the mean duration of the free path, that is, a path without collisions. This result agrees with the relation $\alpha = \nu$ if the duration of individual collisions can be neglected. When we multiply $\frac{1}{\alpha}$ by the mean velocity, we obtain the length of the mean free path (or, more correctly, one of the possible definitions of the mean free path).

The formulas developed can be given a further interpretation when we consider a lattice of continuous probability sequences. The number of sequences itself may be discrete, that is, characterized by a discrete superscript k. If we make the further assumption that the lattice is lattice-invariant, the phase probabilities counted horizontally are realized also in the vertical direction. These conditions correspond to the case of a lattice in which the horizontal sequences represent the world lines of gas molecules. For the analysis of such a lattice it is advisable to use, first, enumeration by sections in the horizontal direction, since we may assume that the sequences are regular-invariant.

If sections of the length t are cut off, the relative frequency of sections lying completely within B is given by $e^{-\alpha t}$, according to (9); they will be called "sections of the first kind". The relative frequency of the other sections, or "sections of the second kind", is given by $1 - e^{-\alpha t}$. On the other hand, we can count the number of changes by counting for each section the number ν^i of one-sided changes; if n is the number of sections, their sum is given by

$$\sum_{i=1}^{n} \nu^i = \nu_\tau \qquad \tau = nt \tag{22}$$

Therefore their average is represented by

$$M(\nu^i)^i = \lim_{n\to\infty} \frac{1}{n} \sum_{i=1}^{n} \nu^i = \lim_{n\to\infty} \frac{1}{n} \nu_\tau = t \cdot \lim_{\tau\to\infty} \frac{1}{\tau} \nu_\tau = t\nu \tag{23}$$

This result is in agreement with the definition of ν, since ν represents the number of changes for $t = 1$. For n sections the number of one-sided changes will be given approximately by $nt\nu$. This differs from the number $n(1 - e^{-\alpha t})$ of the sections of the second kind, because the latter contain partly no change, partly more than one change.

Second, we introduce lattice enumeration, adding the assumption [as for (9, § 34)] that the first element $t = 0$ of all horizontal sequences belongs to B. (The lattice is then not homogeneous.) Because of the lattice invariance, the frequency, counted vertically, of sections of the first kind of the length t, all of which begin with $t = 0$, is given by $e^{-\alpha t}$; the frequency of sections of the second kind is given by $1 - e^{-\alpha t}$. The summation of the changes contained in all the sections does not admit immediately of the interpretation (22) since these sections do not combine into one continuous sequence. But (23) is valid nonetheless, since the summation, because of the lattice invariance gives the same result as the enumeration by sections; thus the number ν can be determined in the vertical direction also. In the relation (19) we must substitute for p the value of the horizontal sequences.

We can now consider a case that is important in many applications, resulting from a degeneration of the lattice. Let the horizontal sequence represent the world lines of radioactive atoms, $\bar{B}$ the disintegration of the atom; then the horizontal sequences are degenerated, because they will no longer return to B after the event $\bar{B}$ has once occurred. The definition (3) therefore supplies $P(A,B^{kt})^{t} = p = 0$. Correspondingly, we infer from (7) that also $\beta = 0$; and (19) gives $\nu = 0$. The lattice is therefore not lattice invariant; but this very circumstance makes it possible that some of the relations derived previously remain valid in the vertical direction. Thus the vertical sequences realize the probability $f(t)$, because the infinitesimal relations (6) remain valid in the vertical direction and determine α, and because the factorization of (8*a*), stated in (8*b*), holds for vertical enumeration. Equations (9) and (10) then represent the familiar law of radioactive disintegration. If there are N atoms present, then during the time t approximately

$$N \cdot (1 - e^{-\alpha t}) \tag{24}$$

atoms will disintegrate, according to (10). At the end of this time only

$$N \cdot e^{-\alpha t} \tag{25}$$

atoms will be left over. The quantity α is called the *constant of disintegration*; it can be determined only by enumeration in the vertical direction. The relation (14) remains valid, too, for enumeration in the vertical direction. The average number of changes for a section of the length t in the vertical direction is identical with the number of sections of the second kind, since in the horizontal direction there can never occur more than one change; and, because of the assumption concerning the first element of the sequences, each section of the second kind must contain at least one change. This number, therefore, is $1 - e^{-\alpha t}$. The number, however, is not identical with the number νt, as defined earlier, since the definition of ν in the horizontal direction leads to $\nu = 0$ and the relation (19) is fulfilled only for this value.

In another example the horizontal sequences may represent the life lines of persons and $\bar{B}$ the death resulting from a traffic accident. Since this example displays the same degeneration as the previous one, α is determined, once more, only by enumeration in the vertical direction. If we put $dt = 1$ year, we have $\alpha = \frac{32,300}{142,673,000} \sim \frac{1}{4,400}$, according to p. 183 (footnote). In this case we may apply the approximation (11): the probability of such an accident in the course of two years is twice as great as for one year. This proportionality, however, holds for short periods only; for long periods the strict law (10) is to be employed. The usual treatment of social statistics by the use of the law of proportionality (11), therefore, provides merely an approximation. The quotients obtained from social statistics must be interpreted, strictly speaking, as transfer coefficients of continuous probability sequences. The necessity of this interpretation is even more clearly visible for nondegenerated cases, such as are supplied, for instance, by the number of attacks of illness.

The examples show that continuous probability sequences are of great practical importance. They supply an instrument permitting us to interpret the continuous lines of causal connection, or *causal chains*, as probability sequences. Through this interpretation the continuous probability sequences, besides their practical applicability, acquire a fundamental importance.

The usual conception of the causal chain as a strictly determined connection between events, the course of which can be predicted with certainty from given initial conditions for any subsequent length of time, has turned out to represent an untenable idealization, which cannot do justice to the facts gathered in natural sciences. Since all the laws of nature possess the form of probability implications, the causal chain must be represented by a continuous sequence in which the later elements are determined by the preceding ones with a certain probability only. The particular form of the determination will be of the type studied here: at every point event the ensuing occurrences are determined with certainty only for an infinitesimal period of time, whereas for a finite length of time they are determined only with a probability that decreases continuously with increasing time. This statement holds for classical as well as for quantum physics, since, even if Heisenberg's uncertainty relations are not taken into account, the concept of causal chain must be described by a process of convergence exhibiting the following form: when we wish to predict at the time t_0 the state of a phenomenon for the later time t with the probability $1 - \delta$ (δ small), we can specify at the time t_0 an observational exactness G that achieves this result; but for the same exactness G a time $t_1 > t$ can be specified for which the probability of the prediction is smaller than a given small value ϵ. Only this double assertion can characterize exhaustively the process of convergence, and we recognize that the idea of determinism seems scarcely justifiable even in classical physics.

The result holds to a much higher degree in Heisenberg's quantum mechanics, in which it is impossible, in general, to increase the probability of prediction as close to certainty as we wish, as soon as finite time intervals are considered, however small they may be. In quantum physics, therefore, not only every actually formulated causal chain is represented by a continuous probability sequence, but this character is also to be ascribed to the limit that the formulated causal chains approach when the exactness of the observation is made as great as possible.[4] The zigzag path of the Brownian motion must therefore be regarded as the prototype of every causal chain.

§ 47. Competition of Chances

The developed methods will now be used for the treatment of some problems of practical importance.

Assume that a man is fishing at a pond. The sequence of events is a continuous sequence of time points; if the man catches a fish, the time point has the attribute B, otherwise $\bar{B}$. We may idealize the problem by assuming that the catching of the fish is an event of so short a duration that it fills only one time point. The probability of getting a fish increases continuously with time; considered for any moment, however, it has the same value, and therefore we may write, in correspondence with (6, § 46),

$$P(A.\bar{B},B^{dt}) = \beta dt \tag{1}$$

By considerations like those applied in (8–10, § 46) we derive the formula

$$P(A,B_{\Delta t}) = 1 - e^{-\beta\Delta t} \tag{2}$$

for the probability of catching a fish during the time Δt.

Assume that there is a second pond at which another man fishes. We shall distinguish the two men by the subscripts $_1$ and $_2$; and, assuming that both men fish during the same period Δt, we shall omit the subscript Δt of B. Thus we write

$$\begin{aligned} P(A_1,B_1) &= 1 - e^{-\beta\Delta t} = p \\ P(A_2,B_2) &= 1 - e^{-\gamma\Delta t} = q \end{aligned} \tag{3}$$

The constants β and γ measure the fishing chances of each man.

Now introduce the assumption that each pond contains only one fish. Formulas (1)–(3) apply to this case as well. For the statistical interpretation of the probabilities we then assume that the fishing experiment is made during many periods Δt. If the fish is caught in one period, no new fish is added during that period; for the following period, however, a new fish is

[4] Further explanations by the author are found in *Erkenntnis*, Vol. I (1930), p. 158; Vol. II (1931), p. 156; Vol. III (1932), p. 32; *Naturwissenschaften*, Vol. 19 (1931), p. 713; and *Philosophic Foundations of Quantum Mechanics* (Berkeley, 1944), § 1.

put into the pond. The probabilities q and p then represent frequencies counted in terms of the periods Δt.

What is the probability that at least one man catches a fish during Δt? Since we can assume independence, we have for this probability the value

$$P(A_1 . A_2, B_1 \vee B_2) = p + q - pq \tag{4}$$

We now turn to a new problem. Assume that there is only one pond with one fish, and that both men fish at the pond. So long as neither man has caught the fish, the chances of catching it are the same as before for each man. But if one man catches the fish, the other can no longer get it; the chances are diminished for each man by the competition of the other. Thus there is a *competition of chances.*

The probabilities (3) must now be denoted in a different way. The probability p exists for the first man only when the second does not fish, and vice versa for the probability q. We therefore must write

$$\begin{aligned} P(A_1 . \bar{A}_2, B_1) &= 1 - e^{-\beta \Delta t} = p \\ P(A_2 . \bar{A}_1, B_2) &= 1 - e^{-\gamma \Delta t} = q \end{aligned} \tag{5}$$

What is now the probability that the fish is caught during the time Δt? We can easily see that the answer (4) remains correct. To show this, let us return to the arrangement with two ponds and assume that, if both men catch the fish during the same period, only that man is given credit for catching who catches the fish first, and that he also gets credit if the other does not catch his fish. The probability that credit is given is then obviously represented by (4), since in the frequency counting of the cases $B_1 \vee B_2$ the case where both B_1 and B_2 are true is counted only once anyway. Now when both men fish at the same pond, the catching of the fish by one man corresponds to the receiving of credit in the other arrangement. As the credit received by one man deprives the other of the chance of getting it, so does the fish caught by one man deprive the other of the chance of getting it. Therefore (4) also represents the probability of the fish being caught if only one pond is used.

It is more difficult to compute the probability $P(A_1 . A_2, B_1)$, that is, the probability that the first man catches the fish if the other man fishes simultaneously at the same pond. Since the events B_1 and B_2 are now exclusive, we have

$$P(A_1 . A_2, B_1) + P(A_1 . A_2, B_2) = P(A_1 . A_2, B_1 \vee B_2) \tag{6}$$

In combination with (4) this relation indicates that

$$\begin{aligned} P(A_1 . A_2, B_1) &< p \\ P(A_1 . A_2, B_2) &< q \end{aligned} \tag{7}$$

We therefore have a reduction of the individual probabilities through the competition of chances.

The value of $P(A_1.A_2,B_1)$ can be found when we return to the arrangement with two ponds and ask for the probability that the first man catches the fish before the other man catches his. The probability can be constructed as follows. Because of the transfer character of the probability sequence and the independence of the two fishing processes, we have, in a generalization of (1),

$$\begin{aligned} P(A_1.A_2.\bar{B}_1^{dt} \ldots \bar{B}_1^{(n-1)dt}.\bar{B}_2^{dt} \ldots \bar{B}_2^{(n-1)dt},B_1^{ndt}) &= \beta dt \\ P(A_1.A_2.\bar{B}_1^{dt} \ldots \bar{B}_1^{(n-1)dt}.\bar{B}_2^{dt} \ldots \bar{B}_2^{(n-1)dt},B_2^{ndt}) &= \gamma dt \end{aligned} \tag{8}$$

The probability that both men catch their fish in the same interval dt then is $= \beta\gamma dt^2$ and is thus of the second order; it may therefore be neglected. Now the case that the first man catches the fish in one of the intervals $dt_1, dt_2, \ldots dt_n$ may be written

$$\begin{aligned} (B_1^{dt} \vee B_1^{2dt} \vee \ldots \vee B_1^{ndt}) \equiv (B_1^{dt} \vee \bar{B}_1^{dt}.\bar{B}_2^{dt}.B_1^{2dt} \vee \ldots \\ \vee \bar{B}_1^{dt} \ldots \bar{B}_1^{(n-1)dt}.\bar{B}_2^{dt} \ldots \bar{B}_2^{(n-1)dt}.B_1^{ndt}) \end{aligned} \tag{9}$$

Since the terms on the right side of (9) are exclusive, the corresponding probabilities may be added, and we have, for the probability r that the first man catches the fish in one of the intervals $dt_1, dt_2, \ldots dt_n$,

$$\begin{aligned} r &= P(A_1.A_2,B_1^{dt} \vee B_1^{2dt} \vee \ldots \vee B_1^{ndt}) \\ &= P(A_1.A_2,B_1^{dt}) + P(A_1.A_2,\bar{B}_1^{dt}.\bar{B}_2^{dt}.B_1^{2dt}) \\ &\quad + \ldots + P(A_1.A_2,\bar{B}_1^{dt} \ldots \bar{B}_1^{(n-1)dt}.\bar{B}_2^{dt} \ldots \bar{B}_2^{(n-1)dt}.B_1^{ndt}) \end{aligned} \tag{10}$$

For the combination terms on the right side we may employ the special theorem of multiplication, since we assume that the two fishing processes are independent. Using (8), we thus have

$$\begin{aligned} r &= \beta dt + (1 - \beta dt) \cdot (1 - \gamma dt) \cdot \beta dt + \ldots + (1 - \beta dt)^{n-1} \cdot (1 - \gamma dt)^{n-1} \cdot \beta dt \\ &= \beta dt \frac{(1 - \beta dt)^n (1 - \gamma dt)^n - 1}{(1 - \beta dt)(1 - \gamma dt) - 1} \\ &= \frac{\beta}{\beta + \gamma - \beta\gamma dt} [1 - (1 - \beta dt)^n (1 - \gamma dt)^n] \end{aligned} \tag{11}$$

The term $\beta\gamma dt$ in the denominator can be neglected, since it goes to zero with $dt \to 0$. Putting $dt = \frac{\Delta t}{n}$ we have

$$(1 - \beta dt)^n = \left(1 - \frac{\beta \Delta t}{n}\right)^n \tag{12}$$

and using the definition of the number e, we have

$$\lim_{n\to\infty}\left(1-\frac{\beta\Delta t}{n}\right)^n = e^{-\beta\Delta t} \tag{13}$$

Therefore (11) assumes the form

$$r = \frac{\beta}{\beta+\gamma}\left[1 - e^{-(\beta+\gamma)\Delta t}\right] \tag{14}$$

Now the probability $P(A_1.A_2,B_1)$, referring to the arrangement with only one pond, is obviously identical with r; in fact, the relation (10) applies likewise to this arrangement. The consideration of the two-pond arrangement acts only as a help in illustrating the existing relations. The independence of the two fishing processes, even when performed at the same pond, is expressed in (8), relations which say that so long as the fish is not caught the chances for each man are not changed by the presence of the other man; only when the fish is caught does one process interfere with the other. We therefore have

$$P(A_1.A_2,B_1) = r = \frac{\beta}{\beta+\gamma}\left[1 - e^{-(\beta+\gamma)\Delta t}\right] \tag{15}$$

By similar considerations we derive

$$P(A_1.A_2,B_2) = s = \frac{\gamma}{\beta+\gamma}\left[1 - e^{-(\beta+\gamma)\Delta t}\right] \tag{16}$$

Now it is easily verified that, with the meaning of p and q defined in (5), the brackets in (15) and (16) have the form $p + q - pq$, so that we finally arrive at the result

$$\left.\begin{aligned} P(A_1.A_2,B_1) &= r = \frac{\beta}{\beta+\gamma}(p+q-pq) \\ P(A_1.A_2,B_2) &= s = \frac{\gamma}{\beta+\gamma}(p+q-pq) \end{aligned}\right. \tag{17}$$

We see that the relations (6) and (4) are satisfied. The probabilities r and s may be called *reduced probabilities*, in comparison to the probabilities p and q; they are reduced by the competition of chances.

Since (5) supplies

$$\left.\begin{aligned} \beta &= -\frac{1}{\Delta t}\log_e(1-p) \\ \gamma &= -\frac{1}{\Delta t}\log_e(1-q) \end{aligned}\right. \tag{18}$$

we can write (17) also in the form

$$P(A_1 . A_2, B_1) = r = \frac{\log_e(1-p)}{\log_e(1-p) + \log_e(1-q)} (p + q - pq)$$
$$P(A_1 . A_2, B_2) = s = \frac{\log_e(1-q)}{\log_e(1-p) + \log_e(1-q)} (p + q - pq) \tag{19}$$

Because of the logarithmic series

$$\log_e(1-p) = -p - \frac{p^2}{2} - \frac{p^3}{3} - \dots \tag{20}$$

we have for small values p and q the approximation

$$r \sim \frac{p}{p+q} (p + q - pq)$$
$$s \sim \frac{q}{p+q} (p + q - pq) \tag{21}$$

For larger values p and q, however, the approximation (21) cannot be used.

The rules of the competition of chances find many applications. When two antiaircraft batteries fire at one plane, formulas (4), (5), and (17) apply. Or if a man suffers from two diseases, the probability of his death is given by (4), whereas the probability of his death by either disease is reduced by the competition of the other disease. In all such problems, the probability of the disjunction is not affected by the competition, whereas the probability of the individual attribute is reduced thereby.

These results may now be applied to the discussion of a numerical problem, which also offers some interest from another angle in that it illustrates the fact that the juristic inferences used in cases of circumstantial evidence can be analyzed in terms of the calculus of probability. The reader may try to solve the following problem for himself before he studies the solution given below.

When Dr. Bergmann returned to the Tyrolese village, he immediately reported to the gendarmerie that his wife had fallen down the 3,000-foot precipice. Her foot had slipped, he said, when they were crossing the narrow crest. A crew of experienced guides was dispatched, accompanied by Dr. Bergmann. After seven hours they returned, carrying the dead woman on a stretcher.

One of the guides expressed astonishment at the fact that Dr. Bergmann had not roped his wife while traversing the crest. He said that guides always roped tourists at that crossing; he estimated that one out of ten persons would slip on the icy stones at that place and was then held only by the rope.

Dr. Bergmann, who was an experienced mountaineer, answered that the crossing had not appeared hazardous to him, and the guides admitted that the danger could not be foreseen by anyone not familiar with the location.

The gendarmerie chief reported the case as an unfortunate accident. The situation changed, however, when sometime later it was learned that Dr. Bergmann, six months before the accident, had taken out a life-insurance policy for $20,000—in the name of his wife. An inquiry at the office of the insurance company brought out the fact that, when a man takes out a large amount of life insurance for his wife in a childless marriage, he does so, in four out of five cases, with criminal intentions. Dr. Bergmann and his wife had no children.

When questioned, Dr. Bergmann ridiculed the suspicion of murder and advanced the objection that within the last few months he had made another equally hazardous mountain trip with his wife. Furthermore, as a doctor he would have had ample opportunity of taking her life. During a recent attack of gallstones he had been obliged to give her a morphine injection, and could easily have administered an overdose if he had intended to murder her.

What is the probability that Dr. Bergmann killed his wife by pushing her down the crest or making her slip in some way?

We denote by A_1 the situation of a person exposed to murder by a man in the situation of Dr. Bergmann during the time Δt of the crossing of the mountain crest. By A_2 we denote the situation of a person exposed to fatal accident on the mountain crest during the time Δt. B_1 may represent death by murder; B_2, death by accident. Since we know that, in either case, the disjunction $B_1 \vee B_2$ is true, the probability sought after has the form

$$P(A_1 . A_2 . [B_1 \vee B_2], B_1) \tag{22}$$

Because of the analytic equivalence

$$([B_1 \vee B_2] . B_1 \equiv B_1) \tag{23}$$

we can write, using the general theorem of multiplication,

$$P(A_1 . A_2 . [B_1 \vee B_2], B_1) = \frac{P(A_1 . A_2, B_1)}{P(A_1 . A_2, B_1 \vee B_2)} \tag{24}$$

The difficulty consists in the correct evaluation of the two probabilities occurring on the right side of (24).

In looking for the numerical values, we must use the given data as best we can, realizing that we can expect to find only crude appraisals, since extensive statistics are not available. The value 0.8 given by the insurance company for the probability of murder is the best value we can use, although a psychological analysis of Dr. Bergmann may lead to the result that for him a much higher, or much lower, value would be appropriate. So long as such an analysis is not known to us we must employ the known data. Insurance murders are usually committed not very long after the insurance is taken out; we shall assume, therefore, that the probability that Dr. Bergmann will murder his wife during the six-month period is $= 0.8$. Most of the time he will have no opportunity to commit murder, since it is not easy to do so without leaving clues. Assume that the three occasions mentioned in Dr. Bergmann's defense exhaust his opportunities for committing murder, and that they can be given equal weight. The probability of 0.8 may then be equally distributed over the three occasions.

For this purpose, we regard the probability of murder as progressing with time according to the form $1 - e^{-\beta t}$, the argument t running, however, only through the three occasions. Equal weights for the occasions may be represented as equality of the time intervals Δt covered by them. Putting $\Delta t = 1$ (this is only a convention for the time unit), we thus have, with $w = 0.8$,

$$w = 1 - e^{-3\beta} = 0.8 \tag{25}$$

which, with the use of (18), gives

$$\beta = -\tfrac{1}{3}\log_e(1 - w) = 0.536 \tag{26}$$

The probability of murder during any of the three occasions, on the condition that no murder was committed on one of the previous occasions, is then given by[1]

$$p = 1 - e^{-\beta} = 0.36 \tag{27}$$

The value (27) may be regarded as giving the probability $P(A_1,B_1)$; but since the statistics of insurance companies were prevalently compiled from cases in which the chance of death was not in competition with the chance of a serious accident, the value (27) may also be interpreted as representing the probability $P(A_1.\bar{A}_2,B_1)$, corresponding to the meaning of p in (5). The two values will not differ very much, since $P(A_1,A_2)$ may be assumed to be small; the reason is that the rule of elimination (2, § 19) gives

$$\begin{aligned} P(A_1,B_1) &= P(A_1,A_2)\cdot P(A_1.A_2,B_1) + P(A_1,\bar{A}_2)\cdot P(A_1.\bar{A}_2,B_1) \\ &= P(A_1.\bar{A}_2,B_1) + P(A_1,A_2)\cdot[P(A_1.A_2,B_1) - P(A_1.\bar{A}_2,B_1)] \end{aligned} \tag{28}$$

[1] It might appear paradoxical that, if for each of the three periods we have the probability 0.36, the sum of the three values is > 1. But these probabilities are not additive. Let A denote the situation of Dr. Bergmann after taking out life insurance for his wife. For the first period we have $P(A,B_1{}^{\Delta t}) = 0.36$; for the second, $P(A.\bar{B}_1{}^{\Delta t},B_1{}^{2\Delta t}) = 0.36$; and so on. That is, the value 0.36 is dependent on the condition that the murder has not occurred before. Seen from the beginning of the total time period, the probability of a murder in the first period is $P(A,B_1{}^{\Delta t})$ and that of a murder in the second period is $P(A,\bar{B}_1{}^{\Delta t}.B_1{}^{2\Delta t})$, since the murder can occur in the second period only if it did not occur in the first. The latter probability is therefore smaller than the first. Seen from the beginning of the second period, however, when we know that the murder has not occurred in the first, the probability of a murder in the second period is $P(A.\bar{B}_1{}^{\Delta t},B_1{}^{2\Delta t})$, and this value is equal to $P(A,B_1{}^{\Delta t})$.

an expression in which the second term may be neglected for small values $P(A_1,A_2)$. The probability $P(A_1.A_2,B_1)$, however, can then be very different from $P(A_1,B_1)$.

For the chance of accident the value 0.1 is given, representing the probability $P(A_2.\bar{A}_1,B_2) = q$, since these statistics were certainly made with persons not exposed to an insurance murder, which was excluded by the presence of guides. Since the accident chance operates during the period $\Delta t = 1$, we derive from (18)

$$\gamma = -\log_e(1 - q) = -\log_e(1 - 0.1) = 0.105 \tag{29}$$

The two chances of death, that by murder and that by accident, are in competition during the passage over the mountain crest, and we therefore have here the conditions expressed in formulas (4) and (17). Applying these formulas to (24), we obtain

$$P(A_1.A_2.[B_1 \vee B_2],B_1) = \frac{\beta}{\beta + \gamma} = \frac{0.536}{0.536 + 0.105} = 0.84 \tag{30}$$

The probability of murder, on the basis of the known facts, is therefore 84%. What makes this value even higher than the antecedent probability of 80% is the fact that the insured person died, and not of a natural death. That the resulting inverse probability is only slightly higher is chiefly a consequence of Dr. Bergmann's excellent defense, which distributes the high antecedent probability of 80% over several periods.

This analysis of the problem corresponds to the weights we would give instinctively to the different factors involved. It is superior to an instinctive appraisal in that it supplies the resultant of all the factors by mathematical computation. The precision of the numerical result, of course, should not be overestimated. Based on rough statistical estimates and omitting any psychological analysis of the man involved, it will represent no more than the best judgment available. And it is scarcely necessary to say that the characters and the numerical values of the story are fictitious.

Chapter 7

THE FREQUENCY PROPERTIES OF PROBABILITY SEQUENCES

THE FREQUENCY PROPERTIES OF PROBABILITY SEQUENCES

§ 48. The Frequency Sequences

The questions to be considered in this section correspond somewhat to the questions on the structure of order, which was discussed in chapter 4. The problems of the theory of order were concerned with questions about the probabilities in subsequences derived in a certain manner from the major sequence or from a lattice of major sequences. It was in terms of these probabilities that the *properties of order* of probability sequences were expressed. A number of types of order were thus defined, by means of which the total domain of probability sequences was subdivided with respect to their structure of order.

In the present chapter the *frequency properties* of probability sequences will be characterized by similar methods, as, once more, certain sequences derived from the major probability sequence are considered. This investigation, however, differs from the theory of order in employing, not subsequences of the major sequence, but certain other sequences derived from it, to be called *coördinated frequency sequences.* In defining this term one must distinguish three kinds of frequency sequences. In order to simplify the discussion, we shall assume the sequence A to be compact.

First, we can coördinate to each element y_n of the sequence the amount

$$f^n = F^n (A,B) \tag{1}$$

The sequence of the amounts f^n may be called the *frequency sequence counted through.* This is the most important of the frequency sequences.

Second, we can coördinate to each element y_i of the sequence the frequency f^n according to (1), but counted only for the preceding segment of the length n, so that the enumeration of the frequencies begins at y_{i-n+1} and ends at y_i. We thus count the frequency within segments overlapping each other. The sequence of values f^n thus constructed will be called a *frequency sequence counted in overlapping segments.*

Third, we can divide the sequence into sections of the length n that do not overlap; we then coördinate to each section the frequency f^n according to (1). Thus an amount f^n is coördinated to each element y_i for which i is a multiple of n. The sequence of the values f^n thus selected will be called a *frequency sequence counted in sections.*

The frequency properties of probability sequences will be characterized by means of certain statements about the *probabilities in coördinated frequency sequences.* For instance, we may inquire into the probability that the frequency assumes certain specified values or that it lies within a certain specified interval. Thus we arrive at probability statements concerning coördinated frequency sequences. It is of great import that such questions can be answered not only in the interpreted but also in the formal calculus of probability.

It is true that in the formal calculus no certainty statements can be made about the structure of coördinated frequency sequences; nevertheless, we can make probability statements about the properties of frequency sequences by calculating the probability of finite sections or segments according to the rules for the probability of combinations. Thus this investigation answers the question concerning the kind of statements about frequencies that can be made in the formal calculus, that is, without the use of the frequency interpretation. It will turn out that these statements assume the character of statements of convergence.

But the results are important also for the interpreted probability calculus. The convergence of frequency sequences, which the frequency interpretation asserts, is not subject to quantitative restrictions stipulating the degree of convergence for a given n; we obtain statements about the quality of the convergence only when we have recourse to probability statements about the convergence of frequency sequences.

Since the question of the probability of frequencies leads to problems of convergence, its answer concerns problems important in practical statistics; the degree of convergence of frequency sequences naturally plays a prominent part in practical applications. Another advantage of the method is that it leads to further classification of types of probability sequences, this time defined in terms of the kind of convergence. As a consequence, questions will arise how far the types that were developed previously coincide with the convergence types. Thus frequency considerations lead to a more profound understanding of the structure of order of probability sequences.

§ 49. The Theorem of Bernoulli

The principal theorem among the results of frequency considerations is the theorem of Bernoulli. Like all the theorems previously presented, it will be developed first in the formal calculus of probability; its meaning can then be stated easily in the interpreted calculus. In this section the derivation will be given only for normal sequences, so that the special theorem of multiplication is applicable. An extension to other than normal sequences will be explained later.

The questions raised by this theorem start with the probability of combinations, and thus it is advisable to develop the theorem first for frequency sequences counted in segments. Later it will be extended to frequency sequences counted in other ways. Assume

$$P(A,B) = p \tag{1}$$

We inquire into the probability of the occurrence of a combination of m elements B with $(n - m)$ subsequent elements $\bar{B}$, that is, a combination of the form

$$B^1 \ldots B^m . \bar{B}^{m+1} \ldots \bar{B}^n \tag{2}$$

The probability of this combination is

$$P(A,B^1 \ldots B^m . \bar{B}^{m+1} \ldots \bar{B}^n) = p^m(1-p)^{n-m} \tag{3}$$

This combination of n elements has the property that the frequency occurring in it is given by

$$f = \frac{m}{n} \tag{4}$$

The notation is to be understood as follows. As in §§ 35 and 43, we distinguish the *individual value* f^i of the frequency from its *possible amounts*. For a given n the amounts are discrete steps determined by the successive values of m, so that we would have to write f_m for the step, in correspondence to the amount u_m in § 35. However, if we regard n as a variable, that is, as representing any chosen integer, it is advisable to regard the possible amount $\frac{m}{n}$ as a continuous variable f (without subscript), similar to the amount u in § 43. In doing so we consider the continuous background of all rational values $\frac{m}{n}$ as representing the range of the variable f.

The combination (3) is not the only one having the frequency (4); the same frequency f is produced by different combinations, provided they have the same total number of elements, and among them an equal number of elements B, that is, the same n and m. How many such combinations are possible? Their number may be designated by k; it is calculated as follows. If we permutate the terms of (2), each keeping its superscript, all the possible permutations of the n terms represent combinations of the same frequency $f = \frac{m}{n}$. How many permutations of this kind exist? The answer is seen from the schema presented. If there is only one term, there are two empty places, designated by points, that may be occupied by a second term. For two terms, therefore, we have $1 \cdot 2$ permutations. If two terms are given, three empty places are available for a third term (second row of the schema), that is, three arrangements are possible. Since this holds for each of the two arrangements in which the two terms may occur, the total number of permu-

$$\begin{array}{c} \cdot\ 1\ \cdot \\ \cdot\ 1\ \cdot\ 2\ \cdot \\ \cdot\ 1\ \cdot\ 2\ \cdot\ 3\ \cdot \end{array}$$

tations of three terms obtains as $1 \cdot 2 \cdot 3$. Continuing the inference, we arrive at the result that for n terms there are $n!$ permutations.

This, however, is more than the desired number k; this number is rather obtained by the permutations of (2) resulting when the superscript does not participate in the permutation. In the permutation of (2) with superscripts, permutations that are distinguished only by the order of their superscripts are counted as different, while they have the same order in respect to B and to $\bar{B}$. Obviously, the number of such arrangements is given by the permutation of the m elements B among themselves and of the $(n - m)$ elements $\bar{B}$ among themselves, that is, by $m!$ and $(n - m)!$ respectively. The number must not be subtracted, however; we must divide by it, since each arrangement without superscript supplies $m!(n - m)!$ arrangements that contribute to the permutations with superscript. Thus, between the k arrangements without superscript and the $n!$ arrangements with superscript, we have the relation

$$km!(n - m)! = n!$$

that is,

$$k = \frac{n!}{m!(n - m)!} = \binom{n}{m} \tag{5}$$

It is an advantage for the mathematical evaluation that k happens to be determined by the binomial coefficients $\binom{n}{m}$.

We now wish to find the probability of obtaining any one of the k arrangements. The question concerns the probability of a disjunction. We introduce the abbreviation

$$\begin{aligned} F^n_m =_{Df} \ & [B^1 \ldots B^m . \bar{B}^{m+1} \ldots \bar{B}^n] \vee \ldots \\ & \vee [\bar{B}^1 \ldots \bar{B}^{n-m} . B^{n-m+1} \ldots B^n] \end{aligned} \tag{6}$$

in which the terms of the disjunction consist of all the combinations that contain m letters B and $(n - m)$ letters $\bar{B}$ in any arrangement. We read F^n_m as "frequency m for B among n elements". To the case F^n_m is then coördinated the amount $f = \frac{m}{n}$. Since all the k terms of the disjunction (6) have the same probability (3), we obtain with the help of (5) the *binomial distribution*

$$P(A,F^n_m) = \binom{n}{m} p^m(1 - p)^{n-m} = w_{nm} \tag{7}$$

This expression is called *Newton's formula.*[1] Although it appears to have a

[1] For a disjunction of r terms having the probabilities $p_1 \ldots p_r$ Newton's formula assumes the extended form of the *multinomial distribution:*

$$w_{n_1 \ldots n_r} = \frac{n!}{n_1! \ldots n_r!} p_1^{n_1} \ldots p_r^{n_r} \qquad n_1 + \ldots + n_r = n$$

where $n_1 \ldots n_r$ are the frequencies, respectively, of the r cases. The proof follows according to the considerations used for (5).

simple structure, its functional properties are difficult to realize; but a graphical representation provides a suitable illustration of the formula. For this purpose we plot, for a given n, the number m as abscissa and w_{nm} as ordinate. Figure 23 presents a diagram of this kind. I have chosen $p = \frac{3}{4}$ and drawn the curve for $n = 4$ and $n = 8$, respectively. The end points of the ordinates are connected by a train of lines. The maximum of w_{nm} is situated for the first case at $m = 3$; for the second, at $m = 6$. That is, in both it corresponds to $f = \frac{3}{4} = p$. But in the second case the maximum is smaller than in the

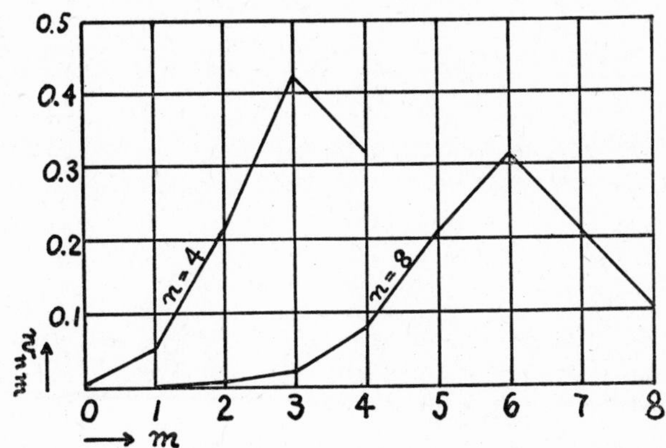

Fig. 23. Graphical representation of Newton's formula (7) for $p = \frac{3}{4}$ and two different values of n.

first case, since in the first we have $(w_{nm})_{\max} = 0.42$, and in the second we find $(w_{nm})_{\max} = 0.31$.

The calculation of the maximum is relatively easy. Consider the quotient of two successive values (assuming that $p \neq 0$ and $1 - p \neq 0$):

$$\frac{w_{nm}}{w_{n,m-1}} = \frac{p}{1-p}\left(\frac{n+1}{m} - 1\right) \tag{8}$$

While m varies from $m = 1$ to $m = n$ (the value w_{n0} in the denominator is included as the case $m = 1$), the expression (8) decreases continually for increasing m; it becomes equal to 1 for

$$m = p\,(n+1) \tag{9}$$

The maximum of w_{nm} is characterized by a change-over of the quotient (8) from values > 1 to values < 1. For the computation we must distinguish the following cases:

1. $p\,(n + 1)$ is an integer > 0, but $< n + 1$. Then the value $m = p\,(n + 1)$ resulting from (9) determines a w_{nm} that, because of (8), is equal to $w_{n,m-1}$; these two values represent the maximum.

2. $p\,(n + 1)$ is a fraction (not an integer) > 1. Then the largest integer m below $p\,(n + 1)$ gives the maximum w_{nm}, since this value represents the largest integer m for which the quotient (8) is > 1. Here there is only one maximum value w_{nm}.

3. $p\,(n + 1) < 1$. In this case the quotient (8) is not > 1 for any possible value m. This means that the maximum lies at $m = 0$.

The cases $p = 0$ and $p = 1$ remain to be considered:

4. $p = 1$. Then the maximum is $m = n$, a result that follows directly from (7).

5. $p = 0$. Then the maximum is $m = 0$, a result that also follows directly from (7).

It is common to all the cases that the desired m is equal either to $p\,(n + 1)$ or to the next smaller integer; we can therefore represent the results collectively in the double inequality for m:

$$p\,(n + 1) \geqq m \geqq p(n + 1) - 1 \tag{10}$$

That integer m or those two integers m that satisfy (10) supply the one or the two maximum values w_{nm}. Further maxima cannot exist, since (8) falls off continuously with increasing m.

For the frequency $f = \frac{m}{n}$ we obtain from (10) the limits

$$p + \frac{p}{n} \geqq f \geqq p - \frac{1 - p}{n} \tag{11}$$

Since the difference between the two limits amounts to only $\frac{1}{n}$, we can assume for large n that this frequency is given approximately by

$$f = p \tag{12}$$

This equation represents the first result: the frequency that corresponds to the probability p has the highest probability.

The probability with which $f = p$ is to be expected can be found by the calculation of the corresponding w_{nm}. For this purpose we must substitute for m in (7) the value pn. The computation is achieved by the help of a method of approximation, based on *Stirling's formula*

$$n! = n^n e^{-n} \sqrt{2\pi n} \tag{13}$$

This formula is often used in the calculus of probability; it converges so well that the deviation from the correct values of the factorial in the case $n = 10$ amounts to only 8 per 1,000, becoming much smaller for larger n. If we replace the factorials in (7) by the expressions given in Stirling's formula (13) and substitute np for m, we arrive easily at the result

$$(w_{nm})_{\max} = \frac{1}{\sqrt{2\pi np(1-p)}} \tag{14}$$

This is the probability for the occurrence of the frequency $f = p$.

This probability, although representing a maximum, is not very large; it even goes toward zero for increasing n. But this is not surprising, since for large numbers a small deviation of f, in percentage, presents a large difference in m. A result in terms of percentage can be very exact, whereas the corresponding m in its absolute value may differ by a large number from the optimum value of m. From the standpoint of frequency, however, the question of the precise value of f is not important; it suffices to answer the question with which probability f lies within an interval from f_1 to f_2. This modification means that we consider not only combinations of n elements that have precisely the same value m, but further combinations possessing somewhat different values m. We thus permit m to vary from m_1 to m_2. Consequently, we consider a *disjunction of disjunctions*, in which each disjunction occurring as a term has the form (6). We introduce the abbreviation

$$F^n[f_1, f_2] =_{Df} F^n_{m_1} \vee F^n_{m_1+1} \vee \ldots \vee F^n_{m_2} \tag{15}$$

$$f_1 = \frac{m_1}{n} \qquad f_2 = \frac{m_2}{n}$$

which may be read: "a frequency between f_1 and f_2 for B among n elements". We then derive from the theorem of addition, because of (7),

$$P(A, F^n[f_1, f_2]) = \sum_{m=m_1}^{m_2} w_{nm} = b_n(f_1, f_2) \tag{16}$$

Again it is advisable to use a graphical representation, which illustrates this quantity as well as the transition from (7) to (16). However, we choose a representation by means of a *histogram*, a staircase-shaped distribution, which permits transition to a continuous distribution. The resulting diagram is presented in figures 24A and B, which are constructed for several values of n, the first two of which are identical with the values n of figure 23 (p. 265). The first difference from figure 23 is that in figures 24A and B the abscissa is f and not m as in figure 23. Therefore, all curves of figures 24A and B extend over the same domain of the abscissa, since for each n the value of f varies between 0 and 1. The second difference from figure 23 is that the ordinate of figures 24A and B represents a probability density, that is, the

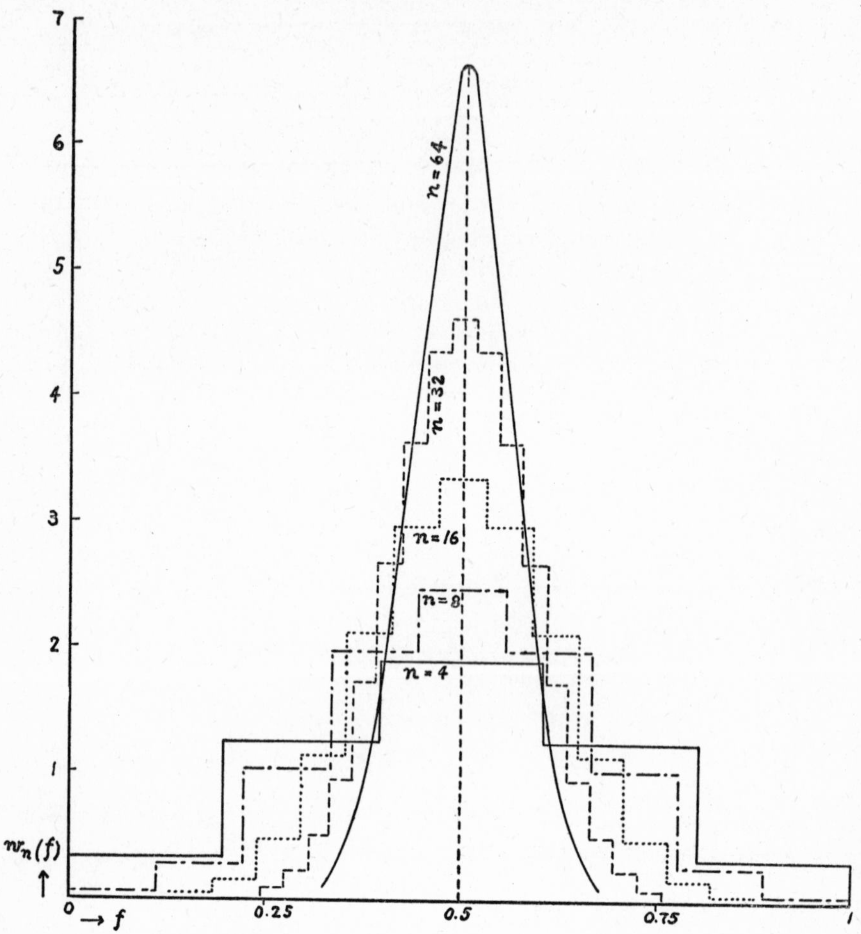

Fig. 24A. Bernoulli curves for $p = \frac{1}{2}$ and different values of n, according to (19) and (21).

area of a rectangular strip of ordinates is made equal to w_{nm}. In other words, we use as ordinate a height $w_n(f_m)$, which is determined by the equation

$$w_n(f_m) \cdot \Delta f = w_{nm} \tag{17}$$

In order to allocate the $n + 1$ possible values of $\frac{m}{n}$ (m varying from 0 to n) to the abscissa, we need $n + 1$ intervals; thus

$$\Delta f = \frac{1}{n+1} \tag{18}$$

and

$$w_n(f_m) = (n + 1) \cdot w_{nm} \tag{19}$$

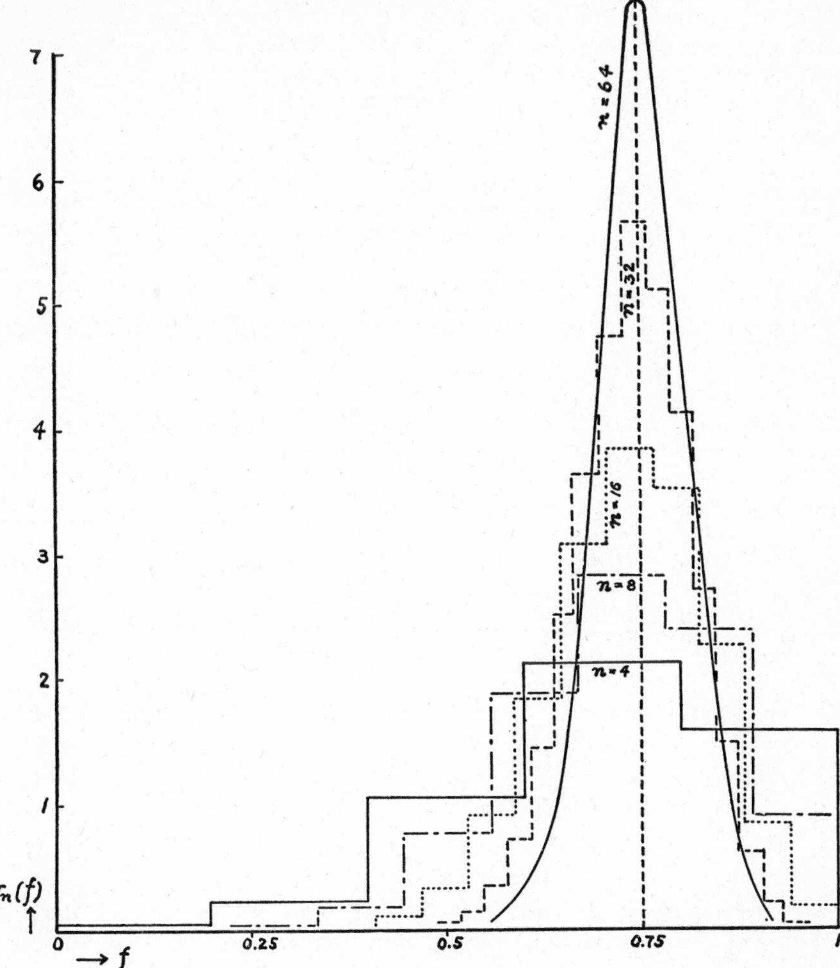

Fig. 24*B*. Bernoulli curves for $p = \frac{3}{4}$ and different values of n, according to (19) and (21).

These ordinates are represented in figures 24*A* and *B*. The value $\frac{m}{n}$ to which an interval belongs lies within the interval at a point that divides the interval in the ratio of m to n. The point $m = 0$ lies at the left end of the first interval; the point $m = 1$ lies $\frac{1}{n}\,\Delta f$ to the right of the left end of the second interval; and so on to the point $m = n$, which lies at the right end of the last interval. The ordinate strips supply the histogram, which takes the place of the oblique lines of figure 23. For the sake of clarity the steps have been drawn in outline only. All staircase-shaped curves have their peaks at $f = p$. Figure 24*A* is drawn for $p = \frac{1}{2}$; figure 24*B*, for $p = \frac{3}{4}$. The histograms are symmetrical only for $p = \frac{1}{2}$; for $p = \frac{3}{4}$ the columns on the right side are higher than those on the left. The diagram for $p = \frac{1}{4}$ would be the mirror image of figure 24*B*.

The quantities $b_n(f_1,f_2)$, too, are illustrated in figures 24*A* and *B*. They would be represented in figure 23 by a sum of neighboring ordinates; in figure 24*A–B* they are given by a sum of neighboring ordinate strips. Thus they are determined by areas delimited on both sides by the ordinates belonging to f_1 and f_2, and on top by the staircase-shaped curve. The farther apart the values f_1 and f_2, the greater is $b_n(f_1,f_2)$. If we put $f_1 = 0$ and $f_2 = 1$, the total area between the curve and the abscissa corresponds to the quantity $b_n(f_1,f_2)$, for which we write $b_n(0,1)$. This quantity represents the probability that f assumes some value between 0 and 1; the probability is equal to 1. We therefore have

$$b_n(0,1) = \sum_{m=0}^{n} w_{nm} = 1 \tag{20}$$

That this relation follows from the definition (7) of the w_{nm} is shown as follows. The expressions (7) are known from algebra as the terms of the binomial expansion (45, § 24) for $q = 1 - p$. Since for this value $p + q = 1$, the expression $(p + q)^n$ and therefore the sum also is $= 1$.

For large n the staircase-shaped train of lines can hardly be distinguished from a smooth curve. In this case f may be regarded as a virtually continuous variable, to which an ordinate $w_n(f)$ is coördinated by the equation resulting from (19):

$$w_n(f) = (n + 1)w_{nm} \quad \text{for } f = \frac{m}{n} \tag{21}$$

In order to extend this formula to noninteger values of m and n, an interpolation is used for the factorials in (7), for instance, by means of Stirling's formula (13).

In the definition of $b_n(f_1,f_2)$ we must then replace the summation by an integration, obtaining

$$b_n(f_1,f_2) = \int_{f_1}^{f_2} w_n(f)df \tag{22}$$

This equation, in which the transition to the continuous distribution is carried through, is strictly valid only for $n \to \infty$; but it represents a very good approximation for finite n, so that for large n it cannot be distinguished practically from the precise value (16). The function $w_n(f)$ is a probability density. The original step nature of f finds its expression in the fact that the usual approximation $w_n(f)df$ supplies a precise value if m and n are integers nd $df = \dfrac{1}{n+1}$. The function $w_n(f)$ will be called *Bernoulli density;* the function $b_n(f_1,f_2)$ will be called *Bernoulli probability.*

In the sense of (21) and (22), the train of lines is replaced by a continuous urve for the case $n = 64$, in figure 24*A–B*. The replacement offers no difficulties for large n, since for $n = 64$ the step would assume a width of only mm. in the scale of the drawing. The quantity $b_n(f_1,f_2)$ is given by an area ounded on the sides by the ordinates of f_1 and f_2 and on top by the curve $w_n(f)$. The peak of the smoothed-out curve corresponds, of course, to the alue $f = p$. Nevertheless, we do not speak of the probability that f is exactly qual to p, but of the probability that f lies within an interval around p, ince the probability assumes a nonzero value only in respect to an interval. Thus $b_n(f_1,f_2)$, not $w_n(f)$, is a probability. That the probability of obtaining frequency f in the neighborhood of p becomes larger with a transition to arger n is expressed by the fact that for larger n the curve contracts to a arrower region around $f = p$. Its total area remains equal to 1, since, according to (20), we have the condition of normalization

$$\int_0^1 w_n(f)df = 1 \tag{23}$$

For large values of n, however, the larger part of the area is concentrated n a narrow vertical strip around the value $f = p$. The peak value of the curve esults from (21) and (14) as

$$w_n(f)_{max} = \frac{n+1}{\sqrt{2\pi\, n\, p(1-p)}} \sim \frac{\sqrt{n}}{\sqrt{2\pi\, p(1-p)}} \tag{24}$$

This value approaches ∞ with increasing n, whereas the area remains inite and equal to 1. The probability $b_n(f_1,f_2)$, considered for an interval from f_1 to f_2 not including the peak, becomes smaller with greater n; but if the nterval includes the peak, the probability $b_n(f_1,f_2)$ becomes larger with inreasing n. The point $f = p$ constitutes, therefore, the *critical point* of a undle of Bernoulli curves. This peculiarity can be characterized as follows: o arrive at an interval including this point becomes more and more probable or increasing n; to arrive at an interval not including the point becomes nore and more improbable for increasing n.

These properties are expressed in the following relations holding for the *Bernoulli probability* $b_n(f_1,f_2)$:

$$\begin{aligned} &\lim_{n\to\infty} b_n(f_1,f_2) = 1 \qquad \text{if } f_1 \leqq p \leqq f_2 \\ &\lim_{n\to\infty} b_n(f_1,f_2) = 0 \qquad \text{if } p < f_1 \text{ or } p > f_2 \end{aligned} \tag{25}$$

These relations will be proved in § 55.

There are no general analytic expressions by which the curves $w_n(f)$ can be represented. However, it is often sufficient to restrict the consideration to the environment of the point $f = p$ and to put $f_1 = p - \delta, f_2 = p + \delta$. We then abbreviate the symbols introduced in (15) and (16) by the definitions

$$\begin{aligned} F^n_\delta &=_{Df} F^n[p - \delta, p + \delta] \\ b_{n\delta} &=_{Df} b_n(p - \delta, p + \delta) \end{aligned} \tag{26}$$

For this domain Laplace has given an approximation for the integral occurring in (22). The result[2] (which is strict for the limit $n = \infty$) gives approximately the equation

$$b_{n\delta} = P(A,F^n_\delta) = \frac{1}{\sqrt{\pi}} \int_{-\gamma}^{+\gamma} e^{-t^2} dt \qquad \gamma = \frac{\delta\sqrt{n}}{\sqrt{2\, p(1-p)}} \tag{27}$$

We see that the Bernoulli curve, in the neighborhood of its peak, is approximated by a Gauss exponential function. Whereas the diagram of figure 24*A*–*B* (pp. 268–269) represents the increase of the probability for a fixed δ by contraction of the curve within given limits of the abscissa, (27) supplies a representation in which the curve e^{-t^2} keeps its form, but in which the limits of the integral are extended. Therefore t is an auxiliary variable, and the dependence of $b_{n\delta}$ on n finds its expression in the fact that n enters into the limits of a definite integral. The construction of this integral may be regarded as a mathematical device to represent a functional relation between $b_{n\delta}$ and n and likewise between $b_{n\delta}$ and p.

By the use of the abbreviation $b_{n\delta}$, the first of the relations (25) can be written

$$\lim_{n \to \infty} b_{n\delta} = 1 \tag{28}$$

The formula is valid for any δ, however small, but δ must remain constant while the transition to the limit is carried out. The meaning of (28) can be paraphrased as follows:

If a certain inexactness δ is specified, the probability that the frequency $f = \frac{m}{n}$ of a segment of the length n will lie within $p \pm \delta$ can be made as closely equal to 1 as desired by making n larger. This holds for any chosen δ, however small, but for small δ a greater value of n is to be employed if we wish to have the same probability for the desired frequency.

Some numerical examples will serve to illustrate the result. In table 4 the heading of each vertical column specifies the value p and the exactness δ^* to which the column belongs; the numbers given under $b_{n\delta}$ refer to the n of the

[2] The proof is found in most of the mathematical texts on probability, for instance, Emanuel Czuber, *Wahrscheinlichkeitsrechnung* (2d ed.; Leipzig, 1908), p. 119. The theorem can also be derived as a special case from the theorem mentioned at the end of § 44. See Richard von Mises, *Vorlesungen aus dem Gebiete der angewandten Mathematik*, Vol. I: *Wahrscheinlichkeitsrechnung* (Leipzig, 1931), pp. 209 ff. The most important consequence, which is expressed in (25), is more easily derivable (see § 55).

respective horizontal rows. The deviation, measured in percentage of p, is designated by δ^*, that is, $p \pm \delta = p(1 + \delta^*)$.

The table shows not only the increase of $b_{n\delta}$ toward 1 but also that the same high value of $b_{n\delta}$ is reached at larger n for smaller δ^*. Thus for $p = \frac{1}{2}$ and $\delta^* = 20\%$ the value $b_{n\delta} = 0.95$ obtains for $n = 100$, whereas for the same p, but with $\delta^* = 10\%$, the same value of $b_{n\delta}$ obtains for about $n = 500$.

TABLE 4

p	¼	¼	½	½	¾	¾
δ^*	20%	10%	20%	10%	20%	10%
n	$b_{n\delta}$	$b_{n\delta}$	$b_{n\delta}$	$b_{n\delta}$	$b_{n\delta}$	$b_{n\delta}$
10	0.29	0.14	0.47	0.24	0.72	0.42
30	0.48	0.24	0.73	0.41	0.94	0.66
50	0.59	0.31	0.84	0.52	0.99	0.78
100	0.75	0.44	0.95	0.68	~1.00	0.92
500	0.99	0.80	~1.00	0.97	~1.00	~1.00

Furthermore, for smaller p, but with the same exactness in percentage, the same value of $b_{n\delta}$ is reached much later. Thus $p = \frac{1}{2}$, $\delta^* = 10\%$, $n = 100$: $b_{n\delta} = 0.68$; $p = \frac{1}{4}$, $\delta^* = 10\%$, $n = 100$: $b_{n\delta} = 0.44$. The difference results from the fact that equal exactness in percentage, for a smaller probability, means a smaller absolute value of δ. Thus $\delta^* = 10\%$ means for $p = \frac{1}{2}$ that $\frac{m}{n}$ lies within 0.5 ± 0.05, whereas it means for $p = \frac{1}{4}$ that $\frac{m}{n}$ lies within 0.25 ± 0.025.

Bernoulli was well aware of the significance of his theorem. He justly emphasized that it cannot be regarded as a matter of course that the probability of observing the ideal frequency converges toward 1, and that this theorem requires "a proof based upon scientific principles".

> Some other circumstances must be taken into account, of which perhaps no one has even thought, so far. It remains to investigate whether with the increase of the number of observations there likewise increases continually the probability of obtaining the true ratio for the number of favorable to the number of unfavorable observations; and whether this happens in such a manner that this probability finally surpasses any chosen degree of certainty, or whether the problem, so to speak, has its asymptote, that is, whether there exists a definite degree of exactness . . . that can never be surpassed, however much we increase the number of observations. . . . This is the problem I intend to publish at this place, after I have carried it around with me for twenty years; its novelty as well as its extraordinary usefulness, connected as it is with an equally great difficulty, will make all other aspects of the theory gain in importance and significance.[3]

The bearing of the theorem upon the calculus of probability will now be investigated in more detail.

[3] Jacob Bernoulli, *Ars conjectandi* (Basel, 1713), Part 4, chap. IV.

§ 50. The Significance of Bernoulli's Theorem

The function of Bernoulli's theorem is to supply frequency statements even within the formal conception of the probability calculus.

Since the quantity $b_{n\delta}$, according to (27, § 49), is defined in the formal calculus of probability by

$$b_{n\delta} =_{Df} P(A,F^n_\delta) \tag{1}$$

the Bernoulli theorem as expressed in (28, § 49) leads to the following result:

Any given probability referring to a normal sequence can be transformed into a probability of a higher kind, the numerical value of which can be made as high as desired, whereas the first probability is interpreted as a frequency. Thus the statement of the first kind, "The event is to be expected with the probability $\frac{1}{2}$", can be replaced, according to Bernoulli, by the statement of the second kind, "Among 500 repetitions the event is to be expected in a frequency between 225 and 275 with the probability 0.97". Here the number $\frac{1}{2}$ has changed from a probability to a frequency; it determines the middle point 250 between the two limits as a function of the total number 500, and a new probability of the value 0.97 is introduced, which is a high probability.

In this transformation the quantities n and δ play the role of arbitrarily assumed parameters. The occurrence of the parameters means that we can coördinate to the statement of the first kind, not only one statement of the second kind, but a class of such statements. Thus in the example we could select as a statement of the second kind, "Among 100 repetitions the event is to be expected in a frequency between 40 and 60 with the probability 0.95". For the first-mentioned statement of the second kind, $n = 500$, $\delta^* = 10\%$, have been chosen; for the second, $n = 100$, $\delta^* = 20\%$. Thus the new probability differs in the two statements. But the transformation is always of the type illustrated; *a probability is transformed into a frequency, and its place is taken over by a new probability, the degree of which can be made as high as desired by a suitable choice of parameters.*

The procedure may be continued: to the probability statement of the second kind we can coördinate a probability statement of a third kind, in which the probability of the second kind is transformed into a frequency, and a new probability of the third kind occurs. The probability of the second kind will then be translated into a frequency within a series of segments, or sections. Such frequencies will presently be encountered in connection with further problems.

One fact, however, should be noticed: even with the help of the Bernoulli theorem it is impossible to eliminate the concept of probability and replace it by a frequency. The reason is that the frequency statement resulting from the Bernoulli transformation, too, contains the probability concept, though at a different logical place. Thus Bernoulli's theorem cannot make the frequency interpretation dispensable. On the contrary, since every statement of the formal probability calculus admits of a frequency interpretation, the Bernoulli theorem likewise can be given a frequency interpretation. It is the probability of the second kind that for this purpose is to be interpreted as a limit of a frequency.

We have only to translate (1) into a frequency statement and then to reformulate (28, § 49). Since (1) represents a frequency in enumeration with overlapping, the corresponding frequency statement will be as follows: if, in a normal probability sequence, we consider a segment of n consecutive elements, which is shifted from element to element with overlapping, and count among the segments those whose "internal frequency" lies between f_1 and f_2, then the frequency of the segments is, in the limit, equal to $b_{n\delta}$. The longer the segment, the higher is the frequency of these segments, and this frequency goes toward 1 with increasing length of the segments. In this interpretation the theorem represents a statement about a frequency sequence counted in overlapping segments.

The Bernoulli theorem can be formulated also as a statement in enumeration by sections. According to (11, § 30), the special theorem of multiplication holds for this enumeration when we deal with normal sequences. Therefore we have

$$P(A\,.\,S^1_{n1},F^n_m) = P(A,F^n_m) = w_{nm} \tag{2}$$

and also, using (15 and 26, § 49),

$$P(A\,.\,S^1_{n1},F^n[f_1,f_2]) = P(A,F^n[f_1,f_2]) = b_n(f_1,f_2) \tag{3a}$$

$$P(A\,.\,S^1_{n1},F^n_\delta) = P(A,F^n_\delta) = b_{n\delta} \tag{3b}$$

In this interpretation, formula (28, § 49) states that a limit statement for the frequency of sections holds in enumeration by sections. This statement corresponds exactly to the limit statement concerning overlapping segments.

Finally, the Bernoulli theorem can be stated by using a lattice enumeration, if it is applied to normal sequences in the narrower sense. In order to construct this interpretation, which makes the meaning of the theorem particularly clear, a few remarks must be made about the symbolism to be used.

In the definition of the abbreviation F^n_m given in (6, § 49) the superscript n represents a phase. If we wish to indicate, apart from the phase, the running superscript i (an indication that is necessary for lattice enumeration, but can be used likewise for enumeration in overlapping segments), we must write, instead of F^n_m, the detailed symbol

$$F_m^{i-n+1/i} =_{Df} [B^{(i-n)+1} \ldots B^{(i-n)+m} . \bar{B}^{(i-n)+m+1} \ldots \bar{B}^{(i-n)+n}] \vee$$
$$\ldots \vee [\bar{B}^{(i-n)+1} \ldots \bar{B}^{(i-n)+(n-m)} . B^{(i-n)+(n-m)+1} \ldots B^{(i-n)+n}] \quad (4)$$

in which the terms of the disjunction consist of all the combinations that contain m times B and $(n - m)$ times $\bar{B}$ in any arrangement. In the detailed notation the superscript i refers not to the class B but to the element y_i; therefore $F_m^{i-n+1/i}$ represents the frequency m for B within that segment of n elements that ends with the element y_i, that is, the frequency m for B from y_{i-n+1} to y_i. If the running superscript is indicated outside the parentheses, the probability referring to this frequency will be written

$$P(A,F_m^{i-n+1/i})^i = w_{nm} \quad (5)$$

Similarly, we define, in analogy to (15, § 49),

$$F^{i-n+1/i}[f_1,f_2] =_{Df} F_{m_1}^{i-n+1/i} \vee F_{m_1+1}^{i-n+1/i} \vee \ldots \vee F_{m_2}^{i-n+1/i}$$
$$f_1 = \frac{m_1}{n} \qquad f_2 = \frac{m_2}{n} \quad (6)$$
$$F_\delta^{i-n+1/i} =_{Df} F^{i-n+1/i}[p-\delta, p+\delta]$$

and write the corresponding probabilities

$$P(F^{i-n+1/i}[f_1,f_2])^i = b_n(f_1,f_2) \quad (7a)$$
$$P(A,F_\delta^{i-n+1/i})^i = b_{n\delta} \quad (7b)$$

The symbols F^n_m and F^n_δ may be regarded as abbreviations of the symbols $F_\delta^{i-n+1/i}$ and $F_\delta^{i-n+1/i}$

In lattice enumeration the superscript k is added. We have for each horizontal sequence k, analogous to (5) and (7a–b),

$$P(A,F_m^{k,i-n+1/i})^i = w_{nm} \quad (8a)$$
$$P(A,F^{k,i-n+1/i}[f_1,f_2])^i = b_n(f_1,f_2) \quad (8b)$$
$$P(A,F_\delta^{k,i-n+1/i})^i = b_{n\delta} \quad (8c)$$

Since normal sequences in the narrower sense are lattice-invariant, according to (7, § 34), we derive

$$P(A,F_m^{k,i-n+1/i})^k = w_{nm} \quad (9a)$$
$$P(A,F^{k,i-n+1/i}[f_1,f_2])^k = b_n(f_1,f_2) \quad (9b)$$
$$P(A,F_\delta^{k,i-n+1/i})^k = b_{n\delta} \quad (9c)$$

The interpretation of these expressions becomes particularly clear when we make $i = n$, that is, when we consider the frequency of the first n elements of the horizontal sequences. We introduce the abbreviations

$$F_m^{kn} =_{Df} F_m^{k,1/n} \tag{10a}$$

$$F^{kn}(f_1,f_2] =_{Df} F^{k,1/n}[f_1,f_2] \tag{10b}$$

$$F_\delta^{kn} =_{Df} F_\delta^{k,1/n} \tag{10c}$$

which are permissible because we no longer need to make the distinction between the phase n and the running superscript i. With $i = n$ the relations (9) then assume the form

$$P(A,F_m^{kn})^k = w_{nm} \tag{11a}$$

$$P(A,F^{kn}[f_1,f_2])^k = b_n(f_1,f_2) \tag{11b}$$

$$P(A,F_\delta^{kn})^k = b_{n\delta} \tag{11c}$$

F_δ^{kn} represents the case that the frequency of the k-th horizontal sequence up to the element y_{kn} lies within $p \pm \delta$, that is, we have simply

$$y_{kn} \epsilon F_\delta^{kn} \equiv [F^n(A,B^{ki})^i = p \pm \delta] \tag{12}$$

where the symbol $F^n(A,B^{ki})^i$ has the meaning introduced in (3, § 16), and the equality sign has the meaning "within" (see 3, § 89).

Because of these simplifications the meaning of (11c) can be stated as follows. We construct for each horizontal sequence the frequency sequence counted through;[1] to each element y_{kn} of the lattice is then coördinated, as amount, the frequency f^{kn}, so that we have

$$f^{kn} = F^n(A,B^{ki})^i \tag{13}$$

The lattice f^{kn} will be called the *coördinated frequency lattice.* The relation (11c) may then be expressed in the following way. We classify (see § 41) the lattice f^{kn} in such a manner that we take as case F_δ all the f^{kn} for which $f^{kn} = p + \delta$ holds, whereas $\bar{F}_\delta$ represents the opposite case; thus an alternative lattice for F_δ results in which the probability $b_{n\delta}$ holds for the vertical sequences, according to (11c).

The probability of F_δ for the horizontal sequences is determined when we assume the frequency interpretation. Since the horizontal sequences y_{ki} approach for B the limit p of the frequency, it follows that from an element

[1] Notice that the frequency sequences counted through are not normal sequences; they rather possess a probability drag, since the frequency can change only by steps from element to element, that is, the value m can be followed only by $\frac{m}{n+1}$ or by $\frac{m+1}{n+1}$.

f^{kn} onward all further f^{kn} $(n > r)$ of the frequency lattice must belong to F_δ, so that we have

$$P(A, F_\delta^{kn})^n = 1 \tag{14}$$

Since the frequency of F_δ in the vertical sequences is given by the Bernoulli probability $b_{n\delta}$, according to (11*c*), and the $b_{n\delta}$, according to (28, § 49), approach with increasing n the limit 1, the f^{kn} classified with respect to F_δ form a convergent lattice. *The Bernoulli theorem, applied to normal sequences in the narrower sense, states that the coördinated frequency lattice, classified with respect to* F_δ, *is a convergent lattice of the probability 1.* Bernoulli's theorem thus assumes the form of a statement about frequency sequences counted through and arranged in a lattice. The classified frequency lattice has a structure of the following kind:

$$\begin{array}{cccccccccccccc}
\bar{F}_\delta & \bar{F}_\delta & \bar{F}_\delta & F_\delta & F_\delta & \bar{F}_\delta & . & . & F_\delta & F_\delta & F_\delta & F_\delta & . & . \to 1 \\
F_\delta & \bar{F}_\delta & F_\delta & F_\delta & F_\delta & \bar{F}_\delta & . & . & . & F_\delta & F_\delta & F_\delta & . & . \to 1 \\
. & . & . & . & . & . & . & . & . & . & . & . & . & . \\
\bar{F}_\delta & \bar{F}_\delta & \bar{F}_\delta & F_\delta & F_\delta & \bar{F}_\delta & . & . & . & . & F_\delta & F_\delta & . & . \to 1 \\
. & . & . & . & . & . & . & . & . & . & . & . & . & . \\
\downarrow & \downarrow & \downarrow & \downarrow & \downarrow & \downarrow & & & & \downarrow & \downarrow & \downarrow & & \\
b_{1\delta} & b_{2\delta} & b_{3\delta} & b_{4\delta} & b_{5\delta} & b_{6\delta} & . & . & . & b_{n\delta} & b_{n+1,\delta} & b_{n+2,\delta} & . & . \to 1
\end{array} \tag{15}$$

This representation by the frequency lattice is restricted, however, to the interpreted calculus of probability, since the frequency statement about F_δ, that is, the relation (14), is not derivable from the Bernoulli theorem, but can be inferred only by means of the frequency interpretation.

These considerations may be illustrated by an example. Assume

$$P(A,B) = p = \tfrac{1}{2} \qquad n = 30 \qquad \delta^* = 20\%$$

We obtain from table 4 (p. 273) the value $b_{n\delta} = 0.73$. This means, according to (4) and (8), that for $\delta^* = 20\%$ the 30th vertical column of the coördinated frequency lattice supplies the value 0.73 as the limit of the frequency. In other words, of all the horizontal sequences, 73% will have attained up to the 30th element a frequency within the interval $\frac{1}{2} \pm 20\%$, that is, will possess elements of the kind B in a number between 12 and 18. The example illustrates the transformation of a probability into a frequency. The statement a_1 of the first kind, "B may be expected with the probability $\frac{1}{2}$", is transformed into the statement a_2 of the second kind, "We may expect with the probability 0.73 that after 30 elements the frequency will lie within the interval $\frac{1}{2} \pm 20\%$". The result is derivable even in the formal calculus.

When we use the frequency interpretation, the new probability 0.73 is also transformed into a frequency; the result is stated in the assertion that in the lattice 73% of all horizontal sequences possess the property explained.

The difference between the two conceptions of the Bernoulli theorem, the formal and the interpreted conceptions, is expressed in the following schema (P = probability):

TABLE 5

THE TWO CONCEPTIONS OF BERNOULLI'S THEOREM

P-statement	Formal conception	Interpreted conception
1st kind	*P* 1st kind: formal meaning	*P* 1st kind: frequency interpretation in an infinite sequence
2d kind	*P* 1st kind: frequency interpretation in a finite section of a sequence *P* 2d kind: formal meaning	*P* 1st kind: frequency interpretation in a finite section of a sequence *P* 2d kind: frequency interpretation in an infinite sequence lattice
3d kind	*P* 1st kind: frequency interpretation in a finite section of a sequence *P* 2d kind: frequency interpretation in a finite section of a lattice *P* 3d kind: formal meaning	*P* 1st kind: frequency interpretation in a finite section of a sequence *P* 2d kind: frequency interpretation in a finite section of a lattice *P* 3d kind: frequency interpretation in an infinite three-dimensional sequence lattice

The frequency interpretation assumed for the probability in the interpreted conception is introduced in the formal conception only in the probability statement of the next higher kind. The frequency occurring in the formal conception, however, is restricted to a finite section of a sequence or of a lattice. The interpreted conception, too, includes frequencies of finite sections, but the frequency of the highest step refers to infinite sequences or to lattices infinite in one direction.

The results show that an inference from the formal to the interpreted conception can be made only if an assumption about the meaning of the probability 1 is introduced. This result will be used in § 65.

In this connection I may point out a fallacy easily committed with respect to Bernoulli's theorem. The supposition offers itself that the property of homogeneity of the lattice, which distinguishes normal sequences in the narrower sense from those in the wider sense, could be derived from the properties of the latter sequences with the help of Bernoulli's theorem if the horizontal sequences are assumed to be independent by combinations [see (6, § 34)]. On this assumption, the probability of a combination of k elements occurring in a vertical column is equal to

$$P(A,F^{*ki})^i = b_{k\delta} \tag{16}$$

F^* stands for the frequency of elements B counted vertically. The $b_{k\delta}$ are identical with the Bernoulli values $b_{n\delta}$, and thus we have

$$\lim_{k\to\infty} b_{k\delta} = 1 \tag{17}$$

The result might be interpreted as meaning that a limit p of the frequency of B in the vertical sequences is to be expected with the probability 1. But although formulas (16) and (17) are correct, this interpretation of them would be incorrect. Its falsehood may be demonstrated by constructing an opposite case: the lattice y_{ki} may be occupied by elements $y_{ki} \in B$ in such a way that on the left side of the lattice diagonal line, given by $y_{11}, y_{22}, y_{33}, \ldots,$ only elements $y_{ki} \in B$ occur, whereas each horizontal sequence is continued on the right side of the lattice diagonal line so as to form a normal sequence, in the wider sense, with the probability p. In such a lattice it is possible to demand also the independence by combinations of the horizontal sequences, so that the relations (16) and (17) are satisfied. Nevertheless, all vertical sequences possess the frequency limit 1 for B. The lattice is not homogeneous, and the quantity $\lim_{k\to\infty} b_{k\delta}$ does not represent the probability that the frequency of a vertical sequence lies within $p \pm \delta$.

Equations (16) and (17) state only the following property: if we consider k horizontal sequences that are completely independent of each other, and count the frequency within the vertical columns, then, if we count the columns in the horizontal direction, those vertical columns will prevail in which the internal (vertical) frequency lies within F_δ. If k is increased, the percentage of vertical columns becomes still greater, if only we count far enough in the horizontal direction. (The latter addition is required, because, if the length of the horizontal sequence remained unchanged, it could happen that an increase in k would cause a diminution in the percentage mentioned.) But it is not possible to make an inference concerning the limit of frequency in the vertical columns.

A corresponding statement must be made in respect to the process of mixing [see (9, § 34)]. A lattice assumption can be dispensed with if we restrict ourselves to a finite number k of horizontal sequences and make probability statements with respect to the horizontal direction only. Then the following conclusions can be drawn with the help of Bernoulli's theorem and the frequency interpretation: if we proceed sufficiently far in the horizontal direction, the percentage of well-shuffled vertical columns approaches a limit, which is determined by k and lies the closer to 1 the larger k is. But if we keep the length of the horizontal sequence unchanged, then an increase in k may effect a diminution of this percentage.

The objection might be raised that in practice we always deal with a finite number k of sequences; but the answer is that also the length n of the sequences in the horizontal direction is always finite. Idealizing the problem by making a transition to the limit $n \to \infty$ means that we wish to make predictions about events that would occur for an increase of a finite n. Correspondingly, we want to make predictions about events that occur when k is increased; in the idealized problem, also, we must therefore make a transition to the limit $k \to \infty$. But then it is necessary to assume lattice invariance for the process of mixing. Applied to a finite number k, the assumption signifies a statement about events that occur when k is increased; the assumption that the sequences are completely independent is insufficient for such statements. It is for this reason that only the lattices introduced in this work, which are infinite in both directions, provide a complete idealization of such problems.

§ 51. The Amplified Bernoulli Theorem

The probability $b_{n\delta}$ of Bernoulli's theorem refers to a combination of n elements for which the frequency counted in the combination lies within the interval $p \pm \delta$. But when such a frequency is reached at the n-th element, we must not infer that the frequency will remain within the interval $p \pm \delta$ when the sequence is continued further; and, in fact, for normal sequences we can never derive a statement that from a specified n the frequency will certainly remain within $p \pm \delta$. The impossibility of such a statement is obvious when we realize that there may occur at any time disturbing runs of elements B that are long enough to push the frequency outside the interval $p \pm \delta$. However, we can inquire at least after the probability that the frequency will remain in the interval $p \pm \delta$ after it has reached this interval at the n-th element.

The question is important for the frequency interpretation. To say that the frequency f^n goes toward a limit p means that for every δ, however small, there exists an element y_n such that f^n lies within $p \pm \delta$ and remains inside this interval for all the following elements. Such an element is called a *place of convergence for* δ. (I do not wish to incorporate into the concept the condition that the element y_n is the first element of this kind; thus there exist, if at all, infinitely many places of convergence for δ, because, if y_n is such a place, all the following elements have the same property.) The question to be studied, therefore, concerns the probability that an element y_n is a place of convergence. We shall inquire later (§ 65) what is gained by these considerations for the frequency interpretation. For the present we shall deal only with the mathematical question of the probability concerned.

We begin with the construction of a simpler probability. Assume the counting of the frequency is continued, after the elements y_n, up to the element y_s $(s > n)$. We ask for the probability that each of the sections $y_1 \ldots y_n, y_1 \ldots y_{n+1}, \ldots y_1 \ldots y_s$ has a frequency within $p \pm \delta$. The problem will be formulated in lattice enumeration, since this way of counting offers the best illustration of a probability of sequences. We introduce the abbreviation

$$F_\delta^{k,n/s} = F_\delta^{kn} \,.\, F_\delta^{k,n+1} \ldots F_\delta^{k,s} \tag{1}$$

The probability that the frequency remains within the interval $p \pm \delta$ of convergence for the section from n to s is then given by

$$P(A, F_\delta^{k,n/s})^k = c_{ns\delta} \tag{2}$$

The computation of $c_{ns\delta}$ cannot be achieved by the multiplication of the quantities $b_{n\delta}, b_{n+1,\delta}, \ldots b_{s\delta}$, since this would result in too small a value, the terms on the right side of (1) not representing independent events. When the frequency f^n has arrived in the interval $p \pm \delta$, there exists a greater probability than $b_{n\delta}$ that it will remain in the interval. The computation of $c_{ns\delta}$ is found in the appendix to the German edition of this book (§ 83); only a short summary of the results will be given here.

From (2) we construct the desired probability by assuming s to go to infinite values. This probability can be written in the form

$$P(A,F_\delta^{k,n\,\cdots})^k = \lim_{s\to\infty} c_{ns\delta} = c_{n\delta} \tag{3}$$

The abbreviation in the second term, which stands for an infinite conjunction, is to be defined by means of an all-operator, according to a remark made in § 6:

$$F_\delta^{k,n\,\cdots} =_{Df} (s)\, F_\delta^{k,n/s} \tag{4}$$

The relation (3) is based on the assumption of the commutativity of the limit operation and the probability operator [see (6, § 41)], since (3) states that the limit of the classes $F_\delta^{k,n/s}$ for $s \to \infty$ has a probability that is the limit of the corresponding probabilities. This proof, therefore, holds only for lattices that satisfy this condition.[1] The restriction, however, is not serious, since the theorem holds only for lattices of a special kind, the proof being valid only for normal sequences. Lattices of the kind required play an important part in the theory of probability.

The computation leads to the result that the limit $c_{n\delta}$ is not $= 0$ but that, although we have, of course,

$$c_{n\delta} < b_{n\delta} \tag{5}$$

the relation

$$\lim_{n\to\infty} c_{n\delta} = 1 \tag{6}$$

holds. This result means: *the probability $c_{n\delta}$, that the frequency at the n-th element lies within the interval $p \pm \delta$ and remains within this interval for the whole infinite remainder of the sequence, has a finite value and converges to 1 when n goes to infinity.* The theorem holds for every δ, however small; but δ must be kept constant when the transition to the limit is made. The result can be stated also in the form: the probability that the n-th element is a place of convergence for a given δ goes toward 1 with increasing n.

This represents an amplification of the Bernoulli theorem, which was proved only recently by György Pólya[2] after the question had been raised

[1] A lattice that does not satisfy the condition can be constructed as follows: we draw the lattice diagonal line passing through the elements $y_{11}, y_{22}, y_{33}, \ldots$. Now assume that none of the horizontal sequences has a limit of the frequency, but that all the horizontal sections to the left of the diagonal line satisfy the relation (2) for every n and s, the probability $c_{ns\delta}$ being counted vertically. Then the probability on the left side of (3) is $= 0$, and the relation (3) is not true.

[2] *Nachr. d. Ges. d. Wiss. z. Göttingen, math.-phys. Kl.*, 1921.

by Paul Hertz. According to the theorem, it is permitted to expect with a probability converging toward 1, not only that the desired exactness is reached at the n-th element, but also that it is adhered to thereafter.

The theorem can be formulated for individual sequences also—both for enumeration in overlapping segments and for enumeration in sections. In the first mode of enumeration a segment of n elements with overlapping is moved through the entire sequence; the frequency is counted from the first element of the segment so that the starting point of the counting is moved along with the segment, and the counting is continued beyond the end of the segment through the whole sequence. We regard as positive the segments for which, first, the frequency f^n at the end of the segment lies within $p \pm \delta$, and second, the frequency at all later places remains within $p \pm \delta$ for the whole infinite remainder of the sequence. The other segments are taken as negative. Even for larger n there will usually exist again and again negative segments; the differences in enumeration result from the circumstance that the frequency f^n is always counted from the beginning of the respective segment and not from the beginning of the sequence. However, the frequency of the segments is counted from the beginning of the sequence. Then the frequency of the positive segments, counted for the whole sequence, is $= c_{n\delta}$. The greater the length of the segments, the larger the number of positive segments.

For enumeration in sections the sequence is divided into consecutive sections of the length n, which do not overlap; the frequency f^n is counted, as before, from the beginning of the respective section and, passing the end of the section, through the infinite remainder of the sequence. The continuations of the enumeration then overlap one another. The counting of the sections is handled as in the preceding case.

§ 52. The Frequency Dispersion

The Bernoulli theorem leads to statements about the convergence of frequency sequences by determining probabilities of a higher kind. It is possible to characterize the convergence in a different, though logically equivalent, manner, which is based on the concept of dispersion. The concept was developed in § 37 in a general way, applicable to any amounts u_k; we can use it for the convergence of frequency sequences when we treat, in particular, frequencies as such amounts. Therefore we speak, more precisely, of *frequency dispersion*; for the sake of brevity the term *dispersion* will be used in the same sense. The concept of dispersion is of particular advantage to practical calculations of a statistical kind.

We begin the investigations with the interpreted conception, by developing the concept of dispersion with the help of the frequency interpretation.

Starting with lattice enumeration, we construct the frequency lattice according to (13, § 50):

$$f^{kn} = F^n(A,B^{ki})^i \tag{1}$$

The value f^{kn} represents the frequency of the initial section of the length n of the k-th horizontal sequence. This time, however, we do not classify in respect to F_δ, but carry out the calculation with the amounts f^{kn} themselves, as in § 39 with the amounts u^{ki}. The lattice has the form

$$\begin{array}{cccccccccc} f^{11} & f^{12} & f^{13} & . & . & . & f^{1n} & . & . & . \\ f^{21} & f^{22} & f^{23} & . & . & . & f^{2n} & . & . & . \\ . & . & . & . & . & . & . & . & . & . \\ f^{k1} & f^{k2} & f^{k3} & . & . & . & f^{kn} & . & . & . \\ . & . & . & . & . & . & . & . & . & . \end{array} \tag{2}$$

Apart from the individual values f^{kn} of the amounts, steps of amounts must also be introduced. Since we want to consider the frequencies in the n-th vertical column, the steps of amounts are given by the possible values of the frequency for n elements. Since the possible amounts are given by the number m of the element B, we designate the step of amount by f^n_m and have

$$f^n_m = \frac{m}{n} \tag{3}$$

The amount f^n_m is coördinated to the case F^{kn}_m, according to (10, § 50); the probability of this case is

$$P(A,F^{kn}_m)^k \tag{4}$$

We can now deal with the lattice f^{kn} as we did previously with the lattice u^{ki}.

First, however, a different conception of the lattice should be mentioned. We can imagine the lattice as resulting from a single sequence that is treated in enumeration by sections. If we cut off sections of the length n within a sequence, writing these sections under one another, then a lattice is produced that is infinite in vertical direction and extends horizontally to the n-th column. The f^{kn} represent the internal frequency of the k-th section of the length n; F^n_m has the meaning given in (6, § 49), and the corresponding probability is to be written as

$$P(A \,.\, S_{nn},F^n_m) \tag{5}$$

where S_{nn} is a regular division of the length n and $\kappa = n$ [see (2, § 30)]. Again f^n_m is the step of amount coördinated to F^n_m. All the following considerations can therefore be regarded as theorems about one sequence enumerated by sections. The disadvantage of such a conception is that the transition to larger n requires a new construction of the lattice. It is even possible to interpret these considerations as referring to enumeration with overlapping, but such an interpretation is not advisable. The f^{kn} would not be mutually independent in enumeration with overlapping for k as running superscript; and we would arrive at an unnecessary increase in segments, by which the accuracy of the consideration would not be improved.

Turning now to the treatment of the f^{kn} lattice, we can derive from the frequency interpretation the relation

$$M(f^{kn})^n = p \tag{6}$$

For the derivation we use the symbol F_δ^{kn}, according to (10, § 50). Although the symbol refers, not to a fixed amount of f^{kn}, but to the interval $p \pm \delta$, we can proceed as follows. According to (14, § 50), $P(A,F_\delta^{kn})^n = 1$, that is, $P(A,\bar{F}_\delta^{kn})^n = 0$; thus the sum occurring in the average, according to (12, § 35), is reduced to a single term. Employing for the step of the amount first the lower limit $p - \delta$, and then the upper limit $p + \delta$, we obtain the inequalities

$$P(A,F_\delta^{kn})^n \cdot (p - \delta) \leqq M(f^{kn})^n \leqq P(A,F_\delta^{kn})^n \cdot (p + \delta) \tag{7}$$

Since this probability is equal to 1, according to (14, § 50), and (7) is to be valid for any δ, however small, (6) follows.[1]

The average in the vertical direction will now be considered. We can use either of the definitions given in § 35: the statistical definition gives the result

$$M(f^{kn})^k = \lim_{s\to\infty} \frac{1}{s} \sum_{k=1}^{s} f^{kn} = M(f^n) \tag{8a}$$

whereas the theoretical definition assumes the form

$$M(f_m^n)_m = \sum_{m=0}^{n} P(A,F_m^{kn})^k \cdot f_m^n = M(f^n) \tag{8b}$$

The two definitions are identical. For enumeration by sections the probability (5) is to be introduced in the expression (8*b*).

The value of (8) cannot, of course, be calculated without further assumptions in regard to the structure of the sequences. We want to investigate, in

[1] Another form of the proof of (6) obtains if, instead of the classification of the f^{kn}-sequence with respect to F_δ, and, in place of the theoretical definition of the average (12, § 35), we employ the statistical definition of the average (1*b*, § 43). Then we can make use of a familiar theorem of the theory of convergent series: from $\lim_{n\to\infty} a_n = a$ we can infer that $\lim \left(\frac{1}{n} \sum_{i=1}^{n} a_i\right) = a$. We have $a_n = f^{kn}$, $a = p$. The relation (6) can be derived only from the frequency interpretation, which is no longer required for the following considerations; so (11) and (15) hold likewise in the formal calculus of probability.

particular, normal sequences, and for such sequences (8) can be evaluated. We have for normal sequences, according to (11a, § 50) and (7, § 49),

$$P(A,F^{kn}_m)^k = w_{nm} = \binom{n}{m} p^m(1-p)^{n-m} \tag{9}$$

The same value holds for the probability (5) when we employ enumeration by sections. Introducing this value into (8), we can carry out the summation when we make use of the relations

$$\binom{n}{m} = \frac{n}{m} \cdot \binom{n-1}{m-1} \tag{10a}$$

$$\binom{n}{m} = \frac{n(n-1)}{m(m-1)} \cdot \binom{n-2}{m-2} \tag{10b}$$

We first use only (10a) and have

$$M(f^{kn})^k = \sum_{m=0}^{n} \binom{n}{m} p^m (1-p)^{n-m} \cdot \frac{m}{n}$$

$$= p \cdot \sum_{m-1=0}^{n-1} \binom{n-1}{m-1} p^{m-1} (1-p)^{(n-1)-(m-1)}$$

The transition to the index of summation $m - 1$ means that in the sum the term with $m = 0$ is dropped; but the term vanishes, since it contains 0 as a factor. The sum now has the form $\sum_{m-1=0}^{n-1} w_{n-1,m-1}$ and is therefore equal to 1, according to (20, § 49). Thus we have

$$M(f^{kn})^k = p \tag{11}$$

in analogy to (6). The f^{kn} lattice of normal sequences is thus homogeneous in respect to the average, since the average is the same for all the horizontal and vertical sequences.

We can also calculate the dispersion, always assuming that we are dealing with normal sequences in the narrower sense. For the dispersion we employ the deviations

$$\delta f^{kn} = f^{kn} - p \tag{12a}$$

$$\delta f^n_m = f^n_m - p = \frac{m}{n} - p \tag{12b}$$

Like the average, the dispersion can be defined in two different ways. The statistical definition, according to (8, § 37), is given by

$$\Delta^2(f^{kn})^k = \lim_{s\to\infty} \frac{1}{s} \sum_{k=1}^{s} \delta^2 f^{kn} = \Delta^2(f^n) \tag{13a}$$

The theoretical definition, according to (6, § 37), supplies the equation

$$\Delta^2(f_m^n)_m = M(\delta^2 f_m^n)_m = \sum_{m=0}^{n} P(A, F_m^{kn})^k \cdot \delta^2 f_m^n = \Delta^2(f^n) \tag{13b}$$

The two expressions are identical, as was demonstrated above. As before, the theoretical definition (13*b*) admits of a transformation. With (9), (11), (12*b*), and (13, § 27) we conclude:

$$\Delta^2(f^n) = M(f^n)^2 - M^2(f^n)$$

$$= \sum_{m=0}^{n} \left(\frac{m}{n}\right)^2 \binom{n}{m} p^m(1-p)^{n-m} - p^2 = \frac{1}{n^2} A - p^2 \tag{14}$$

$$A = \sum_{m=0}^{n} m^2 \binom{n}{m} p^m (1-p)^{n-m}$$

Using the factorization $m^2 = m(m-1) + m$, we obtain

$$A = \sum_{m=0}^{n} m(m-1) \binom{n}{m} p^m (1-p)^{n-m}$$

$$+ \sum_{m=0}^{n} m \binom{n}{m} p^m (1-p)^{n-m}$$

Applying (10*b*) to the first expression, (10*a*) to the second, and making an inference similar to the one above, we find the following result, when we make use of the fact that the sums are equal to 1:

$$A = n(n-1)\, p^2 \cdot \sum_{m-2=0}^{n-2} \binom{n-2}{m-2} p^{m-2}(1-p)^{(n-2)-(m-2)}$$

$$+ np \cdot \sum_{m-1=0}^{n-1} \binom{n-1}{m-1} p^{m-1}(1-p)^{(n-1)-(m-1)}$$

$$= n(n-1)p^2 + np = n^2p^2 + np(1-p)$$

Substitution of the value of A in (14) gives

$$\Delta^2(f^n) = \frac{p(1-p)}{n} \tag{15}$$

$$\Delta(f^n) = \sqrt{\frac{p(1-p)}{n}}$$

This represents the theoretical value of the dispersion for normal sequences. It may be seen from (15) that the dispersion of normal sequences diminishes with increasing n and becomes equal to 0 in the limit $n \to \infty$.

These considerations can be extended to other than normal sequences. It may even happen that (11) holds also for certain nonnormal sequences, though this relation was derived only for normal sequences. For instance,

TABLE 6

	Normal dispersion, sequence (1, § 27)			Supernormal dispersion, sequence (1, § 33)			Subnormal dispersion, sequence (2, § 33)		
k	f^{kn}	δf^{kn}	$\delta^2 f^{kn}$	f^{kn}	δf^{kn}	$\delta^2 f^{kn}$	f^{kn}	δf^{kn}	$\delta^2 f^{kn}$
1	0.6	−0.09	0.01	0.7	−0.14	0.02	0.5	+0.03	0.00
2	0.3	+0.21	0.04	0.2	+0.36	0.13	0.5	+0.03	0.00
3	0.4	+0.11	0.00	0.4	+0.16	0.03	0.7	−0.17	0.03
4	0.6	−0.09	0.01	0.9	−0.34	0.12	0.7	−0.17	0.03
5	0.5	+0.01	0.00	0.5	+0.06	0.00	0.4	+0.13	0.02
6	0.7	−0.19	0.04	0.4	+0.16	0.03	0.6	−0.07	0.00
7	0.7	−0.19	0.04	0.7	−0.14	0.02	0.4	+0.13	0.02
8	0.3	+0.21	0.04	0.7	−0.14	0.02	0.4	+0.13	0.02
	4.1		0.18	4.5		0.37	4.2		0.12
	$M(f^n) = \frac{4.1}{8} = 0.51$			$M(f^n) = \frac{4.5}{8} = 0.56$			$M(f^n) = \frac{4.2}{8} = 0.53$		
	$\Delta(f^n) = \sqrt{\frac{0.18}{8}} = 0.15$			$\Delta(f^n) = \sqrt{\frac{0.37}{8}} = 0.22$			$\Delta(f^n) = \sqrt{\frac{0.12}{8}} = 0.12$		

(11) can be shown to be valid if the horizontal sequences possess probability transfer. There are sequences of other types for which (11) holds. All such sequences are characterized by the statement that their f^{kn} lattice is homogeneous in respect to the average; therefore the relations (12)–(13) remain valid for them, too. However, (15) holds only for normal sequences, since equation (9) was used for the derivation. Therefore (15) represents a criterion of the normal character of sequences.

These results can be applied for practical purposes as follows. In the practice of statistical calculations finite lattices always occur. It suffices to compute the f^{kn} for the column n. Let the number of the horizontal rows be s. We determine the value p as the mean value $M^s(f^{kn})^k$, that is, the summation occurring in (8a) is carried out only as far as s, and the mean value obtained is regarded as a *practical limit*. By the help of this value the deviations δf^{kn} are determined according to (12a); then we calculate the dispersion by the summation according to (13a), likewise carried out only as far as s. Finally, the value thus obtained—the mean square of the deviations—is regarded as a practical limit, and we arrive at the statistical value of the dispersion.

The procedure may be illustrated by an example. In the sequences (1, § 27), (1, § 33), and (2, § 33), which were studied previously, the first $s = 8$ sections of $n = 10$ elements were each counted with respect to B (that is, using enumeration by sections), so that the frequency f^{kn} was ascertained for each section. The results, together with the values δf^{kn}, are reproduced in table 6.

The δf^{kn} column shows that the sequences converge differently. The greatest deviations occur in (1, § 33); the sequence (2, § 33) shows the best convergence. In the dispersion $\Delta(f^n)$ we have, correspondingly, a *measure of convergence,* which is greatest in (1, § 33) and smallest in (2, § 33); in (1, § 27) it has an intermediate value. Had we made the calculation for larger n, that is, for longer sections, the δf^{kn} and therefore the $\Delta(f^n)$ would all have smaller values.

It is possible to calculate the normal theoretical value of the dispersion, that is, the value that the dispersion assumes if the sequences are normal. We then substitute in (15) for p the value $M(f^n)$ that we have regarded as the practical limit. By comparing the normal value of $M(f^n)$ thus calculated with the statistically found value, we are able to judge whether the assumption of a normal character of the sequence is correct. For this purpose the following definitions are introduced:

1. If the statistical dispersion is equal to the normal one, the sequences have a *normal dispersion.*
2. If the statistical dispersion is greater than the normal one, the sequences have a *supernormal dispersion.*
3. If the statistical dispersion is smaller than the normal one, the sequences have a *subnormal dispersion.*

From this viewpoint, consider the example again. Formula (15) furnishes for $n = 10$ and $p = \frac{1}{2}$:

$$\Delta(f^n)_{\text{normal}} = 0.16 \tag{16}$$

Comparison with table 6 shows that the sequence (1, § 27) has a normal dispersion; (1, § 33) a supernormal dispersion; (2, § 33) a subnormal dispersion.

The classification of dispersions is related as follows to the preceding classification, which divides all sequences into normal and nonnormal sequences. If the dispersion is normal we are not certain that the sequence is normal; but if the dispersion is not normal we can conclude that the sequence is not normal. The concept of normal dispersion is more comprehensive than that of normal sequence; it provides a less detailed characterization of the sequence. For practical applications the classification by the normal dispersion is usually preferred, since we can easily ascertain the dispersion, whereas it is rather cumbersome to determine whether the sequence is normal.

§ 53. The Dispersion of Nonnormal Sequences

In § 49 the Bernoulli theorem was derived from the assumption that the sequence is normal. It is possible to show that this is not a necessary condition of a theorem of this kind; sequences of more general character, among them sequences with probability transfer, likewise possess a Bernoulli theorem. By a Bernoulli theorem we mean a theorem according to which the probability that the frequency f^n lies within $p \pm \delta$ converges toward 1 with increasing n. Such sequences satisfy (28, § 49), whereas (27, § 49) holds with a different expression for γ. The type of the sequence has, for all such sequences, an influence only on the kind of convergence; the fact remains that there exists a convergence toward 1.

Since the kind of convergence is best characterized by the frequency dispersion, we shall deal with the problem of dispersion also for sequences with probability transfer. For such sequences (13*b*, § 52) is valid; but the quantity $P(A,F^{kn}_{m})^k$ is not measured by the value w_{nm} of (9, § 52), and therefore (15, § 52) does not hold. Thus (9, § 52) must be replaced by a more complicated relation. These calculations were first carried out by Pólya.[1] In the appendix (§ 82) of the German edition of this book (omitted in this edition) I present a calculation developed by V. Bargmann; it provides for the value $\Delta(f^n)$ an approximation given by the formula

$$\Delta(f^n) = \sqrt{\frac{p(1-p)}{n}} \cdot \sqrt{1 + \frac{2\epsilon}{1-p-\epsilon}}$$

$$= \sqrt{1 + \frac{2\epsilon}{1-p-\epsilon}} \cdot \Delta(f^n)_{\text{normal}} \qquad (1)$$

The formula is valid for enumeration with overlapping, and holds also for enumeration by sections, if the sequences possess a regular domain of invariance. For lattice enumeration the corresponding condition of lattice invariance must be added. ϵ is the quantity that characterizes probability transfer, which was called the degree of transfer. If we restrict the consideration again to the nondegenerate case $1 - p - \epsilon > 0$, which was formulated in (16, § 33), the sign of the second term within the scope of the square root of (1) depends only on the sign of ϵ. We thus arrive at the result

1. For $\epsilon = 0$ the dispersion is normal.
2. For positive ϵ, that is, probability drag, the dispersion is supernormal.
3. For negative ϵ, that is, probability compensation, the dispersion is subnormal.

[1] György Pólya, "Über die Statistik verketteter Vorgänge," in *Zs. f. angew. Math. u. Mech.*, Vol. III (1923), p. 279; and *Sur Quelques Points de la théorie des probabilités* (Paris, 1930).

Theorem 1 is a matter of course; the significant result is formulated in theorems 2 and 3. They can be paraphrased as follows:

The generalization of the sequence type introduced by the concept of probability transfer includes a classification of sequences corresponding to the classification provided by the concept of dispersion. Sequences with probability drag converge less well than, and sequences with probability compensation converge better than, normal sequences.

This fact permits an inference from the dispersion to the sequence type if it is known from other reasons that the sequence has the character of probability transfer expressed in (9, § 33) (that for a selection by predecessors only the first predecessor is relevant), and if it is furthermore known that the sequence has a regular domain of invariance. On these conditions the following classification can be set up:

1. For normal dispersion the sequence is normal.
2. For supernormal dispersion there exists probability drag.
3. For subnormal dispersion there exists probability compensation.

These reversals of the previous theorems follow because, if the sequences have probability transfer, they satisfy formula (1), from which we easily derive

$$\Delta(f^n) = \Delta(f^n)_{\text{normal}} \text{ implies } \epsilon = 0$$

$$\Delta(f^n) > \Delta(f^n)_{\text{normal}} \text{ implies } \epsilon > 0$$

$$\Delta(f^n) < \Delta(f^n)_{\text{normal}} \text{ implies } \epsilon < 0$$

Once the value of $\Delta(f^n)$ is found empirically by the use of the statistical definition (13*a*, § 52), the value of ϵ can be derived from (1), if only we know that the conditions of probability transfer are satisfied.

Formula (1) may be applied to the example of the sequences (1, § 33) and (2, § 33), which were constructed as sequences with probability transfer; we explained that for (1, § 33) we have $\epsilon = +\frac{1}{6}$, and for (2, § 33), $\epsilon = -\frac{1}{6}$. With these values, formula (1) supplies

$$\text{for } (1, \S\, 33)\text{: } \Delta(f^n) = 0.22 \qquad \text{for } (2, \S\, 33)\text{: } \Delta(f^n) = 0.11 \tag{2}$$

These values are in agreement with the statistically found values,[2] which were given in table 6 (p. 288). Conversely, ϵ can be computed from the statistical values given in the table.

In more general sequences, which are not subject to the condition of probability transfer, the correspondence between dispersion and selection by the predecessor no longer exists. For instance, it may happen that, for subnormal

[2] I emphasize once more that the sequences chosen as examples are too short to be used for a reliable calculation of the dispersion. I merely wish to demonstrate by this example the method of calculation.

dispersion, $P(A.B,B^1) > P(A,B)$, as will be shown by an example in § 56. In these cases the sequence type is too complicated to admit of a sufficient characterization by a single quantity, such as is supplied by the dispersion or the degree of transfer ϵ. Classification by dispersion represents for such sequences a relatively rough characterization by which only certain average properties of the sequences are expressed.

For probability sequences of the most general type we cannot even guarantee that a dispersion exists, that is, that the sequence of mean values in (13*a*, § 52) converges toward a limit with increasing s. That this convergence is contingent upon certain conditions is seen from (13*b*, § 52). The formula, it is true, contains only a summation over a finite number of terms, and therefore a dispersion will always exist if the probabilities $P(A,F_m^{kn})^k$ exist. But the existence of the probabilities depends on the existence of the individual phase probabilities (§ 27), or, in lattice enumeration, on the condition that the vertical sequences are of a probability character and possess combinatory probabilities. The type of probability sequence for which a dispersion exists is, therefore, very general, but it is not identical with the most general type of probability sequence.

§ 54. A Simple Interpretation of the Dispersion

In § 43 the probability w_Δ that the deviation remains within the limits given by the linear dispersion was calculated for a Gaussian distribution, or normal curve. It was found that w_Δ is independent of the measure of precision h and very nearly equal to $\frac{2}{3}$. This fact permits a corresponding interpretation of the dispersion, since the Bernoulli distribution can be approximated by a Gauss exponential function, according to (27, § 49). In fact, if we substitute (15, § 52) for δ in (27, § 49), we find for γ the value $\frac{1}{\sqrt{2}}$, which, according to (11, § 43), leads to $w_\Delta = \frac{2}{3}$. Considering, for the present, only normal sequences, we can formulate the following result: *the dispersion determines the limits within which the frequency is to be expected with the probability* $\frac{2}{3}$. (See fig. 16, p. 222.)

The result may be illustrated by representing numerically the value of $\Delta^*(f^n)$ in a table calculated in accordance with (15, § 52). (See table 7.) Like δ^* in table 4 (p. 273), $\Delta^*(f^n)$ stands for the dispersion measured in percentage of p; so we have $p \pm \Delta(f^n) = p(1 \pm \Delta^*(f^n))$.

The tabulation is only a slightly changed and abbreviated reproduction of table 4, which states the probability $b_{n\delta}$ for given limits δ^* as a function of n. Table 7 states the limits $\Delta^*(f^n)$ that exist for a fixed probability $b_{n\delta} = \frac{2}{3}$ as a function of n. We see that the specification of $\Delta^*(f^n)$ represents only a more convenient form of a statement that was made above by the help of the Bernoulli probability $b_{n\delta}$. Instead of saying, "If $p = \frac{1}{2}$, then for 100 cases the frequency lies with the probability 0.95 between 40 and 60", we now say, "If $p = \frac{1}{2}$, then for 100 cases the frequency lies with the probability $\frac{2}{3}$ between 45 and 55". For the latter sentence we use the abbreviation, "If $p = \frac{1}{2}$, then for 100 cases the dispersion $\Delta^*(f^n)$ amounts to 10%". The concept of dispersion contains no new logical problems; its use is logically equivalent to the probability statements of the Bernoulli theorem.

TABLE 7

DISPERSION FOR NORMAL SEQUENCES

p	¼	½	¾
n	$\Delta^*(f^n)$	$\Delta^*(f^n)$	$\Delta^*(f^n)$
10	55%	32%	18%
30	32%	18%	11%
50	25%	14%	8%
100	17%	10%	6%
500	8%	4%	2½%

The approximation of the $b_{n\delta}$ by a Gauss function is not necessarily linked to normal sequences; certain nonnormal sequences have the same property, though with a different measure of precision because of their different dispersion. Thus the relation $w_\Delta = \frac{2}{3}$ also holds for many sequences with supernormal or subnormal dispersion and, in particular, for sequences with probability transfer. Upon this fact rests the great practical value of the dispersion. The statistical calculation of the dispersion represents a procedure that, without a more exact knowledge of the sequence type, permits determination of the limits within which the deviation can be expected with the probability $\frac{2}{3}$.

§ 55. A Simple Derivation of Bernoulli's Theorem

A proof of Bernoulli's theorem will now be presented. It recommends itself by an amazing mathematical simplicity, though it exhibits less clearly the logical problems of the theorem. The derivation is based on the properties of the dispersion and the Tchebychev inequality (18, § 37).

We start with the method developed in (3, § 35) by which the enumeration of classes is reduced to an addition of amounts, a method employing the amounts 1 and 0 supplied by the symbol V. We designate the cases B and $\bar{B}$ by B_1 and B_2 and coördinate to them the amounts $u_1 = u(B_1)$ and $u_2 = u(B_2)$ by the definition

$$u^i = V(y_i \in B_1) \qquad u_k = u(B_k) = V(B_k = B_1) \tag{1a}$$

that is,

$$u_1 = 1 \qquad u_2 = 0 \tag{1b}$$

Furthermore, we define the amount of a combination of consecutive elements in the sense of (1, § 38) by the addition of the separate amounts. Thus we have

$$u(B_i . B_k^1) = u(B_i) + u(B_k) = u_i + u_k \tag{2}$$

A combination of n elements that contains m_l elements B gives the value

$$u(B_{k_1}^1 \ldots . B_{k_n}^n) = u_{k_1} + \ldots . + u_{k_n} = m_l \tag{3}$$

According to (3b, § 38), we have

$$M(u_{k_1} + \ldots . + u_{k_n})_{k_1 \ldots k_n} = M(u_{k_1})_{k_1} + \ldots + M(u_{k_n})_{k_n} = n \cdot M(u) \tag{4}$$

With the help of (1b) we obtain

$$M(u) = \sum_{k=1}^{2} P(A,B_k) \cdot u_k = p \tag{5}$$

and thus find, with (3),

$$M(u_{k_1} + \ldots . + u_{k_n})_{k_1 \ldots k_n} = M(m_l)_l = M(m) = np \tag{6}$$

where m designates the possible values 0, 1, 2, . . . n for m; or, with (13, § 35) and (3, § 52),

$$M\left(\frac{m_l}{n}\right)_l = M(f^n_{m_l})_l = M(f^n) = p \tag{6'}$$

The equation may be explained by the following considerations. Because of the definition (1) for the amounts, taking the average for sections of the length of one element means nothing but counting the elements B; thus $M(u)$, according to (5), is equal to the probability of B. Since the average is additive, we derive with (6) the result that for sections of the length n the average of the relative frequency must also be equal to this probability.

The result was derived without any restricting assumptions concerning the structure of the sequence, since (4), as was shown in § 38, holds for any kind of mutual dependence of events, and the nature of the phase probabilities is irrelevant for (4). However, if we now calculate the value of the dispersion, we must introduce a specializing assumption concerning the structure of the sequence. Assume that the sequence is normal. Since then, according to § 38, the dispersion is additive, we have

$$\Delta^2(u_{k_1} + \ldots + u_{k_n})_{k_1 \ldots k_n} = \Delta^2(u_{k_1})_{k_1} + \ldots + \Delta^2(u_{k_n})_{k_n}$$
$$= n \cdot \Delta^2(u) \tag{7}$$

Now we obtain from (1) and (5)

$$\delta u_1 = u_1 - M(u) = 1 - p \qquad \delta u_2 = u_2 - M(u) = -p \tag{8}$$

and thus

$$\Delta^2(u) = \sum_{i=1}^{2} P(A,B_i) \cdot \delta^2 u_i$$
$$= p(1-p)^2 + (1-p)p^2 = p(1-p) \tag{9}$$

Therefore we have, with (7) and (3),

$$\Delta^2(u_{k_1} + \ldots + u_{k_n})_{k_1 \ldots k_n} = \Delta^2(m) = np(1-p) \tag{10}$$

With (9a, § 37) and (3, § 52) we obtain

$$\Delta^2\left(\frac{m}{n}\right) = \Delta^2\left(\frac{m_l}{n}\right)_l = \Delta^2(f^n_{m_l})_l = \Delta^2(f^n_m)_m$$
$$= \Delta^2(f^n) = \frac{p(1-p)}{n} \tag{11}$$

This is identical with the value of the dispersion for normal sequences derived in (15, § 52). The occurrence of n in the denominator, and thus the dying down of the dispersion toward 0 with increasing n, appears here as the effect of the law of the compensation of the dispersion expressed in (7), according to which the quadratic dispersion increases only with n and not with n^2 [see (8b, § 38)].

Since for normal sequences in the narrower sense the probabilities in lattice enumeration are equal to those holding for enumeration with overlapping or for enumeration by sections, the results can be interpreted also in lattice enumeration, but that interpretation need not be elaborated.

Further conclusions can be drawn by the use of Tchebychev's inequality (18, § 37). When we substitute for $\Delta^2(u)$ the value $\Delta^2(f^n)$, we obtain

$$w_\delta \geqq 1 - \frac{p(1-p)}{n\delta} \tag{12}$$

On account of the definitions (1a) and (1b), w_δ represents the probability that the relative frequency for n elements lies within $p \pm \delta$, and therefore w_δ is identical with the Bernoulli probability $b_{n\delta}$. Since n occurs in the denominator of (12), w_δ goes toward 1 for increasing n, if δ is kept unchanged; thus we have

$$\lim_{n\to\infty} b_{n\delta} = 1 \tag{13}$$

in correspondence to (28, § 49).

This relation is virtually identical with the first of the relations (25, § 49), differing from it only in that the interval $\pm\ \delta$ is assumed symmetrical with respect to p. If p is not in the center of the interval, we can cut off an outer section of the larger part of the interval so as to place p in the center; since the convergence to 1 holds for this smaller interval, it holds also for the original interval. This proves the first of the relations (25, § 49). The second relation (25, § 49) follows from the first in consideration of (20, § 49).

The most important conclusion of Bernoulli's theorem is thus proved. But it is not possible to derive the approximate validity of the Gauss distribution in this manner.

§ 56. Poisson Sequences

A special type of nonnormal sequences was investigated by Poisson. He considers the sequence that results when an event is played for with the successive probabilities $p_1 \ldots p_\lambda$, so that after a period of length λ the probabilities repeat themselves in the same order. In order to find the structure of such sequences we must first translate the given definition into our symbolism.

Poisson's arrangement may be regarded as a regular division of the length λ, for which we have

$$P(A.S_{\lambda\kappa},B) = p_\kappa \qquad \kappa = 1,2, \ldots \lambda \tag{1}$$

Thus p_κ is the probability within the κ-th subsequence. Besides, from the nature of the division we have

$$P(A,S_{\lambda\kappa}) = \frac{1}{\lambda} \qquad \kappa = 1,2 \ldots \lambda \tag{2a}$$

$$\begin{aligned} P(A.S_{\lambda\kappa},S^{\rho}_{\lambda,\kappa+\rho}) &= 1 \\ P(A.S_{\lambda\kappa},S^{\rho}_{\lambda\nu}) &= 0 \qquad \text{for } \nu \neq \kappa + \rho \end{aligned} \tag{2b}$$

(2a) follows from the frequency interpretation, (2b) from axiom II,1 because of $(S_\kappa \supset S^{\rho}_{\kappa+\rho})$. The subscript is again counted cyclically: we put $\kappa - 1 = \lambda$ for $\kappa = 1$, $\kappa + 1 = 1$ for $\kappa = \lambda$. We inquire after the probability $P(A,B) = p$, that is, after the probability in the major sequence. According to the theorem of elimination (21, § 19), we have

$$P(A,B) = \sum_{\kappa=1}^{\lambda} P(A,S_{\lambda\kappa}) \cdot P(A.S_{\lambda\kappa},B) \tag{3}$$

With (2a) and (1) we thus find

$$P(A,B) = p = \frac{1}{\lambda}\sum_{\kappa=1}^{\lambda} p_\kappa \tag{4}$$

This probability is the mean value of the individual probabilities, a result which seems plausible. For the derivation, however, the frequency interpretation was used in (2a). If we do not use this interpretation, but remain in the formal calculus of probability, we cannot prove that the mean value p has the meaning of a probability; but even then it can be shown, in analogy to Bernoulli's theorem, that the mean value takes over the role of the critical point of the Bernoulli curves. Because this proof was the chief aim of Poisson, his theorem is often regarded as an extension of the Bernoulli theorem. Like normal sequences, the Poisson sequences satisfy the relations (25, § 49), but their convergence is quantitatively different. The result will be proved in (17). It is preferable, however, to define the Poisson sequences in the formal calculus by the conditions (2*a*) and (2*b*) in combination with (1); then all the following calculations can be carried out in the formal calculus of probability. But then we lose the possibility of interpreting equation (2) as a regular division.

The inner order of the sequence will be investigated first. Its structure is not yet defined by (1) and (2); we must add Poisson's assumption that the individual subsequences are mutually independent. The assumption is written

$$P(A\,.\,S^{\nu}_{\lambda\kappa}\,.\,B^1_{i_1}\;.\;.\;.\;B^{\nu-1}_{i_{\nu-1}},B^{\nu}_{i_\nu}) = P(A\,.\,S^{\nu}_{\lambda\kappa},B^{\nu}_{i_\nu}) \tag{5}$$

$$i_1\;.\;.\;.\;i_\nu = 1, 2 \qquad B_1 = B \qquad B_2 = \bar{B}$$

Because of (2*b*) we can conclude with the help of (4, § 25) that $S^{\nu+\rho}_{\lambda,\kappa+\rho}$ may be added in the first term; by applying (4, § 25) again, we show that $S^{\nu}_{\lambda\kappa}$ may be dropped in the first term, thus obtaining

$$P(A\,.\,S^{\nu+\rho}_{\lambda,\kappa+\rho}\,.\,B^1_{i_1}\;.\;.\;.\;B^{\nu-1}_{i_{\nu-1}},B^{\nu}_{i_\nu}) = P(A\,.\,S^{\nu}_{\lambda\kappa},B^{\nu}_{i_\nu}) \tag{6}$$

We can now determine the phase probabilities. According to the theorem of elimination (21, § 19), we have, for instance,

$$P(A\,.\,B,B^1) = \sum_{\kappa=1}^{\lambda} P(A\,.\,B,S^1_{\lambda\kappa}) \cdot P(A\,.\,B\,.\,S^1_{\lambda\kappa},B^1) \tag{7}$$

With (5) we obtain

$$P(A\,.\,B\,.\,S^1_{\lambda\kappa},B^1) = P(A\,.\,S^1_{\lambda\kappa},B^1) = P(A\,.\,S_{\lambda\kappa},B) = p_\kappa \tag{8}$$

Now we have, because of (2*b*) and (6, § 25), since we may assume $P(A,B) > 0$,

$$P(A\,.\,B\,.\,S^1_{\lambda\kappa},S_{\lambda,\kappa-1}) = 1 \qquad\qquad P(A\,.\,B\,.\,S_{\lambda,\kappa-1},S^1_{\lambda\kappa}) = 1$$

Using the theorem of multiplication, we derive

$$\begin{aligned} P(A\,.\,B,S^1_{\lambda\kappa}) &= \frac{P(A\,.\,B,S^1_{\lambda\kappa}\,.\,S_{\lambda,\kappa-1})}{P(A\,.\,B\,.\,S^1_{\lambda\kappa},S_{\lambda,\kappa-1})} \\ &= P(A\,.\,B,S_{\lambda,\kappa-1}) \cdot P(A\,.\,B\,.\,S_{\lambda,\kappa-1},S^1_{\lambda\kappa}) \\ &= P(A\,.\,B,S_{\lambda,\kappa-1}) \end{aligned} \tag{9}$$

From Bayes's rule (10, § 21) we obtain

$$P(A.B,S_{\lambda,\kappa-1}) = \frac{P(A,S_{\lambda,\kappa-1}) \cdot P(A.S_{\lambda,\kappa-1},B)}{\sum_{\rho=1}^{\lambda} P(A,S_{\lambda\rho}) \cdot P(A.S_{\lambda\rho},B)}$$

$$= \frac{p_{\kappa-1}}{\sum_{\rho=1}^{\lambda} p_\rho} \tag{10}$$

The relation (7) thus assumes the form

$$P(A.B,B^1) = \frac{\sum_{\kappa=1}^{\lambda} p_\kappa p_{\kappa-1}}{\sum_{\rho=1}^{\lambda} p_\rho} \tag{11}$$

This probability will, in general, differ from $P(A,B)$. With the help of (11) and (4), we obtain for the ratio of these probabilities the value

$$\frac{P(A.B,B^1)}{P(A,B)} = \frac{\lambda \cdot \sum_{\kappa=1}^{\lambda} p_\kappa p_{\kappa-1}}{\sum_{\kappa=1}^{\lambda} p_\kappa \cdot \sum_{\rho=1}^{\lambda} p_\rho} \tag{12}$$

This ratio is, in general, different from 1; the Poisson sequences are thus not free from aftereffect. But the ratio (12) may be smaller or greater than 1; whether the first predecessor induces a tendency to stay or a tendency to change depends on the values and the arrangement of the p_κ. If $\lambda = 2$, (12) is always < 1 (so long as the two p_κ do not have the same value); here there is a tendency to change. For larger λ, however, (12) may assume different values. For instance, if $\lambda = 8$ and $p_1 = p_2 = p_3 = p_4 = \frac{3}{4}$, $p_5 = p_6 = p_7 = p_8 = \frac{1}{4}$, we have $P(A,B) = \frac{1}{2}$, $P(A.B,B^1) = \frac{9}{16}$, that is, the ratio (12) is greater than 1. This result can be explained as follows: the attribute B will result prevalently for elements played with $p_1 \ldots p_4$; the attribute $\bar{B}$, for elements played with $p_5 \ldots p_8$. The B-elements will thus prevalently possess B-elements as successors.

The dispersion is calculated as in § 55; we use for the determination of the frequency the definition of amounts given in (1*a* and 1*b*, § 55). In analogy to (5, § 55), for the first element of the sequence, $M(u)$ is $= p_1$; for the second element it is $= p_2$; and so on. Thus we obtain by addition, as in (4, § 55),

$$M(u[B^1 \ldots B^n]) = \frac{n}{\lambda} \sum_{\kappa=1}^{\lambda} p_\kappa \tag{13}$$

This holds strictly only for a value of n that is a multiple of λ, but it is valid in good approximation for any large n. Furthermore, the dispersion for the first element is $= p_1 \cdot (1 - p_1)$; for the second, it is $= p_2 \cdot (1 - p_2)$; and so on. Thus we obtain by addition, in analogy to (7, § 55),

$$\Delta^2(u[B^1 \ldots B^n]) = \frac{n}{\lambda} \cdot \sum_{\kappa=1}^{\lambda} p_\kappa(1 - p_\kappa) \tag{14}$$

The transition to the relative frequencies, analogous to (11, § 55), gives

$$M(f^n) = \frac{1}{\lambda} \sum_{\kappa=1}^{\lambda} p_\kappa = p \qquad \Delta^2(f^n) = \frac{1}{\lambda n} \sum_{\kappa=1}^{\lambda} p_\kappa(1 - p_\kappa) \tag{15}$$

In order to prove, in analogy to (12, § 55), that the frequency converges toward the mean value p, we use the Tchebychev inequality (18, § 37) and construct the expression

$$w_\delta \geqq 1 - \frac{1}{\lambda n \delta} \cdot \sum_{\kappa=1}^{\lambda} p_\kappa(1 - p_\kappa) \tag{16}$$

With $w_\delta = b_{n\delta}$ (though these quantities are different from the $b_{n\delta}$ of normal sequences) we conclude

$$\lim_{n \to \infty} b_{n\delta} = 1 \tag{17}$$

This relation is equivalent to the relations (25, § 49), as was shown at the end of § 55.

Furthermore, an important result concerning the dispersion follows from (15). We can show that $\Delta^2(f^n)$, according to (15), is always smaller than (in the limiting case, equal to) $\Delta^2(f^n)$, as given by (15, § 52), if the latter dispersion is calculated by using the mean value p, according to (15): *the Poisson sequences always have subnormal dispersion in comparison with a normal sequence of the same frequency.* This conclusion can be drawn by the help of the Schwartz inequality,[1]

$$\left(\sum_{i=1}^{n} a_i b_i\right)^2 \leqq \left(\sum_{i=1}^{n} a_i^2\right) \cdot \left(\sum_{i=1}^{n} b_i^2\right) \tag{18}$$

which for $b_i = 1$ assumes the more special form

$$\left(\sum_{i=1}^{n} a_i\right)^2 \leqq n \cdot \sum_{i=1}^{n} a_i^2 \tag{19}$$

[1] See, for instance, R. Courant and D. Hilbert, *Methoden der mathematischen Physik* (Berlin, 1924), Vol. I, p. 2.

For our purpose we compare the two expressions

$$n \cdot \Delta^2(f^n)_{\text{Poisson}} = \frac{1}{\lambda}\sum_{\kappa=1}^{\lambda} p_\kappa(1 - p_\kappa) = \frac{1}{\lambda}\sum_{\kappa=1}^{\lambda} p_\kappa - \frac{1}{\lambda}\sum_{\kappa=1}^{\lambda} p_\kappa^2 \quad (20a)$$

$$n \cdot \Delta^2(f^n)_{\text{normal}} = p(1 - p) = \frac{1}{\lambda}\sum_{\kappa=1}^{\lambda} p_\kappa - \frac{1}{\lambda^2}\left(\sum_{\kappa=1}^{\lambda} p_\kappa\right)^2 \quad (20b)$$

The first term in each is the same, but the second term, which is to be subtracted, has a larger absolute amount in the first expression because of (19) (it has an equal amount only in the limiting case), and thus we have

$$\Delta(f^n)_{\text{Poisson}} \leqq \Delta(f^n)_{\text{normal}} \quad (21)$$

Since in (18) the equality sign can hold only if the a_i are proportional to the b_i, the equality sign in (21) can result only if all the p_κ have the same value, that is, if the Poisson sequence goes over into a normal sequence.

Because of this property of the dispersion the Poisson sequences represent a case in which classification by the dispersion does not correspond to classification by phase probabilities. The discrepancy originates from the fact that in the Poisson sequences the phase probabilities are not determined by the first predecessor alone. In general,

$$P(A.B.B^1,B^2) \neq P(A.B^1,B^2) \quad (22)$$

Thus we are dealing with a type more general than that of probability transfer. Furthermore, the Poisson sequences are, of course, not regular-invariant, a fact that can be seen from their definition. In another respect, however, the Poisson sequences are more special than sequences with probability transfer, since they can represent only the type of subnormal dispersion and never that of supernormal dispersion.

The latter peculiarity may be illustrated by comparing the Poisson sequences with a type that, in contradistinction to (2), is defined by (1) and the relations

$$P(A,S_{\lambda\kappa}) = \frac{1}{\lambda} \qquad \kappa = 1,2, \ldots \lambda \quad (23)$$

$$P(A.S_{\lambda\kappa_0}.S^1_{\lambda\kappa_1} \ldots S^{\nu-1}_{\lambda\kappa_{\nu-1}}.B_{i_0}.B^1_{i_1} \ldots B^{\nu-1}_{i_{\nu-1}},S^\nu_{\lambda\kappa_\nu}) = P(A,S^\nu_{\lambda\kappa_\nu}) = P(A,S_{\lambda\kappa_\nu}) \quad (24)$$

This definition means that, as before, the main sequence divides into subsequences. The division is not regular, however, but of a random type in the sense of (24). The arrangement can be illustrated as follows. First we draw the ball κ from an auxiliary bowl containing λ balls; then we draw a ball from the bowl $S_{\lambda\kappa}$, in which there are black, B, and white, $\bar{B}$, balls in the ratio p_κ. The procedure is repeated for every element of the sequence. Besides (24), we assume independence, according to (5), in the form

$$P(A \,.\, S^1_{\lambda\kappa_1} \ldots S^\nu_{\lambda\kappa_\nu} \,.\, B^1_{i_1} \ldots B^{\nu-1}_{i_{\nu-1}}, B^\nu_{i_\nu}) = P(A \,.\, S^\nu_{\lambda\kappa_\nu}, B^\nu_{i_\nu}) \tag{25}$$

a condition that is also satisfied by the illustration.

In analogy to (3) and (4) we have

$$P(A,B) = \sum_{\kappa=1}^{\lambda} P(A,S_{\lambda\kappa}) \cdot P(A \,.\, S_{\lambda\kappa}, B) = \frac{1}{\lambda} \cdot \sum_{\kappa=1}^{\lambda} p_\kappa \tag{26}$$

that is, the mean value p represents also in this case the probability of the major sequence. But the phase probabilities result in a different form. We have, for instance,

$$P(A \,.\, B, B^1) = \sum_{\kappa=1}^{\lambda} P(A \,.\, B, S^1_{\lambda\kappa}) \cdot P(A \,.\, B \,.\, S^1_{\lambda\kappa}, B^1) \tag{27a}$$

For the last probability, formula (8) is valid because of (6, § 22); the preceding probability is equal to $P(A,S^1_{\lambda\kappa}) = \frac{1}{\lambda}$ according to (24), and we obtain

$$P(A \,.\, B, B^1) = \frac{1}{\lambda} \cdot \sum_{\kappa=1}^{\lambda} p_\kappa = p \tag{27b}$$

A corresponding conclusion can be derived for any length of groups of predecessors. With (24) and (25) we have

$$\begin{aligned} &P(A \,.\, B_i \ldots B^{\nu-1}_{i_{\nu-1}}, B^\nu) \\ &= \sum_{\kappa=1}^{\lambda} P(A \,.\, B_{i_0} \ldots B^{\nu-1}_{i_{\nu-1}}, S^\nu_{\lambda\kappa}) \cdot P(A \,.\, B_{i_0} \ldots B^{\nu-1}_{i_{\nu-1}} \,.\, S^\nu_{\lambda\kappa}, B^\nu) \\ &= \sum_{\kappa=1}^{\lambda} P(A, S^\nu_{\lambda\kappa}) \cdot P(A \,.\, S^\nu_{\lambda\kappa}, B^\nu) \\ &= \frac{1}{\lambda} \sum_{\kappa=1}^{\lambda} p_\kappa = p \end{aligned} \tag{28}$$

We thus obtain a sequence that is free from aftereffect. It is even a normal sequence if we add to (24) and (25) the assumption that the regular divisions belong to the domain of invariance of these probabilities, that is, (24) and (25) hold likewise in subsequences that are selected by a regular division.

The difference between the Poisson sequences and the sequences defined by (23)–(25) may be illustrated as follows: for the Poisson sequences the individual elements are played with the probabilities $p_1 \ldots p_\lambda$ in a regular order; for the sequences defined by (23)–(25) the individual elements are also played with the probabilities $p_1 \ldots p_\lambda$, but since the selection of the probability for each element is left to chance, there arises a new source of dispersion that makes the dispersion greater than that of the Poisson sequences. Now the dispersion of the sequences (23)–(25) is the normal one. This is apparent, for instance, in (27*a*); since we have there $P(A.B,S^1_{\lambda\kappa}) = P(A,S^1_{\lambda\kappa}) = \frac{1}{\lambda}$, the mean in (27*b*) is taken in the same manner as for the major probability $P(A,B)$ according to (26). In the expression (7), however, the term $P(A.B,S^1_{\lambda\kappa})$ is determined by $p_{\kappa-1}$ according to (9) and (10) because of the regular succession of the $S_{\lambda\kappa}$; it is thus not equal to $P(A,S^1_{\lambda\kappa})$, and so the mean in (11) is formed differently.

For the phase probabilities of the Poisson sequences this may involve a tendency to change or a tendency to stay, depending on the length of the phase with respect to the total period of the p_κ. On the whole, however, these conditions produce a tendency to change; long groups of the same elements will be rare. This result is obvious for $\lambda = 2$. For instance, if $p_1 = \frac{1}{4}$, $p_2 = \frac{3}{4}$, we have $p = \frac{1}{2}$; but since we play alternatively with $\frac{1}{4}$ and $\frac{3}{4}$, $\bar{B}$ is obtained prevalently at one time, B prevalently the next time, with the result that the change from B to $\bar{B}$ is stronger than for the play with constant probability $\frac{1}{2}$. For larger λ, too, a tendency to change must eventually arise. For the example with $\lambda = 8$, groups of 4 successive elements B will still be more frequent than in normal sequences; yet groups of 8 elements B will occur less frequently, since in 8 elements the change of the $p_1 \ldots p_8$ will become noticeable. The regular cycle of the Poisson sequences thus acts in the sense of "a shuffling increased above the normal". It is therefore the probability aftereffect in respect to the $S_{\lambda\kappa}$, expressed in (2*b*), that carries with it the subnormal character of the dispersion of the Poisson sequences.

The conditions of the Poisson sequences are clearly exhibited in lattice enumeration, since the vertical sequences express immediately the p_κ. The resulting lattice is nonconvergent and not lattice-invariant, since because of the independence of the subsequences we have

$$P(A.B^{ki},B^{k,i+\rho})^k = P(A,B^{k,i+\rho})^k \tag{29}$$

in contradistinction to the corresponding phase probability $P(A.B,B^\rho)$. In the lattice representation the definition of the Poisson sequence can be extended even to $\lambda \to \infty$ if we add the condition that the mean value p

$$\lim_{n\to\infty} \frac{1}{\lambda} \sum_{\kappa=1}^{\lambda} p_\kappa = p \tag{30}$$

exists. The amount $n \cdot \Delta^2(f^n)$ then remains finite, as can be seen from (20*a*); therefore the dispersion approaches 0 with increasing n, as in all cases previously considered. Here, too, the dispersion is always subnormal.

The importance of the Poisson sequences for the theory of probability has been overestimated. They have been regarded as supplying a generalization of the Bernoulli theorem; but this generalization has the disadvantage of being one-sided, since these sequences always lead to subnormal dispersion. Furthermore, as shown by (22), the sequences represent a somewhat involved generalization of normal sequences. For these two reasons, sequences with probability transfer are superior in logical significance.

However, the Poisson sequences possess a certain practical importance, particularly in the form extended to infinitely many p_κ according to (30). Consider the transactions of a businessman: his chances of making money will differ from case to case, and are given by probabilities p_κ, which, however, lie between two not-too-distant limits $p^{(1)}$ and $p^{(2)}$ and satisfy (30). Therefore he can expect an average gain p, even with a lower risk than would obtain for constant probability, since the sequence has subnormal dispersion. We shall return in § 72 to these results, which are relevant for the practical value of the frequency interpretation.

§ 57. Bernoulli Sequences

It has been shown in repeated instances that Bernoulli's theorem is satisfied also by sequences other than normal, if the theorem is interpreted as meaning the convergence relations (25, § 49). Conversely, the Bernoulli theorem can be used for the definition of a sequence type of a very general nature.

The definition must be prefaced by some mathematical remarks about the Bernoulli functions b_n and $w_n(f)$, connected by the relation (22, § 49). The latter function was defined in (21, § 49) in terms of the function w_{nm}, which, in turn, was defined by Newton's formula (7, § 49). This formula is restricted to normal sequences and must be abandoned for the general case, whereas the other two formulas can be taken over. We thus assume that for the sequences to be defined there exist probabilities w_{nm}, which, however, may be different for the three modes of enumeration and are given, respectively, by

$$w_{nm} = P(A,F_m^n) \text{ for enumeration with overlapping} \tag{1}$$

$$w_{nm} = P(A\,.\,S_{n1}^1,F_m^n) \text{ for enumeration by sections} \tag{2}$$

$$w_{nm} = P(A,F_m^{kn})^k \text{ for lattice enumeration} \tag{3}$$

The symbols F_m^n and F_m^{kn} have the meanings defined in (6, § 49) and (10a, § 50).

For normal sequences the w_{nm} are determined by the probability p of the sequence. For sequences of more general types the w_{nm} depend on further parameters, such as phase probabilities, probabilities in regular divisions, and lattice probabilities. These parameters may be named $s_1 \dots s_r$, and the Bernoulli density $w_n(f)$ defined in (21, § 49) can therefore be written:

$$w_n(p, s_1 \dots s_r;f) = (n+1) \cdot w_{nm} \qquad f = \frac{m}{n} \tag{4}$$

The semicolon indicates that w_n is a relative probability function. The Bernoulli probabilities b_n depend on the same arguments and may be written

$$b_n(p,s_1 \dots s_r;f_1,f_2) = \int_{f_1}^{f_2} w_n(p,s_1 \dots s_r;f)df \tag{5}$$

They satisfy the relation

$$b_n(p,s_1 \dots s_r;0,1) = \int_0^1 w_n(p,s_1 \dots s_r;f)df = 1 \tag{6}$$

The functions b_n are said to have Bernoulli properties if convergence relations analogous to (25, § 49) hold:

$$\begin{aligned} \lim_{n\to\infty} b_n(p,s_1 \dots s_r;f_1,f_2) &= 1 \qquad \text{for } f_1 \leqq p \leqq f_2 \\ \lim_{n\to\infty} b_n(p,s_1 \dots s_r;f_1,f_2) &= 0 \qquad \text{for } p < f_1 \text{ or } p > f_2 \end{aligned} \tag{7}$$

By means of integrations over the parameters $s_1 \dots s_r$, the following function is constructed:

$$w_n^*(p;f) = \int_{\alpha_1}^{\beta_1} \dots \int_{\alpha_r}^{\beta_r} w_n(p,s_1 \dots s_r;f)ds_r \dots ds_1 \tag{8a}$$

The values α_i and β_i are the end points of the range of s_i and may be functions of p. Correspondingly, a function b_n^* is defined:

$$b_n^*(p;f_1,f_2) = \int_{f_1}^{f_2} w_n^*(p;f)df \tag{8b}$$

This function can be shown to have Bernoulli properties in the form

$$\begin{aligned} \lim_{n\to\infty} b_n^*(p;f_1,f_2) &= b(p) \qquad \text{for } f_1 \leqq p \leqq f_2 \\ \lim_{n\to\infty} b_n^*(p;f_1,f_2) &= 0 \qquad \text{for } p < f_1 \text{ or } p > f_2 \end{aligned} \tag{9}$$

The value 1 of the convergence is here replaced by a function $b(p)$. The relations (9) are derivable from (7) through commutations of the integrations over f and the $s_1 \dots s_r$ and subsequent commutation of the integration over the $s_1 \dots s_r$ and transition to the limit. The admissibility of these commutations is understood.

The assumption is now introduced that the function w_n^* has Bernoulli properties of a second kind, holding for integration over p:

$$k_n(f;p_1,p_2) = \int_{p_1}^{p_2} w_n^*(p;f)dp$$

$$\lim_{n\to\infty} k_n(f;p_1,p_2) = k(f) \quad \text{for } p_1 \leqq f \leqq p_2 \tag{10}$$

$$\lim_{n\to\infty} k_n(f;p_1,p_2) = 0 \qquad \text{for } f < p_1 \text{ or } f > p_2$$

The limit function $k(f)$ is assumed nonvanishing and finite. The convergence relations (10) are derivable from (9) if the functions b_n^* converge *smoothly*, i.e., if, for every interval not containing the critical point, the zero convergence of the area is associated with a uniform zero convergence of the ordinates w_n^*, and if the assumption is added:

$$\lim_{n\to\infty} \int_0^1 w_n^*(p;f)dp = k(f) \tag{11}$$

The relations (10) have the following meaning:

If the cross sections p = const. of the function $w_n^*(p;f)$ have the limit properties of Bernoulli functions for $f = p$ as critical point, the cross sections f = const. have the same limit properties for $p = f$ as critical point. Note that the cross sections p = const. have the staircase shape of histograms, whereas the cross sections f = const. are smooth curves.

The proof of theorem (10) is as follows. Without a specific knowledge of the function w_n^* we cannot say that the cross sections p = const. have a maximum at $f = p$; but from the Bernoulli properties (9) it follows that for increasing n there must arise a maximum close to $f = p$, which converges toward $f = p$ for $n \to \infty$. The ordinate at the critical point is the critical ordinate of the curve; it goes to infinite values for $n \to \infty$, whereas every noncritical ordinate $f \neq p$ goes to 0. For a curve f = const., the ordinate $p = f$ must be the critical ordinate, too, because it is so for the corresponding curve p = const., and thus goes to infinite values. Any other ordinate $p \neq f$ of the curve f = const. is, at the same time, a noncritical ordinate of a certain curve p = const., and thus goes to 0 with increasing n. Since, according to (11), the area between the curve f = const. and the p-axis goes to k with increasing n, the first of the relations (10) follows. The second of these relations is then a consequence.

After these mathematical preparations, I now define Bernoulli sequences as follows:

A sequence of the probability p is a Bernoulli sequence in the wider sense if both for enumeration with overlapping and for enumeration by sections there exist probability functions $w_n(p,s_1 \ldots s_r;f)$ such that the relations (7), (9), and (10) are satisfied.

A lattice of sequences is a lattice of Bernoulli sequences in the narrower sense, or Bernoulli lattice, if each sequence is a Bernoulli sequence in the wider sense, and if lattice probability functions $w_n(p,s_1 \ldots s_r;f)$ exist such that relations (7), (9), and (10) are satisfied.

The definition of the b_n for different modes of enumeration is given by (1)–(3) in combination with (4)–(6). Note that the functions b_n need not be the same.

Using the symbol F_δ^{kn} defined in (10*c*, § 50), we can write the first relation (7) for lattice counting in the form

$$\lim_{n\to\infty} P(A,F_\delta^{kn})^k = 1 \tag{12}$$

According to § 50, this means that the frequency sequences of a Bernoulli lattice form a convergent lattice.

The Bernoulli functions of normal sequences were shown in (20, § 49) and (25, § 49) to have the properties (6) and (7). It can be proved that they have the property (11) also. For normal sequences the function w_n (see 7 and 21, § 49) is identical with w_n^*, since p and f are here the only arguments, and has the form:

$$w_n(p;f) = (n+1)\binom{n}{m} p^m(1-p)^{n-m} \qquad f = \frac{m}{n} \tag{13}$$

This function has the property

$$\int_0^1 w_n(p;f)dp = 1 \tag{14}$$

for every n, so that here $k = 1$. Formula (14) is proved by integration of the function (13) over p. For this purpose we use an auxiliary formula, known from the theory of Euler's integrals; it holds for integers a and b, which are $\geqq 0$, and is derivable by repeated use of integration by parts:

$$\int_0^1 x^a(1-x)^b dx = \frac{a!b!}{(a+b+1)!} \tag{15}$$

By the use of this formula, (14) is easily verified. It follows that the limit properties (10) hold for the Bernoulli functions of normal sequences, in the form:

$$\begin{aligned} k_n(f;p_1,p_2) &= \int_{p_1}^{p_2} w_n(p;f)dp \\ \lim_{n\to\infty} k_n(f;p_1,p_2) &= 1 \quad \text{for } p_1 \leqq f \leqq p_2 \\ \lim_{n\to\infty} k_n(f;p_1,p_2) &= 0 \quad \text{for } f < p_1 \text{ or } f > p_2 \end{aligned} \tag{16}$$

For normal sequences the three kinds of enumeration lead to the same functions b_n, whereas, of course, the functions b_n and k_n have different mathematical forms. The amplified Bernoulli theorem, derived in § 51 for normal sequences, is not included in the definition of Bernoulli sequences and cannot be derived for such sequences without further presuppositions.

The question may be asked whether it is possible to characterize the Bernoulli sequences thus defined in a different manner, for instance, by properties of phase probabilities or by their behavior in respect to regular divisions. No comprehensive answer to the question has yet been found. All we know is that several sequence types belong to the Bernoulli sequences; thus, apart from normal sequences, sequences with probability transfer and Poisson sequences are Bernoulli sequences.

Chapter 8

THEORY OF PROBABILITIES OF A HIGHER LEVEL

THEORY OF PROBABILITIES OF A HIGHER LEVEL

§ 58. Probabilities of a Higher Level

The construction of the probability expressions so far employed represents, in one respect, a special case. Every probability expression contains the probability implication only once: the probability implication stands between terms that are not themselves probability expressions. All the probability expressions previously used are therefore of the type

$$(A \underset{p}{\ni} B) \qquad (1)$$

A and B may be very complicated expressions, but they themselves never contain a probability implication. In the P-notation this restriction is expressed by the fact that within a P-symbol

$$P(A,B) = p \qquad (2)$$

other P-symbols never occur, that is, no P-symbols are contained in A and B.

There are a number of applications that cannot be interpreted by means of probability expressions of this particular form. We find instances in which we do not know for certain which probability exists in a given sequence. We speak, therefore, of *probabilities of the second level;* they are employed in probability statements concerning the existence of a probability. The iteration may be further continued: it is possible to make a probability statement about the existence of a probability of the second level, so that a probability of the third level results, and so on. The operations that are carried out with probabilities of a higher level constitute the *theory of probabilities of a higher level* or the *theory of the hierarchy of probabilities.*

For what problems do we need probabilities of a higher level? The question is linked to the problem of the origin of probability statements and therefore leads to considerations such as are given in § 17. In some cases, for instance, the throwing of dice or the drawing of balls from a bowl, we know the degree of probability before the sequence is produced; we then speak of an *a priori* determination of probability. But a superficial glance tells us that in such cases we do not have a completely reliable knowledge about the degree of probability, if only for the reason that the die may possess an unsymmetrical distribution of weight, or that the mechanism employed may be constructed inaccurately. Closer inspection teaches us that all cases of a so-called *a priori* determination of probability must be analyzed in terms of the theory of levels of probability.

In other cases we speak of an *a posteriori* determination of probability; this is applied when the probability is found by the enumeration of a given sequence. Since we never observe the whole infinite sequence but only a finite initial section of it, it will be impossible to know the limit of its frequency with absolute certainty; so the analysis of the *a posteriori* determination of probabilities and thus of the inductive inference requires the theory of levels of probability. Consequently, the analysis of all probability statements requires the theory of levels; and, in fact, the theory of probabilities of the first level so far developed must be regarded as an approximation, which is applicable when the probabilities of higher levels are very nearly equal to 1. As before, however, discussion of such epistemological considerations will be postponed, and the mathematical theory of probabilities of a higher level will be developed first.

Since a probability of the first level refers to a sequence, a probability of the second level leads to a sequence of sequences, and therefore the lattice is the natural representation of such expressions. To simplify the presentation, it will be assumed throughout that the elements z_{ki} of C or $\bar{C}$ are identical with the elements y_{ki} of B or $\bar{B}$, that is, we have an internal probability implication (see § 9); otherwise we would need two lattices z_{ki} and y_{ki}. The lattice y_{ki} is symbolized in (3)

$$\begin{array}{cccccccccc} x_1 & y_{11} & y_{12} & y_{13} & . & . & . & y_{1i} & . & . \\ x_2 & y_{21} & y_{22} & y_{23} & . & . & . & y_{2i} & . & . \\ . & . & . & . & . & . & . & . & . & . \\ x_k & y_{k1} & y_{k2} & y_{k3} & . & . & . & y_{ki} & . & . \\ . & . & . & . & . & . & . & . & . & . \end{array} \tag{3}$$

The elements y_{ki} belong to B or $\bar{B}$ and in another classification to C or $\bar{C}$, so that we have a probability implication

$$(B^{ki} \underset{p_\rho}{\ni} C^{ki})^i \tag{4}$$

We write p_ρ for the degree of probability in order to indicate that this probability is not constant for the lattice; the probabilities of the horizontal sequences differ from sequence to sequence, that is, the lattice is horizontally inhomogeneous. The implicational form of writing is used because expressions of a new logical structure are to be developed. Later the results will be translated into the P-notation.

Since we wish to construct the probability of a probability, the expression (4) is to represent the second term within a probability implication, and so we must now explain the first term. For this purpose we have added in (3) the column $x_i \ldots x_k \ldots$, the elements of which correspond each to a horizontal row of the lattice; the elements x_k belong to A or $\bar{A}$. The probability to be constructed goes from an element x_k belonging in A to the existence of a probability p_ρ in the corresponding horizontal sequence. Thus we have, in the detailed notation, the following form for the inquired probability:

$$(k)\ (x_k \epsilon A \underset{q_\rho}{\ni} [(i)\ (y_{ki} \epsilon B \underset{p_\rho}{\ni} y_{ki} \epsilon C)]) \tag{5}$$

For the formal calculus of probability the probability p_ρ is a number coördinated to a horizontal sequence, and likewise the probability q_ρ is a number coördinated to a sequence of sequences. In the frequency interpretation (5) means that among the horizontal sequences are found sequences with the frequency limit p_ρ, the frequency of which, counted vertically, goes in the limit toward q_ρ.

As in (2 and 3, § 34), we translate (5) into the abbreviated notation by adding the subscript of the elements as superscript to the class symbols and repeating the running superscript (the bound variable) outside the parentheses:

$$(A^k \underset{q_\rho}{\ni} (B^{ki} \underset{p_\rho}{\ni} C^{ki})^i)^k \tag{6}$$

The technical relation to (1) is apparent: (6) results from (1) if in (1) the expression (4) is substituted for B and the superscripts are added. The addition is necessary because of the lattice form.

We can translate (6) into the P-notation:

$$P(A^k,[P(B^{ki},C^{ki})^i = p_\rho])^k = q_\rho \tag{7}$$

In analogy to (7) a probability of the third level would be written

$$P(A^l,P(B^{lk},[P(C^{lki},D^{lki})^i = p_\rho])^k = q_\rho)^l = r_\rho \tag{8}$$

There are other kinds of probabilities of a higher level, which result when a probability expression is introduced into the first term of another probability expression. We then have expressions of the form

$$(A^k . (B^{ki} \underset{p_\rho}{\ni} C^{ki})^i \underset{r_\rho}{\ni} D^k)^k \tag{9}$$

or, in the P-notation,

$$P(A^k . [P(B^{ki},C^{ki})^i = p_\rho],D^k)^k = r_\rho \tag{10}$$

This expression means that we add to the lattice (3) a vertical column of the elements z_k and consider the probability from the existence of a horizontal sequence with the probability p_ρ to the existence of a corresponding $z_k \epsilon D$. Expressions of this form must be included in the category of probabilities of a higher level.

Probabilities of a higher level must be carefully distinguished from probabilities of a higher kind, which were discussed with respect to the Bernoulli theorem (§ 50). In probabilities of a higher kind there occur within the P-symbol only frequency expressions referring to a finite number of elements, that is, to finite sections of sequences or to finite lattice sections.[1] Their existence can always be verified by the enumeration of a finite number of elements. This fact is clear from table 5 (p. 279), and holds even for the interpreted conception, since throughout it is only the highest probability, that is, the probability of the whole P-symbol, that refers to infinite sequences. Probabilities of a higher kind are, therefore, probabilities of the first level; they contain within the P-symbol only expressions of a *finite reference*. Probabilities of the second level, however, contain within the P-symbol expressions of an *infinite reference*, since the probability occurring inside the symbol refers to an infinite sequence and therefore can be verified only after the enumeration of an infinite sequence.

The transition to probabilities of the second level finds a technical expression in the fact that the lattices are horizontally inhomogeneous, whereas in the preceding chapters they were always homogeneous in the horizontal direction. It is true that we sometimes deal with vertically inhomogeneous lattices; and since vertically homogeneous lattices will be used occasionally in the following considerations, there would be no difference in such cases, since the two lattice directions are mathematically equivalent. For convenience in notation, however, a distinction will be made between horizontally inhomogeneous and vertically inhomogeneous lattices. The sequences that, taken as wholes, are elements of a frequency enumeration of sequences will always be written as horizontal sequences. On this condition, only the horizontally inhomogeneous lattice represents the transition to probabilities of the second level, whereas the horizontally homogeneous lattice belongs to probabilities of the first level. The distinction means no more than the convention not to ask questions concerning the frequency of vertical sequences with respect to the last-mentioned lattice.

§ 59. Constants in Probability Expressions

The introduction of probabilities of a higher type requires an extension of the calculus that will be developed first in the implicational notation. It will be necessary to introduce into probability statements expressions that do not depend on a subscript or a superscript and, therefore, represent *constants*. The introduction is possible if the constant is regarded as a function of the variable i having the same value for all i, a conception that is familiar from

[1] Only in § 51 did we deal with a probability of a different type, since the probability $c_{n\delta}$ considered there is a probability of a higher level.

mathematical constants. As an illustration of a probability expression containing a constant, consider the probability that, when a die is thrown, the "6" shows up *and* Ohm's law holds for electrical currents. This probability has the same value as the probability of the occurrence of face 6 alone, since Ohm's law is always valid. When D is the constant, this probability has the form

$$(A^i \underset{p}{\ni} B^i . D)^i \tag{1}$$

A similar form results if the expression introduced contains a variable k, but does not contain the bound variable i of the probability expression, as in

$$(A^k . B^{ki} \underset{p}{\ni} C^{ki})^i \tag{2}$$

Here A^k plays the same role as D in (1). Furthermore, the expression introduced may contain a bound superscript:

$$(A^i . (B^k)^k \underset{p}{\ni} C^i)^i \tag{3}$$

Here $(B^k)^k$, that is, $(k)(y_k \epsilon B)$, has the meaning of a constant. It is even possible that one term of the probability expression does not contain a variable superscript, so that there result expressions of the form

$$(D \underset{p}{\ni} B^i)^i \tag{4a}$$

$$(A^i \underset{p}{\ni} D)^i \tag{4b}$$

The variable superscript is given by the other term, and the constant term has the value D for all i.

For operations with constants in probability expressions no additional axioms are required; the rules of operation follow from the axioms of the calculus in combination with the general rules of symbolic logic if such expressions as (1)–(4) are admitted as meaningful.

For the interpretation of expressions like (1), (4*a*), (4*b*), we must realize that capital letters in probability symbols represent classes, not propositions. According to a remark made at the end of § 7, however, we can speak of the extension of a proposition; it is either the universal class or the null class, depending on whether the proposition is true or false. This interpretation follows because if d is a proposition we regard the expression $(x)d$ as meaningful and equivalent to d. The constant D in (1), (4*a*), or (4*b*), therefore, represents the universal class or the null class, whatever the proposition d may mean. Since the transition from the class symbol to the propositional symbol is made, in the notation used in this work, by the addition of parentheses, the proposition d is to be written in the form (D), a form that means originally $(i)D$ and thus is the same as d. But if the constant term inside the parentheses is originally a proposition, like the term $(B^k)^k$ in (3), such terms will be considered to represent the corresponding class, that is, the universal class or the null class. That in such cases the notation does not supply an independent distinction between propositions and classes, making this distinction dependent on the occurrence of the expression within parentheses, seems without danger because of the isomorphism between the calculus of propositions and that of classes, although, of course, such a notation would not be expedient for other purposes.

Consider the expression (4*b*). The proposition d, or (D), coördinated to the class D, will be either true or false; correspondingly, we have either $(A^i \supset D)^i$ or $(A^i \supset \bar{D})^i$, a disjunction that expresses merely the *tertium non datur*. Applying axiom II,1 and (9, § 13), we thus derive

$$(A^i \underset{p}{\ni} D)^i \supset [(p = 1) \vee (p = 0)] \tag{5}$$

The alternative $p = 1$ holds if (D) is true; the alternative $p = 0$ holds if (D) is false. Furthermore, we have the tautology $(D \supset (A^i \supset D)^i)$, which is a generalization of the logical formula (8*c*, § 4). By means of (1, § 25) we thus derive

$$(D \supset [(A^i \underset{p}{\ni} B^i)^i \equiv (D.A^i \underset{p}{\ni} B^i)^i]) \tag{6}$$

Because of $(D \supset (B^i \supset D)^i)$ we derive from (4, § 25)

$$(D \supset [(A^i \underset{p}{\ni} B^i)^i \equiv (A^i \underset{p}{\ni} D.B^i)^i]) \tag{7}$$

Putting in (6) the special value $(A^i)^i$ for D, we obtain the expression $(A^i)^i.A^i$ on the right side; when we apply (1, § 25) a second time, we see that this expression may be replaced by $(A^i)^i$, or $(A^m)^m$. We thus derive

$$((A^i)^i \supset [(A^i \underset{p}{\ni} B^i)^i \equiv ((A^m)^m \underset{p}{\ni} B^i)^i]) \tag{8}$$

If the sequence A is compact, we may regard it as a constant if it occurs in the first term of the probability expression.

With (6*d* and 7*a*, § 4) and the help of some simple transformations, we derive from (6) the two formulas

$$(D.(A^i \underset{p}{\ni} B^i)^i \supset (D.A^i \underset{p}{\ni} B^i)^i) \tag{9}$$

$$(D.(A^i \underset{p}{\ni} B^i)^i \equiv D.(D.A^i \underset{p}{\ni} B^i)^i) \tag{10}$$

To (9) we can add

$$(\bar{D} \supset (D.A^i \underset{p}{\ni} B^i)^i) \tag{11}$$

This follows from (7, § 12). Formulas (9) and (11) together give

$$(A^i \underset{p}{\ni} B^i)^i \supset (D.A^i \underset{p}{\ni} B^i)^i \tag{12}$$

The truth of the formula is easily seen when we consider the meaning of D: if D is the universal class, its addition on the right side does not change the reference class A; if D is the null class, the reference class on the right side is empty and the corresponding probability may have any value, so that the implication holds. By similar considerations the meaning of other formulas may easily be explained.

In formulas (9), (11), (12) the expression $(D.A^i \underset{p}{\ni} B^i)^i$ is implied by a stronger expression; in (6) its equivalence to another expression is linked to a condition. We now wish to find an expression that represents an unrestricted equivalence to $(D.A^i \underset{p}{\ni} B^i)^i$. It is supplied by the formula

$$(D \supset (A^i \underset{p}{\ni} B^i)^i) \equiv (D.A^i \underset{p}{\ni} B^i)^i \tag{13}$$

The implication from left to right contained in the equivalence sign is proved by transforming the implication written on the left side into $(\bar{D} \vee (A^i \underset{p}{\ni} B^i)^i)$; since both terms of the disjunction imply the right side of (13) according to (11) and (12), we derive that the implication holds. The implication from right to left is inferred from (10) by the use of the implication from right to left contained in the equivalence sign of (10), the dropping of D on the left side, and the application of (6*d*, § 4).

An important special case of (13) obtains if D is replaced by $(A^m \underset{p}{\ni} B^m)^m$. Then the left side of (13) becomes a tautology, and therefore the right side of (13), taken alone, also represents a formula that is always true:

$$(A^i.(A^m \underset{p}{\ni} B^m)^m \underset{p}{\ni} B^i)^i \tag{14}$$

Here the existence of the probability is incorporated in the first term as condition, so that a tautology results.

Formula (13) may be regarded as an extension of the logical formula

$$a \supset (b \supset c) \equiv a.b \supset c \tag{15}$$

In (13) the second implication sign on the left side of (15) is replaced by a probability implication, and consequently the implication on the right side of (15) also is replaced by a probability implication. We may ask whether a further extension can be constructed for the case in which the first implication sign on the left side of (15), too, is replaced by a probability implication.

The question leads us back to probabilities of the second level (see 6, § 58), and will be discussed in § 60.

The formulas of the present section cannot be expressed exhaustively in the P-notation, since they contain logical symbols outside the parentheses of the probability implication. Formulating the meaning of the symbols in words, the most important formulas for the P-notation are

$$P(A^i,D)^i = \begin{cases} 1 \text{ for } (D) \\ 0 \text{ for } (\bar{D}) \end{cases} \tag{5'}$$

If (D) is true, we have

$$P(A^i,B^i)^i = P(A^i.D,B^i)^i \tag{6'}$$

$$P(A^i,B^i)^i = P(A^i,B^i.D)^i \tag{7'}$$

Formula (13) cannot be written in this manner, but the formula will be used so far as we introduce permission to write the equivalent expression $P(A^i.D,B^i)^i$ in place of the expression occurring on the left side.

§ 60. Operations with Probabilities of the Second Level

Operations with probabilities of the second level are determined entirely by the rules of operations with probabilities of the first level, when the rules for constants in probability expressions are added. Probabilities of the first level occurring inside probability expressions of the second level, taken as wholes, are treated in the same manner as other expressions, and are combined according to the rules of symbolic logic with one another or with other expressions. If the probability expression of the first level contains a free variable k, as in the expression $(B^{ki} \underset{p_\rho}{\ni} C^{ki})^i$ in (9, § 58) or the expression $P(B^{ki},C^{ki})^i = p_\rho$ (10, § 58), it is regarded as the class of all situations x_k to which the expression applies. If it contains no free variable, like the expression $(A^m \underset{p}{\ni} B^m)^m$ in (14, § 59), it is treated according to the rules for constants.

That we do not introduce new axioms for the treatment of probabilities of a higher level leads to the consequence, however, that the possible operations are rather restricted. In probability expressions like (6, § 58), the superscripts i and k indicate two different directions of counting, and no inferences exist that would transform a probability enumerated horizontally into a probability enumerated vertically. The absence of such inferences is made obvious by the frequency interpretation: in dealing with lattice formation (§ 34), we pointed out that the limits of frequencies in the horizontal direction do not determine those in the vertical direction. The convergent lattice was offered as an instance in which the limits in the vertical direction are different from those in the horizontal direction. As explained on p. 280, we can imagine an even more general lattice, in which, for instance, all elements on the left of the lattice diagonal are given as B, and normal sequences are added on the right in the horizontal direction. Then all vertical sequences possess the frequency limit 1; all horizontal sequences, however, have a frequency limit p.

The restriction of possible inferences is apparent when we investigate the question raised at the end of § 59. With the use of (3, § 25) and (10, § 59) we can transform (6, § 58) into

$$(A^k \underset{q}{\ni} (B^{ki} \underset{p}{\ni} C^{ki})^i)^k \equiv (A^k \underset{q}{\ni} A^k . (B^{ki} \underset{p}{\ni} C^{ki})^i)^k$$

$$\equiv (A^k \underset{q}{\ni} (A^k . B^{ki} \underset{p}{\ni} C^{ki})^i)^k \qquad (1)$$

But this is all that can be achieved without further assumptions; the two probabilities p and q cannot, in general, be combined into one probability, because they refer to frequencies counted in different directions.

Only when we consider more special cases is it possible to make inferences from one direction to the other. In earlier sections we studied more special lattices possessing determinate probabilities in both directions. In such lattices the direction of the enumeration of those probabilities of the first level that are counted vertically coincides with the direction of the enumeration of probabilities of the second level. In this case some conclusions can be drawn: we are able to infer relations holding between vertical probabilities of the first level and probabilities of the second level; we can also make inferences from horizontal probabilities of the first level to vertical probabilities of the second level. Such inferences are possible when certain relations between the two directions of enumeration are assumed.

We introduce, first, the simplifying assumption that the sequence A and the lattice B are compact, that is, all x_k belong to A and all y_{ki} belong to B. This simplification does not restrict the generality, since it can always be carried through by a suitable choice of the enumeration. For the P-notation, however, a considerable advantage is thus achieved.[1] Assume, furthermore, the presence of horizontal sequences in the lattice to which belong the horizontal probabilities $p_1 \ldots p_\tau$ in respect to C. We now define a subclass B_ρ of B by the condition that it include those y_{ki} that belong to a horizontal sequence in which the probability of C is $= p_\rho$. A horizontal sequence of the y_{ki} of this kind will be called a *C-sequence of the kind B_ρ*. In symbols

$$(y_{ki} \in B_\rho) =_{Df} (y_{ki} \in B) . (i)(y_{ki} \in B \underset{p_\rho}{\ni} z_{ki} \in C) \qquad (2)$$

[1] In the noncompact lattice B, complications will result if there are horizontal rows in which no element B occurs. According to (7, § 12) the probability $P(B^{ki},C^{ki})^i$ has every numerical value for such rows; therefore we must count such rows in enumerating the probability of the second level, and we must do so for every p_ρ. These conditions do not correspond to what we want to measure, and they create difficulties for the calculation, since the cases B_ρ and B_σ would not be exclusive for $\rho \pm \sigma$.

or abbreviated as

$$B_\rho^{ki} =_{Df} B^{ki}.(B^{ki} \underset{p_\rho}{\ni} C^{ki})^i \tag{2'}$$

Since the sequence B is compact, we have

$$((B^{ki} \underset{p_\rho}{\ni} C^{ki})^i \supset B^{ki})^k$$

With (4, § 25) we now derive

$$(A^k \underset{q_\rho}{\ni} (B^{ki} \underset{p_\rho}{\ni} C^{ki})^i)^k \equiv (A^k \underset{q_\rho}{\ni} B^{ki}.(B^{ki} \underset{p_\rho}{\ni} C^{ki})^i)^k$$

$$\equiv (A^k \underset{q_\rho}{\ni} B_\rho^{ki})^k \tag{3}$$

By means of this transformation the probability of the second level has assumed the form of a probability of the first level in which a free variable i occurs; a formula of this kind is valid for every i. Formula (3) enables us to give a simple representation of the probability of the second level in the P-notation:

$$P(A^k, B_\rho^{ki})^k = q_\rho \tag{3'}$$

This formula means that in every vertical column i the probability of finding an element y_{ki} that belongs to B_ρ is $= q_\rho$. This probability is the same for all columns i because, if one element of a horizontal sequence belongs to B_ρ, so do all. Note that for this reason formula (3′) is not true if the superscript i is chosen as a running superscript. We rather have, because of (2′),

$$P(A^k, B_\rho^{ki})^i = \text{either 1 or 0} \tag{3''}$$

depending on whether the horizontal sequence k is of the kind B_ρ. However, a simplification results for the horizontal sequences. Because of (14, § 59) and with the use of (2) we can write

$$(B_\rho^{ki} \underset{p_\rho}{\ni} C^{ki})^i \tag{4}$$

or, in the P-notation,

$$P(B_\rho^{ki}, C^{ki})^i = p_\rho \tag{4'}$$

The expressions are true for every value k because, if k denotes a sequence for which the horizontal probability is not p_ρ, the reference class is empty and thus any value of the horizontal probability may be asserted. Both (4) and (4′) are probabilities of the second level, since they contain, after the elimination of the abbreviation B_ρ according to (2), a probability of the first level in the first term. The character of a probability of the second level may be indicated by the Greek subscript in B_ρ^{ki}, in contrast to the Latin subscripts used previously. In the following formulas we can return to the use of the P-notation.

In order to make inferences possible, we assume, not only that the horizontal and vertical probabilities p_ρ and q_ρ exist, but also that they satisfy the two conditions:

a. The disjunction $B_1 \vee \ldots \vee B_\tau$ is complete, that is,

$$\sum_{\rho=1}^{\tau} P(A^k, B_\rho^{ki})^k = \sum_{\rho=1}^{\tau} q_\rho = 1 \tag{5}$$

That the disjunction is exclusive follows from the univocality of the probability implication.

b. The sublattice of the C-sequences of the kind B_ρ is homogeneous for every ρ, so that

$$P(B_\rho^{ki}, C^{ki})^i = P(B_\rho^{ki}, C^{ki})^k = p_\rho \tag{6}$$

Assumption *b* is the decisive assumption, enabling us to make an inference that reaches beyond (1). We obtain first, since $(B_\rho^{ki} \supset A^k)^k$ because of the compact character of the sequence A,

$$P(A^k . B_\rho^{ki}, C^{ki})^k = P(B_\rho^{ki}, C^{ki})^k = P(B_\rho^{ki}, C^{ki})^i = p_\rho \tag{7}$$

Furthermore, from the rule of elimination (21, § 19) we derive

$$P(A^k, C^{ki})^k = \sum_{\rho=1}^{\tau} P(A^k, B_\rho^{ki})^k \cdot P(A^k . B_\rho^{ki}, C^{ki})^k \tag{8a}$$

$$= \sum_{\rho=1}^{\tau} P(A^k, B_\rho^{ki})^k \cdot P(B_\rho^{ki}, C^{ki})^k \tag{8b}$$

$$= \sum_{\rho=1}^{\tau} P(A^k, B_\rho^{ki})^k \cdot P(B_\rho^{ki}, C^{ki})^i \tag{8c}$$

$$= \sum_{\rho=1}^{\tau} q_\rho p_\rho = p \tag{8d}$$

The formula contains probabilities of the second level on its right side, and a probability of the first level on the left side. Thus it represents a relation between probabilities of different levels. The relation is possible because the probability expression B_ρ^{ki}, which occurs within the P-symbols on the right side, is eliminated according to the rule of elimination. In the form (8*c*) the formula represents a relation between probabilities counted in different directions; this relation is possible because a corresponding relation is given in the formula (6).

Returning to the implicational notation, we can write the result in the form: if the conditions (5) and (6) are satisfied, the formula holds:

$$(\rho)\,[(A^k \underset{q_\rho}{\ni} (B^{ki} \underset{p_\rho}{\ni} C^{ki})^i)^k] \supset (\exists p)\,(i)\,(A^k \underset{p}{\ni} C^{ki})^k . (p = \sum_{\rho=1}^{\tau} q_\rho p_\rho) \qquad (9)$$

The bound variable ρ runs through the values $1 \ldots \tau$ given in (5). Comparing (9) with (1), we see that, on the conditions stated, two probability implications arranged in series can be replaced by one probability implication the degree of which is given by a mean value constructed from all the probabilities involved. Without the conditions mentioned, however, no such replacement is possible. For the clarification of formula (9) it may be remarked that the all-operators referring to ρ and i are added because the formula cannot be written with the use of the free variables ρ and i.[2] The operators express the facts that only when the left side is true for all values ρ does it imply the right side, and that the value p of (8*d*) holds for all values i, that is, for all columns.

The result (8) bears some resemblance to the result (26, § 56) obtained for the sequences defined by (23–25, § 56), if we write in (8) $S_{\lambda\kappa}$ instead of B_ρ and B for C; the latter sequences differ from (8) only in the fact that all the $P(A,S_{\lambda\kappa})$ are equal. Yet the two expressions are fundamentally different. The symbol $S_{\lambda\kappa}$ denotes an attribute that is attached to all elements of the subsequence, or, in lattice enumeration, to all elements of the lattice sequence. In the illustration given for the assumptions (23–25, § 56), $S_{\lambda\kappa}$ represents the property of being drawn from the bowl $S_{\lambda\kappa}$, an attribute that can be verified directly for every element of the sequence. The sign B_ρ, however, does not represent an attribute that an individual element y_{ki} possesses in virtue of an individual characteristic. In order to find out whether y_{ki} belongs to B_ρ we must know the sequence of elements y_{ki} (i running subscript) completely and we must know also the probability that exists in it with respect to C. This meaning of B_ρ is seen from its definition in (2). Therefore, the classification B_ρ is a characterization with *infinite means*, whereas the classification $S_{\lambda\kappa}$ represents a characterization by *finite means*. This analysis corresponds to the distinction between probabilities of the first and of the second level, which was formulated in § 58 in the statement that the first contain only finite expressions in the P-symbol, whereas the latter include infinite expressions in the P-symbol.

[2] If we omit the all-operators and write ρ and i as free variables, we can put all-operators referring to ρ and i before the whole formula, according to the rule for free variables; but the formula thus resulting does not have the meaning of (9). This follows from the formula (11*d*, § 25) given in *ESL*, p. 135; see also pp. 107 and 144 of that book. Because of the restriction of the values ρ to the values $1 \ldots \tau$ given in (5), the operator referring to ρ is a restricted operator (see *ESL*, p. 162).

The resemblance of the two kinds of probability has an obvious reason: if the classification of the elements of a sequence by simple observational criteria leads to subsequences of such a kind that the probability of a certain attribute within the subsequence is known, the probability of the observational criterion in the major sequence is translatable into the probability of a subsequence of a certain probability and thus into a probability of a higher level. But it is an intrinsic difference whether the definition of the subsequence is given in terms of an observational criterion of its elements or in terms of the probability of the subsequence as a whole. Only in the latter case do we speak of a probability of a higher level. Formally speaking, however, we may use the symbol B_p like a class symbol representing a classification by finite means. But the theory of such probabilities cannot be constructed without a lattice. Only if $S_{\lambda\kappa}$ is an individual attribute are we able to say that an element of the major sequence belongs to the subsequence $S_{\lambda\kappa}$. If we omitted the indication $S_{\lambda\kappa}$ for the elements of the major sequence (23–25, § 56), we would not know into which subsequence a given element was to be incorporated. Therefore, we are always dependent on lattice formation for probabilities of the second level.

Probabilities of the second level are distinguished from those of the first level in a further sense. That we are able to interpret the mean value p in (8*d*) as a probability—namely, as a probability in the vertical direction[3]—derives from the special condition (6) assumed for the lattice. The bowl schema is representative of such a lattice only when the drawing from a bowl satisfies the condition (6), which it usually does. Otherwise we are confronted by a lattice of a different kind, in which the mean value p, computed according to (8*d*), cannot be interpreted as a probability; and the two probabilities of

[3] We might suppose that it is possible to interpret the value p as a probability in another sense if we count the two-dimensional lattice in a one-dimensional arrangement, for instance, in the enumeration

$$z_{11}\ z_{12}\ z_{22}\ z_{21}\ z_{31}\ z_{32}\ z_{33}\ z_{23}\ z_{13}\ z_{14}\ \ldots$$

But from the assumptions mentioned we cannot derive that such an enumeration must result in the limit p for the frequency. This impossibility is easily demonstrated by the construction of an opposite instance: draw the lattice lines running through $z_{11}\ z_{23}\ z_{35}\ \ldots$ and $z_{11}\ z_{32}\ z_{53}\ \ldots$ respectively, and put C's in the places of all the z_{ki} lying between the lines; then all the vertical and horizontal sequences can be supplemented so as to form normal sequences, in the narrower sense, with the frequency limit $p < \frac{1}{2}$. Then we can easily infer that the specified enumeration results in a frequency limit $> p$, since the C-elements in the inner sector always amount to approximately one-half of all elements. Conclusions in regard to the frequency resulting from the enumeration can be drawn only if we introduce assumptions about a uniform convergence of all the horizontal sequences, that is, if for every δ there exists an n such that for all k

$$|\, p - F^n(B^{ki}, C^{ki})^i \,| < \delta$$

holds. But we do not wish to introduce such assumptions (see § 65).

If it should be found objectionable, by the way, that normal sequences in the narrower sense admit of "abnormal" lattices like the one mentioned, the definition of normal sequences in the narrower sense can be further restricted by suitable conditions—for instance, the condition that the same limit of the frequency should exist in every lattice line, that is, in every straight line drawn through the points of the lattice.

different levels occurring in a probability expression of the second level like (6, § 58) cannot be combined in one probability. Here we must be satisfied with the statement of a pair q_ρ, p_ρ, or, when ρ is varied, of a series of such pairs.

But even when the mean value of this pair-series can be interpreted as a probability, it does not provide a complete substitute for the pair-series. Only for the vertical direction does it represent a probability; for the characterization of the horizontal sequences two probabilities, or a pair-series of probabilities, are always required. Probability expressions of the second level are always characterized by a pair, or by a pair-series, of degrees of probability. This necessity is clearly expressed in (1): if we know that A and B exist, what we wish to state about C can be expressed only by the two probabilities p_ρ and q_ρ. If it is true that we look forward to a chance event with a definite feeling of expectation, the intensity of which stands in a certain relation to the magnitude of the probability, we must admit that for probabilities of the second level we have to adjust our feeling of expectation in two different directions.

Such a double adjustment, in fact, is recognizable in our behavior. If, before a horse race, a well-versed expert of the sport tells us that the winning chances of the favorite amount to 80%, and another racing fan, more enthusiastic than expert, claims the same probability for the victory of the favorite, we shall evaluate the two identical statements differently: we place more trust in the statement of the expert. This means that his statement has a higher probability of the second level. It is obvious, however, that the higher probability of the second level is not expressible by a change in the probability of the first level: we must not assume the probability of the victory of the favorite as smaller or greater if our only basis is the information given by the inexperienced fan. What is smaller is solely the probability that the probability of 80% is correct. If we have no better information, we should rather refrain from betting than bet on the basis of a value other than 80%.

§ 61. The Dispersion in a Horizontally Inhomogeneous Lattice

The mean value calculated in (8*d*, § 60)

$$p = \sum_{\rho=1}^{\tau} q_\rho p_\rho \qquad (1)$$

cannot be interpreted as a probability in the horizontal direction. However, for this direction, too, p is a quantity of mathematical import if the dispersion is calculated.

Let δf^{kn} be determined in analogy to (12*a*, § 52) by

$$\delta f^{kn} = f^{kn} - p \qquad (2)$$

Then $\Delta^2(f^n)$ is defined as in (13a, § 52) by

$$\Delta^2(f^n) = M(\delta^2 f^{kn})^k \tag{3}$$

Although, for a lattice that is homogeneous in the horizontal direction, $\Delta(f^n)$ approaches 0 with increasing n, as can be seen from the theoretical value (15, § 52), this is no longer true of horizontally inhomogeneous lattices. For the limit $n \rightarrow \infty$ there appear deviations

$$\delta_\rho = p - p_\rho \tag{4}$$

of the frequencies of the C-sequences of the kind B_ρ; therefore we have, for the limit $n \rightarrow \infty$,

$$\Delta(f^\infty) = \sqrt{\sum_{\rho=1}^{\tau} \delta_\rho^2 \cdot P(A^k, B_\rho^{ki})^k} = \sqrt{\sum_{\rho=1}^{\tau} \delta_\rho^2 q_\rho}$$

$$= \sqrt{\sum_{\rho=1}^{\tau} (p - p_\rho)^2 q_\rho} \tag{5}$$

The mean value p defined in (1) is distinguished from all other values by the fact that $\Delta(f^\infty)$ becomes a minimum if this value is taken as the point of reference; for any other choice of p, formula (5) leads to a larger $\Delta(f^\infty)$. This follows from (12, § 37).

Therefore, replacement of the frequencies of horizontal sequences of an inhomogeneous lattice by the mean value p has a certain meaning: by this replacement we commit the smallest mean error.

The dispersion calculated with the mean value p becomes supernormal for larger n; this follows from the fact that it does not go toward 0 with increasing n. But if we want to infer the inhomogeneous character of the lattice from the dispersion, it is not sufficient to ascertain its supernormal character, since a supernormal dispersion may occur likewise in homogeneous lattices, for instance, in sequences with probability drag. Besides this result, we must therefore investigate statistically whether the dispersion approaches zero with increasing n. This is feasible to the same degree as the ascertainment of limits in general. For the mean value p we may take, for instance, the mean value of the frequencies in the vertical sequences. The investigation of the dispersion thus presents a method by which we can conclude whether the lattice is inhomogeneous.

The dispersion of the total lattice can be calculated for finite n if we assume that for each ρ the sublattice of the C-sequences of the kind B_ρ is normal in the narrower sense. Considering first one of these sublattices we obtain, with (12, § 37) and (15, § 52), as its dispersion referred to p,

$$\Delta_\rho^2(f^n) = \frac{p_\rho(1 - p_\rho)}{n} + (p - p_\rho)^2 \tag{6}$$

The dispersion of the total lattice, likewise referred to p, is the average of the dispersions of the sublattices, that is,

$$\Delta^2(f^n) = M(\Delta^2_\rho(f^n))_\rho = \sum_{\rho=1}^{\tau} q_\rho \cdot \left[\frac{p_\rho(1-p_\rho)}{n} + (p - p_\rho)^2\right] \tag{7}$$

With the help of (1) this result can be transformed, by intermediate calculations, into

$$\Delta(f^n) = \sqrt{\frac{p(1-p)}{n} + \frac{n-1}{n} \cdot \sum_{\rho=1}^{\tau} q_\rho(p-p_\rho)^2} \tag{8}$$

The expression approaches the limit (5) for increasing n; for every finite n the value $\Delta(f^n)$ is greater than $\Delta(f^\infty)$ and, at the same time, is greater than the normal dispersion calculated with reference to p according to (15, § 52). For more general types of lattices, the relation (8) is to be replaced by a different relation; but all such formulas must correspond to (8) so far as they must become identical with (5) for the limit $n \to \infty$.

§ 62. The Inductive Inference

In respect to horizontally inhomogeneous lattices there arises a question that has no analogue for homogeneous lattices. Given a finite initial section of a horizontal sequence (assumed as standing in the k-th row) for which the frequency $F^n(B^{ki},C^{ki})^i$ has the value $f = \frac{m}{n}$, what is the probability that the sequence is controlled by a probability $p = f \pm \delta$, or in the frequency interpretation, that the sequence, upon prolongation, will converge toward a limit of the frequency within the interval $f \pm \delta$? Since the question concerns an inference from a given finite initial section to the infinite remainder of the sequence, it refers to an *inductive inference* (see § 17).

The question cannot be answered unless certain specializing conditions are introduced. The range of possible values for p, which goes from 0 to 1, may be divided into small intervals of the length dp. A sequence of a probability that lies within the interval from p to $p + dp$ will be called a sequence of the kind $B_{p,dp}$, this symbol taking over the function of the symbol B_ρ previously used. If the term $B_{p,dp}$ denotes the attribute class, the formula computed for a precise value p is a probability density and requires multiplication by dp to become a probability. If the term stands for the reference class, the formula computed for a precise value p is a probability and must not be multiplied by dp; a relative probability function can have a precise value in the reference class (see § 45). We thus have, instead of (4′, § 60),

$$P(A.B^{ki}_{p,dp},C^{ki})^i = p \tag{1}$$

Instead of a set of probabilities of the second level q_p, we then have a continuous function $q(p)$ such that

$$P(A,B^{kn}_{p,dp})^k = q(p)dp \tag{2}$$

Assuming that these probabilities exist, we take over condition a, formulated in (5, § 60), in the form

$$\int_0^1 q(p)dp = 1 \tag{3}$$

Instead of condition b, § 60, we introduce the stronger condition:

c. The sublattice of the C-sequences of the kind $B_{p,dp}$ is a lattice of normal sequences in the narrower sense.

On this assumption, the probability of obtaining an initial section of the frequency $f = \frac{m}{n}$, if the horizontal probability is $= p$, is given by

$$P(A.B^{kn}_{p,dp},F^{kn}_m)^k = w_{nm} = \frac{1}{n+1}w_n(p;f) \qquad f = \frac{m}{n} \tag{4}$$

where the symbol F^{kn}_m has the meaning defined in (10*a*, § 50) and the function $w_n(p;f)$, given by (13, § 57), has Bernoulli properties.

The probability sought for is derivable from (2) and (4) by means of the continuous form of the rule of Bayes (8, § 45) and is to be written

$$P(A.F^{kn}_m,B^{kn}_{p,dp})^k = v_n(f;p)dp = \frac{q(p)w_n(p;f)dp}{\int_0^1 q(p)w_n(p;f)dp} \tag{5}$$

The function $q(p)$ takes over the role of the antecedent probabilities. The factor $\frac{1}{n+1}$ drops out.

The expression (5) may first be discussed on the simplifying assumption that $q(p) =$ const., that is, the antecedent probabilities are equal to one another; the term $q(p)$ then drops out. According to (14, § 57), the denominator is then $= 1$; thus we arrive at the result

$$v_n(f;p) = w_n(p;f) \tag{6}$$

This relation may be stated in words as follows. The Bernoulli function $w_n(p;f)$ has the dual meaning of the probability density that the frequency is $= f$ if the probability controlling the sequence is $= p$, and of the probability density that the probability controlling the sequence is $= p$ if the frequency is $= f$. The latter meaning is contingent upon equality of the antecedent probabilities and restricted to normal sequences.

The significance of the property (14, § 57) can now be stated. The forward probability that a certain frequency $f = \frac{m}{n}$ will result, computed as long as the value p is unknown, is given by

$$P(A,F_m^{kn})^k = \int_0^1 q(p)w_{nm}dp = \frac{1}{n+1}\int_0^1 q(p)w_n(p;f)dp \tag{7}$$

If $q(p) = \text{const.}$, (3) gives $q(p) = 1$; because of (14, § 57) the probability (7) is then $= \frac{1}{n+1}$. This means: as long as the value p is unknown, every one of the $n + 1$ possible values of the frequency $f = \frac{m}{n}$ is to be expected with the same probability $\frac{1}{n+1}$. The condition (14, § 57) thus states that, if the antecedent probabilities are equal, the forward probabilities of the possible frequencies are equal to one another.

The probability of a value p between p_1 and p_2 is given by

$$v_n(f;p_1,p_2) = \int_{p_1}^{p_2} v_n(f;p)dp \tag{8}$$

Since the function $w_n(p;f)$ satisfies the convergence relations (16, § 57), we derive from (6) and (8) the same convergence relations for $v_n(f;p)$:

$$\begin{aligned} &\lim_{n\to\infty} v_n(f;p_1,p_2) = 1 \text{ for } p_1 \leqq f \leqq p_2 \\ &\lim_{n\to\infty} v_n(f;p_1,p_2) = 0 \text{ for } f < p_1 \text{ or } f > p_2 \end{aligned} \tag{9}$$

These convergence relations, however, are not restricted to the special case $q(p) = \text{const.}$, but are demonstrable for the general case (5). For this purpose we integrate (5) according to (8):

$$v_n(f;p_1,p_2) = \frac{\int_{p_1}^{p_2} q(p)w_n(p;f)dp}{\int_0^1 q(p)w_n(p;f)dp} \tag{10}$$

The denominator may be divided into three integrals, for the case $p_1 \leqq f \leqq p_2$,

$$v_n(f;p_1,p_2) = \frac{\int_{p_1}^{p_2} q(p)w_n(p;f)dp}{\int_0^{p_1} q(p)w_n(p;f)dp + \int_{p_1}^{p_2} q(p)w_n(p;f)dp + \int_{p_2}^{1} q(p)w_n(p;f)dp} \tag{11}$$

Because of its Bernoulli character, the function $w_n(p;f)$ has a maximum $w_{n,\max}$ in the interval from $p = 0$ to $p = p_1$ and from $p = p_2$ to $p = 1$; $w_{n,\max}$ goes to 0 with increasing n. Replacing the first and the third integral of the denominator by integrals of the form $w_{n,\max} \cdot \int q(p)dp$, we show that these expressions converge to 0 with increasing n, since (3) holds. The integral from p_1 to p_2 does not converge to 0 if the function $q(p)$ does not vanish for $p = f$. The expression (11) converges then to 1 with increasing n. On the condition $q(p) \neq 0$ for $p = f$, the first of the relations (9) is thus derived. The second follows because $v_n(f;0,1) = 1$.

The proof that the convergence relations (9) hold for every form of the function $q(p)$, and are thus independent of the values of the antecedent probabilities, was first given by Laplace. His proof was constructed for normal sequences. Poisson extended the proof to the sequence type that carries his name. Since the proof does not refer to the particular form of the function w_n, but uses only its convergence properties, it can be extended to the general type of Bernoulli sequence defined in § 57. The function $v_n(f;p)$ is then constructed by substituting the function $w_n(p,s_1 \ldots s_r;f)$ in (5) and integrating over the $s_1 \ldots s_r$ with respect to their total range. For the proof of relations (9), the convergence properties (10, § 57) of the function $w_n^*(p;f)$ are used. This proof, which can easily be given by analogy with the discussion of (11), will not be presented in this book.

The general convergence theorem (9) is of great importance for the theory of *induction by enumeration*, or *a posteriori determination* of a probability, which states that the frequency observed for a finite initial section can be identified with the limit of the frequency, that is, with the probability controlling the sequence. The use of formula (9) presupposes only that the probabilities p and $q(p)$ exist, that $q(p)$ does not vanish for $p = f$, and that the sequences form a Bernoulli lattice. On these conditions we conclude from (10) and (9), putting $p_1 = f - \delta$ and $p_2 = f + \delta$:

Given a finite initial section of a probability sequence of the length n and the frequency f, there exists a probability v_n that the observed frequency f represents the probability p controlling the total sequence within an interval of exactness $\pm\, \delta$. It is true that we cannot calculate v_n if we do not know the function $q(p)$; and we do not know whether v_n is the maximum probability or whether the maximum belongs to a different value p. But we infer from (9): the larger n, the larger is the probability v_n that the observed frequency represents the probability p within the interval of exactness $\pm\, \delta$; and v_n goes toward 1 with increasing n.

The convergence theorem by itself does not tell us what number n is large enough to make the inverse probability v_n, formulated in (10), higher than a given value. But such information can be derived when at least a lower bound

$q^* > 0$ is known for the value of the antecedent probability density $q(p)$ i[n] the interval from $f - \delta$ to $f + \delta$. On this condition a value v_n^* can be con[-]structed such that

$$v_n(f;f - \delta, f + \delta) \geqq v_n^* \qquad (12$$

$$\lim_{n\to\infty} v_n^* = 1$$

The value v_n^*, which does not require a knowledge of the function $q(p)$, can b[e] found as follows. We have

$$\int_{f-\delta}^{f+\delta} q(p)dp \geqq q^* 2\delta \qquad (13$$

and because of (3)

$$\int_0^{f-\delta} q(p)dp + \int_{f+\delta}^1 q(p)dp \leqq 1 - q^* 2\delta \qquad (14$$

As before, the largest value of $w_n(p;f)$ in the interval from 0 to $f - \delta$ and fro[m] $f + \delta$ to 1 may be called $w_{n,\max}$. Replacing in the first and the third integra[l] of the denominator of (11) the function $w_n(p;f)$ by $w_{n,\max}$ makes the whol[e] expression smaller, or, at least, not larger. Dividing the numerator in th[e] denominator and then replacing $q(p)$ by q^* in the integral extending fro[m] $f - \delta$ to $f + \delta$ has the same effect. By the use of the inequality (14) thu[s] obtains a value

$$v_n^* = \frac{1}{1 + \dfrac{(1 - q^* 2\delta) w_{n,\max}}{q^* \displaystyle\int_{f-\delta}^{f+\delta} w_n(p;f)dp}} \qquad (15)$$

which has the properties required in (12), because $w_{n,\max}$ converges to 0 an[d] the integral converges to 1.

Some other formulas may be presented that have been derived within th[e] theory of the inductive inference and that concern, not the probability o[f] elements in the whole infinite sequence, but that of the frequency of element[s] in a finite section immediately following the given initial section, or *consecu-tive section*. In spite of the finiteness of the section, the problem concerns a probability of a higher level, because the horizontal sequences may have different probabilities p and the lattice is thus inhomogeneous. Since different probabilities p can produce the same consecutive section, a mean in terms of the possible values p must be computed.

For the computation, the rule of composition (20, § 22) is used, which for lattice counting can be written

$$P(A.D^{ki}, E^{ki})^k = \sum_{\rho=1}^{r} P(A.D^{ki}, B_\rho^{ki})^k \cdot P(A.D^{ki}.B_\rho^{ki}, E^{ki})^k \qquad (16)$$

For B_ρ, which in the problem under discussion refers to the sequence kind, we put $B_{p,dp}$. The first interval dp may have p_1 as its mean value; the last, p_r. The summation then extends from p_1 to p_r. For D^{ki}, which refers to the observed initial section of the sequence, we put F_m^{kn}, replacing i by n also in the other terms.

The computation will be carried through on two specializing assumptions. The first is that the sublattice of the sequences $B_{p,dp}$ is a lattice of normal sequences in the narrower sense. This includes the assumptions of lattice invariance and of independence of predecessors, so that the last probability can be transformed as follows:

$$P(A.F_m^{kn}.B_{p,dp}^{kn},E^{kn})^k = P(A.B_{p,dp}^{kn},E^{kn})^k = P(A.B_{p,dp}^{ki},E^{ki})^k$$
$$= P(A.B_{p,dp}^{ki},E^{ki})^i \qquad (17)$$

The introduction of i in the third probability expression is permissible because this probability, for normal sequences, does not depend on the variable n or i. The transition to the last form follows from lattice invariance. On the insertion of (17), formula (16) assumes the form

$$P(A.F_m^{kn},E^{kn})^k = \sum_{p=p_0}^{p_r} P(A.F_m^{kn},B_{p,dp}^{kn})^k \cdot P(A.B_{p,dp}^{ki},E^{ki})^i \qquad (18)$$

Calling the probability on the left s, using the notation (5) for the second expression, and putting $r(p)$ for the last probability, we can write (18) in mathematical notation

$$s = \int_0^1 v_n(f;p)r(p)dp \qquad (19)$$

The second specializing assumption is that $q(p) = \text{const.}$, that is, the antecedent probabilities are equal. This assumption allows for the use of the value (6) for v_n, and thus (19) can be written

$$s = \int_0^1 w_n(p;f)r(p)dp \qquad (20)$$

This form will be used to answer several questions.

If the given initial section consists of n elements C, whereas no elements $\bar{C}$ occur, what is the probability s_n that the next element will be a C?

For the answer, using (13, § 57) for $m = n$, we put

$$s = s_n \qquad w_n(p;f) = (n+1)p^n \qquad r(p) = p \qquad (21)$$

Formula (20) then gives

$$s_n = (n+1)\int_0^1 p^{n+1}dp = \frac{n+1}{n+2} \qquad (22)$$

This formula is called the *rule of succession*.[1] It determines the probability s_n that an initial section of n elements, all of which are C's, is followed by another element C, if the sequence is normal and the antecedent probabilities for all values p are equal. The probability s_n refers to a frequency in the vertical direction: among the horizontal sequences that begin with n elements C, is a fraction, determined by s_n, in which the $n + 1$-st element is a C. The probability s_n cannot be interpreted as a horizontal frequency because the first probability on the right of (18) cannot be written for horizontal counting; (3″, § 60) shows that such counting would lead to a different value.

The second question concerns the probability $s_{n,n'}$ of obtaining n' elements C after an initial section of n elements C. Leaving the function w_n of (21) unchanged, for the answer we put

$$s = s_{n,n'} \qquad w_n(p;f) = (n+1)p^n \qquad r(p) = p^n \tag{23}$$

Formula (20) thus gives

$$s_{n,n'} = (n+1)\int_0^1 p^{n+n'}dp = \frac{n+1}{n+n'+1} \tag{24}$$

Whereas the value (22) converges to 1 with increasing n, the value (24) does so only if n' is kept constant. If n is constant and n' increases, (24) goes toward 0. The inductive inference based on the observation of n elements C cannot secure a limit of the frequency that is strictly $= 1$, but only a limit close to 1 within a small interval of exactness δ.

The third question concerns the probability $s_{f,f'}$ of obtaining a consecutive section of the frequency $f' = \dfrac{m'}{n'}$ if an initial section of the frequency $f = \dfrac{m}{n}$ is observed. The question requires the substitutions

$$s = s_{f,f'} \quad w_n(p;f) = (n+1)\binom{n}{m}p^m(1-p)^{n-m}$$

$$r(p) = \binom{n'}{m'}p^{m'}(1-p)^{n'-m'} \tag{25}$$

[1] Since the formula is widely used in the literature, a simple derivation may be added that does not make use of (6). The probability that, if the probability of the sequence is $= p$, a run of n elements C occurs, is $= p^n$; therefore the inverse probability of a value p, for equal antecedent probabilities, is given by (12, § 21) as

$$v_n(f;p) = \frac{p^n}{\int_0^1 p^n dp} \qquad \text{for } f = 1$$

The probability of obtaining one more element C, if p is the probability controlling the sequence, is $= p$; the rule of elimination (21, § 19) thus gives

$$s_n = \int_0^1 v_n(f;p)p\,dp = \frac{\int_0^1 p^{n+1}dp}{\int_0^1 p^n dp} = \frac{n+1}{n+2}$$

Formula (20) then leads to the answer

$$s_{f,f'} = (n+1)\binom{n}{m}\binom{n'}{m'}\int_0^1 p^{m+m'}(1-p)^{n-m+n'-m'}dp$$

$$= \frac{(n+1)!\,n'!\,(m+m')!\,(n-m+n'-m')!}{m!\,(n-m)!\,m'!\,(n'-m')!\,(n+n'+1)!} \tag{26}$$

For the computation of the integral, the auxiliary formula (15, § 57) has been used. Approximate values of $s_{f,f'}$ can be found with the help of Stirling's formula (13, § 49). For the values $m = n$, $m' = n' = 1$, (26) is transformed into (22). For $m = n$, $m' = n'$, (26) is transformed into (24). For $m' = n' = 1$, (26) assumes the form

$$s_{f,f'} = \frac{m+1}{n+2} \tag{27}$$

This value means the probability of obtaining an element C after the observation of a section of m elements C and $(n - m)$ elements $\bar{C}$.

In philosophical discussions the simple formulas (22), (24), (26) and (27) have often been regarded as supplying a justification of the inductive inference. Such a conception is incorrect, since the formulas do not possess a general validity; they hold only for normal sequences and even for these only on the condition $q(p) =$ const., that is, on the condition of equal antecedent probabilities. This condition restricts the applicability of the formulas to cases in which some knowledge of the antecedent probabilities has been acquired in other ways, that is, without the use of the formulas. (See the discussion of the rule of succession in §§ 72 and 86.)

Formula (9) is of greater significance, since it is based on weaker assumptions; it presupposes only the Bernoulli character of the sequences, and holds for any values of the antecedent probabilities. Even this formula, however, cannot be regarded as supplying a general justification of the inductive inference, because it, too, is based on special presuppositions. Its significance lies rather in the function that it performs in the further extension of the theory of induction, once the general justification of induction has been given. These logical questions will be discussed again in §§ 86 and 88–90.

Chapter 9

THE PROBLEM OF APPLICATION

THE PROBLEM OF APPLICATION

§ 63. The Problem

The exposition of this book has now reached a point at which the purely logico-mathematical line of thought is discontinued. The investigation turns into a new direction: its objective is the analysis of the relations between the mathematical calculus of probability and knowledge of nature, that is, the application of the calculus of probability to reality. The complex of these questions is called the *problem of application.*[1] In the course of this presentation, numerous examples for the application of the calculus of probability have been given, but without reflection on the legitimacy of the application, which will now be analyzed in detail.

The problem of application may be divided into two groups of questions. The first group includes the question what is meant by the concept of probability as used in applications; we speak here of the *problem of the meaning of probability statements.* The second group deals with the question whether we are justified in applying the rules of probability to physical objects, whether we are entitled to assert probability statements referring to physical objects when the statements are established by means of the rules of the calculus. This group may be called the *problem of the assertability of probability statements.* The two problems are closely connected, since the assertability will depend on the meaning assumed for the probability statement. I speak of assertability instead of truth because the requirement of truth would be too strong. It will be seen that a certain group of probability statements—those, namely, that state numerical probability values—cannot be proved as true, but are assertable on other grounds.

The problem of meaning can be formulated as the problem of the interpretation of the P-symbol. It was explained above that in the formal conception of the calculus of probability the P-symbol remains uninterpreted; the axioms then are conceived as a set of implicit definitions restricting the meaning of the P-symbol by subjecting it to certain formal conditions, without, however, defining the meaning exhaustively. Although the axioms exclude a class of interpretations as inadmissible, they leave open a *class of admissible interpretations.* The problem of meaning consists in selecting, among the admissible interpretations, the one that is suitable to cover the applications of the P-symbol to physical reality.

There remains, of course, the possibility that several interpretations are used, varying with the subject to which the term "probability" is applied.

[1] The term was introduced by Edgar Zilsel in *Das Anwendungsproblem* (Leipzig, 1916).

In fact, the term has two different forms of usage, which call for separate treatment. In the first usage the word "probability" refers to *sequences of events* or of other physical objects; in this usage the word must, without question, be interpreted as referring to a relative frequency, though the precise formulation of the interpretation requires further investigation. In the second usage, however, the term "probability" is applied to *single events* or other single physical objects; we must inquire whether this usage necessitates the introduction of a genuinely different interpretation or whether it is reducible to the frequency interpretation. Postponing the discussion to later sections (§§ 71–72), we shall turn first to an analysis of the frequency interpretation.

§ 64. The Logical Form of Limit Statements

According to the frequency interpretation, introduced in § 16, the probability of a sequence is defined by the limit of the relative frequency. In order to make the meaning of the limit statement clear, I shall formulate it by means of the logical symbolism.

The limit statement may be formulated in two steps, which will be symbolized in the notation for the frequency sequence f^n counted through (see 1, § 48). On the first step we formulate the condition that, throughout the sequence, terms f^n that are as close to the limit p as we wish will always recur. In symbolic notation (see 3, § 89),

$$(\delta)\,(n_o)\,(\exists n)\,(n > n_o)\,.\,(f^n = p \pm \delta) \qquad (1)$$

Translated into words this means: however small we choose δ and however large we choose n_o, there is an element f^n beyond f^{n_o} that is situated within $p \pm \delta$. In this case I shall call p a *partial limit* of the sequence.

On the second step we set up a stronger condition, adding the requirement: for every δ, however small, there is an n_o such that, from f^{n_o} on, all f^n remain within $p \pm \delta$. In symbolic notation this reads

$$(\delta)\,(\exists n_o)\,(n)\,[(n > n_o) \supset (f^n = p \pm \delta)] \qquad (2)$$

Only if (2) is satisfied do we call p the *limit* of the sequence. The limit is distinguished from the partial limit by the requirement of a certain permanence in the approximations attained.

The following relations hold for the assertions (1) and (2). Statement (2) is the stronger assertion: (1) follows from (2), as can be derived from (16c, § 6). However, (1) can be satisfied without (2) being satisfied. The latter case will be exemplified later.

The relation of the limit to the partial limit can be expressed as follows: a sequence that has a partial limit in p contains a subsequence that has a limit in p.

The result of the analysis must now be applied to probability sequences. Both ways of giving, the intensional and the extensional, can be carried through for probability sequences. These terms will be referred only to the attribute class; an intensionally or extensionally given probability sequence is a sequence for which the attribute is given, respectively, intensionally or extensionally. The strictly alternating sequence (1, § 26) represents an example of an intensionally given probability sequence. Here is to be classified, also, Henri Poincaré's[1] example of the last digits of a table of logarithms, among which odd and even numbers are equally frequent. Sequences of throws of the die, social statistics, and so on, are examples of extensionally given probability sequences. Transferring the results of the analysis of limit statements to probability sequences, we arrive at the following conclusion:

With respect to intensionally given probability sequences of an infinite length, statements expressing a frequency interpretation have the usual meaning of mathematical all-statements and existential statements; like the latter, they are strictly verifiable. With respect to extensionally given probability sequences of an infinite length, however, statements expressing a frequency interpretation are not verifiable.

The first result is important for the mathematician. It proves the existence of an interpreted calculus of probability—a calculus of probability dealing with probability as a frequency, in which probability statements have the usual meaning of mathematical limit statements. The convergence, therefore, can be formulated in the usual way. For every δ the n can be calculated for which f^n represents the place of convergence. The interpreted calculus of probability thus constructed is a discipline of the same type as other mathematical disciplines dealing with infinite sequences that are given by defining rules; it must be conceived as a branch of arithmetic.[2]

With regard to extensionally given probability sequences, however, the result is different. We saw that for them the frequency interpretation of probability leads to completely unverifiable statements because the statements include both all-operators and existential operators. An extensionally given sequence can never be realized in its entirety; we know only an initial section of it, and its infinite remainder is a matter of the future. It is obvious, however, that the relative frequency holding for the initial section imposes no restrictions on the limit of the relative frequency existing for the infinite sequence: given a relative frequency $f^n = \frac{m}{n}$ for the initial section, it is pos-

[1] *Calcul des probabilités* (Paris, 1912), p. 313.

[2] This result is due to the elimination of the principle of randomness. The calculus of probability of von Mises cannot be treated on logically equal terms with the other mathematical disciplines because von Mises' collective cannot be given intensionally. My normal sequences, on the contrary, can be so given; see my remarks on Copeland's model of normal sequences in § 30.

sible to imagine a continuation of the sequence of such a kind that the fre quency converges toward any value p selected arbitrarily. In other words let m' and n' be the respective frequencies of elements counted from the plac of the n-th element; then the expression $\frac{m + m'}{n + n'}$ for m' and n' going towar infinite values, will converge toward a limit independent of the constan values m and n. With respect to extensionally given sequences of an infinit length, therefore, the limit statement is not verifiable.

This result makes questionable the meaning of the limit statement fo infinite sequences of the extensional kind. Assuming the verifiability theor of meaning,[3] we demand that if a statement is to be regarded as meaningfu it should be possible to verify it as either true or false (the word "verify" i used here in the neutral sense, denoting a determination as true or as false) Since such verification is not physically possible for extensionally given se quences of an infinite length, we must inquire in what sense the limit state ment can be maintained for them, or what other change of interpretatio could be envisaged. Before entering into such an analysis, we shall inquir whether the interpretation in terms of a limit is indispensable for a frequenc interpretation.

§ 65. The Necessity of the Concept of the Limit for the Frequency Interpretation

In order to find out whether the limit statement can be dispensed with i a frequency interpretation, some other possible frequency interpretations wi be considered. According to the interpretation by a partial limit as intro duced in (1, § 64), the statement of a probability p would require that i should always recur that the relative frequency approaches p to any desire degree of exactness, whereas no statement about a permanence of the con vergence would be added. Because of the rather weak nature of such a postu late, I attempted, in an earlier publication,[1] to develop this interpretatio but in the meantime I found such serious arguments against it that I aban doned the idea.

The first argument is that even statement (1, § 64) is completely unverifi able and therefore cannot relieve the difficulty under discussion. Anothe

[3] See H. Reichenbach, *Experience and Prediction* (Chicago, 1938), §§ 4–8. Hereafter re ferred to as *EP*.

[1] "Der Begriff der Wahrscheinlichkeit für die mathematische Darstellung der Wirklichkeit" (Diss. Erlangen, 1915), pp. 70–71, and in *Zs. f. Philos. u. philos. Kritik*, Vol. 162 (1917) pp. 246–247. The decision to attempt this interpretation was supported, also, by an objec tion to the limit interpretation, which will be discussed presently (p. 346), since the objectio becomes inapplicable for the case of the partial limit. I no longer regard the objection a tenable, however; so this argument for the interpretation by the partial limit is eliminated too.

argument results from the fact that a statement about a permanence from a certain element on, as employed in the definition of a limit, can be conceived logically as the negation of another statement about a partial limit (see § 64). Thus we are compelled to admit a statement of the form (2, § 64) as meaningful if a statement of the form (1, § 64) is accepted; otherwise we should exclude the operation of the negation. In other words, since (4a, § 64) is equivalent to (2, § 64), the statement about the limit can be replaced by another statement containing only the concept of the partial limit and the negation. Therefore, if we admit the concept of the partial limit, we must also admit the concept of the limit unless we intend to renounce the operation of negation. This is the reason why the interpretation by a partial limit does not carry any logical advantage.

We shall now turn to a second attempt to reach a frequency interpretation of a weaker form. We shall construct an interpretation by a frequency statement that is at least unilaterally verifiable. The following axiom of interpretation is introduced:

Axiom of interpretation α. *If an event C is to be expected in a sequence with a probability converging toward 1, it will occur at least once in the sequence.*

This is a rather modest assumption, compared with the interpretation in terms of the limit. That it is of the type of statements that are unilaterally verifiable follows because it requires the occurrence of only one event for its verification. The interpretation so constructed would therefore seem logically preferable to that of the limit. It can be shown, however, that if the interpretation through the axiom α is assumed, the existence of a limit is deducible for normal sequences. We are therefore led back to the original interpretation.

The proof is based on the amplified Bernoulli theorem (§ 51), which is derivable in the formal calculus of probability, without the use of the frequency interpretation. We understand by C the property that f^n is the place of convergence for δ. Then $c_{n\delta}$ is the probability of C; since the $c_{n\delta}$ converge toward 1, C must occur once. If C, however, has occurred for f^n, C holds also for all the following elements according to its definition. That is, f^n and at the same time all the following f^i are situated within $p \pm \delta$; this is the assertion of the limit.

It is remarkable that such a weak assumption as the axiom of interpretation α has so far-reaching a consequence. The result shows that the rejection of the interpretation in terms of the limit, for normal sequences, leads necessarily to the rejection of the axiom of interpretation α, that is, to the assertion that it is possible that an event that is expected with a probability converging toward 1 will never occur. It seems clear that whoever admits this possibility must abandon every attempt at a frequency interpretation. It is the aim of a frequency interpretation to reduce a probability statement to a certainty statement about frequencies; yet a statement demanding less than the axiom α is hardly conceivable. This axiom connects the modality of certainty with the probability 1 by a minimum of requirements. It does not state that the probability 1 is identical with certainty. That would be asking too much. It states only that an event that is to be expected in the limit with the probability 1 will occur at least a single time in the sequence. This assumption, which, taken by itself, would admit even the value 0 for the limit of the relative frequency in regard to such an event, leads, in connection with axioms of the formal calculus of probability, to the interpretation in terms of the limit.

Other assumptions, for instance, the interpretation by a partial limit, can likewise be shown to lead to this consequence. I therefore regard this proof as sufficient evidence that the interpretation of probability in terms of the limit of the frequency cannot be dispensed with. The proof makes evident the great importance of the theorem of Bernoulli: although taken alone this theorem cannot be used for a derivation of the frequency interpretation of probability, it leads to the interpretation in terms of a limit of the frequency as soon as it is combined with an assumption as weak as the axiom α.

In this connection I shall deal with an objection that was raised against the limit interpretation.[2] The attempt was made to construct a contradiction between the limit interpretation and Bernoulli's theorem by the following consideration. If the sequence of the frequency has a limit p, there must exist an n for a given δ such that the frequency beyond f^n does not deviate from p more than δ. From this fact it follows, however, that the elements succeeding the element y_n are in some way restricted. Thus a run of elements B beginning at the place y_n cannot surpass a certain length r. According to Bernoulli's theorem, on the contrary, a certain probability greater than 0 must be allowed for a run of s elements B $(s > r)$. Thus it seems that at the place y_n a section having a certain probability greater than 0 must be regarded as impossible when the limit interpretation is used.

The contradiction is solved as follows. We must distinguish between the probability that a run of s elements B will occur and the probability that a run of s elements B will occur *at the place* y_n. In the frequency interpretation the first probability refers to sections all of which are situated within the same sequence. It is quite possible that such runs occur beyond the element y_n without disturbing the convergence, since for greater n a run of the length s will not cause a transgression of the convergence limit δ. The second probability, however, refers to lattices of sequences, since here the place y_n is specified; for its interpretation we must use a frequency in the vertical direction. Now the following property holds for a lattice of normal sequences in the narrower sense: if all horizontal sections extending from y_n to y_{n+s} are counted in the vertical direction, a certain number of sections containing only

[2] W. Sternberg, in *Zs. f. angew. Math. u. Mech.*, Vol. IX (1929), p. 501.

elements B will be found among them, in correspondence with the Bernoulli theorem. This does not contradict the limit interpretation, because not all the horizontal sequences will have their places of convergence for δ at the same n. The lattice is not required to exhibit a *uniform convergence*: in *each* sequence there is an n for a given δ such that f^n is the place of convergence for δ, but there does not exist *one* such n for *all* sequences. With this solution, which was first offered by von Mises, the contradiction disappears.

With these results we have returned to the limit interpretation. Its necessity is indisputable if a frequency interpretation is to be adopted. We must therefore find a way to eliminate the logical difficulties of the limit interpretation presented in § 64.

§ 66. The Meaning of Limit Statements

We saw the logical difficulties in the fact that the statement about the limit appears meaningless for extensionally given sequences of an infinite length. A limit at a given value p is compatible with every finite beginning of the probability sequence; since we can count the frequency only in a finite initial section, all limit statements must be called nonverifiable and, consequently, meaningless.

The analysis of the problem of meaning with respect to sequences of physical events has suffered from the fact that it has been too closely attached to the mathematical formulation of the calculus of probability. Only in the mathematical version do we find infinite sequences; the sequences of actual statistics, however, are always finite. They are so, not only with respect to the initial section for which the statistics were compiled, but also with respect to the section of the sequence that lies in the future and is not yet observed. In fact, we are interested only in finite sequences because they will exhaust all the possible observations of a human lifetime or the lifetime of the human race. We wish to find sequences that behave, in a finite length of these dimensions, in a way comparable to a mathematical limit, that is, converging sufficiently well within that length and remaining within the interval of convergence. If a sequence of roulette results or of mortality statistics were to show a noticeable convergence only after billions of elements, we could not use it for the application of probability concepts, since its domain of convergence would be inaccessible to human experience. However, should one of the sequences converge "reasonably" within the domain accessible to human observation and diverge for all its infinite rest, such divergence would not disturb us; we should find that such a *semiconvergent* sequence satisfies sufficiently all the rules of probability. I will introduce the term *practical limit* for sequences that, in dimensions accessible to human observation, converge sufficiently and remain within the interval of convergence. Sequences of this kind will include only a subclass of all sequences having a limit and will include also a subclass consisting of semiconvergent sequences. It is with sequences having a practical limit that all actual statistics are concerned.

In applying the rules of probability to such sequences, we shall find that theorems derived from the axioms of the elementary calculus (axioms I–IV, §§ 12–14) hold strictly, since these axioms are strictly valid for finite sequences (see § 18). Theorems derived in the theory of order, however, will be found to hold only approximately, since the axioms V, § 28, hold strictly only for infinite sequences. The conditions of types of order will be regarded as satisfied if they hold to a certain degree of exactness, and the inexactness will become worse for subsequences when the number of elements becomes smaller. For a random sequence, the frequency of groups of successive elements will follow the Bernoulli theorem only if the group is not too long, whereas for long groups the deviations between calculated and observed frequency may surpass any degree.

It would be possible to transform the calculus of probability so that it takes account of finite sequences. We should then speak, not of a limit of the frequency, but of a limiting interval ϵ and should have to stipulate, in every theorem referring to subsequences, which larger interval of convergence δ we are willing to admit. A calculus of this kind would be rather complicated and cumbersome. In comparison, the calculus in its present form must be regarded as an *idealization,* with the advantage, however, that it is much simpler and easier to handle. This is reason enough to prefer it. In this respect we follow the practice of surveyors who apply the ideal notions of geometry to physical elements that satisfy the axioms of geometry only to a certain extent. With respect to all applications, in fact, the notions of a line without width and a point without spatial extension must be regarded as idealizations that are never realized by physical objects. The practical geometry of lines of small width and points of small extension can be identified with the ideal geometry to a certain degree of approximation. We deal with the latter in all computations because its mathematical structure is much simpler. For the same reason we shall not use the concept of a practical limit in the following chapter, but shall carry through all technical analysis with respect to infinite sequences. We know that the results will hold approximately for practical limits.

These considerations will settle the problem of meaning for probability sequences of physical events. To those who insist that every meaningful statement should be strictly verifiable in a finite number of observations we answer that all sequences with which we are concerned are sequences of a practical limit, and thus a *finitization* of the limit statement can be carried through. There can be no doubt that this finitization will satisfy all requirements of the verifiability theory of meaning in its most rigorous form.

The situation is different with extensionally given sequences when an infinite length is admitted, as in the purely mathematical conception of the calculus of probability. If infinite sequences of the extensional kind are used in the sense of an idealization, statements referring to such sequences must be meaningful in some sense, or it would not be possible to translate them into approximative statements holding for finite sequences. An interpreted calculus of probability that deals with infinite sequences not given by rules, for example, the random sequences of von Mises and Church (see § 30), must be classified along with certain chapters of the theory of sets that likewise do not satisfy the requirements of the verifiability theory of meaning in the form assumed by physicists. Thus the axiom of choice maintains the possibility of a selection from an infinite class of classes, though in general we have no means of defining the selection by a rule that is statable in a finite number of terms. Since the present work is concerned primarily with the probability calculus of physical events, an analysis of these logical questions is beyond its province. I may add, however, that a solution seems possible when the meaning of such postulates as the axiom of choice, or of limit statements about infinite sequences not given by a rule, is regarded as being of a type that I have called *logical meaning.*[1] For this type of meaning the logical possibility of verification, as distinct from the physical possibility, is regarded as sufficient. A fact is called logically possible if it is no contradiction to assume that it happens, though the occurrence may be excluded by physical laws.[2] It is called physically possible if there are no physical laws excluding it. Thus an increase of energy in a closed system is logically, though not physically, possible, whereas interplanetary travel is not only logically, but also physically, possible. The enumeration of an infinite sequence can be called only physically impossible; it is not a logical impossibility. In other words, it is no contradiction to assume that an infinite sequence is known element by element, though such knowledge is not physically possible for a human organism. The ultimate answer to all such questions, of course, is closely connected with the problem of a proof of consistency in Hilbert's sense.

If the proposal of logical meaning for statements as described is not accepted, it may be possible to construct a meaning for them on the basis of the fact that they are translatable into approximative statements for finite sequences or classes. Such a meaning, at least, is all that is needed for the theory of probability so far as it is applied to physical events. It may then be called a fictitious meaning. For the purpose of these investigations, however, such questions need not be discussed. In its physical applications the frequency interpretation of probability can be given a *finitist* meaning, satisfying the verifiability theory.

[1] See *EP*, §§ 6–8.

[2] The definition of physical laws, or *nomological statements*, is given in *ESL*, chap. viii.

§ 67. The Assertability of Probability Statements in the Frequency Interpretation

We turn now to the problem of the assertability of probability statements, still restricting ourselves to the frequency interpretation of such statements. Two questions must be answered. The first concerns the *assertability of probability laws*, that is, theorems of the calculus and thus of formulas that do not include assertions of specific numerical values of probability. The second concerns the *assertability of numerical degrees of probability* and therefore may be formulated as the question of the *ascertainment of the degree of probability.*

The question of the assertability of probability laws is easy to answer. It was shown in § 18 that axioms of the calculus of probability follow tautologically from the frequency interpretation. Therefore, if probability is regarded as meaning the limit of a relative frequency, the validity of probability laws is guaranteed by deductive logic. This result is of major importance for the epistemological critique of probability statements. So far as theorems of the calculus of probability are concerned, the application of probability statements to physical objects offers no greater problems than that of theorems of deductive logic. In other words, with respect to theorems of the calculus there is no problem of application that is specific for probability. Now the assertability of theorems of deductive logic is based on the fact that logical formulas are empty, that they do not anticipate properties of the physical world. In the same sense, therefore, theorems of probability must be regarded as empty when applied to relative frequencies. The theorems constitute mere transformations of probability expressions into others, without any addition as to content. The very emptiness, however, supplies the reason for the usefulness of probability laws. If the laws were not empty it would not be permissible to add them to given premises, since only emptiness of the laws can guarantee that the results of transformations do not state more than is supported by the premises. That, nonetheless, the conclusions of probability transformations can be new, in the psychological sense, is obvious to the logician who is familiar with the twofold nature of tautologies as logically empty but full of psychological content.

In this context the construction of the theory of order assumes particular importance. We saw that all special types of probability sequences can be characterized by postulates stating the equality of probabilities in certain subsequences. The treatment of any sequence type, therefore, requires no addition to probability axioms other than statements about the equality of degrees of probability. It follows that the verification of statements classifying a sequence with respect to its type of order can be achieved as soon as methods for the determination of numerical probability values are known. The analysis of such a classification thus finds its place in the discussion of the ascertainment of the degree of probability. We see why a solution of the problem of the assertability of probability laws cannot be given until a general theory of the order of probability sequences is constructed. Without such a theory we should have no proof that all probability laws are tautologous when the frequency interpretation is assumed; we could not exclude the possibility that some unknown synthetic law concerning the type of order is implicitly involved.

The problem of the ascertainment of the degree of probability, in contradistinction to the problem of probability laws, offers serious difficulties. They originate from the difficulties in the verification of the limit statement explained above (§ 64).

In extensionally given sequences—and probability sequences referring to physical reality are of this type—verification of the limit statement is impossible if the sequences are infinite. It has been emphasized that sequences of physical applications are not infinite and that verification by enumeration is possible in principle. Although this result is sufficient to settle the problem of the *meaning* of the limit statement, it does not help so far as the *assertability* of the limit statement is concerned. The reason is that in all practical applications we wish to know the value of the limit before the sequence is completely produced; indeed, all practical use of probability statements consists in the fact that they are applied for the prediction of relative frequencies, and thus cannot be based on counting the total sequence. It is the *predictive nature* of probability statements, therefore, that presents difficulties to a proof that such statements are assertable.

This consideration makes clear that the ascertainment of the degree of probability must be achieved by methods other than counting the relative frequency in the total sequence. In practical applications, two different methods are used. The first is called the *a posteriori* determination of probabilities; the second, the *a priori* determination of probabilities.

The *a posteriori* determination is identical with a procedure known in logic as *induction by enumeration*, if the term is applied in a somewhat wider sense than in traditional logic. It is based on counting the relative frequency in an initial section of the sequence, and consists in the inference that the relative frequency observed will persist approximately for the rest of the sequence; or, in other words, that the observed value represents, within certain limits of exactness, the value of the limit for the whole sequence. This inference is called the *inductive inference;* its formulation is called the *rule of induction.* A strict formulation of the rule will be given later (§ 87). The inductive inference of traditional logic is of a somewhat narrower form; it refers to initial sections in which all elements have the same attribute B and assumes the permanence of B for the rest of the sequence. This inference, which may be called *classical induction*, is a special case of the general form, namely, the case where the observed relative frequency is $= 1$. To distinguish the general rule of induction from the special case of classical induction, it may be called *statistical induction*.

The problematic character of the inductive inference has often been discussed. In § 64 it was pointed out that a relative frequency $\frac{m}{n}$ observed in an initial section is compatible with any value p for the limit of the relative frequency in the whole infinite series. For sequences of a finite length the frequency $\frac{m}{n}$ will impose certain restrictions on the limit; but if the sequence is long enough—and that is true of practical applications—the restrictions are negligible and the value of the limit is virtually independent of the observed value $\frac{m}{n}$. The inductive inference, therefore, is essentially different from a deductive inference; it carries no logical necessity with it. For the present, however, discussion of the logical nature of the inference will be postponed.

It would be a simple way out of the difficulty if it could be shown that other methods than the inductive rule are available for the ascertainment of a degree of probability. Therefore we must now inquire whether an *a priori* determination of a degree of probability is possible.

§ 68. The So-Called *A Priori* Determination of the Degree of Probability

The *a priori* determination of the degree of probability occupies a special position in the historical development of the philosophy of probability. The determination of probabilities on the ground of properties of symmetry, without the use of frequencies, has been regarded by some authors as the nucleus of the problem of probability. It has been contended that every other determination of probabilities must be reduced to the same logical schema. In the pursuit of this idea it was believed that all attempts to solve the philosophical problems of the concept of probability should start at this point.

Mechanisms like the die and the roulette wheel are characterized by the existence of a complete and exclusive disjunction the terms of which are equiprobable. The theorem of addition permits us to infer, from the number r of the terms of the disjunction, that the probability is $= \frac{1}{r}$ for every single term. With respect to such disjunctions it is therefore possible to reduce the determination of the degree of probability to the determination of the concept *equiprobable*. On this fact was based the well-known definition of probability as the ratio of the favorable to the possible cases.

The idea led to the conception that the starting point of every determination of probabilities must be a disjunction consisting of equiprobable terms. Nonequiprobable terms of a disjunction, according to this conception, must be reduced to another disjunction consisting of equiprobable terms, from which the nonequiprobable terms are derived by transformations. The reduction may be illustrated by the computation of the probability of obtaining a number greater than 5 by throwing two dice: here the two nonequiprobable cases of the alternative "greater than 5" and "not greater than 5" are reducible to the original equiprobable cases given by the faces of the die.

It is true that in some cases such a reduction is possible as well as justified. Thus the greater probability of male births as compared with female births may be explained by the fact that in the surroundings of the ovum the number of spermatozoöns producing the male sex somewhat surpass the number of spermatozoöns producing the female sex, and the probability of fertilization is assumed to be equal for both kinds of spermatozoöns. It is not true, however, that reduction of the degree of probability to equiprobability is *always* possible. Theories like the *Spielraumtheorie* of Johannes von Kries,[1] which regard this idea as the solution of the problem of probability, must therefore be rejected as untenable.

A more important objection must be added: even if the degree of probability can be reduced to equiprobability, the problem of probability is only shifted to this concept. All the difficulties of the so-called *a priori* determination of probability, therefore, center around this issue.

In the search for an argument by which the equiprobability can be derived, a principle was established that must be regarded as the foundation of all *a priori* determination of probability: the *principle of indifference*, also called the *principle of no reason to the contrary*. It maintains that events are equiprobable when there is no reason to assume that one should occur rather than another. Thus we have no reason, it is argued, to favor one of the faces of the die; therefore they are equiprobable. Some authors present the argument in a disguise provided by the concept of *equipossibility*: cases that satisfy the principle of "no reason to the contrary" are said to be *equipossible* and therefore equiprobable. This addition certainly does not improve the argument, even if it originates with a mathematician as eminent as Laplace,[2] since it obviously represents a vicious circle. *Equipossible* is equivalent to *equiprobable*. It appears advisable to discuss the principle of indifference in a form that avoids this fallacy, and ask whether the absence of a reason to the contrary can guarantee equal probabilities.

[1] *Die Principien der Wahrscheinlichkeitsrechnung* (Freiburg i. B., 1886).

[2] *Essai philosophique sur les probabilités* (Paris, 1814; rev. ed., Gauthier-Villars, 1921, p. 9).

At first sight the principle seems plausible, since we actually use such inferences in the determination of probabilities. The inference is often associated with a feeling of self-evidence that makes the conclusion seem "logical". But the fact that we use and believe in inferences of this type does not prove that they are valid. On the contrary, a brief reflection on the content of the principle shows that it is invalid. The absence of a reason to the contrary is a condition of our knowledge; equiprobability is a condition holding for physical objects. Why should the occurrence of physical events follow the directive of human ignorance? Perhaps we have no reason to prefer one face of the die to the other; but then we have no reason, either, to assume that the faces are equally probable. To transform the absence of a reason into a positive reason represents a feat of oratorical art that is worthy of an attorney for the defense but is not permissible in the court of logic.

That the principle leads to correct results for mechanisms like the die must be explained by the fact that in such mechanisms more conditions are realized than are formulated in the principle. The principle represents an instance of the *fallacy of incomplete schematization* (§ 21); it does not state all the conditions that must be satisfied for equiprobability. It leads to correct results only where the additional conditions are satisfied; but general usage leads to absurd conclusions. Thus it would follow that every event about the occurrence of which we know nothing is to be expected with the probability $\frac{1}{2}$, an inference that is not improved by the fact that some philosophers maintain the right to use it. Or consider the probability of male births: so long as nothing was known about the mechanism of fertilization, the probability was assumed to be $= \frac{1}{2}$, since there was no reason to expect a male birth rather than a female birth—a result that is now known to be false. If it is argued that there were reasons to expect the contrary, derived from the statistical prevalence of male births, we must answer, first, that the argument represents the use of an inductive inference rather than that of the principle; and second, that if the inductive inference is admitted it will falsify the conclusion that results when the principle is applied without knowledge of birth statistics.

The latter criticism indicates the weakest spot in the *a priori* theory of probability: if both an *a posteriori* and an *a priori* determination of the degree of probability are admitted, their results may contradict each other. The only possible defense of the *a priori* determination, therefore, consists in the attempt to restrict it to a meaning of probability that is not expressible in terms of frequencies and is not verifiable by the inductive rule. The discussion of this attempt will be given in § 71. For the time being, the result may be stated as follows: the principle of indifference is not admissible when the frequency interpretation of probability is assumed.

In § 69 it will be shown that the equiprobability for mechanisms like the die is derivable from certain of their physical properties, among which geometrical symmetry plays an important part. The invalid inference can be replaced, in such applications, by a correct inference. Like other fallacies, the principle owes its apparent plausibility to the fact that in many instances it can be applied successfully, though the correct explanation of such instances is rather involved.

The absurdity of the principle was demonstrated, in particular, for applications to geometrical probabilities, where it leads to the conclusion that, if nothing is known to the contrary, equal areas obtain equal probabilities. This consequence, however, entails contradictions for transitions from one attribute space to another when the two spaces are coupled by a nonlinear measure transformation. To which of the two spaces should the principle be applied? If equal areas possess equal probabilities in one space, they cannot do so in the other.

For instance, assume it is known only that the specific weight of an unknown substance lies between 4 and 6. The principle then leads to the consequence that there is a probability of $\frac{1}{2}$ that the specific weight lies between 4 and 5, the same holding for the interval from 5 to 6. Applying the same inferences to the determination of the specific volume, which is the reciprocal value of the specific weight, we arrive at the result that the specific volume is to be expected with a probability greater than $\frac{1}{2}$ in the interval from $\frac{1}{4}$ to $\frac{1}{5}$, since the difference $\frac{1}{4} - \frac{1}{5}$ is greater than the difference $\frac{1}{5} - \frac{1}{6}$. This result contradicts the previous one.

It should be noticed that the difficulty is inherent in the principle and is not introduced artificially. In fact, if nothing is known about a preference of certain areas in one attribute space, the same must hold for the other, and the contradiction is unavoidable. If the contradiction is to be eliminated, the principle would have to be supplemented by a rule selecting the attribute space to which it is to be applied. No such rule has ever been suggested, and it is hard to see how it could be formulated.

§ 69. Explanation of the Equiprobability in Mechanisms of Games of Chance

The explanation of the equiprobability in mechanisms of games of chance can be given by the use of an idea that was first developed by Henri Poincaré.[1]

[1] *Calcul des probabilités* (Paris, 1912), p. 149. I used the idea in a demonstration showing that the application of the calculus of probability contains no assumptions different from those presupposed in the application of the principle of causality: "Der Begriff der Wahrscheinlichkeit für die mathematische Darstellung der Wirklichkeit" (Diss. Erlangen, 1915), and in *Zs. f. Philos. u. philos. Kritik,* Vol. 161 (1916), p. 209; Vol. 162 (1916), pp. 98, 223. Even though this first among my papers referring to the problem of probability was written under the influence of Kant's epistemology, it seems to me that the result concerning the theory of probability can be stated independently of Kant's doctrine and incorporated in my present views. The present section represents such a restatement of the analysis given in the paper.

The example of the roulette wheel offers itself for the exposition. Assume that the wheel is played in such a manner that before each spinning the indicator is brought into the same initial position. The entire angle ω of the rotation of the indicator may be counted in multiples of 2π, that is, from the initial position to the stop. The probability that a certain value ω will obtain is determined by a probability function $\varphi(\omega)$ (see fig. 25). The following considerations are based on the existence of such a function, whereas the actual form of the function need not be known.

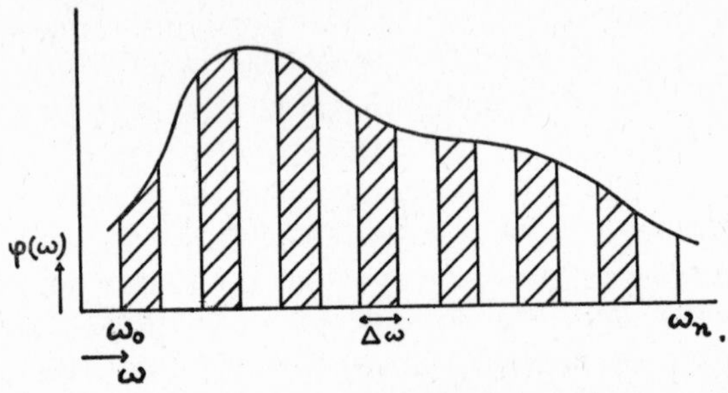

Fig. 25. Probability function for roulette game, used for demonstration of equiprobability of "red" and "black". ω is rotation angle of hand counted in multiples of 2π.

The division by black and red sectors supplies a division by intervals of equal width $\Delta\omega$ on the axis of the abscissa ω. To simplify the problem, assume that no sector is designated for the profit of the bank; so there are black and red sectors only. Every second interval represents the result "red"; the other intervals lead to the result "black". The corresponding probability is given by the respective stripe of ordinates. In figure 25 one of the sets of stripes is shaded; it may represent the result "black". The probability of obtaining "black" will then be represented by the total area covered by the shaded stripes; the probability of obtaining "red" is given by the area of the other stripes.

Now it is easy to show that the sum of the shaded stripes almost equals the sum of the unshaded stripes if the function $\varphi(\omega)$ does not oscillate too much. The two sums of stripes will become equal in the limit $\Delta\omega = 0$ even if the function $\varphi(\omega)$ is submitted to no other condition than that of continuity.

This theorem will be demonstrated first for a finite section extending from ω_0 to ω_n. Consider two successive stripes as forming a pair. If φ' is the smallest, and φ'' the greatest, value of the ordinate φ within the pair, then the difference in the area of the two stripes is not larger than

$$(\varphi'' - \varphi') \cdot \Delta\omega \tag{1}$$

The number of such pairs of stripes is $\frac{n}{2}$. In one of the $\frac{n}{2}$ pairs the difference $(\varphi'' - \varphi')$ has its greatest value, which is called $(\varphi'' - \varphi')_{\max}$. Then the difference between the sum of the shaded and the unshaded areas is not larger than

$$\frac{n}{2} \cdot (\varphi'' - \varphi')_{\max} \, \Delta\omega \tag{2}$$

With increasing n the value $(\varphi'' - \varphi')_{\max}$ approaches 0 because of the continuity of $\varphi(\omega)$, whereas $n \cdot \Delta\omega$ remains always equal to the constant value $\omega_n - \omega_0$. Consequently the expression (2) converges to 0. Thus the proof is given.

The proof can be extended to probability functions the abscissa of which is not bound to the limits ω_0 and ω_n, running through all values from $-\infty$ to $+\infty$. Since the total area of all the stripes is finite because of (6, § 42), it is possible to choose two limits ω_0 and ω_n so that the main part of the area between the curve and the axis of abscissas is situated between these values and the remaining part of the area becomes smaller than a given number ϵ. The foregoing proof can then be given for the main part of the area. By transition to the limit $\epsilon = 0$, the proof is extended to the whole area.

When this consideration is used for the explanation of the game of roulette, the following assumptions must be made:

1. The probability of a rotation angle ω is determined by a continuous probability function $\varphi(\omega)$.
2. The intervals $\Delta\omega$ are equal in size.
3. The size of the intervals $\Delta\omega$ is small with respect to the oscillation of the function $\varphi(\omega)$.

The last assumption is necessary because in a practical case we never deal with the limit $\Delta\omega = 0$.

The assumption 1 represents a very general assumption about the existence of probabilities; it does not contain any *metrical* presuppositions.

The assumption 2 concerns metrical relations; it has no reference to probability, however, since it refers to geometrical relations. In order to confirm it, we have only to measure the size of the sectors of the roulette wheel.

The assumption 3 represents a certain rough appraisal of the metrical properties of a probability function that may, however, remain undetermined within wide limits.

The equiprobability assumed for the game of roulette is easily derived from these three assumptions. But only two of the assumptions, the first and the third, express probability statements. The second has nothing to do with probability. Therefore, we have reduced the equiprobability assumed for the game of roulette to weaker probability assumptions, namely, to the assumptions 1 and 3. They state less than the assertion of equiprobability, since the equiprobability can be derived only if the second assumption is added. If the second assumption is replaced by another, for instance, by the presupposition that the black sectors are twice the size of the red sectors, we can no longer infer the equiprobability of "black" and "red", but shall rather derive a probability relation of 2 to 1. In this sense the theory supplies an *explanation* of the equiprobability. The latter has been derived from more general assumptions.

This reduction represents an inversion of the method of reduction that was attempted in the theory of equiprobable cases. We do not reduce a probability metric of a nonsymmetrical form to an equiprobability, but, conversely, reduce a symmetrical metric, that is, an equiprobability, to a nonsymmetrical probability metric. This way of handling the problem carries the advantage that the principle of "no reason to the contrary" is completely eliminated. The equiprobability does not appear as following from the *absence* of reasons, but as a result of the *existence* of definite reasons, namely, of the occurrence of the facts formulated in the second assumption. Only the question how we arrive at the assumptions 1 and 3 remains to be discussed.

We are dealing here with assumptions that do not specifically concern mechanisms of games of chance, but apply also to physical problems of a very different nature—for instance, problems of the theory of errors, since every evaluation of measurements of physical values will depend on assumptions of this type. Even the rough appraisal of degrees of probability, employed in the assumption 3, is used in many cases that apparently have nothing to do with probability. We always make use of such appraisals in daily life when we regard statements about future events as "practically certain". This result shows that the assumption of the equiprobability concerning mechanisms of games of chance does not contain any assumptions different from those used in probability statements made in everyday life or in science. We must regard the mechanisms of games of chance as mechanical devices by which certain general properties of physical phenomena are transcribed into the special form of equiprobable cases. Mechanisms of this sort express the probability character of nature in an instructive and simplified form, but by no means do they represent the logical archetype of the problem of probability.

With this analysis the seemingly self-evident character of the principle of "no reason to the contrary" finds its explanation. Cases in which this principle is applied always refer to events that, according to the schema of figure 25 (p. 356), can be reduced to a division by small and equal intervals with respect to a probability function. For more complicated mechanisms, for example, the die, we shall use a probability function of several variables; yet the foregoing mathematical considerations can easily be extended to cases of this kind.[2] Our belief in a connection between the symmetry of the

[2] See H. Reichenbach, in *Zs. f. Physik*, Vol. II (1920), p. 163; Vol. IV (1921), p. 448.

terms of the disjunction and the equiprobability originates from a correct estimation of the role that is played in such problems by the assumption 2. The apparently *a priori* inference of symmetry is actually a legitimate inference that leads from an *a posteriori* knowledge, that is, from the assumptions 1 and 3, to an equiprobability by the use of strictly deductive methods as soon as the geometrical symmetry of the cases is assumed. The inference of symmetry is thereby cleared of its mysterious character and becomes a logically justifiable method.

This analysis clarifies the question of nonlinear measure transformations also. When the roulette wheel is surrounded by a square frame and the red and black sectors are projected on the frame, the areas into which the frame is divided will not be equal in size. Yet when the spinning hand is replaced by a spinning light beam, the probability that the light beam comes to rest is equal for red and black areas. However, if the frame is divided into small equal areas, alternately colored, the probabilities of "red" and of "black" will be equal, too, if the areas are sufficiently small. The reason is that the proof expressed in figure 25 (p. 356) is independent of the shape of the curve; the equality of the probabilities derives from a limit property of the curve and is therefore invariant with respect to nonlinear transformations.

These considerations answer the question of the so-called *a priori* determination of the degree of probability. The successful applications of the method can be reduced to probability assumptions of a more general character in combination with physical assumptions of another sort, concerning geometrical or mechanical relations. These inferences replace the fallacious inference in terms of the principle of indifference. What has been regarded erroneously as an *a priori* determination of a degree of probability is thus revealed to be a derivation from other probability assumptions, the validity of which was established *a posteriori*, by means of the rule of induction.

§ 70. The Three Forms of an *A Posteriori* Determination of Degrees of Probability

The analysis presented in the preceding sections can be summed up in the following statement: *there is no a priori ascertainment of a degree of probability; a probability metric can be determined only a posteriori.*

There are three possible ways for *a posteriori* establishment of a probability metric:

1. Degrees of probability can be directly ascertained through induction by enumeration (*statistical probabilities*).
2. A probability metric can be inferred deductively from known probabilities (*deduced probabilities*).
3. A probability metric can be inferred by means of general inductive methods from known observational data (*hypothetical probabilities*).

Method 1 has been pointed out. Method 2 was illustrated in § 69 by the derivation of a probability metric from the assumption of a continuous probability function. The primary probabilities can be ascertained by means of induction by enumeration. The value of method 2 consists in the fact that it can determine a probability rather precisely, even if the original probability values obtained by the inductive rule are known only within wide limits of exactness. The result is demonstrated by the considerations attached to figure 25 (p. 356).

Method 3 has not yet been mentioned. It is carried through as follows: a probability metric is introduced in the sense of a hypothesis; then the observational consequences of the assumption are computed and tested, and thus the truth of the assumption is judged in terms of its consequences. This inductive inference is of the same type as any other inference from observational data to a hypothesis. The fact that in this case the hypothesis concerns a probability metric makes no difference.

An illustration is the probability metric on which the theorems of the kinetic theory of gases are based. The equiprobability of all arrangements of molecules in cells of equal size in the velocity space, and likewise the ergodic hypothesis, are hypothetical assumptions from which observational data concerning thermodynamic phenomena are derived and which, conversely, are tested through these phenomena. The *a posteriori* character of the method is demonstrated by the fact that when, in the realm of low temperatures, observable data were discovered that did not conform to the Maxwell-Boltzmann statistics, the assumed metric was replaced by a different assumption, as expressed in the Einstein-Bose statistics. These statistics correspond to a metric described in § 32 as the metric of nonindividualized combinations. Such inferences offer no particular problems. The determination of a probability metric by these methods is to be incorporated in the general method of indirect evidence and to be discussed in the frame of the analysis of this method.

The method of introducing a probability metric hypothetically has found an elaborate treatment in the method of statistical inference, developed in the last decades by R. A. Fisher, J. Neyman,[1] E. S. Pearson, and A. Wald.[2] The general problem dealt with in these investigations is to find the most plausible probability distribution that comprises a given set of observational data. For instance, a physical quantity u has been measured n times, the values $u_1 \dots u_n$ being the numerical results. What function $d(u)$ represents the most probable assumption for the probability distribution of these values? Problems of this kind cannot be solved without further assumptions. For

[1] "Outline of a Theory of Statistical Estimation Based on the Classical Theory of Probability," in *Philos. Trans. Roy. Soc. London*, Series A, Vol. 236 (1937), p. 333.

[2] *On the Principles of Statistical Inference*, Notre Dame Mathematical Lectures (Notre Dame, Indiana, 1942). This booklet includes a report on the literature in this field.

instance, if it is known that the distribution is a normal curve, its form is ascertained by the methods described in § 43, that is, by the statistical determination of mean and standard deviation. The mathematicians mentioned have developed methods for more general cases, which have acquired great practical significance. The applicability of every such method depends on whether the statistical material satisfies the assumptions presupposed for the method. The validity of the assumptions is tested along with the method; if the method leads to results that conform to further observation, it is accepted in the way any other physical hypothesis is regarded as confirmed through observation.

Of the three methods, the second and third can be carried through only when certain probabilities are known; the first method alone is applicable without such knowledge. The first can, therefore, be used as a *primary method* of finding probabilities, whereas the other two are *secondary methods.* Through both the latter methods a probability is derived deductively from other probabilities. In the second method, the derived probability is the inquired one; in the third method, however, the situation is somewhat involved. The hypothesis inquired is statistical, that is, it states a probability, or a probability distribution; however, the deductive inference does not supply this probability directly, but supplies the probability that the hypothesis is correct and thus the probability that the inquired probability holds. The transition to a determinate probability distribution, or statistical hypothesis, is then made through selection of the distribution for which the maximum probability obtains. For this reason, hypothetically introduced probabilities cannot be derived with certainty from given probabilities, and their use will always include the uncertainty of a hypothesis of which we know only that it is the most probable one. Usually the inference is so devised that it omits the determination of the probability of the second level and supplies directly the distribution that possesses the maximum probability.

With respect to the third method, the necessity of a previous knowledge of some probabilities has not always been recognized, and attempts have been made to construe this method as a primary means of finding probabilities. These attempts break down for two reasons. First, the method presupposes a great deal of inductive knowledge for the selection of the general form of the assumed hypothesis, the choice of which depends on previous experiences in the same field. Second, since the method is meant to supply a probability for a hypothesis, it is dependent on the inference by indirect evidence, which cannot be applied unless certain probabilities are known. This fact is obvious when it is realized that the inference by indirect evidence is a form of the rule of Bayes and therefore determines the probability of a hypothesis—an inverse probability—as a function of certain forward probabilities (§ 21). Among these, the probabilities referring to the observational data, the probabilities $P(A.B_i,C_k)$ of (10, § 21), can often be deductively derived; this will be the case if the B_i represent statistical hypotheses. The antecedent probabilities $P(A,B_i)$, however, must be determined empirically. Often this determination can once more be achieved by means of an inference by indirect evidence; but it is obvious that ultimately inferences by indirect evidence can be made only after some probabilities have been ascertained by the first method, statistical enumeration.

With the intention of evading this consequence, the mentioned attempts at construing the third method as a primary method use inferences that dispense with a statistical determination of antecedent probabilities. The aprioristic school of probability makes use of the principle of indifference for this purpose; unknown antecedent probabilities are regarded as being equal to one another, and Bayes' rule assumes the simple form (12, § 21), which contains no antecedent probabilities. The preceding discussion of the principle of indifference (§ 68) makes it clear that such a procedure is not permissible. Empiricist-minded thinkers, however, have tried to construct an inference by indirect evidence that does not presuppose antecedent probabilities. Of this kind is the *inference by confirmation*, the fallacious character of which was explained at the end of § 21. Another attempt makes use of a *principle of maximum likelihood*, which was introduced by R. A. Fisher[3] with the intention of eliminating the necessity of knowing antecedent probabilities for the establishment of statistical hypotheses. This principle must be given a brief discussion, for which a simple application of the principle may be chosen.

Assume a set of n observations of a random variable u has been made; what is looked for is the probability distribution $d(u)$ that controls the occurrence of these values. This function is regarded as known apart from a parameter s, so that it can be written in the form $d(s;u)$; an extension of the method to more than one unknown parameter is easily given and need not be studied here. The problem is to find the value of s that is made most probable by the n given data. The range of observed values may be divided into r intervals $du_1 \ldots du_r$; the number of observed values within an interval du_i may be $= n_i$. The forward probability w that, if $d(s;u)$ is the distribution, the observed set of values will result, can be computed by the *extended formula of Newton* (see footnote, p. 264), written for probabilities of the form $d(s;u)du$:

$$w = L_n(s;u_1,n_1, \ldots u_r,n_r)du_1^{n_1} \ldots du_r^{n_r}$$

$$L_n(s;u_1,n_1, \ldots u_r,n_r) = \frac{n!}{n_1! \ldots n_r!} d(s;u_1)^{n_1} \ldots d(s;u_r)^{n_r} \tag{1}$$

$$n = n_1 + \ldots + n_r$$

[3] "On the Mathematical Foundations of Theoretical Statistics," in *Philos. Trans. Roy. Soc. London*, Series A, Vol. 222 (1922), p. 309; "Two New Properties of Mathematical Likelihood," in *Proc. Roy. Soc. London*, Series A, Vol. 144 (1934), p. 285.

The function L_n is called the *likelihood function*; the principle of maximum likelihood states that the actual value of s is the one for which the function L_n is a maximum (with respect to variations of s). The function $d(s;u)$ can thus be determined without the use of antecedent probabilities. The maximum of the forward probability replaces the unknown maximum of the inverse probability.

Like all inferences by indirect evidence, this inference must be construed as an application of the rule of Bayes, and in a complete formulation it will include a reference to antecedent probabilities. There must exist an antecedent probability density $q(s)$ for the values of s; the inverse probability density v_n for a distribution $d(s;u)$ is then found through a generalization of (5, § 62), the function w_n of this formula [apart from a constant factor, which drops out in (2)] being replaced by the function L_n:

$$v_n(u_1,n_1, \ldots u_r,n_r;s) = \frac{q(s)L_n(s;u_1,n_1, \ldots u_r,n_r)}{\int_{-\infty}^{+\infty} q(s)L_n(s;u_1,n_1, \ldots u_r,n_r)ds} \tag{2}$$

The product $du_1^{n_1} \ldots du_r^{n_r}$ drops out because it occurs in both numerator and denominator. The distribution of the highest probability is the one for which the function $v_n(s)$ has a maximum (the indication of the constants $u_1 \ldots n_r$ may be omitted). Obviously, the maximum of $v_n(s)$ will in general not coincide with the maximum of $L_n(s)$. The use of the principle of maximum likelihood is therefore restricted to certain conditions. What are these conditions?

A sufficient condition for the coincidence of the maxima of $v_n(s)$ and $L_n(s)$ is the assumption $q(s) = \text{const.}$, that is, the equality of the antecedent probabilities. If the method were dependent on this assumption, the principle of maximum likelihood would be an equivalent of the principle of indifference and subject to the same criticism.[4] But this criticism would be too strict. Only if for all values of s the likelihood were regarded as proportional to the inverse probability would the method presuppose the constancy of $q(s)$; but the maxima of $L_n(s)$ and $v_n(s)$ can coincide within a small interval ds if $q(s)$ is not constant. This is obvious when $q(s)$ has its maximum for the same value s as $L_n(s)$; but even if this is not the case, the shift of the maximum of $v_n(s)$ can be small enough to leave the maximum within the interval ds. It follows, however, that the principle can be used for the determination of the most probable distribution only when something is known about the antecedent probability $q(s)$. Such knowledge is often available; antecedent probabilities are not probabilities of a specific kind, but are accessible to direct statistical determination like all other probabilities.[5] And the situation can be of such

[4] This view is held by M. G. Kendall, "On the Method of Maximum Likelihood," in *Jour. Roy. Stat. Soc.*, Vol. 103 (1940), p. 388.

[5] For an illustration see J. Neyman, "Basic Ideas and Some Recent Results of the Theory of Testing Statistical Hypotheses," in *Jour. Roy. Stat. Soc.*, Vol. 105, Part 4 (1942), p. 298.

a kind that a rather inexact knowledge of $q(s)$ is sufficient to justify the use of the principle of maximum likelihood. This is a general characteristic of all inferences supplying inverse probabilities: the power of the inference by indirect evidence derives from the fact that, if the antecedent probabilities are known within wide limits, the inverse probabilities are determinable within narrow limits. The considerations attached to (12, § 62) show how such results can be achieved. For this reason, the inference by indirect evidence can be used for the establishment of precise quantitative results even when knowledge of the antecedent probabilities is of a merely qualitative nature.

As long as the principle of maximum likelihood is regarded as a principle determining one hypothesis as preferable to others, in the sense of being more likely than others, it is dependent on estimates of antecedent probabilities, like all other forms of the inference by indirect evidence. A way of overcoming this limitation will be discussed in § 88; however, it involves a renunciation of attempts to find the most probable hypothesis and reduces the principle to an improved version of the first method. But even if the principle can be freed from the use of antecedent probabilities, it includes other presuppositions that can be proved by inductive methods only: it presupposes a knowledge of the form of the function $d(s;u)$ and, furthermore, a knowledge that the observed values of u are independent of one another and admit of the use of the special theorem of multiplication for the computation of the function L_n. Corresponding qualifications hold for all similar principles employed in the theory of statistical inference, including Neyman's methods,[6] which avoid the principle of maximum likelihood.

I shall use the term *advanced knowledge* to denote a state of knowledge that includes a sufficient number of probabilities; it then follows that the second and third methods of ascertaining probabilities are applicable only in advanced knowledge. For *primitive knowledge*—a state of knowledge that does not include a knowledge of probabilities—the rule of induction is the only instrument for the ascertainment of probabilities. Incidentally, the rule of induction can also be applied in advanced knowledge and then assumes a somewhat different function, which will be studied in § 86; but for primitive knowledge it is the only available instrument. Many controversies about the legitimacy of probability methods arise from a confusion of these two states of knowledge: in advanced knowledge the whole technique of the calculus of probability is at our disposal and can be adduced to justify the methods employed, whereas primitive knowledge requires other means of justifying inductive inferences.

The given analysis shows that only for statistical induction must a justification be required. The other two methods of ascertaining probabilities are

[6] This is also Neyman's opinion; *Jour. Roy. Stat. Soc.*, Vol. 105, Part 4 (1942), p. 294.

covered by the logical system of the calculus of probability. The axiomatic construction of this calculus is proof that the rule of induction is the only nonanalytic principle necessary for the application of the calculus to reality, if the frequency interpretation of probability is assumed. The rule does not enter into the formal calculus, which is a deductive system like all other mathematical disciplines; however, it is required for the applied calculus of probability, which deals with empirical statistics. The assertability of applied probability statements is thus reduced to the problem of a justification of the rule of induction.

The discussion of the problem must again be postponed, since it cannot be given without the construction of specific logical tools. The result may be emphasized, however, that the justification of induction is the only problem remaining for a theory of probability when the frequency interpretation is assumed.

At this point I must discuss an opinion expressed by some mathematicians who deny the existence of a problem of application and thus wish to evade the difficulties of the problem of induction.[7] According to this opinion, the problem of application in the calculus of probability does not differ from that existing for other mathematical disciplines, in particular, geometry. Mathematically speaking, geometry is a deductive system based on certain axioms; if it is to hold for the physical world, the interpretation of the axioms must be so chosen that the axioms are true. Thus light rays may be chosen as an interpretation of straight lines, solid rods as defining spatial congruence, and so on, when it is proved that these physical objects satisfy the relations postulated in the axioms. The problem of the application of geometry is thus solved by the *method of adequate selection of objects.*

It was mentioned above that similar considerations hold for the application of the axioms of probability; and the result was expressed in the requirement that any interpretation of the P-symbol must be an admissible interpretation. The analysis shows, however, that for the interpretation of the calculus of probability there arises a specific problem that has no analogue in the interpretation of other mathematical disciplines: the selection of an interpretation before we know whether the interpretation is an admissible one. In this form we can state the problem of induction. If the sequence under consideration has a limit of the frequency at p, it certainly constitutes an admissible interpretation of the statement $P(A,B) = p$. What creates the difficulty, however, is the fact that we must use the sequence as an interpretation before we know whether it has a limit of the frequency at p, even before we know whether it has a frequency limit at all. This specific difficulty makes the problem of the application of probability statements unique.

[7] This opinion was stated by R. von Mises, *Vorlesungen aus dem Gebiete der angewandten Mathematik,* Vol. I: *Wahrscheinlichkeitsrechnung* (Leipzig, 1931), p. 21; and by E. Tornier, "Grundlagen der Wahrscheinlichkeitsrechnung," in *Acta Math.*, Vol. 60 (1933), p. 313.

It is true that difficulties of this kind arise in the application of other mathematical disciplines; thus the verification of the axioms of geometry involves measurements that supply only approximative results, and the assumption is made that on repetition the measurements will converge toward a certain limit. But it is clear that this assumption is not of a geometrical nature; it rather represents an inductive inference and thus falls in the domain of investigations concerning probability sequences. The example illustrates the fact that, when any mathematical discipline is applied to physical reality, probability inferences intervene; and that therefore all application problems, in addition to the method of adequate selection of objects, include methods used in the application of the calculus of probability. The solution of the problem of induction is thus shown to be the necessary prerequisite for a solution of any problem of application.

§ 71. Attempts at a Single-Case Interpretation of Probability

After the discussion of the frequency meaning of probability, the investigation must turn to linguistic forms in which the concept of probability refers to an individual event. It is on this ground that the frequency interpretation has been questioned. Some logicians have argued that such usage is based on a different concept of probability, which is not reducible to frequencies. Is the existence of two disparate concepts of probability an inescapable consequence of the usage of language?

At first sight, indeed, it would seem that a probability applied to a single case has nothing to do with a frequency. We say, "It is probable that it will rain tomorrow"; "It is improbable that Julius Caesar was in Britain"; and so on, thus referring probability to a single event. How does it help to know that on a certain percentage of days of certain meteorological conditions it will rain, when we wish to know the probability for rain on one particular day? Similarly, the example of Julius Caesar's stay in Britain has often been quoted as an instance of a probability that makes a frequency interpretation impossible. We must analyze the various meanings that can be suggested for such a second concept of probability.

It was explained above that the formal system of the probability calculus does not distinguish one interpretation from others as the only admissible one; formally speaking, therefore, the possibility of the occurrence of several interpretations in practical applications cannot be denied. The analysis of the problem must be carried through from other viewpoints.

What must be asked is whether an interpretation is adequate to account for the use of probabilities as degrees of reliability or as instruments suitable for an evaluation of predictions. For it is the *predictional value* that makes probability statements indispensable tools of conversational as well as of scientific language. We know that a specific event either will happen or will not. The degree of probability, therefore, will be of no use after the truth about the occurrence of the event is known: probability is used as a substitute for truth so long as the truth is unknown. If the event is to happen in the future, the degree of its probability qualifies the reliability of a prediction; and if the event belongs to the past we can also regard the probability as the measure of the quality of a prediction, namely, of a prediction about possible future verifications of a past event the truth value of which is unknown. The criterion for the justification of an interpretation lies in its adequacy for the purpose of prediction.

This criterion immediately rules out an interpretation like the geometrical interpretation introduced in § 40. That probability can be used as a measure of areas is a logically interesting fact, but no one will be inclined to use the term "probability" instead of "area" for surveying or similar purposes. Only with respect to continuous probability sequences (§ 46) is the measure interpretation of probability important for applications; but the interpretation is then an analogue of the frequency interpretation, replacing a counting of elements by the measuring of a length, and its discussion thus belongs in the frame of the analysis of the frequency interpretation.

The first interpretation of the probability of single events is the *degree of expectation* with which an event is anticipated. The feeling of expectation certainly represents a psychological factor the existence of which is indisputable; it even shows degrees of intensity corresponding to the degrees of probability. Difficulty, however, arises from the fact that the degree of expectation varies from person to person and depends on more factors than the degree of the probability of the event to which the expectation refers. Apart from the probability of an event, emotional associations will influence the feeling of expectation. If it is a desirable event, as, for instance, the passing of an examination, optimistic persons will anticipate it with too-certain expectations, whereas pessimistic persons will think of it in terms of too-uncertain expectations.

The contrary holds for undesirable events. The discrepancy between expectation and degree of probability is expressed in the phenomenon of *fear*. In a thunderstorm there is a certain low probability that we might be struck by lightning; if it strikes our immediate surroundings we have no means of protecting ourselves against the violence of the electrical discharge, which does not keep to prescribed paths. But we need not be afraid, as the probability of this event is extremely low—lower, for instance, than the probability of a traffic accident. If, in spite of this fact, many people are afraid of thunderstorms, their behavior demonstrates that their feeling of expectation is much higher than would correspond to the probability of being struck by lightning. The emotional tension caused by the roaring thunderbolts may make the feeling of expectation grow far too intense. In fact, fear may be defined as the exaggeration of the feeling of expectation concerning the probability of an undesirable event. Of course, it would be a mistake in the other direction to exclude the possibility of being struck by lightning; we should try to reduce the feeling of expectation to the proper degree. But few persons succeed in the attempt.

Corresponding considerations hold for the opposite of fear. *Hope* is the exaggeration of the feeling of expectation concerning the probability of a desirable event. The alluring call of hope often persuades us to reckon seriously with probabilities that would be discounted after sober analysis. The delusion concerns not only events influencing the life of an individual, but anticipations of social conditions, of the progress of civilization, and of a better world in future. We are only too willing to believe that what we desire will come true. This *expectational illusion on emotive grounds* plays a great role in political ideology.

These considerations show why the feeling of expectation does not represent the meaning of the probability concept of applications. We rather conceive probabilities as numerical values holding for the physical world to which our feeling of expectation must be adjusted, as measures of what the intensity of our expectation *should be*, but not of what it *is*. We try to regulate the feeling of expectation to the correct degree by means of training and adaptation. Much of the wisdom of the so-called experienced person consists in being well versed in this art. But if there is an adjustment of expectation to the correct degree, there must exist an interpretation of probability independent of the feeling of expectation.[1]

The second interpretation of the probability of the single case is connected with the principle of indifference. It was mentioned in § 68 that the principle has played a role in the history of the theory of probability. In order to maintain the principle of indifference against criticisms of the kind presented in § 68, an interpretation of probability was developed that defines the meaning of probability in such a way that the principle seems to be legitimate.

The interpretation is based on the logical *principle of retrogression*, which is involved in the theory of meaning. It states that the meaning of a sentence is given by the method of its verification.[2] Since, according to the adherents

[1] Apart from this consideration, there are other reasons why the feeling of expectation cannot be carried through as an adequate interpretation of the meaning of the axiomatic system. Even if we admit that within certain limits the intensity of expectation corresponds to the degree of probability, it can by no means be asserted that the correspondence still holds in dealing with more complicated forms of probability relations. Thus we cannot say that axioms of the calculus of probability represent laws of the feeling of expectation, for example, that the theorem of multiplication may be applicable to the intensity of the feeling of expectation. Psychological laws are much too complicated to fit the schema of mathematical probabilities.

[2] See *EP*, p. 49.

of the principle of indifference, we determine a probability by counting the terms of an exclusive disjunction in which we have no reason to prefer one term, the principle of retrogression supplies the result that the meaning of probability is given by reference to such a disjunction. We thus arrive at a *retrogressive interpretation* of probability.

For instance, that the probability of obtaining face 6 with a die is $\frac{1}{6}$ means, according to this interpretation, that face 6 is a term of a disjunction of six terms and that we have no reason to prefer one of the terms.[3] It is obvious that this interpretation justifies the use of the principle of indifference, since then the probability statement states no more than what is assumed as the premise of that principle. Thus we have here a nonfrequency interpretation of probability that is compatible with the principle of indifference (see p. 354).

It is equally obvious, however, that in this interpretation the probability statement loses its predictional value. Why should we bet on the occurrence of the event "non-6" rather than on the occurrence of "6"? The retrogressive interpretation narrows the meaning of the probability statement in such a way that the assertion of the statement is justified on the basis of the principle of indifference; but in the transition from probability statements to predictions, or advices to action, there reappears the very problem that the retrogressive interpretation was intended to evade and that the principle of indifference cannot solve.

The third interpretation of the probability of the single case was constructed with the intention of escaping the difficulties of both the frequency and the retrogressive interpretation. Its proponents insist that we must renounce defining probability in terms of other concepts. Probability, it is argued, is not reducible to certainty, be it the certainty of frequency statements or that of statements about terms of a disjunction. Instead, probability is regarded as a *primitive concept* that is not capable of further definition. According to this conception, the statement that the probability of an expected event is $\frac{1}{6}$ has a meaning of its own, which is comparable to the meaning of primitive notions of logic; and this meaning cannot be interpreted as a frequency or a report about terms in a disjunction. The conception is sometimes stated in the form that probability is a *rational belief*, that the laws of probability constitute a *quantitative logic* based on a self-evidence comparable to that of ordinary logic.[4] However, adherents of the primitive-concept interpre-

[3] So far as I know, the first to use this interpretation was C. Stumpff, in an article published in *Ber. d. bayer. Akad., philos. Kl.*, 1892. The interpretation has been taken up in our day by some logicians under the influence of ideas of Ludwig Wittgenstein, *Tractatus logico-philosophicus* (London, 1922), p. 113; thus by A. Waisman, in *Erkenntnis*, Vol. I (1930), p. 229.

[4] This concept is represented by the ideas of J. M. Keynes, *A Treatise on Probability* (London, 1921), and was continued by Harold Jeffreys, *Theory of Probability* (Oxford, 1939). I cannot say to what extent it is also present in the ideas of Carnap and others about confirmation (see § 88).

tation do not always clearly distinguish it from the retrogressive interpretation; some logicians vacillate between the two interpretations, depending upon what they wish to prove.

The difficulties of the primitive-concept interpretation appear so overwhelming that it is hard to understand how a logician can commit himself to it. First, the degree of probability remains unverifiable in terms of the event predicted. When the event expected with a probability $\frac{5}{6}$ is observed, does this observation verify the probability statement? Obviously not, since the nonoccurrence of the event, too, is compatible with the probability statement. And how could one distinguish between a probability, say, of $\frac{5}{6}$, and one of $\frac{2}{3}$? The numerical value of a probability cannot be ascertained from one observation. We do not escape the predicament by restricting probability statements to relations of order stating merely that a probability is higher or lower than another (see p. 380); such relations cannot be verified by one observation either.

It is sometimes argued that verification of the degree of probability is obtained, not by observation of the event, but by other methods, such as are used in the principle of indifference. Thus it is said that the probability $\frac{1}{6}$ for the face of the die is verified by the existence of the six faces of the die. But with this argument all the difficulties of the principle of indifference rise anew; it seems incomprehensible that absence of reason to the contrary should be a guaranty of equal probabilities. To use the principle of indifference for verification of statements of probability degrees is not permissible when probability has a meaning of its own; it is permissible only for a retrogressive interpretation. In resorting to the principle of indifference for verification of probability statements, adherents of the primitive-concept interpretation adopt a method of defense that is suitable only for a retrogressive interpretation. They forget that with such a defense the primitive-concept interpretation is abandoned and the probability statement loses its predictional value.

Second—and this point is closely connected with the first—even if the probability statement is regarded as having a meaning of its own, not subject to a retrogressive interpretation, there is no way of explaining its usefulness for predictions. We are dealing here with a question of meaning, not of assertability. Even if it were possible to justify the assertion of a probability statement in this interpretation, there remains the question of whether its meaning is of such a kind that the statement can be used as a guide for predictions. Assume, for a moment, that the truth of the probability statement has been established. Why, then, is it advisable to predict the event of the greater probability? If the probability statement cannot be verified through the occurrence or nonoccurrence of future events, it does not state anything concerning future events and thus cannot be used as an advice for predictions. In other words, what difference does it make if a man, instead of assuming the event B to happen, prefers to assume that the event $\bar{B}$ will happen? Suppose we could prove that the probability of B is $= \frac{3}{4}$. What kind of later experience would demonstrate to a man who had assumed $\bar{B}$ to happen that he had acted unwisely? I do not think that the man would recognize a proof that appealed to rational belief and attempted to evoke his faith in the rules of probability. The only convincing argument would be an experience he might undergo with respect to the occurrence or nonoccurrence of the event. But if some experience proved to him that he was wrong, it would be a verification of the probability statement, the meaning of which would thus no longer be primitive, but would be translated into observable properties of future events.

To state the argument briefly: if probability has something to do with the reliability of predictions, the probability statement must be verifiable in terms of the occurrence of the event predicted; otherwise the statement will be empty so far as predictions are concerned. The frequency interpretation satisfies this condition inasmuch as it verifies a degree of probability through repeated occurrence of the event. If, however, the meaning of the probability statement refers to a single event, it is impossible to verify the statement in terms of the occurrence of the event; and therefore the statement has no predictional value.

Adherents of the primitive-concept interpretation usually maintain that the rules of probability constitute a part of logic, and that if human behavior is to be regarded as reasonable it must follow these laws as well as it follows the laws of deductive logic. This conception must be analyzed more closely. First, the laws of probability will then include a rule for the ascertainment of probability, which is usually constructed as a version of the principle of indifference; second, however, they will include the axioms of the calculus of probability, since these axioms cannot be regarded as analytic if the frequency interpretation is rejected. We thus are faced by a comprehensive system of synthetic statements that are asserted to be a part of logic.

The primitive-concept interpretation, therefore, leads to a conception of logic for which logic is not analytic throughout and therefore is not empty. Logic is conceived, rather, as a science revealing the ultimate laws of the physical world, not by means of sense perception but by a rational insight into the nature of the universe. Such a conception of logic has found its classical expression in Kant's philosophy of the *synthetic a priori*, which asserts that there exists a knowledge that is synthetic and yet is independent of experience and strictly certain. Knowledge of this kind is characterized by *synthetic self-evidence*, analogous to the self-evidence of analytic statements. In fact, the rational belief that is alleged to be the basis of the ascertainment of a probability, or of the laws of probability, represents a revival of the rationalism of Leibniz and Kant, of the *vérités de raison* as well as of the *synthetic a priori*.

This conception of logic cannot be discussed in detail here. Suffice it to say that the claims of a *synthetic a priori* cannot be recognized by an empiricist philosophy; that the development of science since the time of Kant has proved his philosophy of the *synthetic a priori* to be untenable; that such a philosophy must be classified as a remnant of the speculative metaphysics of earlier phases of civilization and is incompatible with modern scientific method. In fact, the admission of synthetic self-evidence would spell the breakdown of scientific philosophy. Trust in synthetic self-evidence means belief that nature must conform to reason; but the history of knowledge has shown that what we call "reason" is the product of a physical environment of rather simple structure and that with the widening scope of empirical knowledge the so-called laws of reason have changed. There is no such thing as synthetic self-evidence; the only admissible sources of knowledge are sense perception and the analytic self-evidence of tautologies.

With this result we must abandon all attempts at finding a satisfactory interpretation of probability statements that restricts their meaning to a single case. That the frequency interpretation can supply an empiricist solution of the probability of the single case will be shown in § 72.

§ 72. The Frequency Interpretation of the Probability of the Single Case

The analysis of meaning has suffered from too close an attachment to psychological considerations. The meaning of a sentence has been identified with the mental images associated with the utterance of the sentence. Such conception leads to meanings varying from person to person; and it will not help to find the meaning that a man would adopt if he had a clear insight into the implications of his words. Logic is interested not in what a man means but in what he *should* mean, that is, in the meaning that, if assumed for his words, would make his words compatible with his actions.

When the meaning of probability statements about single events is analyzed according to this objective criterion, it is found that the frequency interpretation can be applied to this case too. True, we must renounce a reconstruction of subjective psychological intentions; but, since we found that it is not possible to translate the expectation associated with the anticipation of a future event into a logical category, we shall welcome the construction of a logical substitute that can take over the function of a probability of the single case without being such a thing in the verbal sense.

Assume that the frequency of an event B in a sequence is $= \frac{5}{6}$. Confronted by the question whether an individual event B will happen, we prefer to answer in the affirmative because, if we do so repeatedly, we shall be right in $\frac{5}{6}$ of the cases. Obviously, we cannot maintain that the assertion about the individual event is true. In what sense, then, can the assertion be made? The answer is given by regarding the statement not as an assertion but as a *posit*. We do not say that B will occur, but we posit B. We do so if $P(B) > \frac{1}{2}$; otherwise we posit $\bar{B}$. The word "posit" is used here in the same sense as the word "wager" or "bet" in games of chance. When we bet on a horse we do not want to say by such a wager that it is true that the horse will win; but we behave as though it were true by staking money on it. *A posit is a statement with which we deal as true, although the truth value is unknown.* We would not do so, of course, without some reason; we decide for a posit, or a wager, when it seems to be the best we can make. The term "best" occurring here has a meaning that can be numerically interpreted; it refers to the posit that will be the most successful when applied repeatedly.

If we wish to improve a posit, we must make a selection S such that $P(A.S,B) > P(A,B)$.[1] If we now posit B only in the case $A.S$ and omit a posit in the case $A.\bar{S}$, we obtain a relatively greater number of successes than by the original posit. It is even more favorable to construct the selection so that at the same time $P(A.\bar{S},B)$ is $< \frac{1}{2}$. We then always posit $\bar{B}$ in the case $A.\bar{S}$. We can thus make a posit for each element of the original sequence and obtain a greater number of successes. The procedure may be called the *method of the double posit.*

It should be noticed that we cannot improve a posit without knowing a selection S that leads to a greater probability. If we were to posit arbitrarily sometimes B, sometimes $\bar{B}$, we would in general construct a selection S for which $P(A.S,B) = P(A,B) = P(A.\bar{S},B)$, that is, a selection of the domain of invariance. Positing B in the case $A.S$ would then lead to the same relative number of successes as in the main sequence, whereas positing $\bar{B}$ in the case $A.\bar{S}$ would lead to a smaller ratio of successes.

In dealing with exclusive disjunctions $B_1 \vee \ldots \vee B_r$ of more than two terms, if we are compelled to posit only one of the terms (if a posit $B_2 \vee \ldots \vee B_r$ is impossible because of practical reasons), the B_k that carries the greatest probability will be the best posit. In this case, therefore, the probability $\frac{1}{2}$ no longer represents a critical value.

The method of positing serves to utilize probability statements for decisions in regard to single cases. It plays an important role in all practical applications. The merchant who stores a great amount of merchandise for the season, the farmer who wants to get in his crop, the physician who prescribes a cure—all must make decisions, though they know only probability statements about the factors determining success: the merchant about the prospective demand, the farmer about the prospective weather, the physician about the illness that presumably confronts him. They make posits by assuming the occurrence

[1] If we have $P(A.S,B) < P(A,B)$, the selection $\bar{S}$ will have the desired property because it then follows from the rule of elimination that $P(A.\bar{S},B) > P(A,B)$. See the discussion following (11*b*, § 19).

of the events that they consider to be the most probable, according to their experience. Each endeavors to improve his posit by increasing the probability through a more precise analysis of the actual conditions, that is, by making a selection S such that a greater probability will hold for the subsequence determined by S, and such that, if possible, even the method of the double posit becomes applicable. The merchant will explore the market situation more thoroughly, the farmer will study the official weather forecast, the physician may try to analyze the condition of his patient more exactly by taking X-rays. There is no instance in which certainty is reached; only a very high probability is attainable.

It is not necessary for the construction of a sequence that similar cases repeat themselves. If we must make a decision today about the prospective weather, tomorrow about the state of an illness, the day after tomorrow about some financial transaction, and if we always posit the most probable case, our decisions represent a sequence in which the probability changes from element to element, that is, a sequence that belongs to the type of the Poisson sequence (§ 56). Here only one horizontal sequence of the Poisson lattice is realized. But this sequence suffices to obtain a statistical success. Since probabilities usually have values around 1 or 0, we shall be able to apply the method of the double posit. The sequence of the numerous actions of a single day—when we turn on the faucet, hoping that the water will run; when we call the telephone number of a friend, hoping to obtain a connection; and so on—represents a rather long Poisson sequence. The statistical justification of the posit, therefore, is applicable to the actions of a single person.

If we are asked to find the probability holding for an individual future event, we must first incorporate the case in a suitable reference class. An individual thing or event may be incorporated in many reference classes, from which different probabilities will result. This ambiguity has been called the *problem of the reference class*. Assume that a case of illness can be characterized by its inclusion in the class of cases of tuberculosis. If additional information is obtained from an X-ray, the same case may be incorporated in the class of serious cases of tuberculosis. Depending on the classification, different probabilities will result for the prospective issue of the illness.

We then proceed by considering *the narrowest class for which reliable statistics can be compiled*. If we are confronted by two overlapping classes, we shall choose their common class. Thus, if a man is 21 years old and has tuberculosis, we shall regard the class of persons of 21 who have tuberculosis. Classes that are known to be irrelevant for the statistical result may be disregarded. A class C is irrelevant with respect to the reference class A and the attribute class B if the transition to the common class $A.C$ does not change the probability, that is, if $P(A.C,B) = P(A,B)$. For instance, the class of persons having the same initials is irrelevant for the life expectation of a person.

We do not affirm that this method is perfectly unambiguous. Sometimes it may be questioned whether a transition to a narrower class is advisable, because, perhaps, the statistical knowledge regarding the class is incomplete. We are dealing here with a method of technical statistics; the decision for a certain reference class will depend on balancing the importance of the prediction against the reliability available. It is no objection to this interpretation that it makes the probability constructed for the single case dependent on the state of our knowledge. This knowledge may even be of such a kind that it does not determine one class as the best. For instance, we may have reliable statistics concerning a reference class A, and likewise reliable statistics for a reference class C, whereas we have insufficient statistics for the reference class $A.C$. The calculus of probability cannot help in such a case because the probabilities $P(A,B)$ and $P(C,B)$ do not determine the probability $P(A.C,B)$. The logician can only indicate a method by which our knowledge may be improved. This is achieved by the rule: look for a larger number of cases in the narrowest common class at your disposal.

Whereas the probability of a single case is thus made dependent on our state of knowledge, this consequence does not hold for a probability referred to classes. If the reference class is stated, the probability of an attribute is objectively determined, though we may be mistaken in the numerical value we assume for it on the basis of inductions. The probability of death for men 21 years old concerns a frequency that holds for events of nature and has nothing to do with our knowledge about them, nor is it changed by the fact that the death probability is higher in the narrower class of tuberculous men of the same age. The dependence of a single-case probability on our state of knowledge originates from the impossibility of giving this concept an independent interpretation; there exist only substitutes for it, given by class probabilities, and the choice of the substitute depends on our state of knowledge. My thesis that there exists only one concept of probability, which applies both to classes and to single cases, must therefore be given the more precise formulation: there exists only one legitimate concept of probability, which refers to classes, and the pseudoconcept of a probability of a single case must be replaced by a substitute constructed in terms of class probabilities.

The substitute is constructed by regarding the individual case as the limit of classes becoming gradually narrower and narrower. The method is justified in the theory of probability by the fact that, as explained above, we obtain a greater number of successes if we employ the probability of the subsequence $P(A.S,B)$ and not the probability $P(A,B)$ as the basis of our posits. A repeated division of the main sequence into subsequences will lead to progressively better results as long as the probability is increased at each step. According to general experience, the probability will approach a limit when the single case is enclosed in narrower and narrower classes, to the effect that, from a certain step on, further narrowing will no longer result in noticeable improvement. It is not necessary for the justification of this method that the limit of the probability, respectively, is $= 1$ or $= 0$, as the hypothesis of causality assumes. Neither is this necessary *a priori*; modern quantum mechanics asserts the contrary.[2] It is obvious that for the limit 1 or 0 the probability still refers to a class, not to an individual event, and that the probability 1 cannot exclude the possibility that in the particular case considered the prediction is false. Even in the limit the substitute for the probability of a single case will thus be a class probability, and we shall always depend on the method of positing.

Besides choosing a suitable reference class, we must also choose a sequence into which the individual case considered is to be incorporated. This choice is usually less difficult than that of the reference class because the frequency will be the same for most sequences that can be reasonably chosen. We often follow the time order of the events observed or of the observations made. One of the rules to be required is that a knowledge of the attribute of the individual case should not be used for the construction of the order of the sequence. (See the remarks on random sequences in § 30.)

There are, however, instances in which the choice of the sequence is connected with ambiguities, as in lattice arrangements where the horizontal and the vertical sequences converge to different limits. If a particular element y_{ki} of the lattice is considered, the probability assumed for it depends on whether further observations concern the horizontal or the vertical sequence to which it belongs. An illustration is offered by the rule of succession (22, § 62), the probability value $\frac{n+1}{n+2}$ of which refers to a column and is applicable only if a sequence in the vertical direction, supplied by a set of horizontal initial sections of a certain kind, represents the experiences to be envisaged. If, on the contrary, the horizontal sequence to which y_{ki} belongs is continued, the probability value characterized by the maximum inverse probability v_n is $p = 1$, a value which is then preferable to the value $\frac{n+1}{n+2}$. This result means that if many horizontal sequences of this kind are considered, most of them will have a limit of the frequency close to 1, which is thus the most appropriate value to be transferred to the element y_{ki}. The illustration makes it obvious that the probability assumed for the single case depends on the mode of procedure by which the single case is incorporated in a sequence.

The solution offered here for the probability of the single case is essentially different from the interpretations discussed in § 71. I regard the statement about the probability of the single case, not as having a meaning of its own,

[2] See H. Reichenbach, "Das Kausalprinzip in der Physik," in *Naturwissenschaften*, Vol. 19 (1931), p. 716; and *Philosophic Foundations of Quantum Mechanics* (Berkeley, 1944), § 1.

but as representing an elliptic mode of speech. In order to acquire meaning, the statement must be translated into a statement about a frequency in a sequence of repeated occurrences. The statement concerning the probability of the single case thus is given a *fictitious meaning*, constructed by a *transfer of meaning from the general to the particular case.* The adoption of the fictitious meaning is justifiable, not for cognitive reasons, but because it serves the purpose of action to deal with such statements as meaningful.

For a better understanding of the solution, consider the analogous solution of a problem of deductive logic. When we speak of a necessary synthetic or physical implication in an individual case, we mean that the case is an instance of a general law. The statement, "If you press this button, the bell will ring", expresses a physical necessity and thus a reasonable implication. And yet by this classification of the statement we mean only that the same adjunctive implication holds in all similar cases. Physical necessity is interpretable in terms of "always". It was explained in the discussion of the general implication (3, § 6) that this interpretation involves a transfer of meaning from the general to the particular case. A similar transfer is characteristic for probability statements. The statement, "If you press this button, the bomb will hit the target with the probability $\frac{2}{3}$", derives its meaning from a reference to generality, just as does the statement about the bell. The only difference is that the probability statement indicates a frequency relation that does not hold for all cases, but is restricted to a certain fraction of cases.

This interpretation of probability statements is complicated by the following peculiarity. If we have an implication $(A \supset B)$, we can add an arbitrary term in the implicans, that is, we can derive the implication $(A . C \supset B)$. If we have a probability implication $(A \underset{p}{\ni} B)$, however, the addition of a term in the implicans will, in general, lead to a different degree of probability, that is, to a probability implication $(A . C \underset{q}{\ni} B)$ where q is different from p. This is why the choice of the reference class is easily made for a general implication, whereas it is difficult to make it for a probability implication. Once a class A is found such that $(A \supset B)$ holds, we know that if $x_i \epsilon A$ we shall have $y_i \epsilon B$; it does not matter to what other classes the event x_i belongs. For a probability implication there is no such simple relation. We must be aware of the possibility that, if x_i belongs to both A and C, the reference to the common class $A . C$ may lead to a value of the probability different from the one resulting for the reference class A.

Therefore, we can ask only for the best reference class available, the reference class that, on the basis of our present knowledge, will lead to the greatest number of successful predictions, whether they concern hits of bombs, cases of disease, or political events. If no statistics are available for the common class $A . C$, we shall base our probability calculations on the reference class A, and must renounce the improvement in the success ratio that might result from the use of the reference class $A.C$ in combination with the method of the double posit. Such a procedure seems reasonable if we realize that narrowing the reference class means nothing but increasing the success ratio, and that there is no reference class that permits the prediction of a single case. This goal, which could be reached if we had a knowledge of synthetic logical implications combining past and future events, is unattainable if probability implications are all that we have to connect the past with the future.

We must renounce all remnants of absolutism in order to understand the significance of the frequency interpretation of a probability statement about a single case. But there is no place for absolutism in the theory of probability statements concerning physical reality. Such statements are used as rules of behavior, as rules that determine the most successful behavior attainable in a given state of knowledge. Whoever wants to find more in these statements will eventually discover that he has been pursuing a chimera.

§ 73. The Logical Interpretation of Probability

In order to construct a logical form for probability statements concerning single cases, it is advisable to introduce a change in the logical classification of probabilities. In the frequency interpretation, a probability is regarded as a property of a sequence of events. Correspondingly, the statement about the probability of a single case is regarded as stating a property, though fictitious, of an individual event. It is possible, however, to go from events to sentences about events, and to regard a probability, not as a property of the event, but as a property of the sentence about the event. Instead of saying, for instance, that the probability of obtaining face 6 with a die is $= \frac{1}{6}$, we can say that the probability of the sentence, "Face 6 will turn up", is $= \frac{1}{6}$. By this transition, probability is made a rating of propositions; and probability statements belong not in the object language but in the metalanguage.

The dual possibility of conceiving probability was first seen by George Boole,[1] who wrote, "There is another form under which all questions in the theory of probabilities may be viewed; and this form consists in substituting for *events* the propositions which assert that those events have occurred, or will occur". I shall introduce the term *logical interpretation of probability* for this conception; the conception previously used will be called *object interpretation*. The logical interpretation offers the advantage that the probability attached to the single case assumes the form of a truth value of a proposition or, rather, since the proposition can be maintained only in the sense of a posit, of the truth value of a posit. We shall use the term *weight* for the truth value of the posit; the probability of the single case, therefore, is regarded, in the

[1] *The Laws of Thought* (London, 1854), p. 247.

logical interpretation, as the weight of a posit. A posit the weight of which is known is called an *appraised posit.*

When we say that probability assumes the form of a truth value, we use the latter term in a wider sense than usual. Classical logic is two-valued; it knows only the two truth values *true* and *false.* In regarding probability as a truth value we construct a multivalued logic, differing from other such logics in that it is a logic with a continuous scale of truth values ranging from 0 to 1. The formal construction of this *probability logic* will be carried through in chapter 10.

An analysis of language reveals that many of its elements can be understood only from the viewpoint of probability logic. We often use sentences referring to individual events that are not asserted to be certainly true; and we indicate our truth evaluation by words like "probably", "likely", or "presumably". The Turkish language possesses a particular mood of the verb expressing that a statement about a past event is not maintained as certain, but only probable.[2] Such forms of language are expressions of the predicate "weight".

That the weight referred to in such sentences is reducible to a frequency meaning is demonstrable, not only for sentences for which the reference class is obvious—as, for example, "It will probably rain tomorrow"—but also for statements that do not easily lend themselves to statistical interpretation. For instance, the statement that Julius Caesar was in Britain must be regarded as a posit having a certain weight that is translatable into a frequency statement. When we look for the methods by which the weight is ascertained, we usually discover the statistical origin of the weight. Thus, in order to ascertain the reliability of the statement about Caesar's stay in Britain, we investigate the number of chroniclers who report such a fact; and we measure the reliability of the individual chronicler by the number of his reports that are corroborated by reports of other writers.[3]

True, we often prefer an intuitive appraisal of the weight to a statistical enumeration. For instance, we judge intuitively from his presentation whether a writer is a reliable authority. But though an intuitive appraisal may sometimes make a rationalized inference unnecessary, it does not invalidate the inference. Thus the inferences of the meteorologist in regard to the weather of the next day are not made false by the fact that the intuitive appraisal of a sailor may arrive at the same result by simpler methods. We prefer an intuitive appraisal to a statistical determination only if statistics are incomplete. The human mind is fortunate in being endowed with the ability of intuitive appraisal; in many cases the use of this talent leads to a better

[2] See *ESL*, p. 338.

[3] In a more precise analysis the inference must be interpreted through explanatory induction (see § 84); my presentation applies a simplification.

determination of probability values than the compilation of incomplete statistics.

The analysis of mental processes during an intuitive action must be left to the psychologist—the logician can ask only for the rational reconstruction of an action. Such a reconstruction is given in the statistical interpretation. For instance, we must regard the scientific inference of the meteorologist as a rational reconstruction of what the sailor does intuitively. In the same sense, the statistical interpretation of the weight of sentences about individual events, like Caesar's stay in Britain, must be regarded as the rational reconstruction of an intuitive estimate of the weight. That, on the other hand, statistics are necessary not only in ascertaining the weight but also in establishing the meaning of the probability statement is apparent from the fact that we use statistics for further verification of the statement, in the form of a verification of the statistical predictions included in the statement.

It has been argued[4] that probability statements of the kind considered are not quantifiable, that they are intended to state only a relation of order expressible by the terms "more probable" or "less probable". It is true that we often restrict ourselves to the statement of order relations. The verification, obviously, will be easier than that of quantitative relations because the statement of order states less than that of quantity. It would be a serious mistake, however, to believe that the employment of relations of order is a proof that quantitative relations cannot be established. When relations of order are asserted, we are often able to supplement them, at least, by rough estimates of numerical probabilities. This ability is demonstrated in the habit of betting. We bet on the outcome of a boxing match, a scientific experiment, a political election, or a war, expressing the numerical value of the appraised degree of probability by the height of our stakes. The rational reconstruction of such bets would lead to statistical evaluations.

The logical interpretation, which was defined for the probability of a single case, can be extended to the case where a sequence is used as the object of interpretation; it then leads to the consideration of sequences of propositions, or *propositional sequences*. Both interpretations thus have correlates in logical interpretations. The logical interpretation of sequence probabilities, however, has only a theoretical interest, and will be discussed later from that point of view. The practical importance of the logical interpretation springs from its application to the single case, because, in this application, probability assumes the function of a substitute for a truth value.

Because of its analogy with the concept of a statement of known truth value, the concept of appraised posit is indispensable for the understanding of language. It defines the logical category under which probability statements concerning individual cases are to be subsumed, and allots to such statements

[4] By J. M. Keynes, *A Treatise on Probability* (London, 1921), p. 34.

a legitimate place within the body of knowledge. A two-valued logic has no place for unknown truth values; so long as the truth value of a statement is not known, as for statements about future events, classical logic does not allow us to judge the truth of the statement. All it offers is a "wait and see". Our actual behavior, however, does not follow this maxim of passivity. We form judgments about the likelihood of the event and use them as a guide for action. We must do so, because action presupposes judgments about future events; if we had to wait until direct observation informed us about the occurrence or nonoccurrence of an event, it would be too late to plan actions influencing the event.

Probability logic supplies the logical form of a truth evaluation by degrees that is applicable before the occurrence of an event. It allows us to coördinate to the sentence about the individual event a fictitious truth value, derived from the frequency within an appropriate sequence, in such a way that, so far as actions are concerned, the fictitious truth value, or weight, satisfies to a certain extent the requirements that can be asked with respect to a truth value. The logical interpretation repeats the procedure followed in the object-language interpretation of the probability statement about the single case: the metalinguistic conception of the probability of the single case as a weight of a statement, too, is constructed by a transfer of meaning from the general to the particular case. The numerical value of the frequency in the sequence is transferred to the individual statement in the sense of a rating, although the individual statement taken alone exhibits no features that could be measured by the rating.

In spite of the fictitious nature of the rating so constructed, the system of posits endowed with weights can be substituted for a system of statements known as true or false. The essential difference between the two systems consists in the fact that in the substitute system our action is determined, not by a knowledge of the truth value of the statement about the individual event, but by a knowledge of a truth frequency in a sequence. The substitution of this statistical knowledge for unavailable specific knowledge is justified because it offers success in the greatest number of cases. This is why we can act when the truth of the sentence about the individual event remains unknown: the frequency interpretation of probability replaces the unattainable ascertainment of truth by a procedure that accords the best success attainable in the totality of cases.

It has been objected that the frequency interpretation of the probability of the single case does not correspond to what a person actually means when he regards an individual event as probable. My answer may be found in the discussion of meanings at the beginning of § 72: the objection seems irrelevant because logical analysis is not concerned with the description of subjective images and intentions associated with words. The use of the word "probable" with reference to individual events, or as denoting a truth value of sentences about individual events, can be given a meaning in terms of frequencies; if this meaning is assumed, our words will be made compatible with our actions. This result seems to be established beyond doubt. Since there is no other way of determining meanings than by defining interpretations that make language correspond to behavior, I do not see on what grounds the universal applicability of the frequency interpretation of probability can be questioned.

§ 74. Probability Meaning

The interpretation of probability as a truth value permits the introduction of a new category of meaning. The verifiability theory of meaning, in its strict form, makes meaning dependent on verifiability as true or false. The theory can be extended, however, so that a sentence is regarded as meaningful when it is possible to determine a weight for the sentence.[1] By "possible" we understand here "physically possible", that is, "compatible with physical laws". The meaning so defined is called *probability meaning*.

The advantage of the new category of meaning derives from the fact that a determination of weight may be physically possible, whereas a corresponding absolute verification is not physically, but only logically, possible. With respect to simple sentences, for example, sentences concerning the weather of the next day, the distinction is irrelevant; here it is physically possible both to determine a weight in advance and to verify after the occurrence of an event. But this cannot be done with more complicated sentences. Thus, a statement about the temperature of the sun cannot be strictly verified in the sense of a physical possibility of verification, but it is physically possible to determine a weight for it. *Probability meaning*, therefore, represents a wider category than *physical truth meaning*, that is, a meaning defined by the physical possibility of strict verification. But it is a narrower category than *logical meaning*, a meaning defined by the logical possibility of strict verification (see § 66). When the term "verification" is used in a wider sense, to include the determination of a weight, probability meaning represents a *physical* meaning, since it is based on the physical possibility of verification. It can be shown that probability meaning constitutes the very category of meaning that underlies conversational and scientific language, for which physical truth meaning is too narrow and logical meaning too wide.

The meaning of limit statements, analyzed in § 66, may now be reconsidered in the light of the category of probability meaning. It is only logically, not physically, meaningful to speak about the limit of the frequency of an infinite sequence that is extensionally given; therefore a *finitization* of limit statements is required for applications to physical reality. This classification is based

[1] See *EP*, § 7.

on the postulate of absolute verification. If verification in the wider sense is admitted, it is sometimes physically possible to verify the limit statement for an infinite extensional sequence, namely, when it is physically possible to determine a weight for it.

These conditions are realized in a probability lattice. Here the individual horizontal sequence may be regarded as a single case the weight of which is determined by a frequency of sequences counted in the vertical direction. In an arrangement of this kind the limit statement concerning the infinite extensional sequence has probability meaning.

These considerations correspond to actual situations. We make statements about the probability of a limit of the frequency at a certain value p, and we can also compute the probability that for a given degree of convergence ϵ the n-th element of the sequence is a place of convergence. Such results concerning probabilities of a higher level can be derived within the frame of Bernoulli's theorem. If we admit probability meaning, therefore, we can dispense with the requirement of finitization.

This analysis, however, has a restricted value. A computation of probabilities of the higher level is possible only when limits of the higher level are known. Whereas statements about limits of the first level are thus given a probability meaning, those concerning limits of the second level do not have this sort of meaning. True, a probability meaning for statements of the second level can be constructed by computing probabilities of the third level, but then new statements are introduced that do not have a probability meaning. In other words, a probability meaning can be constructed for *every* limit statement, but not for *all*.

The use of probability meaning for the discussion of limit statements is therefore restricted to an *advanced state of knowledge*, a state in which a sufficient number of probabilities is known (see § 70). However, within a *primitive state of knowledge*—a state preceding the determination of probability values and therefore the kind of state on which the determination of the first probability values must be based—probability meaning cannot be used. Since no weight is known for limit statements of the highest level, they cannot be incorporated in the frame of probability meaning and must be subject to the method of finitization explained above (§ 66). The general theory of induction, in particular, must be given with respect to primitive knowledge and, therefore, without the use of probability considerations.

Chapter 10

PROBABILITY LOGIC

PROBABILITY LOGIC

§ 75. The Problem of a Multivalued Logic

We turn now to the technical construction of the probability logic envisaged in § 73, in which the alternative *true-false* is replaced by a continuous scale of truth values. In the history of logic the question has been asked repeatedly whether the dichotomy into *true* and *false* is an ultimate necessity of our thinking, or whether the human mind has the capacity to dispense with it. The very beginnings of two-valued logic in Greek philosophy were accompanied by criticism of this logic, which arose out of a consideration of the uncertainty of the future. In recent times, repeated attempts have been made to construct a generalized logic.[1]

These investigations, however, did not originate in the problem of probability, but rather in the question of the *modalities*—the three categories of necessity, possibility, and impossibility. Consequently, interest centered

[1] As J. Lukasiewicz has stressed, the works of Aristotle contain passages suggesting that he thinks of a generalization of alternative logic into a three-valued logic. Whereas the Stoics, as strict determinists, emphatically maintained a two-valued logic, the Epicureans assimilated Aristotle's idea. Among the more recent investigations, the intuitionism of Brouwer, rejecting the *tertium non datur*, belongs to this line of development, which leads away from alternative logic: L. E. J. Brouwer, *Intuitionisme en formalisme* (Gröningen, 1912), and in *Bull. Amer. Math. Soc.*, Vol. 20 (1913). For further reference see the index of *Erkenntnis*, Vol. II (1931), p. 151.

Other multivalued logics have been treated in the form of a calculus: Hugh MacColl, *Symbolic Logic and Its Applications* (London, 1906); E. L. Post, in *Amer. Jour. Math.*, Vol. 43 (1921), p. 182; J. Lukasiewicz, in *Ruch Filozoficzny* (Lwów), Vol. V (1920), pp. 169–170; and in *Comptes Rendus Soc. des Sciences et des Lettres Varsovie*, Vol. 23 (1930), cl. 3, p. 51; J. Lukasiewicz and A. Tarski, *ibid.*, p. 1. Walter Dubislav, in *Jour. f. d. reine u. angew. Math.*, Vol. 161 (1929), p. 107, used multivalued truth tables for a proof of the consistency of the calculus of functions. O. Becker, in *Jahrb. f. Philos. u. phänomen. Forschung*, Vol. XI (1930), p. 497, treated the problem of modalities within the framework of a multivalued logic. These investigations are connected with the concept of strict implication developed by Clarence Irving Lewis. A presentation is found in C. I. Lewis and C. H. Langford, . . . *Symbolic Logic* (New York and London, 1932), Appendix II. Reports on the problem of multivalued logic are given in: C. I. Lewis, *A Survey of Symbolic Logic* (Berkeley, 1918); J. Jörgensen, in *Erkenntnis*, Vol. III (1932), p. 73; S. Zawirski, in *Revue de Métaph. et de Morale*, Vol. 39 (1932), p. 503. The first application of a three-valued logic to physics was given by H. Reichenbach in *Philosophic Foundations of Quantum Mechanics* (Berkeley, 1944), §§ 32–33, which also contains a presentation of this logic.

Turning to probability, we find the first logical interpretation of the theory of probability in George Boole, *The Laws of Thought* (London, 1854), chaps. xvi–xxi. J. Lukasiewicz, in *Comptes Rendus Soc. des Sciences et des Lettres Varsovie*, Vol. 23 (1930), cl. 3, p. 72, mentions an attempt to construct a probability logic by means of truth tables, but his tables of a continuous logic are not compatible with the calculus of probability. A theory of causality, for which future events are objectively undetermined, was published by H. Reichenbach in *Ber. d. bayer. Akad., math.-phys. Kl.*, 1925, p. 133. At the congress for epistemology of the exact sciences at Prague, in 1929, he read a paper presenting the program of using a logic with a continuous scale of truth values for the treatment of the theory of probability: see *Erkenntnis*, Vol. I (1930), p. 158. The author's first publication of probability logic was made in *Sitzungsber. d. preuss. Akad., phys.-math. Kl.*, 1932, p. 476; that of the theory of induction, including the concept of posit, in *Erkenntnis*, Vol. III (1932), pp. 401–425.

around a three-valued logic corresponding to these categories. Furthermore, a five-valued logic was conceived, including truth and falsehood in addition to the three modalities mentioned above, thus uniting five concepts in one logic. All the investigations were focused on the problem of extending the truth tables to a system of more than two truth values. Yet, since there are many methods for such extension, the systems could not be constructed without arbitrariness. Some of the constructions are merely formal; so the question of their interpretation is left open. But even in attempts that preserved the connection with the meaning of the modalities, no satisfactory correspondence between the constructed system and the actual usage of language was attained.

The present method of investigation has the advantage of connecting the problem of a multivalued logic with the analysis of the calculus of probability. It will be seen that the logicized form given to this calculus greatly facilitates the attempt to construct a logic in which the concept of truth is replaced by the concept of probability. We shall find the laws of this probability logic by transcribing the laws of the calculus of probability into a multivalued logic. This method is free from the arbitrariness that impeded the progress of other investigations, and leads to a logic of a continuous scale of truth values, a *quantitative logic.*

Because of this immediate access to probability logic, we need not construct it by way of a detour, going into a discrete ν-valued logic first and then extending the latter to the case of a continuous scale of degrees of truth. We shall be able to proceed inversely, deriving discrete logics from probability logic by dividing the continuous scale in a suitable fashion. A three-valued logic of modalities will thus be constructed.

The generalization of the concept of truth to be given concerns the truth of empirical statements but not mathematical truth. Thus the objective will not be to construct an interpretation in which such statements as the theorem of Pythagoras are regarded as "merely probable". Such a development is ruled out because mathematics is not concerned with the concept of truth but with the concept of tautology (§ 4). A tautology is a formula that is true whatever be the truth values of the elementary propositions of which it is composed. The term "truth", which occurs twice in the sentence, is the concept used in empirical science. It is a matter of experience to discover whether the elementary components of a compound proposition are true. Only for a tautology will this be irrelevant, since the truth of the tautology is independent of that of its elementary propositions. The investigations of mathematics, therefore, do not concern the truth of the elementary components, but the relation between this truth and the truth of the whole formula. It is the aim of mathematics to construct formulas the truth of which is independent of the truth of their components. The theorems of geometry, for instance, must be judged from this point of view. It is not asserted that the theorem of Pythagoras is true, but that it follows from the axioms, that is, the implication from the axioms to the theorem is a tautology. Mathematics is thus not concerned with truth, but with a certain relation between truth values. Later it will be shown that the concept of tautology can be carried through also in multivalued logic.

A notational remark is in order. Truth is a property of sentences, not of physical objects. Sentences that state the truth of sentences of the object language, therefore, belong in the metalanguage. Thus we should write "V ('a') = 1" for the sentence, "The sentence 'a' is true". For the sake of simplicity, however, the quotation marks will be omitted, and the parentheses, used in combination with the symbol V or similar symbols, will be regarded as a sign that includes the function of the quotation marks. We therefore shall write "$V(a) = 1$" for the sentence above. Similarly, we shall write, "the sentence a", instead of writing, "the sentence 'a' ", allowing the word "sentence" in combination with the succeeding italics to assume the function of quotation marks.

§ 76. A Quantitative Logic of an Individual Verifiability

Before turning to the construction of probability logic, I wish to show, in a preliminary investigation, that a quantitative concept of truth can be coördinated to individual statements in such a way that the degree of truth can be determined for each statement directly, without reference to a frequency within a sequence. A quantitative logic of this kind may be said to be of an *individual verifiability*. It does not lead to a logic of probability, however, since the latter is based on frequency notions and is therefore of a *nonindividual verifiability;* but the logic so constructed may be used to illustrate certain fundamental features of a quantitative logic.

In two-valued logic the two truth values are so defined that certain facts *verify* the statement, whereas others *falsify* it. This means that the statement divides all facts to which it is related into two classes: those that make the statement true, and those that make it false. For example, the statement, "The weather is good", will be called true if the sun is shining and no wind is blowing, but also if the sun is occasionally covered by clouds or a soft wind is blowing. Saying that the statement is false likewise includes several other possible facts, for example, the possibility of a strong wind without rain, that of rain without wind, and so on.

It is possible to go from a dichotomy to a quantitative verification by ordering the facts with respect to the degree to which they satisfy the statement. The example mentioned admits of such an interpretation. The weather can be more or less perfect, that is, it can have all gradations of intermediary forms ending with extremely bad weather. We can therefore ascribe to the statement a degree of truth depending on the observed facts. Such gradations of statements occur in everyday language. We frequently say, "This is true to a certain degree"; "This is half true, half false", and so on.

The method may be analyzed by means of a simpler example. A marksman says, "I shall hit the center" (statement a). After the shot we measure the distance r of the hit from the center; r is a measure of the degree of truth of the statement a (fig. 26).

In order to obtain the values 0 and 1 for the limiting cases, we may take the expression

$$v(a) = \frac{1}{1+r} \tag{1}$$

as a truth value. Let r_1 be the value of the radius of the hit. We then have

$$v(a) = \frac{1}{1+r_1} \tag{2}$$

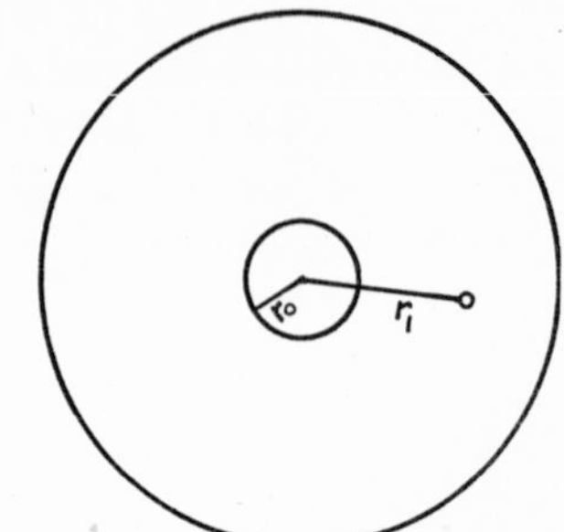

Fig. 26. Hit on a target as example for a logic with continuous scale of truth, according to (2) and (3).

(2) is the truth value of a in the continuous logic.

Obviously it is not necessary, however, to use a quantitative logic in this example. The same fact can be incorporated in two-valued logic in either of two ways:

1. The values of the radius r may be divided into two classes by a demarcation value r_o; and, accordingly, the values v are divided by the demarcation value v_o, corresponding to r_o. We now put

$$V(a) = 1 \text{ for } 1 \geqq v \geqq v_o \qquad V(a) = 0 \text{ for } v_o > v \geqq 0$$
$$\text{or, for } r \leqq r_o \qquad\qquad \text{or, for } r > r_o \tag{3}$$

In this manner the continuous scale of values v has been transformed into the alternative V by a dichotomy. The method has the disadvantage, however, that the truth characterization is diminished in content. If we know only that $V(a) = 1$, we do not know at which point r within r_o the hit is situated, but the expression (2) informs us about this value.

2. We go from the consideration of the degree of truth of the statement a to the consideration of the truth value of the metalinguistic statement

$$v(a) = \frac{1}{1+r_1} \quad \text{abbr. } \grave{a}^{(1)} \tag{4}$$

The addition on the right side means that we abbreviate the statement by the symbol $\grave{a}^{(1)}$. The superscript in the symbol indicates that we deal here with a statement of the first metalanguage, whereas the statement a belongs to the object language. The accent in the symbol $\grave{a}^{(1)}$ indicates that the statement is true. We have, using the second metalanguage,

$$V(\grave{a}^{(1)}) = 1 \qquad (5)$$

This method preserves the full content of the statement. It amounts to the same as the use of the object-language statement, "The shot hit at the distance r_1". It is clear, however, that the statement can be replaced by the metalinguistic formulation above.

The two-valued character thus represents merely a principle of division by which statements are classified. It could be replaced by another principle. When we ask, "Does a hold?" we commit ourselves to the principle of dichotomy, expecting the answer "Yes" or "No". It would, in fact, be possible to ask, "To which degree does a hold?" and to give the answer in the form, "a holds to the degree $\dfrac{1}{1+r}$". The two methods constitute different forms of linguistic classification.

We must not believe, however, that we are bound to a two-valued metalanguage. In place of the true statement $\grave{a}^{(1)}$ we can introduce another statement $a^{(1)}$, which, in its turn, is incorporated in a quantitative logic. Thus we may put

$$v(a) = \frac{1}{1+r} \qquad \text{abbr. } a^{(1)} \qquad (6)$$

If $r = r_1$, $a^{(1)}$ is true and is identical with $\grave{a}^{(1)}$. If $r < r_1$, we have

$$v(a^{(1)}) = \frac{1}{1+\rho} \qquad \rho = r_1 - r \qquad (7)$$

The metalanguage has here a continuous truth scale like the object language. We arrive at a two-valued language only when we go into the second metalanguage, which contains the true statement

$$v(a^{(1)}) = \frac{1}{1+\rho} \qquad \text{abbr. } \grave{a}^{(2)} \qquad (8)$$

The system of the statements a, $a^{(1)}$, $\grave{a}^{(2)}$ is now equivalent to the system of the statements a, $\grave{a}^{(1)}$. This is confirmed by a retranslation: $\grave{a}^{(2)}$ indicates that the value r in $a^{(1)}$ is too small by the amount ρ. If we put for r in $a^{(1)}$ the value $r + \rho$, which is $= r_1$, we obtain the true statement $\grave{a}^{(1)}$ instead of $a^{(1)}$. Using the same procedure, we can also obtain in place of the statement

$$r = r_o \qquad \text{abbr. } a \qquad (9)$$

TABLE 8

COMPARISON BETWEEN TWO-VALUED LOGIC AND A MULTIVALUED LOGIC OF INDIVIDUAL VERIFIABILITY

Two-valued logic on all levels		Multivalued logic on 1st level; two-valued logic on all following levels		Multivalued logic on 1st and 2nd level; two-valued logic on all following levels		Multivalued logic on all levels	
Statement	Abbrev.	Statement	Abbrev.	Statement	Abbrev.	Statement	Abbrev.
$r = r_1$	$\grave{a}$	$r = 0$	a	$r = 0$	a	$r = 0$	a
$V(\grave{a}) = 1$	$\grave{a}^{(1)}$	$v(a) = \frac{1}{1 + r_1}$	$\grave{a}^{(1)}$	$v(a) = \frac{1}{1 + r}$	$a^{(1)}$	$v(a) = \frac{1}{1 + \frac{r_1}{2}}$	$a^{(1)}$
$V(\grave{a}^{(1)}) = 1$	$\grave{a}^{(2)}$	$V(\grave{a}^{(1)}) = 1$	$\grave{a}^{(2)}$	$v(a^{(1)}) = \frac{1}{1 + (r_1 - r)}$	$\grave{a}^{(2)}$	$v(a^{(1)}) = \frac{1}{1 + \frac{r_1}{4}}$	$a^{(2)}$
$V(\grave{a}^{(2)}) = 1$	$\grave{a}^{(3)}$	$V(\grave{a}^{(2)}) = 1$	$\grave{a}^{(3)}$	$V(\grave{a}^{(2)}) = 1$	$\grave{a}^{(3)}$	$v(a^{(2)}) = \frac{1}{1 + \frac{r_1}{8}}$	$a^{(3)}$
. . .	.	. . .	.	. . .	.	. . .	.
. . .	.	. . .	.	. . .	.	. . .	.

which is not completely true, the true statement

$$r = r_1 \qquad \text{abbr. } \grave{a} \tag{10}$$

This statement of the object language is equivalent to the system a, $\grave{a}^{(1)}$.

However, the system a, $a^{(1)}$ is not equivalent to the true statement $\grave{a}$. This results from the fact that the last statement of this system is not true. If we want to cut off a system of multivalued statements after a finite number of levels of language, we must have a true statement as the last one. Language usually contains the implicit convention that an uttered statement has the truth value 1. This principle can be maintained for the last level of finite systems of multivalued statements.

But we are not bound to use finite systems. We can use infinite systems that contain no true statement on any level. For each statement, the degree of its truth will be stated, but the statement concerning the truth will itself be true only to some degree. Using the example, we can construct an infinite system of metalanguages of this kind. We put

$$v(a^{(i-1)}) = \frac{1}{1+\rho_i} \qquad \rho_i = \frac{1}{2^i} \cdot r_1 \qquad \text{abbr. } a^{(i)} \tag{11}$$

According to the principle of retranslation, the true statement $\grave{a}^{(i)}$ coördinated to $a^{(i)}$ is here given by

$$v(a^{(i-1)}) = \frac{1}{1+\rho_i+\sum_{m=i+1}^{\infty}\rho_m} \qquad \text{abbr. } \grave{a}^{(i)} \tag{12}$$

Because of the definition of ρ_i we obtain, for instance,

$$\left.\begin{aligned} v(a^{(1)}) &= \frac{1}{1+\frac{r_1}{2}} \qquad \text{abbr. } \grave{a}^{(2)} \\ v(a) &= \frac{1}{1+r_1} \qquad \text{abbr. } \grave{a}^{(1)} \end{aligned}\right\} \tag{13}$$

The infinite system defined by (11) is thus equivalent to the system a, $\grave{a}^{(1)}$ given by (9) and (4). The system will include (6) and (8) if we put in (6) $r = \frac{r_1}{2}$, in (8) $\rho = \frac{r_1}{2}$.

Table 8 clarifies these relations.

All the systems of statements given in the table are equivalent to each other. They describe the same facts. In the three initial columns, only true statements appear from a certain level on, but the fourth column does not contain any true statements. This system, therefore, cannot be cut off. It is, however, a convergent system, that is, if the system is cut off at $a^{(i)}$, and $a^{(i)}$ is regarded as true, the mistake will be as small as we wish when i is sufficiently great.

The convergent character is not a necessary condition for such systems. We show this by the slightly changed example

$$r = r_2 \qquad \text{abbr. } a \tag{14}$$

$$v(a^{(i-1)}) = \frac{1}{1+\rho_i} \qquad \rho_i = (-1)^i \cdot (r_2 - r_1) \qquad \text{abbr. } a^{(i)}$$

This system has the following properties. When we regard $a^{(1)}$ as true, the true statement corresponding to a will be $r = r_1$; when we regard $a^{(2)}$ as true, the true statement corresponding to a will be $r = r_2$. The two results will change alternatively when we cut off on higher levels. The system (14), therefore, says either that $r = r_1$ or that $r = r_2$, and is equivalent to the statement

$$(r = r_1) \vee (r = r_2) \tag{15}$$

which belongs in a two-valued object language. Incidentally, the truth scale used in (14) is not enclosed in the limits 0 and 1; this, however, is irrelevant.

From these considerations we see that a multivalued logic of individual verifiability can always be carried through. This holds not only for a logic of a continuous scale but also for any ν-valued logic. But statements of a multivalued logic can always be translated into statements of a two-valued logic by one of the two methods described. Conversely, all statements of two-valued logic can be translated into statements of multivalued logic if a suitable rule for the construction of truth values is added, corresponding to the definition (1).

The two-valued character of classical logic is a convention comparable to that used in the decimal system for the notation of numbers. The latter is constructed on the arbitrary basis of the number 10; logic is constructed on the arbitrary basis of the number 2. There are philosophers, indeed, who think that an ultimate truth concerning the structure of the universe is formulated in the statement, "Every proposition must be either true or false"; or in the equivalent statement, "Every fact must either be or not be". But such a statement is no more justified than the statement, "Every number over 99 must be written with at least three digits". The latter sentence is true for the decimal notation, but false if it is asserted for all notations. It says nothing about numbers, but states merely a property of the notation used. Similarly, the preceding proposition concerning truth or existence is true only within two-valued logic, but false if it is asserted without restrictions. In multivalued logic it does not hold, either in the first or in the second version given. Here a fact can "exist halfway"; it will do so when the sentence describing it is half true.

§ 77. Probability as a Property of Propositional Sequences

The logical conception of probability was introduced earlier with reference to individual propositions, but with the qualification that the truth value into which the probability is thus transformed is of a fictitious nature, since it is derived from a frequency. Probability logic, in this conception, is a logic of nonindividual verifiability. It is therefore advisable to attach the technical construction of probability logic, not to individual propositions, but to sequences of propositions. We can then apply the frequency interpretation directly and need not deal with fictitious properties.

Turning to the construction of suitable sequences of propositions, we begin with the implicational form of writing introduced in § 9:

$$(i)(z_i \epsilon A \underset{p}{\ni} x_i \epsilon B) \tag{1}$$

Using the two propositional functions[1]

$$hz_i =_{Df} z_i \epsilon A \qquad fx_i =_{Df} x_i \epsilon B \tag{2}$$

we can write, instead of (1),

$$(i)(hz_i \underset{p}{\ni} fx_i) \tag{3}$$

In the P-notation we have for this:

$$P(hz_i, fx_i) = p \tag{4}$$

All this is object interpretation. Introducing the logical interpretation, we should have to write

$$P(\text{``}hz_i\text{''}, \text{``}fx_i\text{''}) = p \tag{5}$$

The elements of the sequences are now given by individual propositions of the form "hz_i" and "fx_i". The frequency interpretation is constructed by counting the number of true propositions "fx_i" within the subsequence selected by true propositions "hz_i".

Since the notation by quotation marks is inconvenient, these marks will be omitted, and their function will be assumed by parentheses, in correspondence with the rule for the V-symbol introduced at the end of § 75. This means that we shall use (4) instead of (5). Without further indication, probability expressions belong in the metalanguage from now on. This is possible because all rules for metalinguistic expressions correspond strictly to those employed in object-language expressions.

[1] I use here a notation for argument variables that differs from that used above. Furthermore, I omit the parentheses in propositional functions.

A further linguistic obstacle must be overcome if probability is to be regarded as a property comparable to truth. Truth is a property of one sentence, whereas probability is a relation between two linguistic expressions. We can eliminate this difficulty by assuming the first sequence, indicated by the term hz_i in (4), as compact. It is then identical with the sequence of the subscripts and can be omitted. We thus write, adopting the notation of (1, § 24),

$$P(fx_i) = p \tag{6}$$

The sequence of propositions derived from the propositional function f will be called a *propositional sequence*, and is denoted by the symbol (fx_i). The sequence of the elements x_i will be denoted by (x_i). Probability appears here as a property of a propositional sequence and is of the same logical type as truth, which is a property of a proposition.

We can construct for (6) the frequency interpretation, which has a form corresponding to (4, § 16). Since, however, we use the logical interpretation of probability, we must count, not events, but sentences about events. Therefore we write the frequency interpretation of (6) in the form

$$P(fx_i) = \lim_{n \to \infty} \frac{1}{n} \overset{n}{\underset{i=1}{N}} \{V[fx_i] = 1\} \tag{7}$$

The form (7) shows the close analogy between the metalinguistic and the object-language interpretation of probability. For frequencies of events we substitute, in the metalinguistic or logical interpretation, frequencies of sentences about events. All theorems of the object interpretation can therefore be transferred immediately to the logical interpretation; the two interpretations are *isomorphous*. This is why they need not be distinguished by a specific notation and quotation marks are dispensable.

It should be kept in mind that a propositional sequence is not the same as a propositional function. To the propositional function f we must add the ordered sequence of events (x_i); it is this combination that makes up the propositional sequence. We see that a propositional sequence shows a certain analogy to propositions, since the latter can be regarded as a combination of a propositional function and one argument x_1.

These explanations prepare the way for the construction of probability logic. We shall construct it as the logic of propositional sequences—as a logic the elements of which are propositional sequences as wholes. The extension of logic thus envisaged is based on the idea that the logic of propositional sequences can show more general features than the logic of the propositions constituting the elements of these sequences. The two-valued character of the elements need not be transferred to the compound expressions constructed from them.

We may compare the transition from alternative logic to probability logic with the transition from Euclidean geometry to non-Euclidean geometry. Although Euclidean geometry holds for small areas of a Riemann space, it does not determine the structure of large areas, which, in general, is of a more complicated type. Similarly, the alternative character holds for the "logic of smallest domains", namely, of propositions, whereas the continuous scale holds for the "logic of large domains", that is, of propositional sequences.

The comparison may be used for an analysis of the psychological factors involved in the historical predominance of two-valued logic. The corresponding predominance of Euclidean geometry is derived from the fact that our environment satisfies the axioms of Euclid to a high degree of approximation so far as relative positions of solid bodies are concerned. If this environment were of a different structure—if it showed, in the dimensions of our houses, the deviations asserted by Einstein for cosmic dimensions—mankind would have developed from the outset a non-Euclidean conception of space.

Probability logic is in a similar position. In daily life we usually deal with phenomena that have a very high degree of probability. By putting these high probabilities = 1 it is possible to construct a sufficient idealization that knows only the two extreme points. For phenomena of a lower degree of probability, however, such an idealization is inadequate and must be replaced by the construction of a logic of a continuous truth scale.

§ 78. The Probability of Finite Propositional Sequences

For finite sequences (x_i) the frequency definition of probability can be maintained if we understand by the limit the value of $F^n(fx_i)$ for the last term. In this case only the $n + 1$ values $\frac{0}{n}, \frac{1}{n} \cdots \frac{n}{n}$ are possible for the frequency, and we are no longer dealing with a logic of a continuous scale, but with a ν-valued logic for $\nu = n + 1$.

Now it is clear that for finite sequences, in contradistinction to infinite sequences, the order of the elements has no influence upon the limit of the frequency. Consequently, for finite sequences, it is permissible to speak directly of the frequency.

A further difference from infinite sequences consists in that for finite sequences the limit of the frequency can be 1 only if fx_i holds for all x_i. Corresponding considerations hold for the limit 0. With respect to finite sequences, therefore, we need not differentiate between general implication and a probability implication of the degree 1, cases that must be distinguished for infinite sequences (see §§ 12, 18).

By going to smaller and smaller values of n we can accomplish a transition from a ν-valued logic to the two-valued logic. It obtains for the case $n = 1$, that is, for the case that the sequence is reduced to one element. The expression (fx_1) thus means the same as fx_1.

Because the value of the frequency for such sequences can only be 1 or 0, we have here

$$P(fx_1) = 1 \quad \text{or} \quad P(fx_1) = 0 \tag{1}$$

In this case, therefore, the two possible probability values coincide with truth and falsehood, and we have

$$[P(fx_1) = 1] \equiv [V(fx_1) = 1] \tag{2}$$

$$[P(fx_1) = 0] \equiv [V(fx_1) = 0]$$

Truth and falsehood can thus be regarded as the limiting cases of probability resulting when the sequence is reduced to one element.

§ 79. Truth Tables of Probability Logic

Returning now to the probability logic with a continuous scale, I shall show that in this logic truth tables comparable to the truth table 1 (p. 17) can be constructed for propositional operations.

First, operations for propositional sequences analogous to those holding for propositions must be defined. For this purpose the argument sequences (x_i) and (y_i) of the propositional sequences must be coördinated; corresponding elements are indicated by the equality of the subscript. Second, the operation combining two propositional sequences is defined in terms of the corresponding operation combining the elements of the sequences:

$$(fx_i) \vee (gy_i) =_{Df} (fx_i \vee gy_i)$$

$$(fx_i) . (gy_i) =_{Df} (fx_i . gy_i)$$

$$(fx_i) \supset (gy_i) =_{Df} (fx_i \supset gy_i) \tag{1a}$$

$$(fx_i) \equiv (gy_i) =_{Df} (fx_i \equiv gy_i)$$

and correspondingly

$$\overline{(fx_i)} =_{Df} (\overline{fx_i}) \tag{1b}$$

Because of these definitions we can now use the frequency interpretation, in the form (7, § 77), for the probability of the propositional combinations. We have, for instance,

$$P((fx_i) \vee (gy_i)) = P(fx_i \vee gy_i) = \lim_{n\to\infty} \frac{1}{n} \overset{n}{\underset{i=1}{N}} \{V[fx_i \vee gy_i] = 1\} \tag{2}$$

Similar relations hold for the other propositional operations.

We must now introduce a new propositional operation that will allow us to write degrees of relative probabilities. Since probabilities of this kind are written in the form $P(fx_i, gy_i)$, the content of the parentheses in this expression can be regarded as a compound proposition, resulting from the two components by a propositional operation, denoted by the comma. We shall call it *operation of selection*, or *comma operation*. We can also put the comma between propositional sequences. We then define, by analogy with (1*a*),

$$(fx_i),(gy_i) =_{Df} (fx_i, gy_i) \tag{3}$$

The frequency interpretation of $P(fx_i, gy_i)$ is given by the expression

$$P(fx_i, gy_i) = \lim_{n \to \infty} \frac{\overset{n}{\underset{i=1}{N}} \{V[fx_i . gy_i)] = 1\}}{\overset{n}{\underset{i=1}{N}} \{V[fx_i] = 1\}} \tag{4}$$

This form makes clear why we speak of the operation of selection. The proposition fx_i selects the subsequence in which we count the frequency of gy_i. Since (3) allows us to regard the comma on the left side of (4) as standing between the propositional sequences, we can also say that the comma operation represents a selection from one sequence by another sequence.

Turning to the construction of the truth tables, we now meet an intrinsic difference between probability logic and two-valued logic. The truth table 1 (p. 17) contains as arguments the two truth values of the elementary propositions, so that in the adjunctive interpretation the two truth values determine the truth value of the compound proposition. In probability logic, on the contrary, two arguments are not sufficient; we need a third argument when the probability value of the compound propositional sequence is to be determined. This follows from the considerations leading to (7′, § 21), which show that for the determination of the probabilities of compound events three fundamental probabilities are required. The three values $P(A,B)$, $P(A,C)$, $P(A.B,C)$ were chosen for this purpose. In (9, § 24) the notation that omits the reference class A was applied. Transcribing these results for the case of probability logic, we regard the three probabilities

$$P(fx_i) \qquad P(gy_i) \qquad P(fx_i, gy_i) \tag{5}$$

as the arguments of the truth tables; the probabilities of other combinations are determined by these values. According to the remarks added to (7′, § 21), we could replace the value $P(fx_i, gy_i)$ by that of any other of the propositional combinations, for instance, by the value $P(fx_i . gy_i)$. It is a matter of convenience that we prefer to use the three values (5) as the independent parameters.

TABLE 9

Truth Tables of Probability Logic

A. NEGATION

$P(fx_i)$	$P(\overline{fx_i})$
p	$1-p$

RESTRICTIVE CONDITIONS

1. $\frac{p+q-1}{p} \leqq u \leqq \frac{q}{p}$

2. $P(fx_i,fx_i) = 1$

B. BINARY OPERATIONS

$P(fx_i)$	$P(gy_i)$	$P(fx_i, gy_i)$	$P(fx_i \vee gy_i)$	$P(fx_i . gy_i)$	$P(fx_i \supset gy_i)$	$P(fx_i \equiv gy_i)$	$P(gy_i, fx_i)$
p	q	u	$p+q-pu$	pu	$1-p+pu$	$1-p-q+2pu$	$\frac{pu}{q}$

C. BINARY OPERATIONS FOR THE LIMITING CASES 0 AND 1

1	1	1	1	1	1	1	1
1	0	0	1	0	0	0	?
0	1	?	1	0	1	0	0
0	0	?	0	0	1	1	?

The construction of the truth tables is now easily achieved by the use of formulas (7, § 13), (8, § 20), (3, § 14), (22, § 20), (23, § 20), and (6, § 21). The results are compiled in table 9, *A* and *B*.

The table includes a restrictive condition, holding for the numerical values of the three arguments. As pointed out in § 19, the three fundamental values (5) are subject to numerical restrictions, which were formulated in (15, § 19). This condition can be written in the form

$$\frac{p+q-1}{p} \leqq u \leqq \frac{q}{p} \tag{6}$$

This inequality leaves the three values p, q, u independent of each other within wide limits. For the case $p = 1$, however, we derive from (6) that $u = q$. In this case, therefore, we have only two independent values. For $p = 0$ there is no such mutual dependence. The second restrictive condition must be added to the table because the value 1 of u for the identical functions f and g cannot be derived from the table (see 4, § 82). This condition has a similar function as axiom II,1 (§ 12).

It can now be shown that probability logic is a generalization of two-valued logic, or, more precisely speaking, that the truth table 1 (p. 17) of two-valued logic is contained as a special case in the truth table 9 of probability logic. This is proved as follows. Two-valued logic can be regarded as the special case that the numerical values of the probabilities are restricted to 0 and 1. The first two columns in table 9*B*, on this restriction, can assume only the four combinations 1,1; 1,0; 0,1; 0,0; the combinations are presented in table 9*C*. Since for $p = 1$ we have $u = q$, the third column in table 9*B* loses its independent character. This means that the third column of table 9*C* becomes a function of the two preceding columns; the vertical double line separating arguments from functions thus goes to the left by one column. For the case $p = 0$ the indeterminacy in the value of u has been indicated by a question mark. When we now insert, in the expressions of the other columns, the respective values of p, q, and u, we obtain table 9*C*, the columns of which refer to the respective headings of table 9*B*. Thus, putting $p = 1$, $q = 1$, $u = 1$, we obtain for the disjunction the value $1 + 1 - 1 \cdot 1$, which is $= 1$; and so forth. It turns out, and is easily verified, that thereby the indeterminacy indicated by the question mark drops out for all the operations used in two-valued logic. Thus the combination $p = 0$, $q = 1$, $u = ?$, gives for the disjunction the result $0 + 1 - 0 \cdot ?$, which is $= 1$, since the question mark means any finite number, and thus its product by 0 will give 0. Only for the operation (gy_i,fx_i) will the question mark reappear, namely, in the cases $P(gy_i) = 0$. We thus obtain the analogue of table 9*C*.

When we compare table 9*C* with the truth table 1 of two-valued logic (p. 17), we see that the tables are identical with respect to the operations of disjunction, conjunction, implication, and equivalence. The values 1 and 0 of 9*C* correspond to the values *T* and *F* of tables 1. It is clear that, in a similar way, table 9*A* of the negation can be shown to be identical with table 1*A* when the value of p is limited to 1 and 0.

That the tables of two-valued logic are contained in those of probability logic as a special case need not surprise us. In fact, this is a consequence of the previous result (§ 18), according to which the axioms of probability follow from the frequency interpretation and are strictly satisfied even by finite sequences. Referring to the remarks in § 78, we can, therefore, regard two-valued logic as the probability logic holding for propositional sequences having only one element.

The third column of table 9*C* can be regarded as the definition of the *individual operation of selection*, holding between propositions. What distinguishes this truth table from those of the other operations is the occurrence of a question mark, that is, of cases where the truth value is not determined. Truth tables of this kind may be called *defective;* and the individual operation of selection, which holds between propositions, may be called a *defective operation.* In this application of the table, the indeterminacy indicated by the question mark must be interpreted, not as meaning that there exists no truth value, but as meaning that all real numbers are assigned as truth values to the expression. This follows from the corresponding property for empty reference classes (see 7, § 12).[1] For the question-mark cases, the individual operation of selection thus leads to a compound expression which is both true and false, i.e., an expression that is not a proposition. For these reasons, the compound established by the comma operation differs in kind from the other operations and must not be substituted for elementary variables or used as a term in a propositional operation. [For an exception see (31, § 82).] A restricted rule of substitution will be introduced later (see 24, § 82).

This conception of the individual operation of selection will meet the requirements of the frequency interpretation when the latter is introduced in the general form

$$P(hx_i) = \lim_{n \to \infty} \frac{\overset{n}{\underset{i=1}{N}} \overline{\{V[hx_i] = 0\}}}{\overset{n}{\underset{i=1}{N}} \{\overline{V[hx_i] = 1} \vee \overline{V[hx_i] = 0}\}} \tag{7}$$

This expression will, in general, be identical with (7, § 77). But when the functional hx_i has the special form $[fx_i,gy_i]$, the denominator will be given

[1] For a sequence of infinite length in which $p = 0$, the value u need not be indeterminate, although it is not determined by p and q. This is the case if the functional fx_i is not always false, although the limit of the frequency is $= 0$. In the subsequence selected by fx_i, the relative frequency of gy_i has then a definite value. The question mark in table 9*B* means, therefore, that either the value of u is unknown or that u has all real numbers as its values. The latter case results when fx_i is false for all x_i and is thus realized for a finite sequence in which $p = 0$.

only by the cases in which this functional has one determinate truth value and is thus either not true or not false. These are the cases where fx_i is true. Similarly, the numerator will count only the cases where $V[hx_i] = 1$, excluding the question-mark cases. This is precisely the frequency interpretation given in (4).

Several critics have raised the objection that the probability logic presented here is not extensional. When "extensional" is taken to mean a logic in which the truth value of a binary operation is determined by the individual truth values of its two elements, the criticism is formally correct. But it is materially misleading, since it uses a very narrow definition of the term "extensional". In § 4 it was explained that the term "extensional" should rather be replaced by the term "adjunctive". Only the latter term denotes the properties that are relevant for a logic of this kind and which constitute the background of the somewhat vague term "extensional". A logic is adjunctive if its truth tables can be read from left to right, in the direction from the arguments to the function, and not only from right to left, or in the direction from the function to the arguments. It is clear that the truth tables of probability logic are adjunctive. If the term "extensional" is to have a reasonable meaning it should be identified with the term "adjunctive". Therefore it may be said that probability logic is extensional in this wider sense.

The peculiar form of probability logic explains why this generalization of two-valued logic was not found so long as logicians were seeking such a generalization in terms of the two-valued truth tables, disregarding the calculus of probability. Probability logic involves a generalization in which a function of two arguments is replaced by a function of three arguments; two-valued logic appears as the degenerate case in which the third argument becomes a function of the other two. The degenerate case was erroneously assumed as determining the form of every truth-functional logic. However, two propositional components determine, not two, but three truth values: one for each component and a third for the combination of the components. Of this kind are the three arguments of probability logic: two are coördinated to the individual propositional sequences and the third is coördinated to the pair as a whole. The third probability value may be regarded as a measure of the *degree of coupling* existing between the two sequences.

§ 80. Truth Tables of the Logic of Modalities

Two-valued logic is contained in the truth tables as the special case $n = 1$. Similarly, every other metrical ν-valued logic is contained in the tables as the special case $n = \nu - 1$. By metrical logic we understand a logic the truth values of which can be interpreted as probabilities in the sense of the frequency interpretation. Two-valued logic, also, is a metrical logic. But a brief analysis shows that only in the latter logic are the truth values of the operations fully determined by the two individual truth values of their components. For $\nu = 3$, that is, $n = 2$, there are three possible truth values: 0, $\frac{1}{2}$, 1. Here the values of the operations are not determined for the row $p = \frac{1}{2}$, $q = \frac{1}{2}$, and the value of u must be regarded as an independent parameter for this row. When ν increases, the number of rows for which u must be independently given becomes larger and larger.

The construction of metrical logics of a finite number of truth values, other than two-valued logic, is of no particular interest, because we prefer to replace such logics by the logic of the continuous scale. We do so for the same reason that we replace finite sequences of observations by infinite sequences (see § 66). We arrive, however, at an important form of three-valued logic when we turn to the *modalities*. This logic is not a metrical logic, as will be made clear presently.

Like probabilities, the modalities must be regarded as properties not of individual propositions but of propositional sequences. Only in a fictitious sense can they be transferred to individual sentences, in analogy to the concept of weight; but, for the present, such a conception will not be envisaged. By the use of the abbreviation *Nc* for *necessity*, *Ps* for *possibility*, *Im* for *impossibility*, and the symbol $M(f\hat{x}_i)$ for the phrase, the *modality of* $f\hat{x}_i$, the modalities are defined as follows (for the circumflex notation see § 6):

$$[M(f\hat{x}_i) = Nc] =_{Df} (x_i)f(x_i)$$
$$[M(f\hat{x}_i) = Ps] =_{Df} (\exists x_i)f(x_i) . (\exists x_i)\overline{f(x_i)} \qquad (1)$$
$$[M(f\hat{x}_i) = Im] =_{Df} (x_i)\overline{f(x_i)}$$

In this way of writing, the modality is regarded as a property of a situational function. In another conception the modality is considered to be a property of a propositional function. This conception results from (1) when the symbol M is interpreted as a metalinguistic symbol, like the symbol V.

The term "possible" defined here has the meaning of "merely possible", that is, it excludes necessity. This notation is convenient for logical purposes. For finite sequences, necessity and impossibility coincide, respectively, with the probabilities 1 and 0, whereas possibility corresponds to probability values between the two limiting cases. For infinite sequences, no such correspondence exists. Here the limit of the frequency, and therefore the probability, can be $= 1$ when the statement $(x_i)f(x_i)$ is not true; and similarly the probability can be $= 0$ when the statement $(x_i)\overline{f(x_i)}$ is not true. The logic of modalities, therefore, is not strictly identical with that of the three probability values $p = 0$, $0 < p < 1$, $p = 1$. But it turns out that the truth tables resulting for the latter three cases are identical with those holding for the categories defined by (1). These tables are given in table 10 (p. 406). They follow from table

9 (p. 400) when we construct the probability values of the operations on the assumption that p and q, respectively, represent one of the foregoing three cases. The tables can be established also by reference to the meanings of the three categories defined in (1). Thus if (fx_i) is always true and (gy_i) is always true, we see from the structure of the sequences that their disjunction must be always true, and so forth, for the other operations.

Table 10 shows that in general the modalities of the compound propositional sequence are determined by those of the elementary propositional sequence, except for the middle row, where commas are used to separate the possible values. Thus, if both (fx_i) and (gy_i) are possible, their disjunction will be either possible or necessary, but it cannot be impossible. If the ambiguity of the middle row is to be eliminated, we must go back to probabilities. Then the value of the three parameters p, q, u will determine the value of the disjunction.

The ambiguity of the middle row of table 10 has sometimes been regarded as a disadvantage. It is certainly possible to construct other forms of three-valued truth tables[1] containing no such ambiguity, but it seems impossible ever to construe the modalities in such a way. In any interpretation of the modalities, the ambiguity of the middle row is a necessary concomitant, since it corresponds to the usage of these concepts. If two events are possible, it is undetermined whether their conjunction is also possible. They might be mutually exclusive events each of which is possible. Corresponding considerations hold for the disjunction. If two events are possible, their disjunction can be either possible or necessary. Thus if we cast two dice simultaneously, it is possible that face 1 of the first die turns up, and it is possible that face 1 of the second die turns up; here it is merely possible that one or the other of the faces turns up, as well as it is merely possible that both turn up. However, if we toss a coin it is possible that heads turn up, and it is possible that tails turn up; but it is necessary that heads or tails turn up, whereas it is impossible that both turn up.

The definitions (1) can be applied to functional compounds of the form $[f\hat{x}_i, g\hat{y}_i]$, with the qualification that the cases where fx_i is false are canceled for the operator expressions. In these cases, for which the truth table 9C furnishes a question mark, the individual compound is not a proposition and the operators thus cannot be applied. By the use of this rule a modality can be determined for the function $[f\hat{x}_i, g\hat{y}_i]$, except for the case that $f\hat{x}_i$ is empty. This modality is given in the third column of table 10B; it shows an indeterminacy in the middle row, like the other columns. In contradistinction to table 9B, a knowledge of the modality of $[f\hat{x}_i, g\hat{y}_i]$ would not enable us to determine the modalities of the other compounds. This is shown by the

[1] Of this kind are the tables constructed by Post and by Lukasiewicz and Tarski, which the author used for the interpretation of quantum mechanics. See footnote, p. 387.

TABLE 10

Truth Tables of the Logic of Modalities

A

$M(fx_i)$	$M(\overline{fx_i})$
Nc	*Im*
Ps	*Ps*
Im	*Nc*

B

$M(fx_i)$	$M(gy_i)$	$M(fx_i,gy_i)$	$M(fx_i \vee gy_i)$	$M(fx_i \,.\, gy_i)$	$M(fx_i \supset gy_i)$	$M(fx_i \equiv gy_i)$	$M(gy_i,fx_i)$
Nc	*Nc*	*Nc*	*Nc*	*Nc*	*Nc*	*Nc*	*Nc*
Nc	*Ps*	*Ps*	*Nc*	*Ps*	*Ps*	*Ps*	*Nc*
Nc	*Im*	*Im*	*Nc*	*Im*	*Im*	*Im*	?
Ps	*Nc*	*Nc*	*Nc*	*Ps*	*Nc*	*Ps*	*Ps*
Ps	*Ps*	*Im,Ps,Nc*	*Ps,Nc*	*Im,Ps*	*Ps,Nc*	*Im,Ps,Nc*	*Im,Ps,Nc*
Ps	*Im*	*Im*	*Ps*	*Im*	*Ps*	*Ps*	?
Im	*Nc*	?	*Nc*	*Im*	*Nc*	*Im*	*Im*
Im	*Ps*	?	*Ps*	*Im*	*Nc*	*Ps*	*Im*
Im	*Im*	?	*Im*	*Im*	*Nc*	*Nc*	?

examples above. If heads turn up, it is impossible that tails turn up; thus for the two faces of the coin we have the modality of impossibility for $[f\hat{x}_i, g\hat{y}_i]$. The same holds when we regard two faces of the same die. But in the first case the modality of the disjunctive compound event is necessity; in the second case it is possibility.

This peculiarity of the logic of modalities is the reason that other attempts to construct a multivalued logic for the modalities did not succeed. Two-valued logic is governed by the principle of adjunctivity, according to which the truth value of the compound proposition is determined by the truth values of its components. The same principle holds for probability logic, in which, however, three truth values concerning the two components are required for the determination. The assumption that the principle must hold also for the logic of modalities made it impossible to construct this logic. We saw that the modalities of the two components do not determine that of the compound expression; but we saw also that the introduction of a third argument would not remove this difficulty. It is therefore impossible to construct a table of modalities that can be read throughout from left to right; only the direction from right to left can always be used. The logic of modalities does not conform to the principle of adjunctivity; this logic can be understood only when it is viewed with probability logic as its background.

The logic of modalities admits of a further application, which has no analogue in probability logic. The definition (1) of the modalities shows that we are concerned here with properties of sequences that, even in the case of an infinite number of elements, are independent of the order of the sequences. It follows that the definitions can be applied also when we choose as the sequence of elements the total range of objects for which the propositional function is meaningful, no matter in what order. (It is even irrelevant whether the range constitutes a denumerable class.) Since the range is determined by the propositional function alone, we can ascribe a modality, according to (1), to the propositional function itself without referring to a special sequence of arguments. Propositional functions of one variable can thus be conceived as three-valued expressions within the logic of modalities.[2]

The use of the modalities defined in (1) is subject to certain restrictions. The modalities do not fully correspond to the use of the respective categories in conversational language. Thus it may happen that no professor will ever enter a certain classroom of a university on Sundays; but we shall not speak here of impossibility. For the application of the modalities we demand not only that the all-statements used in the first and third of the definitions (1) be true, but that they have some further properties. In particular, we

[2] This conception, which was prepared by Russell, was used by Walter Dubislav, in *Jour. f. d. reine u. angew. Math.*, Vol. 161 (1929), p. 107, for the construction of three-valued truth tables for propositional functions, with which the truth tables 10 (p. 406) are identical. These tables are used in *ESL*, § 23.

TABLE 11

Truth Tables of the Logic of Weight

A. NEGATION

$P(a)$	$P(\bar{a})$
p	$1 - p$

RESTRICTIVE CONDITIONS

1. $\dfrac{p + q - 1}{p} \leqq u \leqq \dfrac{q}{p}$

2. $P(a,a) = 1$

B. BINARY OPERATIONS

$P(a)$	$P(b)$	$P(a,b)$	$P(a \vee b)$	$P(a.b)$	$P(a \supset b)$	$P(a \equiv b)$	$P(b,a)$
p	q	u	$p + q - pu$	pu	$1 - p + pu$	$1 - p - q + 2pu$	$\dfrac{pu}{q}$

C. BINARY OPERATIONS FOR THE LIMITING CASES 0 AND 1

$P(a)$	$P(b)$	$P(a,b)$	$P(a \vee b)$	$P(a.b)$	$P(a \supset b)$	$P(a \equiv b)$	$P(b,a)$
1	1	1	1	1	1	1	1
1	0	0	1	0	0	0	?
0	1	?	1	0	1	0	0
0	0	?	0	0	1	1	?

may require that the statements be derivable from the general laws of physics. A definition of modalities of this kind, which requires the metalinguistic interpretation, is given elsewhere[3] under the name of *nomological modalities*. The modalities defined by (1) are called *extensional modalities*. The general laws of physics are defined at the same place under the term *nomological formulas*. The nomological modalities of necessity and impossibility are subclasses of the corresponding extensional modalities; the extensional modality of possibility is a subclass of nomological possibility. Table 10 (p. 406) applies also to nomological modalities and thus represents the general form of the logic of modalities.

The difference between physical and logical modalities, explained with reference to possibility at the end of § 66, cannot be formulated in terms of extensional modalities. It requires nomological modalities because it is based on the distinction between synthetic and analytic nomological formulas.

§ 81. The Logic of Weight

Probability logic has been constructed in § 79 as a logic of propositional sequences. The probability logic of individual propositions, or *logic of weight*, is constructed by a fictitious transfer of the truth properties of propositional sequences to individual propositions. We write

$$P(a) = p \tag{1}$$

thus admitting individual propositions inside the probability functor. The number p measures the weight of the individual proposition a. It is understood that the weight of the proposition was determined by means of a suitable reference class, but once the determination is completed, the indication of this class will be omitted and the notation (1) will be used.

We can also construct compound propositions and determine their weights. Thus we can write

$$P(a \vee b) = r \tag{2}$$

The weights of compound propositions are determined by truth tables resulting from table 9, A and B (p. 400), when we replace the expressions fx_i and gy_i by the symbols a and b. The tables of the logic of weight are presented in table 11, A and B. Like the logic of probability, this logic requires three arguments for the determination of the weight of compound propositions; that is, the truth tables have three argument columns. The value

$$P(a,b) = u \tag{3}$$

is employed as the third argument value.

[3] *ESL*, §§ 23, 65.

Using the implicative notation, we would write (3) in the form

$$a \underset{u}{\ni} b \tag{4}$$

Thus we have an *individual probability implication.* This concept is as fictitious as that of weight; it results from a transfer of meaning from the general to the particular case. The fictitious nature finds its expression in the way of writing the general probability implication:

$$(i)\ (fx_i \underset{p}{\ni} gy_i) \tag{5}$$

Here the degree p of probability, existing for the whole sequence, is regarded as holding for every individual case of the sequence, independently of the truth values of fx_i and gy_i. This conception is expressed by the use of the all-operator in front of the expression (5). It is clear that, when we regard the expression resulting from cancellation of the all-operator as meaningful, the conception of an individual probability implication has only a fictitious meaning.

It is different with the P-notation, when the latter is interpreted by sequences of two-valued propositions. When we write, instead of (5),

$$P(fx_i, gy_i) = p \tag{6}$$

the expression, in this interpretation, must not be regarded as an all-statement the individual cases of which have the form (3). The individual statements of this propositional sequence have the form

$$a,b \tag{7}$$

and must be regarded as propositions that are either true or false, depending on the truth values of a and b, according to the truth tables of the operation of selection given in table 11C. Exception is made for the question-mark case, where the compound (7) is no proposition. Going from this individual operation to statement (6), we cannot use an all-operator, but must use a method of counting the individual truth values, formulated in (4 and 7, § 79). We thus can regard the two-valued operation of selection as a substitute for an individual probability implication,[1] which has the advantage that it is not of a fictitious nature. Like other propositional operations, this operation is given a probability in the logic of weight, written in the fictitious form (3).

In every logic we employ, besides the terms *true* and *false*, the term *assertion.* Whereas "truth" and "falsehood" are semantical terms, "assertion" is a pragmatic term, since it refers to the sign-user. In two-valued logic there is a simple

[1] In the German edition of this book (1935), p. 379, I identified the two operations, a method that I now regard as incorrect. See also my article, "Ueber die semantische und die Objektauffassung von Wahrscheinlichkeitsausdrücken," *Jour. of Unified Science* (*Erkenntnis*), Vol. VIII (1939), pp. 61–62.

relation between these categories: a true statement can be asserted. In probability logic, however, we must, in general, use statements that are not known to be true. It is here that the term *posit* applies, which is a pragmatic term of the nature of the term *assertion*: we assert the best posit. We see that assertability is achieved, in probability logic, by the use of posits.

Even when we restrict assertability to statements of the probability 1—and we shall do so in §§ 82–83—we cannot dispense with posits. If the limit of the frequency is $= 1$, there may yet be false statements in the propositional sequence. When we deal with each of the statements as true, we can do so only in the sense of a posit.

Although positing is a behavior of the sign-user and thus a pragmatic affair, the phrase "best posit" is a semantical term. It is defined by reference to the truth of statements, namely, as the statement that will be true in the greatest number of cases. In probability logic, therefore, it occupies a place similar to that of the term "true statement" in two-valued logic. By the act of positing, we select a certain class of statements as a basis for action. From the plurality of statements of various degrees of weight, we single out a unique group that serves the same purpose as the group of asserted statements of two-valued logic.

Since the structure of probability logic is complicated, we often prefer methods that permit the use of two-valued logic instead of probability logic, thus equating truth and assertability. For a transition from probability logic to two-valued logic the following methods can be used.

1. *Method of division.* By a procedure similar to the first method described in § 76, we divide the scale of probability in two domains, the first running from 0 to $\frac{1}{2}$, and the second from $\frac{1}{2}$ to 1. We regard statements of the first domain as false, those of the second domain as true. More often we use a trichotomy instead of a dichotomy: we choose within the scale of probability the two values p_1 and p_2 so that p_1 is close to 0 and p_2 close to 1, and then call statements of the domain 0 to p_1 *false*, those of the domain p_2 to 1 *true*. Statements of the domain between p_1 and p_2 are omitted, that is, they are regarded as statements of unknown truth value.

It is obvious that this method represents but another form of positing, since the statements defined as true can be maintained only in the sense of posits. The method has the advantage that the statements defined as false can be transformed into assertable statements by the use of the negation. However, the two-valued logic so introduced has the character of an approximation, a result that is visible in certain peculiar consequences of the method of division. Assume that the weights p and q of two statements are situated in the domain from p_2 to 1; then the two statements will be called true. If, furthermore, the two statements are independent of each other, their conjunction will have the weight pq. Now if p and q are not much larger than p_2, it may happen that $pq < p_2$; the conjunction of the two true statements will then not be true; and if p_2^2 is smaller than p_1, the conjunction may even be false.

Cases of this kind may actually happen. Thus, if a judge is considering the testimony of an individual witness whose reliability has not been questioned, he will regard the testimony as true; but he may hesitate to regard the statements of *all* witnesses as true. A way out of the situation, which is contradictory for two-valued logic, is given by returning to the probability nature of the weights assigned to the testimonies. Similar paradoxes will arise for a disjunction; if two or more statements are each false in the sense defined by the division, their disjunction may be true, or at least not false. In spite of these paradoxes, the method of division is widely used.

2. *Method of transition to the metalanguage.* By a method similar to the second method described in § 76, we can go from a statement a concerning an event to the metalinguistic statement $P(a) = p$, according to which the probability of the first statement is $= p$. This method is often used. It is necessary, in particular, when we must consider the weight of statements that cannot be posited because their weight is too low, or when the weight of a posit becomes relevant. The whole calculus of probability must be incorporated in this method when probabilities are regarded as properties of individual statements, not of events, since the assertions made in this calculus concern probabilities.

It should be noticed, however, that this method requires the use of the first method so far as probability statements—statements of the form $P(a) = p$—can never be proved to be strictly true. Such statements are maintained as posits of the second level (see § 89). When we regard such statements as true, we use the method of division.

3. *Method of reduction to two-valued elements.* This method is identical with the frequency interpretation. We substitute here propositional sequences for the individual propositions within the P-symbol, and regard a probability as a frequency of two-valued statements. The ultimate elements of this logic, therefore, are two-valued propositions. Using this method, however, we have abandoned the logic of weights, and so have omitted a probability evaluation of the single case. We have thus returned to the logic of propositional sequences that was developed in § 79. The use of this method, therefore, is restricted to cases in which only large numbers of objects are considered, as in all statistical investigations.

This method, like the second, requires a simultaneous use of the method of division. When we regard the individual statements of statistics as true, we can do so only in the sense of an approximation, since all that can be asserted is a high weight; the same holds for the statement about statistical frequencies when the statement is meant to include future observations.

In contradistinction to the second method, the third method is capable of two linguistic interpretations, as was explained in § 73. We can regard it, like the second method, as a transition to the metalanguage; then the expression (6) is meant to have the form (5, § 77), and thus refers to propositional sequences. But we can also regard (6) as referring to sequences of events. The object language is two-valued in both. In the first, it is so because the ultimate elements of the object language are two-valued propositions the frequency relations of which are expressed in (6). In the second, it is two-valued because it states frequency relations holding for events in a two-valued language.

§ 82. Derivations and Tautologies in Probability Logic

The method of derivation developed for the calculus of probability can be transferred to probability logic. Since the method deals with formulas containing the P-symbol, however, it represents a method of derivation, not within the object language of probability logic, but within the metalanguage corresponding to it, and may be called an *indirect* method of derivation. Its advantage consists in the fact that, since the metalanguage of probability logic is two-valued, the procedure of derivation is identical with the derivative methods of two-valued logic. The presentation of a direct method of derivation will be postponed to a later section (§ 83).

For the analysis of the indirect method, the derivable formulas can be divided into two categories. The first includes formulas stating that the probability of a certain expression is $= 1$, for example,

$$P(a \vee \bar{a}) = 1 \tag{1}$$

In the second category are formulas expressing relations between probability values, for example,

$$P(b) = P(a) \cdot P(a,b) + [1 - P(a)] \cdot P(\bar{a},b) \tag{2}$$

which corresponds to the rule of elimination (2, § 19), resulting from it by the omission of the first term A and a renaming of variables.

Although the derivation of the formulas is carried through in the two-valued metalanguage, there is a certain difficulty connected with it. As long as formulas like (1) and (2) are derived in the calculus of probability, not only the language of the derivation but the language of the symbols inside the parentheses, that is, the object language, is two-valued. This permits the use of logical transformations inside the parentheses; for instance, in the treatment of the expression $P(a.b \vee \bar{a}.b)$ we are allowed, applying the rule of replacement, to write the expression in the form $P(b)$. This replacement is used in the derivation of (2). When, however, the derivation refers to probability logic, the expressions inside the parentheses are subject to the rules of probability logic; the signs of the propositional operations are then defined by the truth tables of probability logic, and we are not allowed, without a specific proof, to apply logical transformations to the expressions. Derivations in the metalanguage of probability logic, therefore, although performed in a two-valued language, are not legitimate until a proof of the applicability of logical transformations inside the parentheses is given.

Turning to the construction of this proof, we shall show that the object language of probability logic admits of the same logical transformations that are used in two-valued logic. For this purpose the problem of *tautologies* must be studied first. In probability logic, tautologies are defined as formulas that have the probability 1 for all probability values of their constituents. When we set up the rule that formulas of the probability 1, and only such formulas, can be asserted, tautologies will constitute the assertable formulas of probability logic.

We shall first consider only derivations of formulas of type (1). It will be shown that every tautology of two-valued logic is, at the same time, a tautology of probability logic. Such formulas will be referred to as tautologies of the first kind, since, as we shall see later (§ 83), probability logic includes a second kind of tautologies that have no analogues in two-valued logic; such tautologies are constructed from formulas of type (2). It will be shown also that tautologies of the first kind may be used for transformations of the expressions inside the P-symbol. This result guarantees that for every formula that is derivable in the calculus of probability there exists an analogue in the metalanguage of probability logic, including formulas of type (2).

To illustrate the problem, consider the formula

$$P(a\,.\,a \equiv a) = 1 \tag{3}$$

When we regard the equivalence sign and the period sign as defined by the two-valued truth tables, the formula is demonstrable in the calculus of probability. For this purpose we would have only to rewrite the axioms of the calculus by the omission of the first term A (which means, in the frequency interpretation, that we restrict our consideration to sequences that are compact in A). Since the probability of every compound expression can be written as a function f of the probabilities of the elementary expressions, and since the probability of every two-valued tautology is $= 1$, we shall find that f is identical with 1 for all probability values of the elementary expressions.

This inference cannot be applied immediately to (3) when the logical signs are regarded as expressing operations of probability logic. Instead, (3) must be proved independently; for this proof it cannot be assumed that $a\,.\,a$ can be replaced by a, since such replacement has not yet been shown to be permissible for the operations of probability logic. However, we can give for (3) an independent proof, which shows, at the same time, why the formula $P(a,a) = 1$ must be added to the truth tables of probability logic:

$$P(a.a) = P(a) \cdot P(a,a) \quad \text{(col. 5, table 11}B\text{)}$$

$$\frac{P(a,a) = 1 \qquad \text{(restrictive condition 2, table 11)}}{P(a.a) = P(a)} \tag{4}$$

Looking at column 7 of table 11*B* (p. 408), we see that (4) is not sufficient to establish (3); we must first prove that $P(a.a,a) = 1$, since $P(a.a,a)$ corresponds to the value u of the column. The proof is given as follows:

$$P(a,a) = 1$$

$$P(a.b,a.b) = 1 \text{ (substitution of } a.b \text{ for } a) \tag{5}$$

$$P(a.b,a.b) = P(a.b,a) \cdot P(a.b.a,b) \text{ (col. 5, table 11}B\text{)}$$

Since the latter product contains two factors that cannot be larger than 1, and since the product is $= 1$ because of the preceding line, each of the two factors must be $= 1$, that is, we have

$$P(a.b,a) = 1 \tag{6}$$

$$P(a.b.a,b) = 1 \tag{7}$$

Substituting a for b in (6), we arrive at

$$P(a.a,a) = 1 \tag{8}$$

(8) in combination with (4) proves (3).

By similar inferences it is possible to prove the commutativity and the associativity of the "and" and the "or", but these proofs will not be given here.

However, we shall prove the rule of replacement (see 3, § 5), which for probability logic assumes the form: if the formula

$$P(b \equiv c) = 1 \tag{9}$$

is derivable, that is, a tautology, it is permissible to replace b by c in all probability expressions. The result enables us to use the tautological equivalences of probability logic for transformations inside the P-symbol.

We first prove that if (9) holds we have

$$P(b) = P(c) \tag{10}$$

Employing the notation

$$P(b) = p \qquad P(c) = q \qquad P(b,c) = u \qquad P(c,b) = v \tag{11}$$

we obtain, from the seventh column of table 11B,

$$1 - p - q + 2pu = 1 \tag{12}$$

$$u = \frac{p + q}{2p} \tag{13}$$

The second inequality of restrictive condition 1 of table 11 gives the result

$$u \leqq \frac{q}{p} \tag{14}$$

which, by the substitution of (13), leads to the inequality

$$p \leqq q \tag{15}$$

Since the values of all probabilities are subject to the condition of normalization, that is, must lie between 0 and 1, limits included, we derive from $u \leqq 1$, by means of the last column of table 11B, that

$$u = \frac{vq}{p} \leqq 1$$

$$v \leqq \frac{p}{q} \tag{16}$$

Inserting the value $u = \frac{vq}{p}$ in (12), we derive

$$1 - p - q + 2qv = 1 \tag{17}$$

$$v = \frac{p + q}{2q} \tag{18}$$

Combining (16) and (18), we arrive at

$$q \leqq p \tag{19}$$

(15) and (19) lead to the conclusion that

$$p = q \tag{20}$$

a result that proves (10). Although the quantity p occurs in the denominator of (13), the relation (20) holds also for the case $p = 0$. Namely, if $p = 0$, we derive from (13) that also $q = 0$, since otherwise u could not be $\leqq 1$.

We now prove that if (9) holds we have

(21a) $\qquad P(b,c) = 1 \qquad P(c,b) = 1 \qquad$ (21b)

This results from (13) and (18) with $p = q$. The proof, however, depends on the condition $p > 0$.

Furthermore, we prove that if (9) holds we have

$$P(a,b) = P(a,c) \tag{22}$$

We first prove that if $P(d) = 1$ we have also $P(a,d) = 1$ for every $P(a) > 0$. With the notation

$$P(d) = p \quad P(a) = q \quad P(d,a) = u \quad P(a,d) = v \tag{23}$$

we derive from the restrictive condition 1 of table 11 that for $P(d) = 1$ we have $P(d,a) = P(a)$. The last column of table 11B then furnishes $P(a,d) = 1$. Applying the result to (9), we derive

$$P(a,b \equiv c) = 1 \tag{24}$$

In order to apply the truth table 11 to expressions like (24), in which the propositional operation stands in the second place of a comma expression, we introduce the following addition to the rule of substitution:

Rule α. If a relation between probabilities $P(x_1)$, $P(x_2)$, . . . , is derivable, it is permissible to substitute a,x_i for every x_i, provided that the same variable a is used in all expressions. Expressions of the form $a(x,y)$ resulting from such substitutions are replaced by the form $a.x,y$.

Applying this rule, and using the notation

$$P(a,b) = p \quad P(a,c) = q \quad P(a.b,c) = u \quad P(a.c,b) = v \tag{25}$$

we can now write (24) in the form (12), since we can insert the first term a in column 7 of table 11B. Applying to (12) the same methods as above, we derive (22).

By means of the last column of table 11B we now derive from (22) that if (9) holds we have

$$P(b,a) = P(c,a) \tag{26}$$

This proof, however, is bound to the conditions $P(b) > 0$ and $P(c) > 0$.

Employing the results obtained, it is now easy to show that if (9) holds we have

$$P(d.b \equiv d.c) = 1 \tag{27}$$

$$P(d \vee b \equiv d \vee c) = 1 \tag{28}$$

The proof need not be given here. In a similar way, all further conditions for the rule of replacement are derivable.[1]

The proofs so far have been restricted to the condition that probabilities of expressions occurring in the first term, like $P(b)$ in (21a), are > 0. It will now be shown that the results can be made independent of this condition. The proof is given by showing that when we apply the frequency interpretation the condition can be dropped. In the formal system of probability logic, then, we shall introduce a specific rule stating that the condition can

[1] *ESL*, p. 60.

be dropped; the proof guarantees that the formal system so extended still admits of a frequency interpretation.

When we apply the frequency interpretation to finite sequences, a probability value $P(A) = 0$ can occur only if the class A is empty. (We write capitals because we are now dealing with classes.) For this case we have set up the rule that $P(A,B)$ is not univocal, but has all real numbers as its value. A relation derived for $P(A) > 0$ will, therefore, hold also for this case, since the value of $P(A,B)$ determined by the relation will then certainly be among the values of $P(A,B)$. Only for infinite sequences must we make an exception, since for them $P(A)$ can be $= 0$, although the class A is not empty (only the limit of the frequency need be $= 0$). For a class A of this kind, for instance, $P(A,D)$ need not be $= 1$ when $P(D) = 1$.

It is not necessary, however, to consider such cases. All formulas that are derivable must hold for both finite and infinite sequences; this follows because the same holds for the formulas given by the truth tables of probability logic. In particular, if a formula $P(D) = 1$ is derivable, it holds strictly for every n of a sequence, since the formula holds for all probability values of the constituents of D and thus is true whatever be these values for the n considered. The formulas consist in the statement of an equality of certain frequencies. Now if the equality holds for every initial section of the length n of the sequence—and this follows from its validity for finite sequences—it must hold also for the limit of the frequency for $n \to \infty$. It is therefore impossible to derive a formula $P(A,B) = \ldots$ that is false for infinite sequences when $P(A) = 0$. In order to express in the formal system of probability logic this result derived from the frequency interpretation, we introduce the rule, returning to small letters:

Rule β. If a formula is derivable for $P(a) > 0$, it holds also for $P(a) = 0$.

As a consequence of this rule, the relations (10), (21a), (21b), (22), (24), (26), (27), (28) are valid for all probability values, on the condition that (9) is a *derivable* formula (and not only a true formula).

Using the rule of replacement, we can now derive all tautologies of the first kind of probability logic, tautologies that are identical with tautologies of two-valued logic. They are derived in the metalanguage, in the form of a statement that the probability of the formula is $= 1$. In the process of derivation we can use the tautologies already derived for logical transformations. It is easily seen that in this way every tautology of two-valued logic is made a tautology of probability logic.[2]

[2] For the treatment of compound expressions like $P(a \supset b, \bar{a} \vee b)$ the following method is used. The truth tables allow us to divide the expression after the comma, that is, to write the probability considered as a function of $P(a \supset b, \bar{a})$, $P(a \supset b, b)$ and $P([a \supset b].\bar{a}, b)$. The last column of table 11B then permits us to determine these probabilities as functions of the reversed ones. We thus arrive at probabilities with a single term before the comma and a compound term after the comma. The latter term is then divided by the method used before. By repetition of these methods the compound probability is broken down to probabilities containing only single terms before and after the comma.

It should be realized, however, that, although every tautology of two-valued logic supplies probability logic with a tautology of identical form, the meaning of the two tautologies is not the same. The corresponding signs of propositional operations have different meanings, since they are defined by different truth tables. Similarly, the fact that the truth values of propositional variables have different ranges leads to a difference in meaning. The meaning of a logical formula cannot be separated from the semantical rules holding for the logic to which the formula belongs; though these rules are statable only in the metalanguage, they control the meaning of the object language formula, being implicitly included in its symbolic form. The relation between probability logic and two-valued logic, therefore, must be stated as an isomorphism rather than as an identity of parts; the tautologies of two-valued logic are isomorphous to a subclass of the tautologies of probability logic. Only when the range of probability values is restricted to the values 0 and 1 will the tautologies of this subclass assume the meaning of the corresponding two-valued tautologies. In other words, two-valued logic is a special case of probability logic resulting for a certain restriction of its semantical rules.

The derivation of formulas of type (2) can now be carried through easily. In this derivation we can use the tautologies of the first kind for logical transformations inside the parentheses; furthermore, we shall use rules α and β. It is obvious that in this way every formula of the calculus of probability is derivable in identical form in the metalanguage of probability logic.

So far, the metalanguage has been used for both the derivation and the assertion of tautologies. It is easily seen that the assertion, at least of tautologies of the first kind, can be transferred to the object language if the rule is introduced:

RULE OF ASSERTION. A formula of the probability 1 may be asserted, except for the case that it has the probability 0 simultaneously.

The *probability 1* thus takes over the function of the *truth* of two-valued logic, although the first concept is slightly wider than the second. The rule permits us to assert all tautologies of two-valued logic unchanged within the frame of probability logic. For instance, we may write the tautologies

$$a \vee \bar{a}$$

$$\overline{a \vee b} \supset \bar{a} \tag{29}$$

The way of writing does not indicate the probability frame, which is understood.

A tautology of the first kind that does not occur in two-valued logic is the formula

$$a,a \tag{30}$$

which holds because of the second restrictive condition of table 11 (p. 408). Exception is to be made for the case $P(a) = 0$, for which (30) is not assertable because of the indeterminacy arising for the comma operation. An assertable formula containing a comma may be connected with another assertable formula by means of an "and". For instance, the formula

$$[a \vee a].[a,a] \tag{31}$$

is a meaningful expression for $P(a) > 0$ and has the probability 1.

§ 83. The Quantitative Negation

The assertion of tautologies of the second kind in the object language of probability logic requires some further technical means, which at the same time make it possible to carry through the process of derivation in the object language. In order to develop these means, a way must be found to express the degree of probability, or the weight, within the object language. This aim can, in fact, be reached. For this purpose a specific instrument, namely, a *quantitative negation*, will be constructed.

In two-valued logic the negation serves as an instrument to express truth values in the object language. Instead of saying, in the metalanguage, the sentence a is false, we say, in the object language, $\bar{a}$. By means of the negation we thus coördinate a true sentence to a given false sentence, and, asserting the true sentence, we express the fact that the original sentence is false.

A similar method is used in multivalued logics, which for such purposes possess a *cyclical negation*. For example, in three-valued logic[1] there are three truth values t_1, t_2, t_3, the "highest" of which, say t_3, may be regarded as truth. The cyclical negation, written as $\sim a$, is defined by the truth table 12. As in two-valued logic, only statements having the highest truth value are assertable. When we wish to say that a statement a has the truth value t_1, usually interpreted as falsehood, we write

$$\sim a \tag{1}$$

and when we wish to say that a has the middle truth value t_2, we write

$$\sim\sim a \tag{2}$$

Thus the rank of the truth value is identical with the number of negation signs placed before the proposition. The truth value t_3 is given by

$$\sim\sim\sim a \tag{3}$$

This, however, is the same as a; the three negation signs in (3) thus can be canceled.

[1] See, for instance, the presentation of three-valued logic in H. Reichenbach, *Philosophic Foundations of Quantum Mechanics* (Berkeley, 1944), § 32.

TABLE 12

TRUTH TABLE OF THREE-VALUED CYCLICAL NEGATION

a	$\sim a$
t_3	t_2
t_2	t_1
t_1	t_3

This procedure can be used for any n-valued logic. Probability logic, however, has an infinite number of truth values arranged in a continuous scale, and the cyclical negation must therefore be constructed as a *quantitative negation*. We write this negation by the addition of a numerical variable $\ulcorner w \urcorner$ in half-brackets before the sentence, and define it by the truth table 13.

TABLE 13

TRUTH TABLE OF QUANTITATIVE NEGATION IN PROBABILITY LOGIC

a	$\ulcorner w \urcorner a$
p	$p - w + \delta_{p-w}$ $\delta_{p-w} = \begin{cases} +1 \text{ for } p - w \leqq 0 \\ 0 \text{ for } 0 < p - w < 1 \\ -1 \text{ for } p - w = 1 \end{cases}$

The *negation variable* w is a real number between 0 and 1, limits included.

Table 13 may be illustrated by figure 27. The range of probability from 0 to 1 is indicated by the circle, running clockwise. The sentence a has a probability p, indicated by its position on the circle. The negation w runs counterclockwise. For $w < p$ the sentence $\ulcorner w \urcorner a$ has a smaller probability. For $w = p$ the probability of $\ulcorner w \urcorner a$ jumps discontinuously to 1, as a consequence of the rules for the δ-symbol set up in the table; and for $w > p$ the sentence $\ulcorner w \urcorner a$ has a probability greater than $P(a)$.

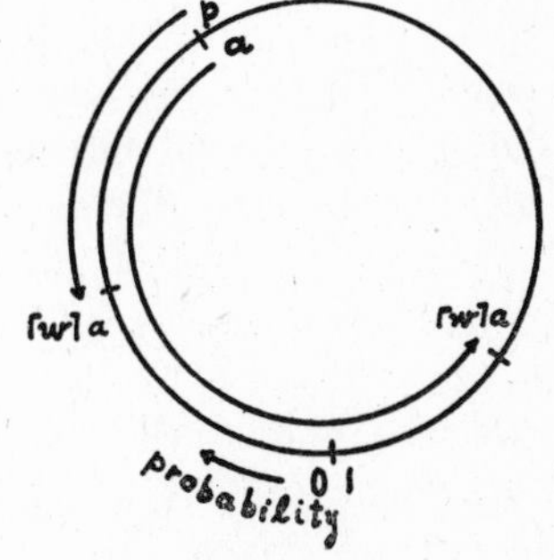

Fig. 27. Diagram of quantitative negation.

By means of this negation we can coördinate a sentence of any degree q of probability[2] to a given sentence of the probability p. When we choose, in particular, $w = p$ for the negation variable, the coördinated sentence has the probability 1.

[2] With the exception of the value $q = 0$, which can be assumed only for $p = 1$ and $w = 0$. For p, however, the 0-value is not excepted; that is, to a sentence of the probability 0 we can coördinate a sentence of any other value.

It is easily seen from the table that this is the only case in which $q = 1$, that is, the statement $\ulcorner w \urcorner a$ has the probability 1 *if and only if* $w = P(a)$. We see, furthermore, that a negation to the degree $w = 1$ leaves the truth value of the statement unchanged. A negation to the degree $w = 0$, in general, also leaves the truth value unchanged, except for the case $p = 0$ or $p = 1$, where the negation reverses the truth value.

We now apply the rule of assertion (§ 82), according to which propositions of the probability 1, and only those, may be asserted. The quantitative negation, consequently, allows us to state the degree of probability in the object language. In order to express the metalinguistic sentence $P(a) = p$, we write in the object language the true sentence

$$\ulcorner p \urcorner a \tag{4}$$

The symbol $\ulcorner p \urcorner$ before the sentence has the same function as the number of negation signs in the expressions (1) and (2). Because of the continuous character of the truth scale, however, we deny a sentence to a certain degree. We state a true sentence by denying the sentence a to the degree p.

The sign $\ulcorner w \urcorner$ can be put before compound sentences also. Thus the meaning of $\ulcorner w \urcorner (a \vee b)$ is determined by the truth table 13 when we put there $a \vee b$ for a. Similarly, the meaning of $\ulcorner w \urcorner (a,b)$ is determined by table 13.

Although the quantitative negation was introduced in the logic of weight, a frequency interpretation of this operation can be given. In a propositional sequence containing the sentences b_1, b_2, b_3, and so on in repeated occurrence, the negation determines for every sentence whether it *satisfies* the sentence $\bar{b}_1$, namely, it determines all sentences b_2, b_3, and so on, as being of this kind. In contradistinction to this procedure, the quantitative negation does not determine individual sentences as being of the form $\ulcorner w \urcorner b_1$. We are free, rather, to regard any sentence of the form b_1, b_2, b_3, and so on, as being of the form $\ulcorner w \urcorner b_1$, the only condition being that the total number of sentences $\ulcorner w \urcorner b_1$ conform to the degree of probability supplied by table 13 (p. 421). If $P(\ulcorner w \urcorner b_1) = 1$, all sentences of the sequence will have the form $\ulcorner w \urcorner b_1$; or, for infinite sequences, at least so many sentences that the limit of the frequency is $= 1$.

Because of the arbitrary distribution of the sentences satisfying $\ulcorner w \urcorner b_1$ in the sequence, the value $P(a, \ulcorner w \urcorner b_1)$ is not determined in terms of $P(\ulcorner w \urcorner b_1)$, but must be given independently. Similarly, the value $P(\ulcorner w \urcorner b_1, c)$ is independent of $P(\ulcorner w \urcorner b_1)$. For the frequency interpretation of the expression $P(\ulcorner w \urcorner (a,b))$ we use only the subsequence selected by a, that is, we regard only sentences chosen from this subsequence as having the form $\ulcorner w \urcorner (a,b)$. This means that we count the frequency of $P(\ulcorner w \urcorner (a,b))$ in the subsequence, like that of $P(a,b)$. The reason for this interpretation of $P(\ulcorner w \urcorner (a,b))$ is that this probability is a function of $P(a,b)$, according to table 13.

The interpretation makes clear that the quantitative negation constitutes a negation relative to a certain sequence as reference sequence. If a probability expression contains no comma, the reference sequence is given by the main sequence tacitly assumed for all probability expressions; if it contains a comma, the sequence given by the first term of the probability expression is the reference sequence. The application of the quantitative negation to individual sentences in the logic of weights, therefore, has the same fictitious character as the application of degrees of probability to individual sentences.

Using the quantitative negation for the indication of the degree of probability, we can now transfer all derivative methods of the metalanguage into the object language. In order to carry through this program, we must first introduce tautologies of a second kind, different from those that are identical with two-valued tautologies and that were discussed in § 82 under the name of tautologies of the first kind.

Tautologies of the second kind contain a quantitative negation and constitute the analogues of P-formulas of the second category derivable in the metalanguage of probability logic, like (2, § 82), that is, of formulas stating the equality of certain degrees of probability. For instance, the formula

$$P(a) + P(\bar{a}) = 1 \tag{5}$$

can be written in the form

$$[P(a) = p] \supset [P(\bar{a}) = 1 - p] \tag{6}$$

The implication is the two-valued implication of the metalanguage. Using the quantitative negation, we can construct the object-language equivalent of (6) in the form

$$\lceil p \rceil a \rightarrow \lceil 1 - p \rceil \bar{a} \tag{7}$$

We must discuss the implication indicated by the arrow. It is not permissible to identify it with the implication defined in the sixth column of the truth table 11B (p. 408) of probability logic. For obvious reasons we can also establish in (7) an implication going from right to left; the two implications, then, would constitute an equivalence in the sense of the seventh column of table 11B. But it is easily seen that the expressions on the two sides of (7), in general, do not have the same probability value. This follows when we apply table 13 of the quantitative negation, which shows that equal probability values result only for $P(a) = p$, or, what is the same, for $P(\lceil p \rceil a) = 1$. Now we want to regard (7) as a tautology, as a formula having the probability 1 for all probability values $P(a)$. In other words, we want (7) to hold also for values other than $P(\lceil p \rceil a) = 1$. It follows, then, that the implication in (7) requires a definition different from that given in table 11B.

For these reasons we introduce the truth table 14, which defines the *alternative implication*, expressed by the arrow. This implication corresponds to the two-valued implication in that it is true whenever the implicans is not completely true (third row of the table). It is obvious that, because of this property of the implication, formula (7) will have the probability 1 for all values of $P(\lceil p \rceil a)$, that is, it will be a tautology. What (7) asserts, in fact, is only that the probability of the implicate must be $= 1$ when the probability of the implicans is $= 1$, whereas any assertion concerning other probability values is omitted. For this reason the existence of the inverse implication in (7), going from right to left, does not lead to an equivalence; the double-arrow implication states the equality of truth values only for the value 1, leaving open any statement about other values.

TABLE 14

TRUTH TABLE OF ALTERNATIVE IMPLICATION IN PROBABILITY LOGIC

$P(a)$	$P(b)$	$P(a \rightarrow b)$
1	1	1
1	< 1	0
< 1	$\leqq 1$	1

The name *alternative implication* has been chosen in order to indicate that the implication is capable only of the two truth values 1 and 0, as shown by the third column of table 14. To use an alternative implication in tautologies of the second kind seems appropriate because such tautologies correspond to formulas of the two-valued metalanguage, like (6). We thus arrive at formulas of the object language that can only be true or false. The formula

$$\lceil p \rceil a \twoheadrightarrow \lceil p \rceil \bar{a} \tag{8}$$

for instance, is false for all values $p \neq \frac{1}{2}$ and $P(a) = p$, and it would not make sense to regard this formula for such values as probable to a certain degree. That it is possible to construct two-valued formulas within a multi-valued logic is known from the study of three-valued logic;[3] the alternative implication of table 14, in fact, corresponds to an operation of the same name employed in three-valued logic.[4]

By means of the alternative implication we can transcribe all formulas of the second category derivable in the metalanguage of probability logic, like (2, § 82), in formulas of the object language. The transcribed formulas contain a quantitative negation and represent tautologies of the second kind. Thus, formula (2, § 82) supplies the tautology:

$$\lceil p \rceil a \,.\, \lceil u \rceil (a,b) \,.\, \lceil v \rceil (\bar{a},b) \twoheadrightarrow \lceil pu + (1-p)v \rceil b \tag{9}$$

[3] See H. Reichenbach, *Philosophic Foundations of Quantum Mechanics* (Berkeley, 1944), pp. 154, 159. [4] *Ibid.*, p. 151.

For this formula, of course, the arrow cannot be reversed, as in (7). There exist, however, inverse formulas—formulas resulting from (9) by exchanging the right side with one of the units of the left side. For the derivation of such formulas the rule of existence is to be assumed, as was explained with respect to (2, § 13). We shall not go into the formalization of this procedure, since we did not do so for the calculus of probability. Note that the comma expressions in (9) are meaningful only if the probability of the reference term does not vanish.

The existence of tautologies of the second kind shows that probability logic is richer in tautologies than two-valued logic. Every tautology of two-valued logic is isomorphous to a tautology in probability logic, but not vice versa.

After the construction of tautologies of the second kind, we can turn to the consideration of the direct method of derivation. The rules of derivation, to be used for derivations in the object language, are the same as those of two-valued logic. The rule of substitution obviously holds; it was demonstrated above that the rule of replacement is applicable. The rule of inference (*modus ponens*) holds in the two forms

$$\begin{array}{c} a \\ a \rightarrow b \\ \hline b \end{array} \tag{10}$$

$$\begin{array}{c} a \\ a \supset b \\ \hline b \end{array} \tag{11}$$

The justification of (10) follows from the definition of the alternative implication in table 14 (p. 424); if $P(a) = 1$ and $P(a \rightarrow b) = 1$, we must have $P(b) = 1$. The implication in (11) is the implication defined in column 6 of table 11*B* (p. 408) of probability logic. It is easily derivable from the value $1 - p + pu$ of this column that $u = 1$ when the premises of (11) have the probability 1; this leads to $P(b) = 1$.

Besides the fundamental rules (10) and (11), every tautological implication can be employed for the construction of a secondary rule of inference. Thus (9) leads to the inferential schema

$$\begin{array}{c} \ulcorner p \urcorner\, a \\ \ulcorner u \urcorner\, (a,b) \\ \ulcorner v \urcorner\, (\bar{a},b) \\ \hline \ulcorner pu + (1 - p)v \urcorner\, b \end{array} \tag{12}$$

This schema is the object-language analogue of the inferential schema of the metalanguage:

$$\frac{\begin{array}{l} P(a) = p \\ P(a,b) = u \\ P(\bar{a},b) = v \end{array}}{P(b) = pu + (1 - p)v} \tag{13}$$

The inferential schema (12) can be regarded as a generalization of the *modus ponens* for premises that are not true but have only certain degrees p,u,v of probability. A special form of (12) results for $p = 1$; we then have

$$\frac{\begin{array}{c} a \\ \ulcorner u \urcorner \ (a,b) \end{array}}{\ulcorner u \urcorner\, b} \tag{14}$$

This schema resembles more closely the *modus ponens*.

I do not wish to suggest the notation in terms of the quantitative negation for practical use. The P-notation in the metalanguage is technically superior for derivations. But it is important that probability logic can be made complete, so that it includes direct methods of derivation.

Chapter 11

INDUCTION

INDUCTION

§ 84. The Various Forms of Induction in Empirical Science

The analysis of probability statements presented in chapter 9 has led to the result that the meaning of the word "probable" is always reducible to a frequency meaning. This result includes the logical interpretation of probability, although by the concept of weight this interpretation acquires a certain independence and assumes the form of a logic in which the reference to frequencies is not explicitly stated. The problem of the assertability of probability statements finds a simple solution so far as the laws of probability are concerned, since the laws are made tautological by the frequency interpretation. There remains only one problem, which is inherent in the frequency interpretation: the problem of the ascertainment of the degree of probability, which has been shown to be the same as the problem of induction.

The word "induction" is usually employed in a sense more comprehensive than the one so far envisaged in this work. The rule by which we infer that the frequency observed in an initial section will persist for the whole sequence is regarded as a special case of induction, often called *induction by enumeration.* Postponing the discussion of induction by enumeration, we shall consider first the other forms of induction employed in scientific method.

Francis Bacon, who was the first to emphasize the need for inductive inference in scientific method, regarded induction by enumeration as a poor instrument of prediction and attempted to devise inductive methods superior to it. His tables of presence, of absence, and of degrees were constructed for that purpose. More than two centuries later, they were adopted by John Stuart Mill, who reformulated them as canons of induction and believed that in them he had constructed the ultimate form of scientific inference.

When we consider these improved forms of induction critically, we find that they contain three additions to induction by enumeration. The first is a trivial use of *deduction*: Bacon's table of absence, or Mill's canon of difference, calls for a collection of instances in which a factor A is not connected with a factor B—instances, therefore, which prove that the conclusion "All A are B" is not true. Here deduction is employed in a trivial form to rule out impermissible inductive conclusions. The method, of course, is applicable only when classical induction (§ 67) is concerned; it cannot be applied to statistical induction. The second addition is the emphasis on *large numbers*, illustrated particularly by Bacon's long tables on the phenomena of heat. The third addition is clearly stated by Bacon, but not by Mill, though it obviously underlies his theories, too: the rule that the instances collected must constitut a *fair sample.* Bacon[1] refers to it in his story of the shipwreck:

> It was a good answer that was made by one who when they showed him hanging in temple a picture of those who had paid their vows as having escaped shipwreck, and woul have him say whether he did not now acknowledge the power of the gods—Aye, asked h again, but where are they painted that were drowned after their vows?

The disregard of the rule of the fair sample may be called the *fallacy of biase statistics.* Mill assumes the rule, in particular, in his canon of concomitan variations, for such variations can establish an inductive relationship onl when they are taken at random. In Charles Peirce's investigations of induc tion, the rule of the fair sample plays an important part.

Another method of improving induction—though it is not mentioned b Bacon, Mill, or Peirce—may be called *cross induction.* For example, th inference that, because all the swans so far observed have been white, all th swans in the world are white, has turned out to be a bad inference, sinc black swans were found in Australia. Even before the discovery of black swans however, the conclusion could have been questioned by the following con sideration: it is a general rule for biological species that color is not a constan characteristic within a species; therefore, one should not have inferred tha all swans are white.[2] Here an inference of induction by enumeration, applie to one sequence, is criticized by means of another inductive inference tha refers to sequences as elements.

The inference may be illustrated by the schema

$$\begin{array}{cccccccccc} B & B & \bar{B} & B & \bar{B} & B & B & . & . & . \\ C & \bar{C} & C & C & \bar{C} & \bar{C} & C & . & . & . \\ . & . & . & . & . & . & . & . & . & . \\ . & . & . & . & . & . & . & . & . & . \\ S & S & S & S & S & S & S & . & . & . \end{array} \qquad (1)$$

For each of the first horizontal sequences an inductive inference leads to the result that both positive and negative instances will always occur; for the last horizontal sequence the inductive conclusion is that only positive instances will occur. Making an inductive inference in the vertical direction, for which each horizontal sequence is an element, we infer that, since all the other sequences show both kinds of instances, the last horizontal sequence will do the same if it is sufficiently continued. We thus cancel an individual inductive inference by means of a cross induction.

The schema is frequently applied. It is given in the inference that, although carbon has not yet been melted, it will melt at higher temperatures because

[1] *Novum organum scientiarum* (published 1620), aphorism 46.

[2] André Lalande, . . . *Les Théories de l'induction et de l'expérimentation* (Paris, 1929).

all other substances do so.[3] The inference that all men are mortal, too, has this form when we regard the lives of individuals as sequences of days; that the sequences of days of persons now living will terminate in a day of death is inferred by a cross induction.

Whereas all these forms of inference clearly show their relation to induction by enumeration, there is another form of induction that, at first sight, seems to be of a different nature. It is based on the application of causal explanation and may therefore be called *explanatory induction.* It consists in the inference from certain observational data to a hypothesis, or theory, from which the data are derivable and which, conversely, is regarded as being made probable by the data. Such inferences are used in the establishment of scientific theories. They are applied also in the investigations of a detective who, in the search for the perpetrator of a crime, constructs an explanation on the basis of observational findings. In the physical sciences the explanation is carried through by means of mathematical methods, a fact showing that the inference from the hypothesis to the data may be of the deductive type. The advancement of science in the last centuries, in fact, is due to the application of explanatory induction; and the critics of John Stuart Mill were right when they insisted that no theory of induction is satisfactory unless it includes an account of explanatory induction.

The various forms of induction presented include a common feature: they all constitute inductive inferences made in an advanced state of knowledge (see § 70), that is, inferences based, not on new observational data alone, but also on the results of previous inductive inferences. In fact, it virtually never happens that an inductive inference is made in isolation; the success of induction is based on a method of *concatenation,* which combines many inductive inferences in a network of inference. This fact has usually been overlooked in the theories of induction. If a great deal of knowledge is taken for granted, the inductive inference assumes particular forms that are justifiable only on the basis of tacit assumptions. Presuppositions of the specific form, however, were overlooked, and theories of induction were constructed that took one or the other specific form, without mention of the conditions of its applicability, as representative of the inductive inference in general. The resulting theories illustrate the fallacy of incomplete schematization (§§ 21, 68).

So far as explanatory induction is concerned, a fallacious interpretation was discussed at the end of § 21, in the analysis of the so-called *inference by confirmation.* This inference has also been called the *hypothetico-deductive* method, the term being meant to indicate the deductive relation from the hypothesis to the observational data. The fallacy consists in the belief that an inductive relation holds for the reversed direction, or, more precisely

[3] See *EP*, p. 365.

speaking, that the implication $a \supset b$ entitles us to regard a as probable when b is given.

Explanatory induction must not be interpreted in the same sense. Only for a superficial consideration does explanation have the form of the hypothetico-deductive method, or of the inference by confirmation. In deeper analysis it reveals a much more complicated structure. Explanatory induction must be regarded, not as an inference in its own right, but as a combination of probability inferences such as are formulated by the rule of Bayes. The complicated nature of the inference is made clear by the fact that the inference is applied only when much more is known than the occurrence of the consequences of an assumption; without an estimate of the antecedent probabilities the inference is never made.

For scientific theories, estimates of the antecedent probabilities are often given in the form of considerations about the plausibility or "naturalness" of a theory, that is, by arguments that make the theory credible, independent of the observed confirmation. For instance, Newton's law of gravitation has a high antecedent probability because of the term r^2 in the denominator, which corresponds to the decrease of a force spread over the surface of a sphere. The law thus appears as an expression of the three-dimensionality of space, since only in a three-dimensional space does the surface of a sphere diminish with the square of the radius.

For detective cases the antecedent probabilities appear in the form of a discussion of the motives of the crime. Simple examples, in which the computation of the probability of the explanation can actually be carried through, were given in the exercises appended to chapter 3. Another illustration is found in the analysis of a problem of circumstantial evidence in § 47, in which the probability of an insurance murder is computed numerically. The misunderstandings of the inference by indirect evidence and its interpretation as an inference by confirmation may perhaps be psychologically explained as an oversight of the role which the antecedent probabilities play in the inference. These probabilities are easily overlooked because they often need not be known otherwise than in the form of crude estimates, while the result of the inference can be very precise (see §§ 62, 70).

The thesis may be generalized: all inductive inferences that do not have the form of induction by enumeration must be construed in terms of the theorems of the calculus of probability. In fact, the calculus of probability contains the key to a theory of induction in advanced knowledge. Philosophers who believe that a philosophical theory of induction is to be developed independent of the statistical methods employed in the sciences make the mistake of overlooking the existing mathematical methodology: all the questions concerning induction in advanced knowledge, or *advanced induction*, are answered in the calculus of probability. While logicians were vainly looking for an inductive logic that could account for scientific method, mathematicians constructed a mathematical system that covers all forms of probability inference and thus of scientific inference—a system that can even be transcribed into a system of logic, as was shown in chapter 10. The logician of our day who is aware of the fallacies of the philosophy of rationalism abandons all attempts at a construction of an inductive logic from pure reason. The inductive method presented by the calculus of probability is a much more powerful instrument than any substitute devised under the name of rational belief; moreover, it admits of an empiricist interpretation that rejects all forms of synthetic self-evidence.

The probability character of explanatory induction is also indicated by the fact that the merging of scientific theories of different domains leads to an increase in reliability. Thus Newton's combination of Galileo's theory of falling bodies and Kepler's laws of planetary motion led to a theory of gravitation that was superior in reliability to either of the theories included in it. Some logicians have regarded the unification of theories as the expression of a tendency to logical elegance, or economy. Such an interpretation, however, seriously misrepresents the nature of scientific method. The unification of theories is an instrument for connecting scientific results in such a way that the combination obtains a higher probability than each of its parts taken separately. The schema of these inferences can be understood when it is interpreted in terms of the theorems of the calculus of probability. Such an analysis makes clear that the theory of advanced induction is identical with the theory of probability.

Since the axiomatic construction of the calculus of probability leads to the result that, when the frequency theory is assumed, all probability inferences are reducible to deductive inferences with the addition of induction by enumeration, it follows that all inductive inferences are reducible to induction by enumeration. This thesis was at the basis of Hume's theory of induction—though he thought only of classical induction—but he had no proof for it. The proof can be given only by the axiomatic construction of the calculus of probability.

The thesis, furthermore, must not be oversimplified to the statement that all inductive inferences can be construed directly as induction by enumeration; the reduction is possible only indirectly through the reducibility of the axioms of probability to the frequency interpretation. The thesis may be compared to a result of axiomatic constructions of mathematics, according to which all mathematical operations are reducible to the operation of addition by one, a reduction that, too, can be claimed only in principle, but cannot actually be carried through.

The thesis is further obscured by a confusion between the context of discovery and the context of justification, if I may be allowed to use certain terms that I have introduced elsewhere.[4] The finding of explanation belongs in the context of discovery and can be analyzed only psychologically, not logically; it is a process of intuitive guessing and cannot be portrayed by a rational procedure controlled by logical rules. Rationalization belongs in the context of justification; it can be applied only when given inductive conclusions are to be judged appropriate to given facts. It is in this context that the thesis belongs. Testing the relations between given observational data and given inductive conclusions is a procedure expressible in terms of theorems of the calculus of probability; and the inductive inferences of the test procedure are, therefore, ultimately reducible to induction by enumeration.

I do not maintain, by this thesis, that the finding of inductive explanation could be achieved by enumerating observations and simple generalization, such as Bacon hoped to attain in his tables; nor do I claim to have better methods. I refuse to answer the challenge of setting up rules of a logic of discovery. There are no such rules. Philosophers who believe that induction could become a sort of philosopher's stone, supplying methods that automatically transform facts into theories, misunderstand the task of logical analysis and burden the theory of induction with an unsolvable problem. Like deductive logic, the logic of induction concerns, not the psychological process of finding solutions, but the critical process of testing given solutions; it applies to the rational reconstruction of knowledge and thus belongs in the context of justification, not in the context of discovery.

§ 85. The Probability of Hypotheses

The thesis that explanatory induction can be construed in terms of the theorems of the calculus of probability and is therefore reducible to induction by enumeration is attacked by the argument that the probability of hypotheses is not interpretable as a frequency. Although the general discussion of nonfrequency probabilities (§ 71) covers this case also, and shows that the probability of a hypothesis, like that of any other single case, must be interpreted as a relative frequency, I should like to add some remarks on how the interpretation can be carried through—how, in particular, the reference class of a hypothesis is to be determined.

Scientific hypotheses are all-statements: they assert that for all things of a certain kind, at all times and places, a certain relation holds. So we begin this inquiry by studying the probability of all-statements.

When a scientific law is stated in the form of a general implication, symbolized as

$$(x)[f(x) \supset g(x)] \tag{1}$$

the formulation must be regarded as a schematization, introduced because

[4] See *EP*, pp. 6–7.

the logical treatment of all-statements in a two-valued logic is much simpler than the use of probability implications within the framework of probability logic. What can be proved by inductive methods is only that a probability implication of a high degree exists; the transition to the general implication is bound to certain conditions the neglect of which leads to paradoxes.

Such paradoxes appeared in the theory of confirmation when it was applied to the establishment of implications like (1), as was shown by C. G. Hempel.[1] The theory employs, as confirming cases, observations of the kind $f(x) . g(x)$; the larger the number of such cases, it is argued, the better is the general implication confirmed. But since, according to the rule of contraposition (6*c*, § 4), (1) is tautologically equivalent to the form

$$(x)[\overline{g(x)} \supset \overline{f(x)}] \tag{2}$$

we must regard as confirming cases observations of the form $\overline{g(x)} . \overline{f(x)}$ also. This consequence seems absurd. Returning to the example of the swans, we would have to regard as confirming cases of the statement, "All swans are white", not only observations of white swans, but also of anything that is not white and not a swan, for instance, of red flowers.

The paradox seems to be unsolvable within the theory of confirmation. It disappears, however, as soon as probabilities are introduced. For a probability implication, even of the degree 1, contraposition does not hold, and thus the two forms corresponding to (1) and (2)

$$(x_i)[f(x_i) \underset{1}{\twoheadrightarrow} g(x_i)] \tag{3}$$

$$(x_i)[\overline{g(x_i)} \underset{1}{\twoheadrightarrow} \overline{f(x_i)}] \tag{4}$$

are not equivalent. This fact can also be made clear as follows. If (3) were manifestly false and we had only a probability implication of a low degree, (4) might remain virtually true. The reason is that only a small number of things that are not white will be swans. Consequently, if the truth of (4) is established to a high degree of probability, (3) need not be true, and we cannot use an establishment of (4) as a proof for the validity of (3). From the standpoint of probability theory, a general implication represents a degenerate case. It can be substituted for a probability implication of a high degree only after both the relations (3) and (4) have been established independently. Conversely, the use of all-statements must be interpreted as indicating, not that the degree of probability is assumed as strictly $= 1$ (all that empirical evidence can prove is a probability within a small interval $1 - \delta$), but that both the relations (3) and (4) have been verified practically within

[1] A purely syntactical definition of confirmation, in *Jour. of Symbolic Logic*, Vol. VIII (1943), p. 128.

an interval δ of exactness. The analysis shows that the method of inductive verification must be attached to probability statements and not to the schematized form of knowledge in which such statements are replaced by two-valued statements.

Apart from contraposition, a further difference between general implications and probability implications of the degree 1, or of a high degree of probability, is given in the fact that from $(A \supset B)$ we can derive $(A . C \supset B)$ for every C; whereas if $(A \underset{p}{\ni} B)$ holds, with p almost equal to 1, there exist always classes C such that $(A . C \underset{q}{\ni} B)$, where q is a low probability.[2] When we use a general implication as a schematization for a probability implication of a high degree of probability, we usually require, therefore, that at least no class C *be known* such that $P(A . C,B)$ is small. If such a class C *is* known we can derive that $P(A . \bar{C},B) > P(A,B)$ [see the remarks following (11b, § 19)], and we then reformulate the all-statement by the use of $A . \bar{C}$ as reference class, that is, by excluding the known exceptions from the implicans. In other words, we require that no rule be known by means of which exceptions to the all-statement could be predicted. This usage of language makes it evident that the use of all-statements in place of probability implications of high degrees depends on more conditions than are given by the existence of a high degree of probability.

In this connection, an objection by E. Nagel may be discussed. It concerns the question why scientific all-statements are usually conceived as so strictly valid that even one exception would be regarded as a sufficient reason to renounce the all-statement. If what is meant by an all-statement is only a high probability, occasional exceptions should not be regarded as evidence to the contrary.

This criticism can be answered in various ways. First, if the limits of exactness are narrowly drawn, there will always be exceptions to scientific all-statements; that such exceptions are called observational errors does not change the fact that the all-statement is not strictly satisfied. Second, it is true that for wide limits of exactness, or merely qualitative statements, a case of one exception is regarded as incompatible with the all-statement. For instance, in scientific language we would not say that all human beings have hearts, if one exception were known. This attitude can be explained in two ways. First, the degrees of probability for such all-statements are usually so high that one exception, in fact, must be regarded as a noticeable diminution of the degree of probability. Second, one exception proves that the strict all-statement is false, and we dislike using an all-statement as a schematization if it is known that the all-statement is false. If a statement is used as a schema

[2] This is true even if $p = 1$, though in this case $q \neq 1$ is possible only if $P(A,C) = 0$; see (6, § 25). See also the discussion of this peculiarity of probability implication at the end of § 72.

ization, it should at least be compatible with the existing observational vidence to assume that the schematization is verbally true.

We shall turn now to the question how to discuss schematized statements ike (1), in the sense of approximations, within the frame of probability tatements. Before the schematization, we have, instead of (1), a statement f the form[3]

$$P[f(x),g(x)] = p \tag{5}$$

Ve shall consider also the probability of the statement (5) and thus a statenent of the form

$$P\{P[f(x),g(x)] = p \pm \delta\} = qdp \tag{6}$$

where $dp = 2\delta$ (for the notation $p \pm \delta$ see footnote, p. 462). The value q s a probability density; for $\delta = 0$, that is, a precise value p, the probability 6) would be $= 0$.

We begin with the discussion of (5). The statement can be transformed o that it informs us about the probability of individual implications of the orm $f(x) \supset g(x)$. To demonstrate the procedure we use the tautological quivalence

$$(A \supset B \equiv A.B \vee \bar{A}) \tag{7}$$

which is easily derivable from the truth table of implication, and have

$$\begin{aligned} P(A \supset B) &= P(A) \cdot P(A,B) + P(\bar{A}) \\ &= 1 - P(A) \cdot [1 - P(A,B)] \end{aligned} \tag{8}$$

f $P(A) = 1$, we have $P(A \supset B) = P(A,B)$. If $P(A) < 1$, the value of $P(A \supset B)$ must be closer to 1 than that of $P(A,B)$, since the brackets in (8) hen are multiplied by a factor smaller than 1. Therefore we have the general nequality

$$P(A \supset B) \geqq P(A,B) \tag{9}$$

The equality sign holds only when $P(A) = 1$ or $P(A,B) = 1$. If $P(A,B) = 1$, we have also $P(A \supset B) = 1$.

Because of (9) we can always replace a probability implication of the legree p by a logical implication, an individual adjunctive implication, with he qualification that the probability of the resulting statement is $\geqq p$. Thus, when we find that 95% of all swans are white, we can express the esult in the form that the probability of the statement, "A swan is white", s $\geqq$ 95%. The degree of the probability implication, therefore, appears as lower limit of the probability of the corresponding logical implication. This interpretation offers itself when p is close to 1; instead of a probability mplication, we then apply a logical implication with the qualification that he probability of the statement does not quite attain certainty.

[3] We omit the subscript i because in the P-notation the order of the elements x is not xpressed.

We now proceed to the discussion of (6). When p is practically $= 1$ and the conditions of a transition to an all-statement are satisfied, (6) assumes the form

$$P\{(x)[f(x) \supset g(x)]\} = q dp \tag{10}$$

From the form of (6) it is clear that we must employ a lattice for the interpretation of this probability, though the lattice superscripts are omitted in (6); the probability of all-statements, consequently, must be defined as a second-level probability in a probability lattice. The occurrence of the factor dp on the right of (10) shows that the probability of a strict all-statement would be $= 0$. Only if the all-statement is regarded as admitting of a small inexactness $dp = \delta$, such that $p \geqq 1 - \delta$, will its probability be > 0.

With respect to all-statements, therefore, we can distinguish two kinds of probabilities. Probabilities of the first level of the lattice, that is, probabilities of the form (5), represent probabilities of individual implications comprised by the all-statement, to be used when the all-statement is not strictly verified but serves only as an approximation. The probability of the second level supplies the probability of the all-statement itself.[4]

Newton's law of gravitation may be used as an example. It states that for all bodies, at all times and in all places, the relation

$$f = k \cdot \frac{m_1 m_2}{r^2} \tag{11}$$

holds, where f is the force of attraction, m_1 and m_2 the respective masses of the bodies, r their distance, and k a constant. The abbreviation x may denote a set of individual conditions, including a specification of the bodies involved and the time and space coördinates. Then, if (11) holds for the individual conditions to a certain degree δ of exactness, we write $\varphi(x)$, otherwise $\overline{\varphi(x)}$. Assume that the relation (11) has been tested for various positions of the planet Mars; we can write the results in the form of a sequence of terms $\varphi(x)$ or $\overline{\varphi(x)}$. Doing the same for other planets, the moon, and other tests of Newton's law (for example, Cavendish's experiment with a torsion balance) we arrive at a lattice

$$\begin{array}{cccccc} \varphi(x_{11}) & \varphi(x_{12}) & \overline{\varphi(x_{13})} & \varphi(x_{14}) & . & . \\ \varphi(x_{21}) & \overline{\varphi(x_{22})} & \varphi(x_{23}) & \varphi(x_{24}) & . & . \\ . & . & . & . & . & . \end{array} \tag{12}$$

Each row belongs to one planet or other test object. The negative cases are usually said to result from errors of observation; for us they are indications

[4] This treatment of the probability of hypotheses was first developed by H. Reichenbach in *Erkenntnis*, Vol. V (1935), pp. 274–278. The present treatment, however, includes some additions and clarifications.

hat observation can never strictly establish a general implication, but only probability implication of a high degree.

We now establish the degree of probability for each row by means of a posit ased on the inductive rule. Assuming the posits to be true, we count in the ertical direction and thus construct the probability of the second level olding for the statement that the probability of a row is $= p$. This prob-bility, again, is stated in the form of a posit. For the probability of the first evel the reference class is rather easily constructed. It is the same class hat is used in implications of the all-statement if such a statement is intro-luced in the sense of a schematization. For the probability of the second evel the definition of the reference class is not unambiguous and thus offers he usual complications combined with this definition. In principle, however, he definition of the reference class follows the same procedure as in all other uch problems. In this example, the class is constructed by reference to other nstances where the law applies. The second-level probability, then, expresses he probability of an all-statement restricted to one planet.

In the astronomical tests of Newton's law, the observation was made that he planet Mercury does not satisfy the law to the same degree of exactness s the other planets. This example illustrates an exception to an all-statement hat was formerly regarded as true without exception. Since the exception s restricted to one planet, it is regarded as an example in which one negative ase is sufficient to disprove a physical law. This interpretation is not entirely orrect. The measurements of the orbit of Mercury cover a large number of ndividual observations, and in schema (12) the exception would be repre-ented by one row that does not converge toward the same limit as the others. But it is at least true that one negative row is regarded as sufficient to lisprove the all-statement extended in the vertical direction. The law has herefore been restricted to planets that are not too close to the sun, that is, o the center of attraction. This is an instance of a transition from a reference lass A to a reference class $A . \bar{C}$, as explained above (see also the discussion f the independence of the single case at the end of § 89). That the reference lass C in which we incorporate the exception is, in this case, assumed to be he class of planets near the sun is based on other inductions, including those alidating Einstein's theory of general relativity.

Instead of constructing an individual sequence for each planet, we can nclude in the first horizontal row all observed instances of the law, regardless f the individual planet or test object to which they belong. Then the limit f the frequency posited for this row determines the probability of the first evel for the general case, that is, for an individual implication not restricted o one planet. In order to define the probability of the second level and thus he probability of Newton's law in general, not restricted to one test object, ve must construct a reference class by filling out the other rows with observations pertaining to other physical laws. For instance, for the second ro we can use the law of the conservation of energy; for the third, the law o entropy; and so on. The reference class employed corresponds to the wa in which a scientific theory is actually judged, since confidence in an individua law of physics is undoubtedly increased by the fact that other laws, too, hav proved to be reliable. Conversely, negative experiences with some physica laws are regarded as a reason for restricting the validity of other laws tha so far have not been invalidated. For instance, the fact that Maxwell's equa tions do not apply to Bohr's atom is regarded as a reason to question th applicability of Newton's or Einstein's law of gravitation to the quantur domain.

These considerations show that the probability of a hypothesis or a scientifi theory can be defined in terms of frequencies. Applied to the individua hypothesis, the probability assumes the character of a weight; all that wa said about the use of a weight for statements of single cases holds likewis for the weight of hypotheses. In fact, speaking of the probability of an indi vidual hypothesis offers no more logical difficulties than speaking of th probability of an individual event, say, the death of a certain person.

It is sometimes argued that in cases of the latter kind the choice of th reference class is easily made—that, for example, the reference class "all pe sons in the same condition of health" offers itself quite naturally. But critics (the frequency interpretation of the probability of theories forget how muc experience and inductive theory is invested in the choice of the referenc class of the probability of death. Should we some day reach a stage in whic we have as many statistics on theories as we have today on cases of diseas and subsequent death, we could select a reference class that satisfies th condition of homogeneity (see § 86), and the choice of the reference clas for the probability of theories would seem as natural as that of the referenc class for the probability of death. In some domains we have actually bee witnesses of such a development. For instance, we know how to find a referenc class for the probability of good weather tomorrow; but before the evolutio of a scientific meteorology this reference class seemed as ambiguous as tha of a scientific theory may seem today. The selection of a suitable referenc class is always a problem of advanced knowledge.

The method described for the statistical definition of the probability of theory, though it is of a schematized form, is not very different from th procedure actually used. What is different, however, is that we do not directl observe instances of the form $\varphi(x)$, but use other observations from whic we infer that the form $\varphi(x)$ holds. Thus we do not directly observe, for certain position of a planet, the force of attraction; we observe, insteac successive positions from which we infer the acceleration, and identify with the force of attraction. The distance between the planet and the sur too, is computed by complicated inferences. Apart from this difference, however, the inductive inference is expressed by schema (12). Thus, when a certain position and acceleration of the planet confirm the law, we assume that for this instance the numerical values of the force of attraction, of the distance, and so on, satisfy the relation φ; but we do not regard a single instance as proof that φ will always be satisfied. There might exist a different law φ', which, however, is so constructed that, for the special case observed, the numerical values of φ and φ' coincide. That such an assumption is false, and that φ always holds, is a result based on the great number of instances in which φ is satisfied.

We can explain why we prefer, for the establishment of the general law, a variation of instances from planet to planet rather than a repetition of instances for the same planet. By previous inductions we have established a general rule that the various positions of one planet, or the various states of one body, are controlled by a simple law; therefore, when we have tested a law for the various states of one body, we assume it to hold for all states of the body, so that further observation virtually does not supply new information. Observations of other planets, on the contrary, must be regarded as independent observations.

The logical analysis of this inference is as follows: we have a probability of the second level telling us when the horizontal sequence is long enough to justify a posit of its persistence, whereas the posit in the vertical direction requires independent evidence and is not justifiable by an analogy with the posit in the horizontal direction. What is called the *weight of an evidence* is to be interpreted as the result of previous inductions, all of which are ultimately reducible to induction by enumeration. The analysis shows that the actual inferences by which the probability of a theory is established include the results of a great many previous inductions, and that any reconstruction of the method will remain a simplified schematization.

A further difference from induction by enumeration results when we ask for the probability, not that the law φ holds, but that it holds *when* verifying observations of a certain kind have been made. This question concerns a lattice inference applied to schema (12) and is answered by the probability density $v_n(f;p)$ of (5, § 62). The answer is constructed in terms of the rule of Bayes and presupposes a knowledge of the antecedent probability density $q(p)$. The latter function can be found, in principle, by counting sequences vertically in schema (12); a division of the range 0 to 1 for the possible values p of the horizontal probability by small but finite intervals dp leads to an approximate determination (the precise analysis of this method is given in § 89). The probability $v_n(f;p)dp$ thus ascertained, in which f measures the frequency of conforming instances of the law φ in the observed initial section, possesses Bernoulli properties and satisfies Laplace's convergence theorem

(9, § 62). Whatever be the antecedent probabilities, if n is large enough, v_n is close to 1 for an interval $p \pm \delta$ that includes the value $p = f$.

In the theory of confirmation, which is intended to supply the probability $v_n(f;p)dp$, the inference is falsely construed as governed by laws that are not included in the calculus of probability. The analysis presented shows that the calculus possesses all the means to account for the inference, and that recourse to laws of an independent "inductive logic" is unnecessary. The use of a knowledge of antecedent probabilities for the inference is visible in the fact that the scientist usually knows when the frequency n of observed instances is large enough to warrant the conclusion of its persistence. His judgment about the number n may be construed as following computations as explained with respect to (12, § 62).

The considerations presented make it evident that the probability of hypotheses offers no difficulties of principle to a statistical interpretation. That, in most practical instances, the statistics cannot be carried through numerically because of insufficient data, and that, instead, crude estimates are used, do not constitute objections to a theory that claims to embody only the rational reconstruction of knowledge, not knowledge in its actual procedure.

§ 86. Induction by Enumeration in Advanced Knowledge

The theory of induction by enumeration in advanced knowledge was presented in § 62. The inductive inference was shown there to refer to a probability lattice and to be reducible to an application of the rule of Bayes. Assuming that the antecedent probabilities are known and that the sequences are normal sequences, the following questions can be answered.

1. Given an initial section of a sequence with the relative frequency f^n, what value $p \pm \delta$ is the best posit for the limit of the frequency?

2. What is the second-level probability v_n that a limit at $p \pm \delta$ will be reached?

If the antecedent probabilities are not known but the sequences at least are known to be of the Bernoulli type, we can prove the following theorem:

3. The greater n, the greater the second-level probability v_n that the limit will be at $p = f^n \pm \delta$; and v_n converges to 1 with $n \to \infty$.

We see that the theory of induction by enumeration, in advanced knowledge, is as complete as can be required; the theory tells us what value to posit, how good our posit is, and that it will become better and better with larger numbers.

In discussions of induction the question has been asked, What is a large number? In fact, the conception of a large number varies greatly with the field in which the induction is applied. For the test of a new medicine a few hundred cases may be sufficient; insurance companies count millions of cases;

and physicists dealing with problems of the kinetic theory of gases do not speak of large numbers unless they are compelled to use a mode of writing in terms of powers of the number ten. It is obvious that the definition of a large number belongs in advanced knowledge: it is a number so large that the probability of the second level for a sufficient convergence from that number on is high enough. A "sufficient convergence" and a "high-enough probability" are matters of definition and depend on what is attainable; the large number is a function of these definitions, computable or appraisable only within advanced knowledge (see 12, § 62).

A large number is not always necessary for an inductive inference. Sometimes the inference can be based on only one instance. Such inferences occur when previous inductions have created a situation in which one experiment, called a *crucial experiment,* is sufficient for an induction. Instances of crucial experiments are found in what a physician calls *differential diagnosis.* One Wassermann test may be regarded as sufficient evidence of a case of syphilis, because previous inductions have established a relation between the test and the disease and have shown that repeated applications of the same test to the same person usually lead to the same result. The existence of crucial experiments has been misconstrued as evidence against an inductive interpretation of scientific methods, but it can be incorporated without difficulty in an interpretation for which all inductive inferences are reducible to induction by enumeration.

A further condition that can be satisfied only in advanced knowledge is the condition of a *homogeneous reference class.* A class of tuberculosis cases is a homogeneous class. But it would seem unwise to compile death statistics in a class of persons with different diseases or in a class including both human beings and animals. The definition of the predicate "homogeneous" depends on the state of our knowledge. An inhomogeneous class can be defined as a class for which we know methods by means of which the class can be so subdivided, without the use of the attribute considered (see § 30), that subclasses of different frequencies for the attribute result. The subdivision will sometimes be achievable by reference to other attributes, such as are given in the example of different diseases, or biological species. However, it can be achieved also by dividing the total sequence, or ordered class, into consecutive sections. The latter method amounts to a determination of the dispersion (see § 52). For an inhomogeneous class the probability of the second level concerning the persistence of the observed frequency is lower than for a homogeneous class. It is obvious that such considerations are restricted to advanced knowledge.

The relevance of the fair-sample condition, too, can be demonstrated in advanced knowledge. If the observed initial section presenting the frequency f is the result of random methods, the formulas (5 and 9, § 62) can be applied, which lead to a high degree of the inverse probability that $p = f \pm \delta$. But if the sample is selected so as to contain a certain attribute by preference, the procedure that supplies it is not the same as the one that determines further observations and is thus not represented by the probability function $w_n(p;f)$ of the formulas mentioned. Therefore, Laplace's convergence formula for the inverse probability cannot be applied. This explanation makes it clear why the fair-sample condition is attached to inductive inferences only when they belong in advanced knowledge; for induction in primitive knowledge there is no fair-sample condition.

In advanced knowledge the inference of induction by enumeration is justified by the theorems of the calculus of probability. Statements about the limit of the frequency and about the reliability of the inductive conclusion can be given a probability meaning (§ 74); in other words, within the frame of advanced knowledge the inductive inference is an appropriate instrument for operations controlled by a logic of probability. There is no question of the legitimacy of induction in advanced knowledge.

It is in advanced knowledge, too, that such formulas as the rule of succession (22, § 62) find their places. The equality of antecedent probabilities on which the applicability of the rule depends must be known before the rule can be used. The argument that equality may be assumed in the absence of knowledge to the contrary makes use of the principle of indifference, but it was shown (§ 68) that the principle is untenable. Logic cannot supply a probability metric; only experience and observation can inform us about degrees of probability or about the equality of such degrees. But even experience and observation can supply such knowledge only if the observational results are linked and carried on by inductive inferences. There is no circularity involved in such a procedure if it is used in establishing special forms of inductive inferences; and the rule of succession, therefore, occupies a legitimate position in advanced knowledge. But it would be circular to base the general use of the inductive inference on formulas that presuppose the knowledge of a probability metric. The ultimate justification of induction must be given by other means than formulas that are derivable in the calculus of probability; the problem falls entirely within the province of primitive knowledge.

§ 87. The Rule of Induction

We turn now to the consideration of induction in primitive knowledge. So long as no probabilities have been established, the inductive rule cannot be based on theorems of the calculus of probability; therefore we cannot prove that the inductive rule leads to the posit of greatest weight, nor do we know how probable it is that the limit posited will be reached. We cannot even prove that the posit becomes better with a greater number of observed instances. In spite of our ignorance, however, we must use the inductive rule, since otherwise we could not establish any probability values and could never proceed to the advanced state of knowledge in which the theorems of probability take over the functions of a guide in inductive method.

To facilitate the discussion of induction in primitive knowledge, or *primitive induction*, we shall proceed by steps. We shall not begin with the analysis of a state in which nothing is known about the progress of sequences, but shall leave the discussion of that question to a later inquiry (§ 91). Rather, we shall introduce the assumption that the sequences under consideration have

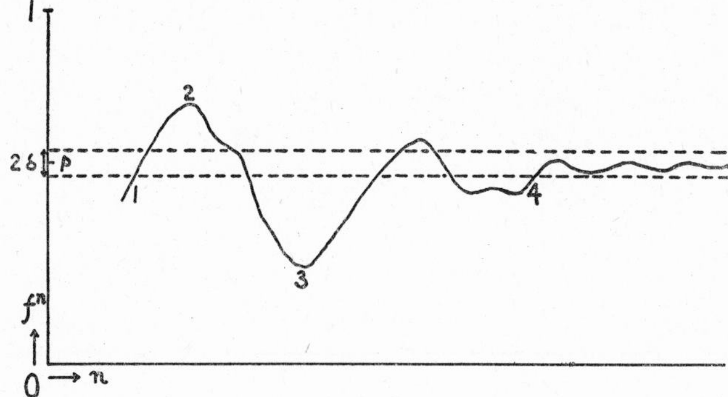

Fig. 28. Frequency curve of a sequence converging to a limit.

a limit of the frequency, although the limit is unknown. Let us see to what extent this assumption can help in the solution of the inductive problem.

Again the concept of *posit* will be used for the interpretation of the statements to be considered. The statement that the observed frequency will persist can be maintained only in the sense of a posit, since it is obvious that we cannot prove it to be true. But it is not an appraised posit, since we have no weight for it. In what sense, then, can the inductive posit be justified if we have no proof that the posit will lead to the greatest number of successes?

To answer the question, we must analyze the way in which the rule of induction is used. The inductive posit is not meant to be a final posit. We have the possibility of correcting a first posit, of replacing it by a new one when new observations have led to different results. From this point of view, the following analysis of the inductive procedure can be made. If the sequence has a limit of the frequency, there must exist an n such that from there on the frequency f^i $(i > n)$ will remain within the interval $f^n \pm \delta$, where δ is a quantity that we can choose as small as we like, but that, once chosen, is kept constant. Now if we posit that the frequency f^i will remain within the interval $f^n \pm \delta$, and if we correct this posit for greater n by the same rule, we must finally come to the correct result. The inductive procedure, therefore, represents a *method of anticipation*; in applying the inductive rule we anticipate a result that for iterated procedure must finally be reached in a finite number of steps. We thus speak here of an *anticipative posit.* In contradistinction to an *appraised posit,* the weight of which is known, it may also be called a *blind posit,* since it is used without a knowledge of how good it is; the term "blind" is meant to express the fact that it is a posit without a rating.

Figure 28 will make the method clear. The abscissa is given by the number n of the elements of the sequence; as ordinates, the relative frequencies f^n are plotted. When the sequence has a limit of the frequency, its oscillations will die down. If the sequence is known only to place 1, we posit the corresponding f^{n_1}. When we continue the observation of the sequence, the next posit may be made at place 2, then at places 3 and 4, and so on. At each place we use the frequency observed as the best posit. We see that, in going from place 1 to place 2, we even make our posit worse; but finally we must reach a place (in the diagram, place 4) where the posit is correct within the interval 2δ and will remain so for the rest of the sequence. The inductive procedure, therefore, has the character of a method of *trial and error* so devised that, for sequences having a limit of the frequency, it will automatically lead to success in a finite number of steps. It may be called a *self-corrective method,*[1] or an *asymptotic method.*

The method of the anticipative posit may be formulated as follows:

Rule of induction. *If an initial section of n elements of a sequence x_i is given, resulting in the frequency f^n, and if, furthermore, nothing is known about the probability of the second level for the occurrence of a certain limit p, we posit that the frequency f^i $(i > n)$ will approach a limit p within $f^n \pm \delta$ when the sequence is continued.*

The distinction between appraised and anticipative posits leads to two different kinds of posit. It is a common feature of both that their use is not justified for the individual case, but only in repeated applications. With respect to the grounds of their use, however, the two posits must be distinguished. The appraised posit is justified by the *principle of the greatest number of successes.* This kind of posit, therefore, can be used only when the corresponding weight is known. The anticipative posit cannot be justified

[1] The self-corrective nature of induction was emphasized by C. S. Peirce, who mentioned "the constant tendency of the inductive process to correct itself," in *Collected Papers* (1878; Cambridge, Mass., 1932), Vol. II, p. 456; see also *ibid.*, p. 501, and Vol. V, p. 90. I have not been able, however, to find a passage in Peirce's work where he clearly states a reason for his contention. The fact that he constantly connects the problem of induction with that of a fair sample, that is, with the use of random sequences, seems to indicate that he bases the self-corrective nature of induction on Bernoulli's theorem. This interpretation is supported by his exposition of the increase in the reliability of induction (*ibid.*, Vol. II, p. 428). Such an argument is invalid, of course, since the justification of induction must be given before the use of probability considerations. As to my own relations to Peirce, with whose ideas I was not acquainted when I wrote the German original of this book, see my remarks in *The Philosophy of John Dewey* (ed. by P. Schilpp; Evanston, Ill., 1939), pp. 188–190.

by a maximum principle. It involves another form of justification based on the *principle of finite attainability*. If the sequence has a limit, the anticipative posit is justified because, in repeated applications, it leads to any desired approximation of the value of the limit in a finite number of steps.

This argument may be called an *asymptotic justification*. It includes an explanation why the value f^n found for the last observed element of the sequence is preferable to any earlier value. If after 100 elements we find $f^n = \frac{1}{2}$, after 200 elements, $f^n = \frac{2}{3}$, we do not claim that $\frac{2}{3}$ is a better value than $\frac{1}{2}$ in the sense that it is more probable. Such a proof is impossible in primitive knowledge and can be given only in advanced knowledge (see § 86). But if the procedure of going through all elements successively can be justified, we know, at least, that in selecting the f^n of the later element we are closer to the end of the procedure. The choice of the last f^n is therefore a matter of economy.

The posit f^n is not the only form of anticipative posit. We could also use a posit of the form

$$f^n + c_n \tag{1}$$

where c_n is an arbitrary function, which is so chosen that it converges to 0 with n increasing to infinite values. All posits of this form will converge asymptotically toward the same value, though they will differ for small n. We shall prefer the inductive posit f^n, for which $c_n = 0$. To do so we can, however, adduce only grounds of descriptive simplicity;[2] that is, the inductive posit is simpler to handle.

Whereas the principle of asymptotic convergence determines a class of inductive rules as equally justified, a distinction between the members of this class—the selection of a particular function c_n—can be achieved in advanced knowledge. For instance, the method of cross induction (§ 84) can be regarded as an instrument for finding a function c_n such that the value $f^n + c_n$ supplies an earlier convergence within an interval δ than does the value f^n. This method and others will be discussed in §§ 88–90.

These results must now be extended to the concept of practical limit, which was introduced in § 66. The concept refers to a sequence which reaches sufficient convergence after a fairly large number of elements, but which may diverge in later parts that lie beyond the reach of human experience. It is obvious that the rule of induction is justified, too, when the condition of the limit is replaced by that of a practical limit. The justification, in fact, will be improved, since finite attainability then means an attainability for human capacities. A sequence that converges so late that human observers

[2] Descriptive simplicity is a property of a description that has no bearing upon its truth. It must be distinguished from inductive simplicity, which classifies descriptions leading to different predictions. See *EP*, § 42. In the same book, on p. 355, I tried to give other reasons for preferring the posit f^n. Dr. Norman Dalkey has since convinced me that they are invalid. It is, however, sufficient for the theory of induction that the posit f^n is descriptively simpler.

cannot experience the convergence has, for all practical purposes, the character of a sequence without a limit. In the following discussions we should therefore regard the limit condition as referring to a practical limit. Since it seems unnecessary to mention this interpretation on all occasions, we shall speak throughout simply of the *limit condition.*

When we use the logical conception of probability, the rule of induction must be regarded as a *rule of derivation,* belonging in the metalanguage. The rule enables us to go from given statements about frequencies in observed initial sections to statements about the limit of the frequency for the whole sequence. It is comparable to the rule of inference of deductive logic (see § 5), but differs from it in that the conclusion is not tautologically implied by the premises. The inductive inference, therefore, leads to something new; it is not empty, like the deductive inference, but supplies an addition to the content of knowledge. It is a consequence of the synthetic nature of inductive inference that the conclusion cannot be asserted as true, but can be asserted only in the sense of a posit. Since we could show that all inductive inferences are reducible to induction by enumeration, in the wider sense of statistical induction (§§ 67, 84), the rule of induction is the only rule of derivation that distinguishes *inductive logic* from *deductive logic.* In other words, inductive logic contains all the rules of derivation of deductive logic with the addition of the inductive rule.

A so-called paradox of inductive logic was constructed by N. Goodman.[3] He assumes that an initial section of n_o elements of a sequence has been observed and that all elements observed have the attribute B. He now defines an attribute C as follows: an element has the attribute C if it is one of the first n_o elements or has the attribute $\bar{B}$. It is obvious that all the n_o elements of the initial section have the property C; the rule of induction therefore advises us to expect C for the next following element $n_o + 1$. But since this element does not satisfy the first part of the disjunctive property C, it must satisfy the second part, and we have inferred by induction that the next element will have the attribute $\bar{B}$.

To regard this consideration as an objection against the rule of induction would reveal a misunderstanding of inductive method. The rule of induction, applied in primitive knowledge, leads only to posits that are justified asymptotically. We cannot expect it to supply correct predictions for every individual element. This justification includes the case of the property C. Assume the total infinite sequence to consist of elements B; then, applying the rule of induction to the property C, we shall first make bad posits, but while going on will soon discover that the following elements do not have the property C.

[3] "The Problem of Counterfactual Conditionals," in *Jour. of Philos.*, Vol. 44 (1947), p. 128. I present the paradox in a somewhat changed version in order to make it applicable to the rule of induction. In the form presented by Goodman it refers to aprioristic inferences that are illegitimate anyway.

We shall thus turn to positing $\bar{C}$ and have success. When we count the frequency m for $\bar{B}$ and the number n of elements in such a way that we begin after the first n_o elements, the paradoxical inference can be regarded as a rule of induction that posits the frequency of $\bar{B}$ as given by $\frac{m + n_o}{n + n_o}$. Since this value converges with growing n to the value $\frac{m}{n}$, it is included in the class of justifiable rules of induction. The inferiority of this particular rule in the example considered cannot be demonstrated in primitive knowledge; in fact, if the sequence had nothing but elements $\bar{B}$ after the n_o-th element, the rule would be superior to the usual rule.

It is only in advanced knowledge that the rule can be criticized; and in advanced knowledge the inference can be shown to be inferior because it violates the rule, "Use the narrowest common reference class available". The property C, by its definition, is identical with $\bar{B}$ from the $(n_o + 1)$-st element; since the reference class $\bar{B}$ is narrower than C, it should be used as a basis for the inference (in other words, the property with respect to which the first n_o elements should be counted is the property $\bar{B}$). Using the property C as reference class means determining the probability of the next element with respect to an initial section that is not specified, since every initial section of n_o elements has the property C.

But the rule of using the narrowest reference class for which there exist reliable statistics belongs in advanced knowledge. It can be applied only when it is known what statistics can be called reliable. If ten cases of cholera are observed, eight of which have a lethal issue, we assign a death probability of 80% to cases of cholera, but would refuse to assign this death probability to cases of disease in general. That we are entitled to prefer the narrower class of cholera cases to the wider class of disease cases is justified by the large amount of statistical material, but it cannot be justified *a priori*. Cholera might be a sort of disease much less dangerous than the average disease, and the eight observed death cases might be a misleading exception.

The choice $c_n = 0$ in (1) cannot be shown to be better than any other choice. The illustration shows that even the rule of putting $c_n = 0$ does not always lead to the same conclusion. It leads to contradictory conclusions according as it is applied to the property $\bar{B}$ or C. This fact offers no difficulties, because inductive conclusions are never final, but are used only as posits and are canceled when other conclusions lead to their abandonment. The choice between contradictory conclusions is made in terms of additional rules, which are developed in advanced knowledge. Another instance of this kind was studied in the method of cross induction (§ 84): the rule of induction is first applied in the horizontal, then in the vertical, direction, and the secondary conclusions may lead to the cancellation of some of the primary conclusions.

Additional evidence, though not contradictory to the original evidence, may change the conclusion—a result unthinkable for deductive logic. With respect to the requirement of consistency, inductive logic differs intrinsically from deductive logic; it is consistent, not *de facto*, but *de faciendo*, that is, not in its actual status, but in a form to be made.

Some remarks may be added in order to clarify the nature of the rule of induction as a rule of primitive knowledge. The rule is an instrument for finding probabilities, but it does not express a probability inference. If an event B has been observed with respect to the reference class A in m out of n cases, the rule advises us to expect the event B, when A occurs, with the probability $\frac{m}{n} \pm \delta$; but it does not say that the grounds on which the advice is based confer a probability on B. The rule says:

I. *If an initial section* F^n_m *is observed, posit*

$$P(A,B) = \frac{m}{n} \pm \delta \tag{2}$$

But it does not state a probability relation between the observations F^n_m and the event B; that is, it does not say:

II. *Posit that*

$$P(A\,.\,F^n_m,B) = \frac{m}{n} \pm \delta \tag{3}$$

The two formulations are not equivalent. In formulation I the observed frequency is regarded as the ground for the assertion of the posit and therefore is included in the content of the rule. In formulation II the observed frequency is included in the content, not of the rule, but of the posit, since it is regarded as determining the reference class of the probability expression. As the rule is one linguistic level higher than the posit, the *ground-for-assertion relation* is formulated in the metalanguage of the language to which the posit, or probability expression, belongs; the ground-for-assertion relation is thus one linguistic level higher than the probability relation.

Now it can easily be seen that only formulation I is permissible, since only this formulation is capable of a justification. To regard the observed frequency as a ground for the assertion of the probability (2) is justified because the latter can be asserted in the sense of an anticipative posit; but we are not entitled to assume that the observed frequency makes the event B probable to a certain degree. That formulation II may lead to mistakes can be illustrated by the use of sequences with aftereffect: for such sequences (2) would be correct, whereas (3) would be false.

That the two formulations are not both legitimate distinguishes inductive from deductive inference. For deductive logic the two corresponding formulations are:

III. If a is true, and $a \supset b$ is true, assert b.

IV. Assert: $a . (a \supset b) \supset b$.

They differ, of course, as to content: formulation III expresses the rule of inference; formulation IV expresses the existence of a corresponding implication in the object language. But both are justifiable, which means here that both lead to true statements. The deductive rule of inference, therefore, has an object-language correlate in an implication. The inductive rule of inference, on the contrary, does not possess an object-language correlate.

The decision for formulation II as the rule of induction derives from the confusion of two relations: the *ground-for-assertion relation* and the *probability relation*. When we say that the observed frequency is the ground for the assertion of the probability (2), we state a relation between our knowledge and a prediction, to be formulated in the metalanguage. This relation, the ground-for-assertion relation, is not a relation of degree or of order. It selects a certain statement as assertable, but it does not include any quantitative measure of assertability.

The absence of degrees of assertability is illustrated by the fact that the asymptotic convergence, which justifies the rule of induction, holds likewise for the class (I) of rules; the ground-for-assertion relation determines, therefore, a class of different posits as assertable. In advanced knowledge these posits would have different weights, but in primitive knowledge they are not ordered by degrees. The ground-for-assertion relation cannot assign ratings to posits; it merely gives a permit of assertability.

For these reasons, formulation I is the only admissible form for the rule of induction. The justification of the rule in this form has so far been given on the assumption of the existence of a limit of the frequency. In a later investigation (§ 91) we shall free ourselves from this last presupposition.

§ 88. Anticipative Posits in Advanced Knowledge

The method of anticipative posits, although it is the natural instrument of primitive knowledge, can be applied also in advanced knowledge. Instances of such application are found in the theory of statistical inference, used for the hypothetical introduction of a probability metric (see § 70). These inferences can be regarded, on the one hand, as methods of discrimination between posits of the class (1, § 87) and thus of determining a function c_n that improves the convergence. On the other hand, these methods again apply the principle of anticipative posits.

If a large statistical material is given, it can be used for a direct determination of the distribution (see § 42); for this determination, statistical induction is applied for every interval du of the attribute u. If the material is insufficient, however, the method, applied to small intervals du, would lead to unreasonable results; the distribution would be an irregular zigzag curve, the shape of which would not indicate the form to which it would converge asymptotically with increased material. A better anticipation of the final curve for incomplete statistical material is afforded by the methods of statistical hypotheses, developed by Fisher, Neyman, and others. These methods can be regarded as determining for every interval du a function c_n such that $f^n + c_n$, rather than f^n, represents the best posit for the limit of the frequency.

The way in which this improvement is achieved must be studied more closely. The principle of maximum likelihood, developed by Fisher, may serve as the basis of the analysis. Formula (2, § 70) determines the inverse probability $v_n(s)$ of a value s of the parameter of the distribution $d(s;u)$. I shall make the assumption that the function $L_n(s)$ of this formula has the Bernoulli properties (10, § 57):

$$k_n(s_1,s_2) = \int_{s_1}^{s_2} L_n(s)ds$$

$$\lim_{n\to\infty} k_n(s_1,s_2) = k \qquad \text{for } s_1 \leqq s_o \leqq s_2 \tag{1}$$

$$\lim_{n\to\infty} k_n(s_1,s_2) = 0 \qquad \text{for } s_o < s_1 \text{ or } s_o > s_2$$

The critical point s_o is the value of s for which, in all intervals du_i, the limit of the frequency $\frac{n_i}{n}$ coincides with the probability, that is, s_o is determined for finite n by the relation

$$\frac{n_i}{n} \sim d(s_o;u_i)du_i \tag{2}$$

The sign $\sim$ denotes approximate equality; for increasing n this approximate equality improves, if the assumption that the distribution conforms to $d(s;u)$ is correct. The convergence toward k stated in (1) holds for every value du, however small, but du is kept constant during the transition to the limit. The Bernoulli properties make it possible to apply the inferences used for (11, § 62) and to show that the maximum of the inverse probability converges asymptotically with the maximum likelihood. If nothing is known about the antecedent probability $q(s)$, the principle of maximum likelihood can therefore be justified as a method of asymptotic convergence, in the same way that the rule of induction is justified. It leads to anticipative posits, which must eventually lead to the correct result if this result is attainable.

The improvement of induction through a term c_n is here achieved by means of an asymptotically convergent method, which does not presuppose a knowledge of antecedent probabilities. In this sense, Fisher's method has, in fact, attained its aim of eliminating the prerequisite of antecedent probabilities. This is possible, however, only if the Bernoulli conditions (1) are satisfied; the last of these conditions, in particular, is a rather strong condition, since it virtually eliminates all but one hypothesis. These conditions presuppose the applicability of the special theorem of multiplication, or some other rule for combinations, and must be proved for the function $d(s;u)$ under consideration, which determines the dependence of L on s.

Furthermore, the method presupposes an inductive knowledge of another kind inasmuch as it presupposes the knowledge of the functional form $d(s;u)$. This knowledge excludes a direct empirical determination of the distribution, as would obtain by the successive application of the rule of induction to all intervals du; for small numbers n the distribution thus resulting would not have the form $d(s;u)$. For incomplete observational material the hypothetical form $d(s;u)$ is preferred to the direct empirical distribution. Fisher's principle can be regarded as an extension of the method of anticipative posits to cases in which previous knowledge has restricted the admissible hypotheses to a one-parameter class (or a class depending on more parameters); the rule of positing that selects among this class is then constructed in the same manner as the rule of induction. It is an interesting fact that the method of anticipative posits, originally the instrument of primitive knowledge, finds renewed application on an advanced level of knowledge. In this application it may be called a method of *restricted anticipative posits*. It is an improved version of statistical induction, supplying a set of simultaneous inductive posits connected by a restrictive condition. The posits refer to the various intervals du and are made in such a way that the individual posits are balanced against each other by means of a concatenation derived from previous knowledge.

The bearing of previous knowledge upon anticipative posits may be illustrated by an example in which only one inductive posit is made. Assume it is known that the limit of the frequency of a certain sequence must be a multiple of $\frac{1}{5}$, that is, one of the values 0, $\frac{1}{5}$, $\frac{2}{5}$, $\frac{3}{5}$, $\frac{4}{5}$, 1; then one would not posit the observed frequency to persist, but would select the nearest multiple of $\frac{1}{5}$ for the posit. In this example, too, advanced knowledge leads to a deviation from the rule of induction.

The restriction of posited hypotheses to a certain class holds likewise for statistical inferences other than those based on the method of maximum likelihood, for instance, for Neyman's methods. All such methods, if they dispense with a knowledge of antecedent probabilities, are justifiable through asymptotic convergence. This includes the usual method of determining a normal curve through statistical ascertainment of mean and standard deviation (§ 43). The class of admissible hypotheses is here the two-parameter class of normal curves. The condition that the distribution must be known to belong to a certain class can be generalized in the following sense: if it is known that for the limiting case of an infinite observational material the distribution belongs to a certain class, the use of this class for an asymptotic convergence is justifiable. Such a generalization was studied by Wald.

The justification of the method of anticipative posits in advanced knowledge differs from the one concerning primitive knowledge so far as it cannot be proved that the method must lead to success if success is attainable. It can be proved only that success is probable if it is attainable. The limitation to a probability originates from the merely probable character of the inductive assumptions entering into the method. It may be false that the distribution belongs to the class considered; then there might still exist a limiting distribution, but it would not be found through the method. In primitive knowledge the method of anticipative posits is free from this limitation.

The methods of restricted anticipative posits share the property of the general method in determining, not one rule of positing, but a class of such rules, all equally justified. For instance, instead of selecting for the posit the value s_o of the parameter s determined by the rule of maximum likelihood, it is permissible to select a value $s_o + c_n$, where c_n is a function converging to 0 with increasing n. It is impossible to distinguish one of these values as preferable to the others unless something is known about the antecedent probabilities. Without such knowledge the decision for a certain form of c_n, for instance, $c_n = 0$, must be regarded as a matter of convenience, as was explained with respect to the rule of induction. Although advanced knowledge leads to a restriction of functions c_n as compared with primitive knowledge, it still leaves open a certain class if it is dependent on anticipative posits.

The term *likelihood* was selected by Fisher in order to indicate that his principle does not determine an inverse probability; he believes that the selection of statistical hypotheses is governed by rules other than those of probability. I cannot agree with his conception. The only legitimate system for the control of hypothetical assumptions is the calculus of probability in combination with the frequency interpretation; it is the only predictive system that is justifiable. The logical function of the likelihood method is not to supply a second kind of probability but to supply a rule of assertability which, though differing from the usual rule, is still justifiable in terms of the calculus of probability. When the usual rule, "Assert the most probable hypothesis", is inapplicable because of lack of knowledge of antecedent probabilities, the likelihood method replaces it by the rule, "Assert a hypothesis which, if continuously adjusted to an increasing statistical material, converges with the most probable hypothesis". This rule is applicable when the forward probabilities have the asymptotical properties of Bernoulli probabilities.

The asymptotical rule, however, cannot provide degrees of preference for inductive hypotheses; it merely distinguishes admissible series of hypotheses from inadmissible ones, without discrimination among the admissible hypotheses. A reasonable measure of preference, or expectation, would be given by the inverse probability. The likelihood cannot take its place, for the following reasons:

1. The order of likelihood, in general, does not correspond to the order of inverse probability. If the inverse probability is known, the likelihood must therefore be disregarded.

2. If the inverse probability is unknown, the order of likelihood does not supply an order of assertability; hypotheses of different degrees of likelihood, if they belong to a certain class, are assertable with equal rights.

It was explained in § 87 that the use of the inductive rule in primitive knowledge is not based on a probability relation but on the ground-for-assertion relation. A similar result holds for anticipative posits in advanced knowledge. The method of maximum likelihood, and with it all other asymptotical rules of statistical inference, are forms of the ground-for-assertion relation. They do not convey a probability, or a degree of assertability, to the assertion designated by them, but they offer legitimate grounds for the choice of these assertions. Asymptotic convergence cannot supply a graduated scale for the rating of hypotheses, but only a yes-or-no criterion.

The term *hypothesis* is usually meant to imply a certain credibility of the conjecture. If this connotation is accepted, the method of maximum likelihood, applied without a knowledge of antecedent probabilities, supplies, not a *hypothesis*, but an *assumption*, the latter term being free from such connotations. The distinction between hypothesis and assumption then repeats, for a comprehensive thesis, the distinction between appraised and blind posit. The establishment of hypotheses is left to the inference by indirect evidence, which supplies a degree of probability, or an estimate of it, to a hypothesis. Anticipative posits, even in advanced knowledge, are forms of statistical induction and can establish assumptions only. In application to cases of unknown antecedent probabilities, Fisher's principle must therefore not be classified as supplying hypothetical probabilities, i.e., as an instance of the third method mentioned in § 70, but as an improved version of statistical induction, i.e., of the first method.

Incidentally, practical applications that at first sight appear to have the form of the anticipative method (including, presumably, most actual applications of the theory of statistical estimation) will often turn out, on closer examination, to be inferences by indirect evidence. A very inexact knowledge of antecedent probabilities may be sufficient to transform the anticipative method into an inference by indirect evidence, and thus into a method that supplies a probability estimate for a hypothesis. The situation resembles the one holding for the inductive inference by enumeration, which in most practical cases includes advanced knowledge and therefore admits of an estimate of its reliability (see § 86).

The method of maximum likelihood has recently found an application in one of the attempts to construct a theory of confirmation of hypotheses by deductive means. This theory, which was developed by O. Helmer, C. Hempel, and P. Oppenheim,[1] concerns propositions that are built up in terms of a finite number of one-place predicates $B_1 \ldots B_t$. Since the number of possible combinations of such predicates in affirmative or negative form is finite, the theory is concerned with finding a probability metric that coördinates a degree of probability to each "cell" of a logical space, that is, to each predicate combination Q_i, such as $Q_i = B_1.\bar{B}_2.\bar{B}_3.B_4.\bar{B}_5 \ldots B_t$. The number of individuals x_k to which the predicates apply need not be finite.

Other theories of confirmation, like R. Carnap's theory,[2] introduce a metric of this kind on *a priori* grounds, and it is difficult not to classify them along with aprioristic methods like the principle of indifference. The theory of Helmer-Hempel-Oppenheim claims to get around this difficulty by making the metric dependent on observational material. They assume that certain cells of the logical space are occupied, that it is known of a finite number of individuals which predicates pertain to them, or that it is known, at least, whether some cells are or are not empty. They then determine the most suitable probability metric for their space by the principle of maximum likelihood.

The latter method degenerates for this problem into a simpler form. The probability distribution is not continuous, but consists of a finite set of probabilities $p_1 \ldots p_t$. Consequently, a functional form $d(p)$ need not be assumed, and the $p_1 \ldots p_t$ may be regarded as the parameters of the distribution. The method then consists in determining the combination $p_1 \ldots p_t$ that makes the observed occupation of cells the most probable result. The only synthetic assumption entering into the method is the assumption of independence, that is, the use of the special theorem of multiplication, without which (or some equivalent) the method of maximum likelihood cannot be carried through. This assumption restricts the applicability of the method to advanced knowledge, if the probabilities used in the theory have a frequency meaning. The assumption of independence is then testable through observations in the logical space.

Some numerical instances may illustrate the method. If only one predicate B is considered and n observations have been made, m of which fall into the

[1] O. Helmer and P. Oppenheim, "A Syntactical Definition of Probability and of Degree of Confirmation," in *Jour. of Symbolic Logic*, Vol. X (1945), p. 25. C. Hempel and P. Oppenheim, "A Definition of Degree of Confirmation," in *Philos. of Science*, Vol. XII (1945), p. 98.

[2] "On Inductive Logic," in *Philos. of Science*, Vol. XII (1945), p. 72.

class B, the probability of B is $\frac{m}{n}$, in accordance with the frequency interpretation. Differences arise if the observational knowledge has a disjunctive form. If it is known that out of six observations either two or four fall into B, the theory provides the probability $\frac{1}{2}$ for B. This result seems rather strange: the direct application of the rule of induction would supply the answer that either $\frac{1}{3}$ or $\frac{2}{3}$ should be taken for the probability. The intermediate value $\frac{1}{2}$ is the result of a different rule for the selection of the anticipative posit, based on the method of maximum likelihood in combination with the assumption of independence; but since the method cannot adduce reasons for the preference of its posits, the inverse probability being unknown, the value $\frac{1}{2}$ is a blind posit in the same sense as the posit "$\frac{1}{3}$ or $\frac{2}{3}$". The difference drops out when larger n are considered, since then the theory shows two maxima of the likelihood to exist, which means a disjunctive answer.[3]

The discussion shows that the authors are mistaken if their theory is meant to supply more than an asymptotic rule and could give answers beyond the reach of the rule of induction. Like all other forms of statistical inference, the theory is an extension of the method of anticipative positing to advanced knowledge; it can be applied if a previous use of the rule of induction has shown that the assumption of independence presupposed for it is true. It is obvious that the theory cannot establish an inductive logic; it depends for its justification on the meaning of probability and the methods developed for the frequency interpretation.

Theories like those of Carnap and Helmer-Hempel-Oppenheim may be said to develop a *deductive conception* of probability, because they derive the values of probabilities from observational material by deductive methods alone. I should like to summarize my criticism of all such methods by comparing them with the *inductive conception* of probability, expressed in the frequency interpretation, for which the derivation of a probability value presupposes the use of inductive inferences.

Given a sequence of events or other objects in which the frequency $\frac{m}{n}$ of a certain attribute converges toward a limit p, we can coördinate to it a set of numbers p_n such that

$$\lim_{n \to \infty} p_n = \lim_{n \to \infty} \frac{m}{n} = p \tag{2}$$

We require that the p_n be defined without a knowledge of the value p; the convergence then must be demonstrable from the functional form of the p_n on the assumption that the limit p exists. The interpretation of such *asymptotic indices* p_n is the subject of the controversy.

---

[3] In order to secure this result for all cases of this type, the theory must be corrected so as to admit likelihood maxima of different heights as assertable metrics.

Asymptotic indices, which are derived from observational material by deductive methods, are an indispensable instrument for both the deductive and the inductive conceptions. In the deductive conception the p_n are called probabilities; I will call them *deductive probabilities*. They must be carefully distinguished from deduced probabilities (see § 70), which are deductively derived from other probabilities and thus presuppose the use of inductive inferences at some place. Deductive probabilities are single-case probabilities because their values do not express frequencies; the values p_n vary, in general, from element to element. The inductive conception reserves the name of probability for the limit p of the p_n, or of $\frac{m}{n}$, and thus employs only *inductive probabilities*, that is, probabilities the values of which can only be asymptotically determined through observational material. Such probabilities are class probabilities; if they are applied to individual cases, the transfer represents only a mode of speech translatable into frequency statements. Besides inductive probabilities, the inductive conception uses asymptotic indices p_n derived deductively from observational material; however, it does not call the p_n probabilities, but regards them as substitutes for the probability p, to be used as long as the limiting value p is unknown. The simplest form of a rule determining values p_n is the rule of induction, which puts $p_n = \frac{m}{n}$.

So far, the difference between the two conceptions is merely terminological. It does not matter whether the p_n are called probabilities of the individual element or whether the limiting value p is regarded as the probability of each element, for which the value p_n is a substitute. A material difference arises, however, with respect to the choice of the rules determining the p_n.

The inductive conception regards the decision for a particular rule of this kind, if the rule is set up in primitive knowledge, as a convention. Since the applicability of the rule is controlled only by the asymptotic criterion (2), the value of an individual p_n is arbitrary and without objective significance; it does not describe a property of the sequence. The different sets of values p_n are asymptotically indistinguishable. It is for this reason that the p_n are not called probabilities in the inductive conception; the name is used only for the limit p, which is an objective characteristic of the sequence. Adherents of the deductive conception, however, select among the admissible rules for the p_n one rule as the best, asserting that they can prove its superiority. For this reason they regard an individual p_n as objectively significant and call it a probability. Being derived from observational material, this probability is usually construed as holding relative to the material, so that an expression denoting the observational material enters into the reference term of the probability functor. Inductive probabilities do not include such a reference, because they are independent of the observational material from which they are inductively inferred; they express an objective relation holding for a sequence.

Since the asymptotic convergence cannot supply a criterion of selection among the rules, deductive theories must resort to aprioristic reasoning. Some of these theories, in fact, are openly advanced with the appeal to a *synthetic a priori*, such as the theories based on the principle of indifference or on rational belief. Others, like those of Carnap or Helmer-Hempel-Oppenheim, are intended to use only analytic principles for the determination of the p_n. It is obvious that such theories must break down, since analytic reasoning cannot prove any set p_n satisfying (3) to be superior to the others. If it is argued that such a theory is justified because its probabilities converge asymptotically with the one supplied by the inductive theory,[4] the argument is easily seen to be invalid: asymptotic convergence justifies all such rules alike, but cannot be adduced if differences between the rules are under discussion.

The rules for the selection of the p_n are sometimes based on the assumption of a metric $q(p)$ for the antecedent probability (Carnap), sometimes on the principle of maximum likelihood (Fisher, Helmer-Hempel-Oppenheim), a method which, as far as the selection of values p_n is concerned, coincides with the metric $q(p)$ = const. (see § 70). If the aim of the method is to construct an asymptotic rule, any such metric is permissible, not because the metric $q(p)$ is derivable from a "plausible" principle like the principle of indifference, but because the metric is irrelevant. Laplace's convergence relations (9, § 62) guarantee the asymptotic convergence for every metric $q(p)$, if it is known, at least, that the Bernoulli relations (10, § 57) are satisfied. But this validation of deductive methods for the selection of a set p_n is very different from the intentions of their authors: every deductive method of this kind is admissible because it leads to a rule of induction of the general form (1, § 87), and there are no arguments that favor any of them. The search for a distinction among the class of admissible inductive rules in primitive knowledge concerns a pseudoproblem.

My objection to deductive theories, therefore, is not that their sets p_n are different from the one introduced by my inductive rule. I have emphasized repeatedly[5] that the justification of induction refers to the class of rules (1, § 87) and is not restricted to the inductive rule in the narrower sense. My criticism concerns the claim of deductive theories to select one set p_n as superior to others by methods for which no justification is given. By justification I understand a proof, not that some people believe in these methods or behave as though they believed in them, but that it is advantageous to use them.

---

[4] This argument is used by R. Carnap, "On Inductive Logic," in *Philos. of Science*, Vol. XII (1945), p. 97.

[5] For instance, in *EP*, pp. 355–356. For a correction of my earlier views see footnote, p. 447.

It was mentioned above (§ 71) that it is possible to introduce a concept of probability for which the principle of indifference is analytic. Similarly, it is possible to define a concept of probability for which the rules determining the p_n are analytic. Of this kind is Carnap's concept of a nonfrequency probability. But this probability concept, and with it every deductive probability, is empty because it has no predictional value. To say that $p_n = \frac{2}{3}$ is as justifiable as saying that $p_n = \frac{1}{2}$ because both values fit the asymptotic rule. And both values are compatible to the same extent with whatever happens. Assume that p_n was chosen $= \frac{1}{2}$, but that observation of further s elements shows the frequency to converge to $\frac{2}{3}$. Then the value $p_n = \frac{1}{2}$ is not regarded as falsified, although a new probability p_{n+s} close to $\frac{2}{3}$ is now introduced; this probability is said to have different observational material in the reference term. Through this evasion of verification by future observation, deductive probabilities lose any predictional value. If the statement $p_n = \frac{1}{2}$ is true, it does not help to know it, because it has no implications for the future.

This is why the inductive conception does not call the p_n probabilities. If the value p_n is regarded as the measure of a probability relation between the observational material and a future event, the probability relation is ambiguous; there is no most advisable value for p_n in primitive knowledge. There is a ground for asserting p_n; but the ground-for-assertion relation determines without discrimination a class of values p_n in terms of the observational material. It is not the use of an individual p_n but the progressive use of the total set that can be shown to be advisable; such progressive use will lead to a prediction of the frequency if there is a limit of the frequency. To this limit of the frequency the inductive conception applies the name of probability. The degree of probability is thus made a number uniquely determined for classes of observable objects, independent of the state of our knowledge. The distinction between (2 and 3, § 87) expresses the difference between the two conceptions. What is called a probability relation in the deductive conception is actually a ground-for-assertion relation inaccessible to quantitative measure.

I do not wish to say that there are no methods of selecting a particular set p_n as better than others; on the contrary, the inductive conception has developed such methods. But it constructs them only in advanced knowledge. This means that it bases the selection of a set p_n on more material than is given by the initial section of the sequence up to the place n; and for the selection it uses the results of inductive inferences referring to other sequences. For instance, if, in advanced knowledge, the special theorem of multiplication is to be used, it must first be proved that the condition of independence holds; this proof presupposes the use of the rule of induction for a certain material, since it amounts to showing that the limits of two frequency sequences are equal. Or, if the anticipative posits are restricted to a certain form of the distribution $d(s;u)$, as for Fisher's method of maximum likelihood, the choice of the form must be based on previous inductive inferences referring to further observational material. If methods of this kind lead to regarding the individual p_n as probabilities, this means that the p_n have been shown to be the limits of the frequencies in certain other sequences. For instance, the p_n can be the limits of vertical columns in a lattice built up of horizontal sequences, such as the probability values given by the rule of succession (22, § 62). The justification of all methods of improved convergence is therefore reducible to the justification of statistical induction and thus to the use of asymptotic methods.

The problem of selecting a set p_n that leads to a better convergence, which the deductive conceptions cannot solve, is thus made accessible to a solution free from aprioristic methods. The solution consists in demonstrating that the use of asymptotic rules for a wider material can lead to a selection among asymptotic rules for a narrower material. The following sections (§§ 89–90) will show in what sense this proof can be given.

§ 89. The Method of Correction

Since the rule of induction is the only primary instrument of finding probabilities, it must be possible to show how advanced knowledge is built up from primitive knowledge by means of the rule of induction.

By means of the inductive rule we set up posits concerning the limit of the frequency in a sequence and thus establish probability values. The probabilities so constructed can be used as the weights of certain other posits; we are thus able to construct appraised posits by means of anticipative posits. The appraised posits can even be identical with some of the anticipative posits; in other words, we can transform an anticipative posit into an appraised posit. Since the weight thus constructed can be used for a change in the posited value of the limit, we speak here of the *method of correction*.

The general method of correction is based on the combined use of a great number of posits; the totality of posits is so evaluated that certain individual posits can be corrected. In this procedure we start with primary posits, posits of the first level, and then construct from them secondary posits, posits of the second level, which supply weights of the primary posits. By this method a weight can be constructed for every posit of the first level. The statement about the weight, however, will itself represent an anticipative posit. In the same manner we can construct a weight for the secondary posits, and so on. Proceeding to higher and higher levels, we can find a weight for every posit on any level and thus can transform every anticipative posit into an appraised posit, though there will always remain anticipative posits in the system, namely, those of the last level.

The procedure, which represents the transition from the primitive to the advanced state of knowledge, follows the schema that we called cross induction (§ 84); its mathematical structure corresponds to the methods developed in § 62. A finite number of finite sections of the sequences, that is, a finite initial part of the lattice (3, § 58), is given:

$$\begin{matrix} y_{11} & y_{12} & . & . & . & y_{1n} & f^{1n} \\ y_{21} & y_{22} & . & . & . & y_{2n} & f^{2n} \\ . & . & . & . & . & . & . \\ . & . & . & . & . & . & . \\ . & . & . & . & . & . & . \\ y_{s1} & y_{s2} & . & . & . & y_{sn} & f^{sn} \end{matrix} \tag{1}$$

The frequency

$$f^{kn} = F^n(A.B^{ki},C^{ki})^i \tag{2}$$

has been determined for every horizontal section. As in § 62, the range from 0 to 1 is divided into τ intervals $\eta_1 \ldots \eta_\tau$ of the width 2δ; the middle value of an interval η_ρ may be called p_ρ or f_ρ. If f^{kn} lies within η_ρ, we write[1]

$$f^{kn} = p_\rho \pm \delta \tag{3}$$

The system of primary posits is constructed by putting

$$\lim_{n\to\infty} f^{kn} = p_\rho \pm \delta \quad \text{if } f^{kn} = p_\rho \pm \delta \tag{4}$$

In order to obtain the system of secondary posits, we count for each η_ρ the horizontal sections of the sequence for which $f^{kn} = p_\rho \pm \delta$. Thus we determine frequencies $\overset{sn}{g_\rho}$, which we define by

$$\overset{sn}{g_\rho} = \frac{1}{s} \overset{s}{\underset{k=1}{N}} (f^{kn} = p_\rho \pm \delta) \tag{5}$$

From here we go in two steps to the secondary posits. We first posit that the $\overset{sn}{g_\rho}$ represent for the limit $n \to \infty$ the frequencies of the s sequences for the intervals η_ρ. Using the notation

$$\overset{s}{g_\rho} = \frac{1}{s} \overset{s}{\underset{k=1}{N}} (\lim_{n\to\infty} f^{kn} = p_\rho \pm \delta) \tag{6}$$

we thus posit

$$\overset{s}{g_\rho} = \overset{sn}{g_\rho} \pm \epsilon \tag{7}$$

where ϵ is a given interval of exactness of this posit. On the second step we posit that the values $\overset{sn}{g_\rho}$ also represent the limits for $s \to \infty$, that is, we posit

$$q_\rho = \lim_{s\to\infty} \overset{s}{g_\rho} = \overset{sn}{g_\rho} \pm \eta \tag{8}$$

[1] In this notation the expression $p \pm \delta$ is regarded as an indefinite description (see *ESL*, p. 264), so that (3) means: there is a value within the interval from $p - \delta$ to $p + \delta$ such that f^{kn} is equal to this value.

This is the posit by which we obtain probabilities of the second level q_ρ from he finite section (1) of the lattice. q_ρ is the probability that the probability of the first level is $= p_\rho \pm \delta$.

The probability of the second level q_ρ does not immediately represent the ating of the posit p_ρ of the first level; q_ρ is the antecedent probability of p_ρ, out not its rating based on the observed initial section. This rating is supplied by the inverse probability v_n, which was computed in § 62 as a certain function of the q and p. The treatment used here, which is meant to portray the pproximative methods of primitive knowledge, is distinguished from the one ised in § 62 in that the intervals η_ρ are finite and a transition to the limit 0 s not made. Correspondingly, summations take the place of integrations. The precise frequency f is replaced by the frequency interval $f \pm \delta$, and the nverse probability density $v_n(f;p)$ of (5, § 62) is replaced by a density $_n(f \pm \delta;p)$. Corresponding to a remark made in § 45, a relative probability unction having a small interval in the reference class is approximately equal o the function written for a precise value in the same place; the value (5, § 62), herefore, is approximately valid.

For the construction of the exact value, the following notation will be used. The inverse probability that, if f is within an interval η_ρ, p is within the inerval η_μ, will be written $v_{n\rho\mu}$. The forward probability that, if p is within an nterval η_σ, the frequency f lies within the interval η_ρ, will be written $b_{n\sigma\rho}$; or normal sequences this is the Bernoulli function (16, § 49). Thus, instead f (5, § 62) and (9, § 62), the following relations result:

$$v_{n\rho\mu} = \frac{q_\mu b_{n\mu\rho}}{\sum_{\sigma=1}^{\tau} q_\sigma b_{n\sigma\rho}} \qquad \lim_{n\to\infty} v_{n\rho\mu} = \begin{cases} 1 \text{ for } \rho = \mu \\ 0 \text{ for } \rho \neq \mu \end{cases} \tag{9}$$

The proof of the convergence to 1 or 0, respectively, follows because, accordng to (25, § 49), $b_{n\rho\rho}$ goes to 1 with increasing n, whereas $b_{n\sigma\rho}$ goes to 0 for $\neq \rho$; for large n the denominator reduces virtually to the term $q_\rho b_{n\rho\rho}$.

The presentation up to (8) has shown how the antecedent probabilities q_ρ an be determined from statistical material in the form of posits. In order o determine $v_{n\rho\mu}$ we must know, furthermore, the functions $b_{n\sigma\rho}$. But the etermination of these functions presupposes a knowledge of the normal haracter of the sequence, or some other knowledge that enables us to compute he probability of combinations, like the knowledge that the sequences have robability transfer. I shall therefore indicate, as an illustration for the etermination of sequence structure in primitive knowledge, by what methods he normal character of the sequences can be ascertained.

The method consists in the determination of probabilities in subsequences nd thus is not different in principle from the methods so far presented. We rst examine whether a given sequence is free from aftereffect. We construct a subsequence from the given section of the sequence by the use of a selection by the first predecessor; in this subsequence the probability is determined by a posit according to (4). We repeat the same procedure for all subsequences that result from a selection by a group of two predecessors, and so on, up to r predecessors. Now we conceive the probabilities thus found in t subsequences as a new probability sequence, and judge about its continuation for $t \rightarrow \infty$, that is, for a transition to longer and longer groups of predecessors, by means of a posit. Assume, for instance, that the probabilities of the subsequences considered are, in the majority, equal to the probability of the main sequence within the interval of exactness $\pm \delta$; then we regard the coincidence of the two probabilities as a property C and posit that the frequency of C will go toward 1 for the limit $t \rightarrow \infty$. The rule of induction is thus applied to obtain a probability of the second level and leads to the posit that for all subsequences there exists a coincidence between their probabilities and the main probability.

By means of a corresponding procedure we can determine also the probabilities of the subsequences resulting from regular divisions (§ 30). Corresponding considerations hold for properties of the lattice. The methods that determine the type of the sequence from the dispersion can also be included in the procedure explained above. Thus we see that the ascertainment of any special type of sequence can, in principle, be carried through by applying the rule of induction, whether we deal with normal sequences in the narrower or in the wider sense, or sequences with probability transfer, or Bernoulli sequences, and so on. No *a priori* assumptions are required.

That the inductive rule is all we need follows from the fact that we could construct the entire calculus of probability from axioms that are logically derivable from the frequency interpretation. We could characterize every special case by the property that some probabilities are numerically equal (as in normal sequences), or that they approach a limit (as in Bernoulli sequences). The occurrence of such a property can always be established by means of the rule of induction. We thus confirm the previous result that all applications of the calculus of probability to physical reality can be carried through by using the rule of induction as the only nondeductive principle.

Once the $b_{n\mu\rho}$ have been determined within the ascertainment of the type of the sequence, the probability $v_{n\rho\mu}$ can be found from (9) in the form of a posit. The values q_μ and $b_{n\mu\rho}$ to be inserted in (9) are posited, not ultimate values; thus the value $v_{n\rho\mu}$ can be posited only within a certain interval of exactness ϑ:

$$v_{n\rho\mu} = \frac{q_\mu b_{n\mu\rho}}{\sum_{\sigma=1}^{r} q_\sigma b_{n\sigma\rho}} \pm \vartheta \qquad (10)$$

When the weights $v_{n\rho\mu}$ are determined, they can be used for the correction of the primary posits. Should it turn out, for instance, that for some of the sequences the maximum probability is not given by $v_{n\rho\rho}$, but by $v_{n\rho\mu}$, where $\rho \neq \mu$, we shall correct the original posit correspondingly. We can thus transform the original anticipative posit into an appraised posit. Such a case will occur, for instance, when all sections (1) have a frequency within $f_\mu \pm \delta$, except one; for this sequence we shall then posit the limit also within $f_\mu \pm \delta$, though this does not correspond to its frequency f^{kn}, which is within $f_\rho \pm \delta$. We have thus constructed the method of cross induction, introduced in § 84.

We now see the importance of the convergence relation added in (9), according to which $v_{n\rho\rho}$ goes toward 1, no matter what the values of the q_ρ are. If it turns out, in a prolongation of the sections, that the frequency remains within $f_\rho \pm \delta$, the method of correction must finally lead to the result that $v_{n\rho\rho} > v_{n\rho\mu}$. We meet here with a peculiar interrelation of the totality and the individual case. At first the totality is more "powerful" than the individual case given by one of the sequences, and may correct it. With sufficient prolongation of the one sequence, however, the individual case can finally become more "powerful" than the totality. Even if only one sequence of a deviating frequency occurs, we assume for a very great n that we are actually concerned with a deviating individual case. We refer to the illustration given by the deviation of the planet Mercury from Newton's law (see § 85). This is the version given by the theory of probability to the specific tension that exists between *individual case* and *totality* in all empirical inquiry. An exception from a hitherto known law will at first be interpreted as an *error*. If it recurs frequently, however, it will be acknowledged as a *fact*, and the general law will be altered correspondingly. In the theory of probability this *independence of individual facts* finds its expression in the convergence relation stated in (9).

§ 90. The Hierarchy of Posits

I shall summarize the results so far obtained. It has been shown that, if we know the limit of the frequency in a sequence, this value can be regarded as the weight of an individual posit concerning an unknown element of the sequence. The weight may be identified with the probability of the single case, assuming the character of a truth value. In order to find the limit of the frequency we use an anticipative posit; its weight is unknown. In order to determine this weight we must make an anticipative posit on a higher linguistic level; the former anticipative posit is then transformed into an appraised posit, that is, a posit of known weight.

The procedure can be extended to higher and higher linguistic levels. It is obvious, however, that we shall never reach a true statement in this way; each of the metalanguages in this infinite hierarchy follows a probability logic. The infinite system may be compared with the system constructed in the fourth column of table 8 (p. 392). There is, however, an important difference. The latter infinite system is given by a mathematical rule, whereas the probability system must be constructed step by step, by the use of empirical methods. We must stop on a finite level without any knowledge about higher levels; in particular, we do not know whether the probability system is convergent. When we thus cut off the probability system on a finite level, we do not know how great the mistake so introduced will be. The posit on the last level will be an anticipative posit, the weight of which is unknown.

The fact that we must cut off the hierarchy of posits at a posit of unknown weight endows the structure of probability logic with a peculiar uncertainty. The question offers itself: What is gained by the transition to posits of a higher level? Why do we prefer to use appraised posits instead of anticipative posits on the lower levels, if all the weights are obtained ultimately by anticipative posits? What is the logical significance of the structure of the levels if on the last level we remain with anticipative posits?

In order to answer these questions, assume that the system is cut off at the secondary posits, and apply to the secondary posits the principle that was used for the justification of the anticipative posit on the first level. This justification was based on the self-corrective nature of induction; we showed that the anticipative posit leads to the correct value of the limit within the given interval of exactness $\pm \delta$ if the sequence has a limit. The same principle is applicable to secondary posits. But we must now compare the convergence of primary and secondary posits.

Consider the determination of the secondary probabilities g_ρ that was given in § 89. The primary posit used there is formulated in (4, § 89); the two steps necessary for obtaining the secondary posits are given in (7 and 8, § 89). The following theorems are derivable from the definition of the limit, on the assumption that the sequences have a limit of the frequency:

1. If s is kept constant, there is an $n = n_1$ such that the primary posits (4, § 89), concerning the horizontal sequences, become correct for all s sequences and remain correct for all greater n from there on.

2. If s is kept constant, there is an $n = n_2$ such that the secondary posits (7, § 89), concerning the vertical frequencies up to s sequences, become correct and remain correct for all greater n from there on.

If we understand by n_1 and n_2 the smallest values having these properties we can prove the inequality

$$n_2 \leqq n_1 \tag{1}$$

This relation follows from the fact that the number g_ρ^{sn} can correspond to g_ρ within the interval of exactness $\pm \epsilon$ even when the correct value of the horizontal limit has not yet been found for all s sequences. In some horizontal sequences the primary posits may still be false, although the frequencies counted vertically, for the determination of g_p^s, may have arrived within the interval of exactness $\pm \epsilon$. If the primary posits, however, are correct in all s sequences, the secondary posits (7, § 89) must also be correct. [Only in exceptional cases will the equality sign in (1) hold; but we cannot make use of this qualification for the following considerations, since it is based on an inductive inference from our experience. It cannot be derived from the definition of the limit.]

Adding the transition to $s \to \infty$, we can derive the following theorems on the assumption that all the sequences have a limit:

3. There is an $s = s_3$ and an $n = n_3$ such that all the posits (8, § 89) are correct.

To prove this theorem, assume that there exists a limit of the frequency for vertical counting with respect to horizontal sequences as elements. Then there must be an $s = s_3$ such that the frequency g_p^s, for the definition of which we have assumed that the limits in the horizontal sequences are known, corresponds within the interval of exactness $\pm \zeta$ to the value q_p resulting for $s \to \infty$. Now we do not know the precise value of g_p^s because the horizontal sequences are known only up to n elements. But since, according to theorem 2, the value g_p^{sn} found statistically will correspond to g_p^s within the interval of exactness ϵ, the value g_p^{sn} will be close to q_p within the interval $\pm(\epsilon + \zeta)$. It can thus be brought as close to q_p as we wish by a suitable choice of s_3 and n_3. The posits (8, § 89) will be correct when $\epsilon + \zeta \leqq \eta$.

4. Generally, there is no $n = n_4$ such that for all s the posits (4, § 89) are correct. Yet if there is such an $n = n_4$, that is, if we deal with the special case of a uniform convergence (§ 65), the inequality

$$n_3 \leqq n_4 \tag{2}$$

holds. This follows by considerations as given for theorem 2.

The four theorems express an intrinsic difference between primary and secondary posits, which can be formulated in the following statements:

a. The secondary posits are independent of the individual primary posits. An individual primary posit may be false, though the secondary posit is true [inequalities (1) and (2)].

b. If all the sequences have a limit of the frequency, the correctness of all the secondary posits can be reached in a finite number of steps, though the primary posits to which they refer are not made in an infinite number of sequences. This holds whether or not there is a uniform convergence.

These results, which are derived from the assumption of the limit of the frequency alone, formulate the superiority of secondary posits over primary ones. Briefly, they say that secondary posits are independent of primary posits; that, increasing the degree of approximation by going to greater values of n, we may count upon the possibility that secondary posits will be correct at an earlier stage than primary ones. The transition to secondary posits makes possible a better convergence. This mathematical result justifies us in regarding secondary posits as better than primary posits.

We are now in a position to recognize the significance of the structure of levels in probability logic. The order of levels does not lead to a system that as a whole is equivalent to a two-valued true statement, as in the multivalued logic developed in § 76. Neither can we claim that the infinite system of probability logic will strictly converge, as does the infinite system of column 4, table 8 (p. 392). But we can make a statement of convergence in a different form: when we improve the approximation by extending the primary sequences, the posits of higher levels will, in general, become correct at an earlier stage—in any case, never later than the posits of lower levels. In this sense the multivalued system of probability logic, which has an infinite number of levels of language, is a convergent system. The higher the level on which it was cut off,[1] the sooner we may expect the convergence of this system with increasing n of the primary sequences. The only assumption on which the theorem is based is that the sequences converge to a limit of the frequency. Such a system will be called a *quasi-convergent system.*

These results will make clear the use of probabilities for the verification of limit statements, a procedure that is applied within an advanced state of knowledge (see § 74). Only within a system of two-valued logic is it impossible to verify statements about infinite sequences or practically infinite sequences before the sequences are completely observed. Within the framework of probability logic, however, such a verification by degree is possible. The statement of verification itself is imbedded in a metalanguage that is governed by probability logic. It is true that we must cut off this system on a finite level and stop with a posit of unknown weight. But we know that if the system is quasi-convergent, verification by levels has a distinct advantage: it opens the possibility of a better convergence than would be attainable by the use of primary posits alone. The statement that the posits so appraised are the best posits is itself a posit, and an anticipative one. When we use the weights so constructed for the correction of primary posits, we may make mistakes

[1] G. H. von Wright, in his thesis, "The Logical Problem of Induction," in *Acta Philos. Fennica*, Part 3 (1941), p. 185, quotes this passage from the German version of this book (p. 407) and adds the following criticism: "Reichenbach does not seem to realize that this 'may expect' is not in the slightest way justified by his system". Apparently von Wright did not see that the passage quoted from me is merely a paraphrase of my inequalities (1) and (2), which he does not even mention, and is therefore based on a mathematical proof. It should be clear that by the words "may expect" I do not mean that some *probability* of an earlier convergence exists, but that I speak here only of the *possibility* of an earlier convergence, expressed in the sign $\leqq$ occurring in (1) and (2). In some of the preceding passages this is stated explicitly. See also the remarks on p. 477 which I have added in this edition. Von Wright's misunderstanding of my theory of induction is shown also in his remark that my theory was anticipated in C. S. Peirce's writings. I refer to my remarks about Peirce in footnote 1, p. 446.

But on the assumption of quasi-convergence we can prove: continuing the sequences observed to greater n and increasing their number to greater s, we shall reach an $n = n_3$ and an $s = s_3$ such that our corrections will be right in most cases. The values n_3 and s_3 can occur at a stage where a great number of primary posits are still incorrect. The system of posits appraised by posits finds its justification, therefore, as an instrument for improving the degree of convergence—if such improvement is attainable at all.

When we connect our results with the considerations of § 85, we see that the procedure developed as a method of correction constitutes a simplified schema for the logical operations performed in scientific inquiry. The system of science, therefore, must be regarded as a system of posits, incorporated in the framework of probability logic and including posits of lower and higher levels. We cut this system off on a certain level of language; the higher the level, the better will be the convergence attained if the system of posits is quasi-convergent.

Although the form of the system is given by probability logic, we know that this logic does not require any specific assumptions when probabilities are reduced to frequencies, since probability logic is then reducible to two-valued logic (see method 3, § 81). The result follows because the axioms of probability are derivable from the frequency interpretation. The statement that the system is quasi-convergent requires, as its only basis, the existence of limits of the frequency. The assumption that such limits exist is sufficient, not only to justify individual anticipative posits, but also to justify the method of correction, that is, the transition to posits of higher levels.

We have thus reduced scientific method to two presuppositions: the validity of the rules of two-valued logic and the assumption that the sequences under consideration have a limit of the frequency. The first presupposition need not be discussed. In a theory of empirical science the validity of two-valued logic may be taken for granted. It is the second presupposition, concerning the existence of a limit, that we must now discuss. The use of this assumption represents the last obstacle to be removed before a satisfactory theory of probability, and thus of scientific method, can be given.

§ 91. The Justification of Induction

In traditional philosophy the problem of induction was restricted to the discussion of classical induction (§ 67), of the inductive inference referring to a sequence consisting throughout of events of the same kind, for which, therefore, we have a relative frequency $f^n = 1$. This book always uses the wider form of statistical induction, for which the relative frequency f^n may have any value between 0 and 1. The critical objections that have been raised against classical induction, however, apply likewise to statistical induction, and we shall therefore consider these objections.

The first to criticize the inference of induction by enumeration and to question its legitimacy was David Hume.[1] Ever since his famous criticism, philosophers have regarded the problem of induction as an unsolved riddle precluding the completion of an empiricist theory of knowledge. In Hume's analysis it does not appear as a problem of probability; he includes it, rather, in the problem of causality. We observe, Hume explains, that equal causes are always followed by equal effects. We then infer that the same effects will occur in future. On what grounds do we base this inference? Hume's criticism gave two negative answers to the question:

1. The conclusion of the inductive inference cannot be inferred *a priori*, that is, it does not follow with logical necessity from the premises; or, in modern terminology, it is not tautologically implied by the premises. Hume based this result on the fact that we can at least *imagine* that the same causes will have another effect tomorrow than they had yesterday, though we do not believe it. What is logically impossible cannot be imagined—this psychological criterion was employed by Hume for the establishment of his first thesis.

2. The conclusion of the inductive inference cannot be inferred *a posteriori*, that is, by an argument from experience. Though it is true that the inductive inference has been successful in past experience, we cannot infer that it will be successful in future experience. The very inference would be an inductive inference, and the argument thus would be circular. Its validity presupposes the principle that it claims to prove.

Hume did not see a way out of this dilemma. He regarded the inductive inference as an unjustifiable procedure to which we are conditioned by habit and the apparent cogency of which must be explained as an outcome of habit. The power of habit is so strong that even the clearest insight into the unfounded use of the inductive inference cannot destroy its compelling character. Though this explanation is psychologically true, we cannot admit that it has any bearing on the logical problem. Perhaps the inductive inference is a habit—the logician wants to know whether it is a good habit. The question would call for an answer even if it could be shown that we can never overcome the habit. The logical problem of justification must be carefully distinguished from the question of psychological laws.

Up to our day the problem has subsisted in the skeptical version derived from Hume, in spite of many attempts at its solution. Kant's attempt to solve the problem by regarding the principle of causality as a synthetic judgment *a priori* failed because the concept of the synthetic judgment *a priori* was shown to be untenable. I may add that Kant never attempted to make use of his theory for a detailed analysis of the inductive inference. In the

[1] *An Enquiry Concerning Human Understanding* (1748).

empiricism of our time the problem has come to the fore, overshadowing all other problems of the theory of knowledge. It has held this place persistently without changing the skeptical form that Hume gave it.

A few philosophers tried to escape Hume's skepticism by denying that a problem of the justification of induction exists. Various reasons were given for such a conception. It was said that the rule of induction does not belong to the content of science; that Hume's criticism concerns only a linguistic problem; that the problem of justification is a pseudoproblem; and so on. It is hardly comprehensible that such arguments could ever have been seriously maintained. They misuse an important modern discovery—the distinction between levels of language—for the purpose of contesting the legitimacy of an old problem, upon which, however, this distinction has no bearing.

It is true that the rule of induction belongs, not in the object language of science, but in the metalanguage. It is a *directive* for the construction of sentences, since it tells how to proceed from verified sentences to predictive sentences. I have therefore called it a rule of derivation (§ 87), the only one that inductive logic requires in addition to the rules of derivation of deductive logic. Such rules, however, are admissible within a scientific language only when they can be justified, that is, when they can be shown to be adequate means for the purpose of derivation. Such a justification is easily given for the rules of derivation of deductive logic: it can be shown that the rules always lead to true sentences if the premises are true. In a systematic exposition of deductive logic this justification of the rules of derivation must be formally given.[2] For the rule of induction such a proof is not possible; that is why the problem of its justification is so involved that it requires a comprehensive analysis.

The frequency theory of probability, with its interpretation of probability statements as posits, makes it possible to give a justification of the rule of induction. The problem will be discussed with respect to the wider form of statistical induction; the results will then include the special case of classical induction. The generalization expressed in the use of statistical induction is relevant because it weakens the inference. Whereas classical induction wishes to establish a rigorous inference that holds for every individual case, statistical induction renounces every assertion about the individual case and makes a prediction only about the whole sequence.

There is another sense in which the statistical version involves a different interpretation of the problem. The classical conception entails the question whether the rule of induction leads to true conclusions, but the statistical version deals only with the question whether the rule of induction leads to a method of approximation, whether it leads to posits that, when repeated, approach the correct result step by step. The answer is that this is so if the

[2] See § 5 above, and *ESL*, §§ 12, 14.

sequences under consideration have a limit of the frequency. The inductive posit anticipates the final result (§ 87) and must eventually arrive at the correct value of the limit within an interval of exactness.

The method of anticipation may be illustrated by an example from another field. An airplane flies in the fog to a distant destination. From two ground stations the pilot receives radio messages about his position, ascertained by radio bearings. He then determines the flight direction by means of a map, adjusts the compass to the established course, and flies on, keeping continuously to the direction given by the compass. In the fog he has no other orientation than to follow the adopted course. After a while, however, he inquires again of the ground stations for another determination of his position. It turns out that the airplane was subject to a wind drift that has carried the ship off its course. The pilot, therefore, establishes a new course that he follows thereafter.

This method, repeatedly applied, is a method of approximation. The direction from the position ascertained to the destination is not the most favorable one because of wind currents; but the pilot does not know the changing currents and therefore at first *posits* this direction. He does not believe that he has found the final direction. He knows that only when he is very close to his destination will the straight line be the most favorable flying direction—but he acts as though the coincidence of the most favorable flying direction and a straight-line connection were reached. He thus anticipates the final result. He may do so because he uses this anticipation only in the sense of a posit. By correcting the posit repeatedly, always following the same rule, he must finally come to the correct posit and thus reach his destination.

The analogy with the anticipative method of the rule of induction is obvious. In the analysis of Hume's problem we thus arrive at a preliminary result: if a limit of the frequency exists, positing the persistence of the frequency is justified because this method, applied repeatedly, must finally lead to true statements. We do not maintain the truth of every individual inductive conclusion, but we do not need an assumption of this kind because the application of the rule presupposes only its qualification as a method of approximation.

This consideration bases the justification of induction on the assumption of the existence of a limit of the frequency. It is obvious, however, that for such an assumption no proof can be constructed. When we wish to overcome Hume's skepticism we must eliminate this last assumption from our justification of induction.

The traditional discussion of induction was dominated by the opinion that it is impossible to justify induction without an assumption of this kind, that is, without an assumption stating a general property of the physical world. The supposedly indispensable assumption was formulated as a postulate of the uniformity of nature, expressed, for instance, in the form that similar event patterns repeat themselves. The principle used above, that sequences of events converge toward a limit of the frequency, may be regarded as another and perhaps more precise version of the uniformity postulate. So long as logicians maintained that without a postulate of this kind the inductive inference could not be accounted for, and so long as there was no hope of proving such a postulate true or probable, the theory of induction was condemned to remain an unsolvable puzzle.

The way out of the difficulty is indicated by the following considerations. The insistence on a uniformity postulate derives from an unfortunate attempt to construct the theory of inductive inference by analogy with that of deductive inference—the attempt to supply a premise for the inductive inference that would make the latter deductive. It was known that the inductive conclusion cannot be asserted as true; but it was hoped to give a demonstrative proof, by the addition of such a premise, for the statement that the conclusion is probable to a certain degree. Such a proof is dispensable because we can assert a statement in the sense of a posit even if we do not know a probability, or weight, for it. If the inductive conclusion is recognized as being asserted, not as a statement maintained as true or probable, but as an anticipative posit, it can be shown that a uniformity postulate is not necessary for the derivation of the inductive conclusion.

We used the assumption of the existence of a limit of the frequency in order to prove that, if no probabilities are known, the anticipative posit is the best posit because it leads to success in a finite number of steps. With respect to the individual act of positing, however, the limit assumption does not supply any sort of information. The posit may be wrong, and we can only say that if it turns out to be wrong we are willing to correct it and to try again. But if the limit assumption is dispensable for every individual posit, it can be omitted for the method of positing as a whole. The omission is required because we have no proof for the assumption. But the absence of proof does not mean that *we know that there is no limit*; it means only that *we do not know whether there is a limit.* In that case we have as much reason to try a posit as in the case that the existence of a limit is known; for, if a limit of the frequency exists, we shall find it by the inductive method if only the acts of positing are continued sufficiently. Inductive positing in the sense of a trial-and-error method is justified so long as it is not known that the attempt is hopeless, that there is no limit of the frequency. Should we have no success, the positing was useless; but why not take our chance?

The phrase "take our chance" is not meant here to state that there is a certain probability of success; it means only that there is a possibility of success in the sense that there is no proof that success is excluded. Furthermore, the suggestion to try predictions by means of the inductive method is not an advice of a trial at random, of trying one's luck, so to speak; it is the proposal of a systematic method of trial so devised that if success is attainable the method will find it.

To make the consideration more precise, some auxiliary concepts may be introduced. The distinction between necessary and sufficient conditions is well known in logic. A statement c is a *necessary* condition of a statement a if $a \supset c$ holds, that is, if a cannot be true without c being true. The statement c will be a *sufficient* condition of a if $c \supset a$ holds. For instance, if a physician says that an operation is a necessary condition to save the patient, he does not say that the operation will save the man; he only says that without the operation the patient will die. The operation would be a sufficient condition to save the man if it is certain that it will lead to success; but a statement of this kind would leave open whether there are other means that would also save him.

These concepts can be applied in the discussion of the anticipative posit. If there is a limit of the frequency, the use of the rule of induction will be a sufficient condition to find the limit to a desired degree of approximation. There may be other methods, but this one, at least, is sufficient. Consequently, when we do not know whether there is a limit, we can say, if there is any way to find a limit, the rule of induction will be such a way. It is, therefore, a necessary condition for the existence of a limit, and thus for the existence of a method to find it, that the aim be attainable by means of the rule of induction.

To clarify these logical relations, we shall formulate them in the logical symbolism. We abbreviate by a the statement, "There exists a limit of the frequency"; by b the statement, "I use the rule of induction in a repeated procedure"; by c the statement, "I shall find the limit of the frequency". We then have the relation[3]

$$a \supset (b \supset c) \qquad (1)$$

This means, $b \supset c$ is the *necessary* condition of a, or, in other words, the attainability of the aim by the use of the rule of induction is a necessary condition of the existence of a limit. Furthermore, if a is true, b is a *sufficient* condition of c. This means, if there is a limit of the frequency, the use of the rule of induction is a sufficient instrument to find it.

It is in this relation that I find the justification of the rule of induction. Scientific method pursues the aim of predicting the future; in order to construct a precise formulation for this aim we interpret it as meaning that scientific method is intended to find limits of the frequency. Classical induction and predictions of individual events are included in the general formulation

[3] The implications occurring here must be regarded as nomological operations: the first as a tautological implication, the second as a relative nomological implication. See *ESL*, § 63.

as the special case that the relative frequency is $= 1$. It has been shown that if the aim of scientific method is attainable it will be reached by the inductive method. This result eliminates the last assumption we had to use for the justification of induction. The assumption that there is a limit of the frequency must be true if the inductive procedure is to be successful. But we need not know whether it is true when we merely ask whether the inductive procedure is justified. It is justified as an attempt at finding the limit. Since we do not know a sufficient condition to be used for finding a limit, we shall at least make use of a necessary condition. In positing according to the rule of induction, always correcting the posit when additional observation shows different results, we prepare everything so that if there is a limit of the frequency we shall find it. If there is none, we shall certainly not find one—but then all other methods will break down also.

The answer to Hume's question is thus found. Hume was right in asserting that the conclusion of the inductive inference cannot be proved to be true; and we may add that it cannot even be proved to be probable. But Hume was wrong in stating that the inductive procedure is unjustifiable. It can be justified as an instrument that realizes the necessary conditions of prediction, to which we resort because sufficient conditions of prediction are beyond our reach. The justification of induction can be summarized as follows:

Thesis θ. The rule of induction is justified as an instrument of positing because it is a method of which we know that if it is possible to make statements about the future we shall find them by means of this method.

This thesis is not meant to say that the inductive rule represents the only method of the kind described. In (1, § 87) were formulated other forms of posits that also must lead to the limit if there is one. Let us investigate whether the rule of induction constitutes the *best* method of finding the limit.

In order to answer the question we must divide the possible methods in two classes. In the first class we put all rules of the form (1, § 87), rules that differ from the rule of induction only inasmuch as they include a function c_n that is formulated explicitly so as to converge to 0 with increasing n. In the second class we put all other methods that will lead to the limit of the frequency. These methods will also converge asymptotically with the rule of induction; but they differ from the form (1, § 87) because they do not state the convergence explicitly.

As to the first class, it was explained in §§ 87–88 that we cannot prove the rule of induction to be superior to other methods included in the class. There may be, and in general will be, forms of the function c_n that are more advantageous than the function $c_n = 0$. If we knew one of these forms, we should prefer it to the rule of induction. The method of correction (§§ 89–90) may be regarded as an instrument for finding such forms. When, on the contrary, we know nothing, we can choose what we like. The rule of induction has the advantage of being easier to handle, owing to its descriptive simplicity. Since we are considering a choice among methods all of which will lead to the aim, we may let considerations of a technical nature determine our choice.

In regard to the second class the situation is different. If a method is presented with the assertion that it is a method of this class, the difficulty arises of how to prove the assertion. Of course, there may be such a method. Every oracle or soothsayer maintains that he has found one. Such a method is usually presented in the form of a prediction of individual events. This is included in our theory as the case in which the probability, or the frequency limit, is $= 1$. We may, therefore, generalize the problem so as to concern the prediction of any value of the limit of the frequency. Assume that a clairvoyant asserts that he is able to predict only the probability of an event—to predict the limit of the frequency in a sequence. We shall not be willing to believe him until we have checked his abilities. He might well be following a method that will never lead to the limit of the frequency. Such methods are certainly possible. For instance, if we were to posit that the limit is *outside* the interval $f^n \pm \delta$, we should certainly never reach the limit by continued application of this rule. The inadequacy of the methods of oracles and soothsayers is not so clearly apparent. But how can such methods be tested?

Obviously, there is only one way—to test these methods by means of the rule of induction. We would ask the soothsayer to predict as much as he could, and see whether his predictions finally converged sufficiently with the frequency observed in the continuation of the sequence. Then we would count his success rate. If the latter were sufficiently high, we would infer by the rule of induction that it would remain so, and thus conclude that the man was an able prophet. If the success rate were low, we would refuse to consult him further. It is true that in the latter case the soothsayer may refer us to the future, declaring that on continuation of the sequence his prediction of a limit may still come true. Although clairvoyants favor such an attitude, finally even the most ardent believer no longer places any faith in them. In the end the believer submits his judgment to the rule of induction. He must do so because the rule of induction is a method *of which he knows that it will lead to the aim* if the aim is attainable, whereas *he does not know anything* about the oracle and the clairvoyant.

We see, by the way, that with this subordination to induction the oracle in all its forms loses its mystical glamor. Like other methods of prediction, it is subject to a scientific test. It was explained above that science itself is at work to find methods of better convergence by the construction of a network of inductions in the form of the method of correction. There is no need, therefore, to ask the help of oracles or clairvoyants in order to improve our methods of approximation.

We thus come to the result that the rule of induction can by no means be maintained to be the best method of approximation. But with its help it is possible to find better methods of approximation. Scientific method makes use of this fact to a great extent. The concatenation of empirical results in a scientific system is the way to improve the method of approximation. The rule of induction, or one of its equivalents, is the only method that can be used in the test of other methods of approximation, because it is the only method *of which we know* that it represents a method of approximation.

In discussing the method of correction we have presented methods that offer ways to a better approximation. However, a proof that the approximation will, in fact, be better cannot be given; only the possibility of a better approximation can be proved. We have no means of excluding the equality sign in the inequalities (1 and 2, § 90) so long as we abstain from the use of inductive inferences. When we regard the use of posits of higher levels as a better method of convergence, the result must be strictly formulated as follows: in employing posits of higher levels we carry through the necessary conditions for obtaining a better convergence. The justification of the method of correction is therefore given in the same way as that of the inductive inference in general.

A remark about the limit condition must be added. It was stated earlier that success by the inductive method is possible only if the sequences under consideration have a limit of the frequency. This statement requires qualification. For the inductive method to have success, it is not necessary that all the sequences considered have a limit. It is possible that some have no limit of the frequency and that we shall discover this fact by using other sequences that do have limits. Assume, for instance, that in continuing a sequence we find its frequency to oscillate between the values $\frac{1}{4}$ and $\frac{3}{4}$. We then regard a frequency value close to $\frac{1}{4}$ as an event B_1, and a frequency value close to $\frac{3}{4}$ as an event B_2. When we now consider the sequence of events B_1 and B_2, we may find that it has a limit of the frequency for each of these events. Using the rule of induction in the latter sequence, we thus find that the former sequence has no limit of the frequency.

Since it is mathematically possible to construct sequences that have no limit of the frequency, it seems reasonable to assume that there are sequences of natural events having this property. It is clear, in any case, that we have no right to assume that all sequences of natural events have a limit of the frequency. The question has been asked whether we cannot prove, at least, that there must exist some sequences of natural events having a limit of the frequency. A theorem of this kind is presumably demonstrable; that is, it seems plausible that, given any system of sequences without a limit, we shall always be able to construct from them another sequence that has a limit. For the inductive problem, however, the question is irrelevant. In making inductive posits it does not help to know that there are sequences with a limit of the frequency; what we must know is whether the sequence under consideration has one. Since the question cannot be answered *a priori*, the justification of induction must be given independently of a limit assumption, as it is achieved in thesis θ.

Returning to a consideration of this thesis, we shall analyze the kind of justification it affords for induction. To clarify the analysis, we shall refer to the distinction between formulations I and II of the inductive rule (see p. 450). Formulation II postulates the existence of a probability relation between an observed frequency and future events, whereas formulation I does not. This formulation, we said, expresses the conception that between an observed frequency and the statement of probability merely the ground-for-assertion relation holds.

We are now able to give the final explanation why this relation can be maintained and thus why the rule of induction can be justified. The title to employ the inductive rule is based on a logical relationship between the aim of knowledge and a method the applicability of which constitutes a necessary condition of success. This relation, which can be formulated only in the metalanguage and is expressed in (1), may be abbreviated as the *condition relation*. The justification, therefore, is made possible only because we use formulation I, which differs from formulation II in that it does not identify the ground-for-assertion relation with a probability relation. For anticipative posits, the ground-for-assertion relation is derived from the condition relation.

Attempts at identifying the ground-for-assertion relation with a probability relation, and thus at basing the justification of induction on a probability relation, are doomed to failure because they lead to a conception in which the rule of induction asserts an object-language relation between an observed frequency and future events as soon as the object interpretation of probability is used (§ 87). The rule would then constitute a synthetic statement of the object language, depending for its support on arguments that are derived from a rational belief, from an *a priori* insight into the structure of the physical world—claims that cannot be taken seriously by anyone who is accustomed to apply the gauge of scientific truth to logical analysis. It was explained above (p. 372) that there is no such thing as synthetic self-evidence; self-evidence can be admitted as a criterion of truth only for analytic statements. That adherence to this fundamental principle of empiricism does not exclude a justification of induction is shown through thesis θ; by means of the condition relation we can construct a justification of induction that is free from all forms of synthetic self-evidence.

The condition relation (1), which is formulated in the implication that constitutes the second part of thesis θ, is a tautology; this relation follows from the definition of a limit. Therefore, an empirical assumption is not used for the justification; this avoids the fallacy analyzed in Hume's second result (p. 470). But Hume's first result is not contradicted: it is not maintained that the inductive conclusion is tautologically implied by its premise. A synthetic inference is justified by means of a tautology. Such a procedure involves no contradiction: whereas the relation between the premise and the conclusion of the inductive inference is synthetic, the relation between the inductive procedure and the aim of knowledge, that is, the condition relation, is analytic. The recognition that a tautological justification of a synthetic inference can be given makes the solution of the problem of induction possible.

This solution presupposes a moderation of the requirements to be satisfied by a justification: it involves the renunciation of a proof that the inductive conclusion is true or probable. To be sure, it would be a superior justification if we could prove that predictions must come true, or that there is some probability of their coming true. But if such a proof is impossible we shall be glad to have, at least, a method that we know will lead to success if success is possible. This logical relation may be illustrated by the example used in the Preface. When Magellan planned to find a passage through or around the Americas to the Pacific, he did not know whether there was one, and he did not even have a probability for the assumption of its existence. But he knew that if there was one he would find it by sailing along the coast—so his enterprise was justified.

Such illustrations suffer from the fact that they belong, not in primitive, but in advanced, knowledge. Thus the implication, "If there is a passage he will find it by sailing along the coast", is synthetic and therefore established by means of inductive inferences. In other examples the attainability of the aim can be judged in terms of probabilities; if the probability of reaching the aim turns out to be very low, the realization of the necessary conditions may seem scarcely advisable. Such a situation corresponds practically to a case where it is known that the aim is not attainable, at least not until further conditions are satisfied. Thus if a man wants to be a millionaire he must have a bank account; but taking out a bank account is usually not associated with the hope of ever being able to write out six-figure checks.[4] In all such instances, inductive inferences are applied and the legitimacy of induction is taken for granted. Only in the ultimate justification of induction itself must we renounce the use of inductive methods; the justification must be given within primitive knowledge, and therefore we have no other means at hand than considerations concerning necessary conditions, not supported by an estimate of the attainability of the aim.

---

[4] Similar considerations apply to a problem known as Pascal's wager, which has sometimes been wrongly compared to my justification of induction; see my answer, "On the Justification of Induction," in *Jour. of Philos.*, Vol. 37 (1940), pp. 101–102.

Some critics have called my justification of induction a weak justification. Such judgments originate from a rationalistic conception of scientific method. In spite of the empiricist trend of modern science, the quest for certainty, a product of the theological orientation of philosophy, still survives in the assertion that some general truths about the future must be known if scientific predictions are to be acceptable. It is hard to see what would be gained by the knowledge of such general truths. As was pointed out earlier, if we knew for certain that sequences of natural events have limits of the frequency, our situation in the face of any individual prediction would not be better than it is without such knowledge, since we would never know whether the observed initial section of the sequence were long enough to supply a satisfactory approximation of the frequency. It is no better with other forms of the uniformity postulate. How does it help to know that similar event patterns repeat themselves, if we do not know whether the pattern under consideration is one of them? In view of our ignorance concerning the individual event expected, all general truths must appear as illusory supports.

The aim of knowing the future is unattainable; there is no demonstrative truth informing us about future happenings. Let us therefore renounce the aim and renounce, too, the critique that measures the attainable in terms of that aim. It is not a weak argument that has been constructed. We can devise a method that will lead to correct predictions if correct predictions can be made—that is ground enough for the application of the method, even if we never know, before the occurrence of the event, whether the prediction is true.

If predictive methods cannot supply a knowledge of the future, they are, nevertheless, sufficient to justify action. In order to analyze the applicability of the inductive method as a basis for action, we must inquire into the presuppositions on which an action depends.

Every action depends on two presuppositions. The first is of a volitional nature: we wish to attain a certain aim. This aim can, at best, be reduced to more general volitional aims, but it cannot be given other than volitional grounds. A man who likes to exercise may justify his volitional aim by stating that he wants to retain a healthy body—but thereby the special volitional aim is only reduced to a more general one. The second presupposition is of a cognitive nature: we must know what will happen under certain conditions in order to be able to judge whether they are adequate for the attainment of the aim. If, for instance, I set up the general volitional aim of a healthy body, I can derive from this aim the usefulness of athletics only when I know that exercise makes the body healthy. Thus for every individual action I must know a statement about the future if the action is intended to contribute to the achievement of a general volitional aim. Only the combination of the two presuppositions, the volitional aim and knowledge about the future, makes purposive action possible. When the physician induces the patient to take an anodyne, he must know, first, whether the patient wants to get rid of his pain, and second, whether the drug will relieve it. When a politician advocates a new law, he wants to reach some goal and assumes that the law will attain it. The two presuppositions for action are of this kind.

The first presupposition, the volitional decision, need not be discussed here. Within the boundaries of a logical analysis we investigate the second presupposition for action, that is, the cognitive presupposition. Now it is clear that, though the inductive rule does not supply knowledge of a future event, it supplies a sufficient reason for action: we are justified in dealing with the anticipative posit as true, not because we can expect success in the individual case, but because if we can ever act successfully we can do so by following the directive of induction.

The justification of induction constructed may, therefore, be called a *pragmatic* justification: it demonstrates the usefulness of the inductive procedure for the purpose of acting. It shows that our actions need not depend on a proof that the sequences under consideration have the limit property. Actions can be made in the sense of trials, and it is sufficient to have a method that will lead to successful trials if success is attainable at all. It is true that this method has no guaranty of success. But who would dare to ask for such a guaranty in the face of the uncertainty of all human planning? The physician who operates on a patient because he knows that the operation will be the only chance to save the patient will be regarded as justified, though he cannot guarantee success. If we cannot base our actions on demonstrative truth, we shall welcome it that we can at least take our chance.

That is a rational argument. But who refers to it when he applies the inductive method in everyday life? If asked why he accepts the inductive rule, he answers that he believes in it, that he is firmly convinced of its validity and simply cannot give up inductive belief. Is there a justification for this belief?

The answer is a definite "No". The belief cannot be justified. As long as such a "No" was averred by a philosophy of skepticism, it constituted a negative judgment on all human planning and acting, which it seemed to prove utterly useless. It is different for the philosophy of logical analysis, which distinguishes between justification of the belief and justification of the action. Actions directed by the rule of induction are legitimate attempts at success; no form of belief is required for the proof. He who wants to act need not believe in success; it is a sufficient reason for action to know how to prepare for success, how to be ready for the case that success is attainable. Belief in success is a personal addition; whoever has it need not give it up. For his actions it is logically irrelevant: whether or not he believes in success, the same actions will follow.

I say "*logically* irrelevant", for I know very well that, psychologically speaking, belief may not be irrelevant. Many a person is not able to act according to his posits unless he believes in their success, since few have the inner strength to take a possible failure into account and yet pursue their aim. Nature seems to have endowed us with the inductive belief as a measure of protection, as it were, facilitating our actions, though without it we would be equally justified, or obliged, to act. It is difficult, indeed, to free oneself from such a belief; and Hume was right when he called the belief in induction an unjustified but ineradicable habit. But since Hume could not show that even without this belief action is justified when it follows the rule of induction, there remained for him only skeptical resignation.

The logician need not share this negative attitude. He can show that we must act according to the rule of induction even if we cannot believe in it. This result may be the reason why it is easier for him to renounce the belief; with the loss of the belief he does not at the same time lose his orientation in the sphere of action. We do not know whether tomorrow the order of the world will not come to an end; tomorrow all known physical laws may be invalidated, the sun may no longer shine, and food no longer nourish us—or at least our own world may come to an end, because we may close our eyes forever. Tomorrow is unknown to us, but this fact need not make any difference in considerations determining our actions. We adjust our actions to the case of a predictable world—if the world is not predictable, very well, then we have acted in vain.

A blind man who has lost his way in the mountains feels a trail with his stick. He does not know where the path will lead him, or whether it may take him so close to the edge of a precipice that he will be plunged into the abyss. Yet he follows the path, groping his way step by step; for if there is any possibility of getting out of the wilderness, it is by feeling his way along the path. As blind men we face the future; but we feel a path. And we know: if we can find a way through the future it is by feeling our way along this path.

INDEX

INDEX

(The numbers refer to pages)